U0170752

学术引领系列

国家科学思想库

中国学科发展战略

软凝聚态物理学（下）

国家自然科学基金委员会
中 国 科 学 院

科 学 出 版 社
北 京

图书在版编目（CIP）数据

软凝聚态物理学.下 / 国家自然科学基金委员会，中国科学院编. —北京：科学出版社，2020.1
（中国学科发展战略）
ISBN 978-7-03-062147-4

Ⅰ. ①软… Ⅱ. ①国… ②中… Ⅲ. ①凝聚态-物理学 Ⅳ. ①O469

中国版本图书馆 CIP 数据核字（2019）第181914号

丛书策划：侯俊琳 牛 玲
责任编辑：张 莉 崔慧娴 / 责任校对：韩 杨
责任印制：徐晓晨 / 封面设计：黄华斌 陈 敬
联系电话：010-64035853
E-mail: houjunlin@mail.sciencep.com

科学出版社 出版
北京东黄城根北街 16 号
邮政编码：100717
http://www.sciencep.com

北京虎彩文化传播有限公司 印刷

科学出版社发行 各地新华书店经销
*
2020年1月第 一 版 开本：720 × 1000 B5
2021年7月第三次印刷 印张：28 1/2 插页：12
字数：500 000

定价：198.00元

（如有印装质量问题，我社负责调换）

中国学科发展战略
联合领导小组

组　　长：丁仲礼　李静海

副 组 长：秦大河　韩　宇

成　　员：王恩哥　朱道本　陈宜瑜　傅伯杰　李树深

杨　卫　汪克强　李　婷　苏荣辉　王长锐

邹立尧　于　晟　董国轩　陈拥军　冯雪莲

王岐东　黎　明　张兆田　高自友　徐岩英

联合工作组

组　　长：苏荣辉　于　晟

成　　员：龚　旭　孙　粒　高阵雨　李鹏飞　钱莹洁

薛　淮　冯　霞　马新勇

中国学科发展战略·软凝聚态物理学（下）项目组

（以姓名汉语拼音为序）

曹　毅　曹文彬　陈　诚　陈　虎　陈　文　陈雷鸣
陈祎璇　陈征宇　程晓辉　程正迪　邓海游　邓林红
范昊翔　方海平　冯西桥　高　翔　郭坤琨　侯　旭
厚美瑛　黄吉平　黄巧玲　贾　亚　江　雷　蒋　滢
冷劲松　黎　明　李宝会　梁　琴　梁好均　梁永日
林乃波　林友辉　刘　锋　刘　军　刘冬生　刘立武
刘雳宇　刘明杰　刘如川　刘向阳　刘彦菊　刘艳辉
刘昭明　吕雄飞　马红孺　欧阳钟灿　邱　东　沈红斌
史安昌　舒咬根　帅建伟　司铁岩　孙洪广　唐建新
唐睿康　涂展春　王　威　王　炜　王琼华　王晓晨
王宇杰　魏志祥　温维佳　巫金波　吴晨旭　吴艺林
肖石燕　刑向军　许文祥　严　洁　严大东　杨　扬
杨朝晖　杨光参　于伟东　於东军　俞燕蕾　袁军华
张　洁　张　阳　张何朋　张平文　张天辉　张文彬
张晓华　张泽新　赵　坤　赵亚溥　郑　鹏　周如鸿
周永丰　朱智超　Holger Merlitz

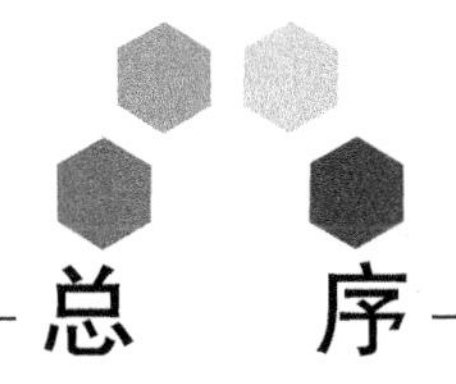

总　序

白春礼　杨　卫

17 世纪的科学革命使科学从普适的自然哲学走向分科深入，如今已发展成为一幅由众多彼此独立又相互关联的学科汇就的壮丽画卷。在人类不断深化对自然认识的过程中，学科不仅仅是现代社会中科学知识的组成单元，同时也逐渐成为人类认知活动的组织分工，决定了知识生产的社会形态特征，推动和促进了科学技术和各种学术形态的蓬勃发展。从历史上看，学科的发展体现了知识生产及其传播、传承的过程，学科之间的相互交叉、融合与分化成为科学发展的重要特征。只有了解各学科演变的基本规律，完善学科布局，促进学科协调发展，才能推进科学的整体发展，形成促进前沿科学突破的科研布局和创新环境。

我国引入近代科学后几经曲折，及至上世纪初开始逐步同西方科学接轨，建立了以学科教育与学科科研互为支撑的学科体系。新中国建立后，逐步形成完整的学科体系，为国家科学技术进步和经济社会发展提供了大量优秀人才，部分学科已进入世界前列，有的学科取得了令世界瞩目的突出成就。当前，我国正处在从科学大国向科学强国转变的关键时期，经济发展新常态下要求科学技术为国家经济增长提供更强劲的动力，创新成为引领我国经济发展的新引擎。与此同时，改革开放 40 多年来，特别是 21 世纪以来，我国迅猛发展的科学事业蓄积了巨大的内能，不仅重大创新成果源源不断产生，而且一些学科正在孕育新的生长点，有可能引领世界学科发展的新方向。因此，开展学科发展战略研究是提高我国自主创新能力、实现我国科学由“跟跑者”向“并行者”和“领跑者”转变的一项基础工程，对于更好把握世界科技创新发展趋势，发挥科技创新在全面创新中的引领作用，具有重要的现实意义。

学科发展战略研究的核心是结合科学技术和经济社会的发展需求，在分析科学前沿发展趋势的基础上，寻找新的学科生长点和方向。在这个过程中，战略科学家的前瞻引领作用十分重要。科学史上这样的例子比比皆是。在 1900 年 8 月巴黎国际数学家代表大会上，德国数学家戴维·希尔伯特发表了题为“数学问题”的著名讲演，他根据过去特别是 19 世纪数学研究的成果和发展趋势，提出了 23 个最重要的数学问题，即“希尔伯特问题”。这些“问题”后来成为许多数学家力图攻克的难关，对现代数学的研究和发展产生了深刻的影响。1959 年 12 月，美国物理学家、诺贝尔奖得主理查德·费曼在加利福尼亚理工学院举行的美国物理学会年会上发表了题为“物质底层大有空间——一张进入物理新领域的请柬”的经典讲话，对后来出现的纳米技术作出了天才的预见。

学科生长点并不完全等同于科学前沿，其产生和形成不仅取决于科学前沿的成果，还决定于社会生产和科学发展的需要。1841 年，佩利戈特用钾还原四氯化铀，成功地获得了金属铀，可在很长一段时间并未能发展成为学科生长点。直到 1939 年，哈恩和斯特拉斯曼发现了铀的核裂变现象后，人们认识到它有可能成为巨大的能源，这才形成了以铀为主要对象的核燃料科学的学科生长点。而基本粒子物理学作为一门理论性很强的学科，它的新生长点之所以能不断形成，不仅在于它有揭示物质的深层结构秘密的作用，而且在于其成果有助于认识宇宙的起源和演化。上述事实说明，科学在从理论到应用又从应用到理论的转化过程中，会有新的学科生长点不断地产生和形成。

不同学科交叉集成，特别是理论研究与实验科学相结合，往往也是新的学科生长点的重要来源。新的实验方法和实验手段的发明，大科学装置的建立，如离子加速器、中子反应堆、核磁共振仪等技术方法，都促进了相对独立的新学科的形成。自 20 世纪 80 年代以来，具有费曼 1959 年所预见的性能、微观表征和操纵技术的仪器——扫描隧道显微镜和原子力显微镜终于相继问世，为纳米结构的测量和操纵提供了“眼睛”和“手指”，使得人类能更进一步认识纳米世界，极大地推动了纳米技术的发展。

作为国家科学思想库，中国科学院（以下简称中科院）学部的基本职责和优势是为国家科学选择和优化布局重大科学技术发展方向提供科学依据、发挥学术引领作用，国家自然科学基金委员会（以下简称基金委）则承担着协调学科发展、夯实学科基础、促进学科交叉、加强学科建设的重大责任。继基金委和中科院于2012年成功地联合发布“未来10年中国学科发展战略研究”报告之后，双方签署了共同开展学科发展战略研究的长期合作协议，通过联合开展学科发展战略研究的长效机制，共建共享国家科学思想库的研究咨询能力，切实担当起服务国家科学领域决策咨询的核心作用。

基金委和中科院共同组织的学科发展战略研究既分析相关学科领域的发展趋势与应用前景，又提出与学科发展相关的人才队伍布局、环境条件建设、资助机制创新等方面的政策建议，还针对某一类学科发展所面临的共性政策问题，开展专题学科战略与政策研究。自2012年开始，平均每年部署10项左右学科发展战略研究项目，其中既有传统学科中的新生长点或交叉学科，如物理学中的软凝聚态物理、化学中的能源化学、生物学中的生命组学等，也有面向具有重大应用背景的新兴战略研究领域，如再生医学，冰冻圈科学，高功率、高光束质量半导体激光发展战略研究等，还有以具体学科为例开展的关于依托重大科学设施与平台发展的学科政策研究。

学科发展战略研究工作沿袭了由中科院院士牵头的方式，并凝聚相关领域专家学者共同开展研究。他们秉承“知行合一”的理念，将深刻的洞察力和严谨的工作作风结合起来，潜心研究，求真唯实，“知之真切笃实处即是行，行之明觉精察处即是知”。他们精益求精，“止于至善”，“皆当至于至善之地而不迁”，力求尽善尽美，以获取最大的集体智慧。他们在中国基础研究从与发达国家“总量并行”到“贡献并行”再到“源头并行”的升级发展过程中，脚踏实地，拾级而上，纵观全局，极目迥望。他们站在巨人肩上，立于科学前沿，为中国乃至世界的学科发展指出可能的生长点和新方向。

各学科发展战略研究组从学科的科学意义与战略价值、发展规律和研究特点、发展现状与发展态势、未来5～10年学科发展的关键科学问题、发展思路、发展目标和重要研究方向、学科发展的有效资助机制与政策建议等方面进行分析阐述。既强调学科生长点的科学意义，也考虑其重要的社会价值；既着眼于学科生长点的前沿性，也兼顾其可能利用的资源和条件；既立足于国内的现状，又注重基础研究的国际化趋势；既肯定已取得的成绩，又不回避发展中面临的困难和问题。主要研究成果以“国家自然科学基金委员会-中国科学院学科发展战略”丛书的形式，纳入“国家科学思想库-学术引领系列”陆续出版。

基金委和中科院在学科发展战略研究方面的合作是一项长期的任务。在报告付梓之际，我们衷心地感谢为学科发展战略研究付出心血的院士、专家，还要感谢在咨询、审读和支撑方面做出贡献的同志，也要感谢科学出版社在编辑出版工作中付出的辛苦劳动，更要感谢基金委和中科院学科发展战略研究联合工作组各位成员的辛勤工作。我们诚挚希望更多的院士、专家能够加入到学科发展战略研究的行列中来，搭建我国科技规划和科技政策咨询平台，为推动促进我国学科均衡、协调、可持续发展发挥更大的积极作用。

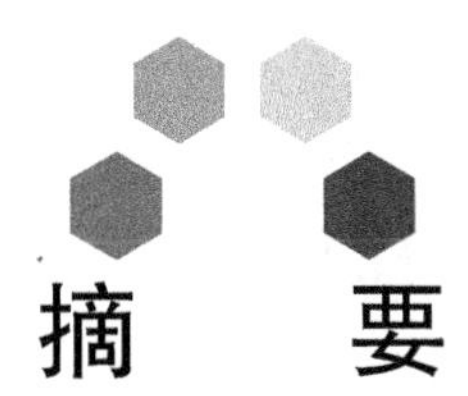

摘　要

一、软物质物理学的科学意义与战略价值

软物质（soft matter）泛指处于固体和理想流体之间的复杂凝聚态物质，主要共同点是其基本单元之间的相互作用比较弱（约为室温热能量级），因而易受温度影响，熵效应显著，且易形成有序结构，处于这种中间状态的常见体系包括胶体、液晶、高分子及超分子、泡沫、乳液、凝胶、颗粒物质、玻璃、生物体系等。软物质不仅广泛存在于自然界，在人类的生产生活中也得到广泛应用，如液晶显示、塑料、橡胶、清洁剂、护肤品等。软物质自19世纪中后期正式进入科学家视野以来，已经发展成为一个高度交叉的研究方向，软物质物理学也已成为物理学的重要组成部分。这一方面是物理学自身走向复杂体系的必然趋势，另一方面也是由巨大的社会需求所推动的。

（一）来自物理学自身发展的内部需求

自1991年德热纳正式命名“软物质”以来，软物质物理学发展极为迅猛，这不仅极大地拓展了物理学的研究对象，还对物理学基础研究尤其是与非平衡过程（如生命现象）密切相关的物理学提出了重大挑战。2005年，著名学术期刊《科学》（*Science*）在创刊125周年之际提出了125个世界性科学前沿问题，其中13个直接与软物质交叉学科相关；2007年，美国物理学会凝聚态物理委员会（Committee on CMMP 2010）发布报告《凝聚态与材料物理：我们身边的科学》，列出未来十年物理学面临的6个重大课题，其中4个直接与软物质和生命系统相关；2011年，美国白宫报告文件专门提到了软物质材料科学研究的重大意义，同年美国能源部在宣布启动的关键材料创新中心的文件中也特别强调加强软凝聚态材料的计算科学研究；2013 年，以 John Hemminger 教授为首的基础能源科学咨询委员会在给美国能源部的一份报告中强调了物质介观尺度的复杂性和重要性，在这个尺度上，经典物理与量子物理产生了碰撞，这将对未来几十年的研究产生深远影响，而软物质的结构特征正好体现在该尺度上。

（二）来自社会、经济的外部需求

软物质在人们的生产生活中发挥着越来越重要的作用。液晶显示作为全球市值逾千亿的平板显示主流技术，仍在不断发展蓝相液晶、柔性液晶等新型显示技术。表面活性剂是软物质中应用最广泛的体系，它渗透到了从日化到石油开采、农业、卫生、环境等几乎所有的经济技术部门，目前全球市值高达数百亿美元。20 世纪末，高分子材料总产量已达 20 亿吨，被广泛应用于科学技术、国防建设和国民经济等各个领域，与金属、陶瓷并列为三类最重要的材料，目前高分子材料正向高性能化、功能化和生物化方向发展。智能型软材料的开发和应用是当前材料科学的重要内容，例如磁流变液已被用作汽车等设备的智能减震器，而电流变液被认为是最有希望在高铁、国防、军工等方面用来制作性能优良的减震器的材料。高性能新材料、新器件的研发，环保智能型建筑材料的开发以及生物医药相关领域的创新，也是当前软物质研究发展的驱动力来源。颗粒物质作为新兴的软物质体系，与很多工程实际问题密切相关，例如河床和河道的土壤失稳和流失、大型水利枢纽工程结构稳定性等，具有重要科学意义和工程应用价值。

综上所述，软物质研究已经成为当代物质科学的重要组成部分，对材料、能源、环境、医疗等人类面对的重大问题有着深远的影响，对我国国计民生具有重大的战略价值。

二、软物质物理学的发展规律和研究特点

1861 年，英国科学家本杰明·格雷厄姆引入“胶体”（colloid）

一词，标志着近代软物质科学的兴起。20世纪初，爱因斯坦、朗之万等提出的胶体颗粒的布朗运动理论可以说是早期软物质物理学的巅峰。由于对布朗运动的精细实验研究，法国物理学家佩兰、瑞典物理化学家斯维德伯格分获1926年诺贝尔物理学奖、化学奖。此外，席格蒙迪因胶体溶液异相性质的工作荣获1925年诺贝尔化学奖，朗缪尔因表面吸附的工作荣获1932年诺贝尔化学奖。

同期软物质领域的另一重大事件是“高分子”的诞生，这一概念由德国科学家施陶丁格于1920年正式提出，1938年得以证实，施陶丁格于1953年获得诺贝尔化学奖，其同胞齐格勒与纳塔因高分子催化合成方面的研究荣获1963年诺贝尔化学奖。美国科学家弗洛里因为在高分子聚合反应机制、高分子构象统计、橡胶弹性等课题上的系统工作荣获1974年诺贝尔化学奖。法国物理学家德热纳因为横跨超导、液晶、高分子等不同领域的深刻理论成果而独获1991年诺贝尔物理学奖。这些科学家的工作充分展现了软物质研究基础性和实用性并重的特点，将软物质整个领域的研究和应用推到了一个新的高度。

液晶是另一种重要的软物质。对液晶光学电学性质的研究促使美国科学家黑尔麦乐、德国物理学家海尔弗里希等在20世纪六七十年代先后提出了液晶作为显示器元件的设想，他们也因此于2012年分享了被誉为工程界诺贝尔奖的德雷珀奖（Charles Stark Draper Prize for Engineering）。海尔弗里希还在1973年开创了液晶生物膜的理论研究，我国理论物理工作者欧阳钟灿等在此基础上做出了一系列有广泛影响的工作，建立了液晶生物膜的完整理论。

从20世纪60年代起，生物体系也逐渐被当作软物质来研究，而其中最为复杂多样的当属生物大分子及其形成的微纳有序结构，催生了不少新的实验和计算技术，如多尺度计算模拟技术（2013年诺贝尔化学奖），超高分辨率显微技术（2014年诺贝尔化学奖）。20世纪90年代出现的单分子技术，将生物学的研究推进到单分子水平，极大地拓宽了软物质研究的层次和范围。

近十余年来出现了另一种重要的软物质体系——超分子凝胶，在分子机器、分子传感器、气体吸附、纳米反应器、化学催化、药物传输、基因传输、生物成像、组织工程等领域被广泛研究，已经产生了重大影响。

总的来看，软物质研究100多年以来，人们认识了越来越多的体系，也总结出某些共性，如热涨落显著、多重亚稳态、介观多尺度自组装结构、熵致有序、宏观柔性、“小作用，大响应”、强非线性等。这些特性通常难以从它们的微观结构（如原子或分子成分）上推测出来，而更多地取决于介观尺度的自组装结构。尽管对这些共性已经有所了解，但软物质物理尚未建立起统一的研究范式，这仍将是未来软物质物理研究的主要内容。

三、学科的发展现状与发展态势

由于软物质物理在基础研究中的重大意义及在经济社会需求中的战略价值，很多国家都建立了相关研究基地，大力推进这方面的研究。

近十多年来，发达国家许多大学的物理系和研究机构已纷纷建立软物质物理的研究方向，如哈佛大学、普林斯顿大学、洛斯阿拉莫斯国家实验室、剑桥大学、牛津大学、海德堡大学、巴黎高等师范学院、代尔夫特理工大学、昆士兰大学、京都大学等。特别值得一提的是，英国爱丁堡大学的Michael Cates教授由于在软物质方面的贡献，于2015年被剑桥大学授予卢卡斯数学教授席位，将这一具有悠久历史和崇高地位的荣誉职位授予软物质科学家，充分体现了国际科学界对这一领域的重视。以软物质为基础的国际复杂自适应性物质组织（ICAM）现在已有72个成员单位。世界范围内的学术出版也反映了软物质研究的发展趋势。从20世纪90年代开始，国际物理学界很多著名的专业学术期刊［如《物理学评论快报》（*Phys. Rev. Lett.*）］纷纷着手开辟软物质专栏或出版软物质专刊。相应地，近十年来也涌现出大批软物质物理的专著和教材。

我国对软物质物理的研究可追溯到20世纪80年代，中国科学院理论物理研究所、中国科学院生物物理研究所、中国科学技术大学、南京大学、北京师范大学等单位率先开展了生物软物质方面的研究。国家自然科学基金委员会（以下简称国家基金委）和中国科学院曾部署并大力资助软物质和生物物理等方面的研究。《国家中长期科学和技术发展规划纲要（2006—2020年）》明确将“软凝聚态物质”列为科学前沿问题。国家基金委从2010年开始设立“可控自组装体系及其功能化”重大研究计划，科学技术部纳米研究重大科学研究计划也对软物质材料尤其是生物软物质提供了大力资助。国内许多大学和研究机构在最近十多年来也纷纷引进国际著名的软物质科学家，同时成立专门的软物质研究实验室，如中国科学院软物质化学重点实验室、中国科学院软物质物理重点实验室、厦门大学生物仿生与软物质研究院、北京大学软物质科学与工程中心、南京大学生物物理研究所、北京航空航天大学软物质物理与应用研究中心、浙江大学软物质科学研究中心、苏州大学软凝聚态物理及交叉研究中心、上海交通大学软凝聚态物理实验室等。此外，从20世纪90年代初国内就开始举办软物质物理的专门学术会议、讲习班和暑期学校多达数十次，编著、翻译出版的相关教材和专著也有10余种。

目前，软物质物理正迅速成为国际上物理研究的潮流之一。我国在这方面已经具备了一定基础，强化这方面的部署与研究，不仅可能使我国在基础研究上产生重大突破，也必将对国民的生产生活产生重大影响。

四、未来5～10年学科发展的关键科学问题、发展思路、发展目标和重要研究方向

软物质物理学的科学内涵和目标主要体现在以下几个层面：观测和发现软物质体系的复杂物理现象；针对不同软物质系统建立精确描述其运动规律的模型和理论；探讨软物质体系的一般运动规律。下面从多个方面简要介绍软物质在未来几年的关键科学问题和重要研究方向。

（一）软物质的基础理论

弹性理论、相变理论、标度理论、自洽场理论等仍将是软物质体系理论研究方法的主流。关于颗粒物质、黏性系统、扩散系统、

界面系统等的动力学问题，有望发展一些新的理论和计算模型。针对复杂软物质体系，仍需要发展新的多尺度计算模拟方法。

（二）软物质的实验方法

软物质相关的实验手段和技术仍需要在信噪比、时间和空间分辨率等方面大幅提高。单分子操控技术还将实现更高的力学精度。单分子荧光技术仍需要解决时间分辨率、系统稳定性等问题。在单分子水平上对活体细胞进行显微成像，未来还需要提升图像快速扫描技术。小角散射未来也有很大提升空间，例如建立更精确的结构因子及不依赖理论模型的数据解析方法等。

（三）软物质介观体系

1. 超分子凝胶

超分子凝胶是近来颇受关注的新体系，它已经对很多领域产生了重大影响。发展具有良好生物相容性、无毒性和可降解的超分子凝胶材料是近年关注的焦点。对超分子凝胶材料与生物体系的相互作用进行深入研究，研制具有更好适用性的多功能超分子凝胶是另一重要发展方向。

2. 聚合物

聚合物是最具有实用价值的材料，未来亟待解决的基础问题包括超越平均场的理论、非平衡系统的动力学等。前沿研究课题包括分相与结晶、玻璃化转变、不同尺度上的非平衡相转变、功能（光、电、磁）材料中的物理特性与结构间的关联、受限聚合物/纳米粒子复合体系的组装机制等。智能软聚合物材料是近年研究热点之一，未来需发展复合改性设计和多功能设计，研究电、热、光、磁、溶剂等智能聚合物复合材料的记忆效应驱动新机制和形状恢复行为规律，以及多场耦合等物理场下的响应行为和感应规律等。

3. 液晶

在液晶的基础理论研究方面，未来需要进一步发展精度更高的分子模型。受限液晶体系相关的问题，例如曲面上液晶分子结构的自洽场模拟、受限体系中液晶分子的指向分布等，也将得到更多关注。液晶显示在未来仍有丰富的研究内容，如三维显示、蓝相显示、柔性显示等。液晶弹性体作为新型智能材料在未来将展现出巨大的应用潜力，如制备动态键盘、柔性器件、微型透镜、微流体设备等。

4. 颗粒物质

近十年来，对颗粒物质的研究已经在颗粒输运性质、颗粒流相转变机制等方面取得了重要的进展，但人们对这类体系的特性和机制的了解还十分有限。未来需要着力开展的前沿课题包括如下四方面：①颗粒气体，致力于揭示类似于平衡态统计力学中的速度分布律；②颗粒流体，主要关注动力学研究，例如，对颗粒流中的激波结构的研究将有助于我们对颗粒流的耗散特性及颗粒流中相转变的理解；③颗粒固体，主要关注力在颗粒介质中的沿非线性力链结构传播的研究，大尺度颗粒介质的本构关系的建立等；④各向异性颗粒及带电颗粒的相互作用。

（四）软物质低维与界面体系

1. 膜

膜是准二维体系，其中最具代表性的是生物膜。1973 年海尔弗里希提出的自发曲率模型奠定了生物膜弹性理论的基础，我国学者在其中也做出了许多重要工作。近年来，生物膜弹性理论研究仍在发展，例如膜方程的解析解、分相膜的理论与应用；另一个重要方向是基于该理论研究膜与蛋白质的相互作用。

2. 表面

超浸润表面具有广泛的应用前景和长期的研究积累，但仍有若干问题亟待解决：提取具有特殊性能的生物界面的仿生学原理，揭示其微观结构、组成和功能之间的内在本质联系；基于仿生原理的仿生功能分子设计、材料表面微纳结构的精细调控与表征；通过调控优化仿生智能多尺度界面材料的结构和组成参数，实现对自然界特殊功能的模拟；发展规模化制备技术，实现仿生智能多尺度界面材料在资源、能源、环境、生物医学等领域的应用。

3. 胶体

胶体广泛存在于日常生活中的各个场合，得益于新技术的发展，胶体的物理化学研究仍在不断深入，并且出现了利用胶体系统研究基础物理问题的新趋势。未来需要发展制备具有复杂外形和复杂相互作用的胶体体系，以研究更复杂的相变行为，完善发展各种控制手段以获取更复杂的晶体结构，建立自驱动胶体体系以研究非平衡自组织行为，研究胶体颗粒与各种有机分子混合体系的相变和自组装行为。此外，表面活化剂和其他软物质构成的复合胶体系统也是未来需要着力研究的方向。

4. 微观尺度下的水

近年来，人们逐渐认识到许多有关水的技术难以取得突破，是由于对微观尺度下水的特殊性质理解还不够，仍有许多问题未解决。例如，纳米尺度管道中准一维水的高流通特性背后的物理本质，离子参与下纳米碳管等碳材料中水的流通特性，水由高流通性向普通流通性转变时水的结构变化；微观界面水的性质及其对浸润和流动输运性能的基本物理机制；自“生物水”的概念被提出后，水对生物大分子的结构稳定性影响目前未清楚；水环境对材料结构与性质的影响未完全解决。

（五）软物质生物体系

1. 生物软物质物理

生物软物质物理是物理学与生物学最重要的交叉前沿，研究范围涵盖分子生物物理学、细胞生物学，主要利用和借鉴凝聚态物理、统计物理、计算物理、数学、信息学、计算机科学方面的概念、理论和技术手段。当前的前沿课题包括：生物分子的相互作用及结构和功能动力学；非编码序列、非编码基因和非编码 RNA；生物膜相关的结构和动力学；细胞骨架自组装、聚集态结构和动态行为等；单分子生物学，包括各种新型技术的研发；生物网络的拓扑结构、动力学；生物神经系统；细菌生物物理学；生物学启发的物理和工程问题等。

2. 纳米颗粒和蛋白及细胞膜的相互作用

纳米材料在生物医药卫生领域的应用是纳米科学的热点，国内外已在积极地开展纳米材料的生物学效应研究。目前的研究重点已经由生物整体效应逐步深入纳米材料与细胞和功能性生物大分子的相互作用。亟待解决的前沿问题包括：纳米粒子进入细胞的驱动力是什么？小于100nm或更小的纳米粒子是否存在新的跨膜过程和机制？纳米粒子与生物大分子的相互作用的分子机制是什么，有什么特异性？这些特异性是否能解释不同纳米材料的特异生物兼容性与安全性？

3. 生物信息大数据

面向生物大数据理解分析的智能生物信息学理论与方法研究是当前的国际学术前沿，研发生物数据特征驱动的先进算法和方法是当前的重要方向。这方面的课题包括：面向海量生物数据的动态学习新框架；多源异质生物数据迁移学习算法，特别是针对多类型多介质生物数据、不同层次生物数据和不同物种生物数据，发展不同的迁移学习机制和相应的智能模型构建方法；面向生物数据层次隐含信息挖掘的深度学习理论与方法；多源异质生物数据挖掘平台的建设等。

4. 生物软物质医学

最近十几年来兴起的癌症生物物理研究开创了癌症研究的新局面，它着重于侵袭转移过程中癌细胞与微环境的关系、细胞群体侵袭的途径和模式，及细胞间的协作与共进化等方面。软物质物理学在肿瘤的侵袭和转移研究中，可能在以下方向做出突破性贡献：细胞内分子信号通路耦合的各种生物力学信号的定量研究，精确可控的三维组织微环境平台构建。而围绕癌细胞的理论和模拟方面，未来可能的研究方向包括：关于细胞信号转导机制的研究，关于基因表达调控机制的研究，关于生物网络结构、动力学和功能的研究。

（六）其他交叉领域

1. 软物质微流变和微流控器件

微流控是一种精确控制和操控微尺度流体的技术，在生物、化学、医学等领域具有巨大的应用潜力，其重要前沿研究包括微通道的设计制造，及制备材料选择、微流控系统的集成封装与移动、微尺度下对流体的输运控制等。此外，许多新的应用目前仍处于实验室研究阶段，如智能仿生材料的设计和开发、多孔介质的模拟、连续流动化学有机合成和微纳米颗粒的制备等。未来微流控研究将在海水淡化、油水分离、药物及生物分子筛分检测和复杂流体中的高效除气技术等领域带来巨大的经济效益。

2. 软智能材料

智能材料是继天然材料、合成高分子材料、人工设计材料之后的第四代材料，拥有传感功能、反馈功能、信息识别与积累功能、响应功能、自诊断能力、自修复能力和自适应能力七大功能，是材料领域目前最前沿的研究领域。例如，巨电流变液被认为是现在最有希望在高铁、国防、军工等方面获得广泛应用的软物质智能材料。除了电、磁作用智能软物质材料外，具有热、光及其他作用的智能软物质材料也是目前材料研究的前沿领域。相关基础理论问题包括：电、磁流变体固-液相的匹配及分散相的选择多样性；纳米颗粒材料的表面修饰，及表面活性剂对电、磁流变液性能的提高的基础物理机制；固相纳米材料颗粒沉淀及极限应用下物理特性评价；软物质智能材料在不同的应用条件过程中的服役、失效问题。

五、有利于学科发展的有效资助机制与政策建议

软物质物理学是高度交叉的学科，研究课题较多样、分散，短期内难以形成一条明确的主线。尤其是在我国，目前从事这方面研究的物理学家绝对数量并不多，但由于学科属性很不明确，难以从固定渠道获得持续支持。这既不利于凝练现有学术队伍、促进学术交流和合作，也不利于青年后备人才的持续培养。为推动我国软物质物理学稳定的发展，我们提出如下建议。

（一）从政策层面设立专门的资助

目前，在国家基金委等国家级科研支持计划中，还没有单列的软物质物理申请代码，与此形成鲜明对照的是，传统领域（如粒子

物理、凝聚态物理等）都有明确的申请代码，而软物质相关的基金申请只能分散地挤到凝聚态物理、统计物理等方向中。政策上的先天倾斜妨碍了软物质研究者获得稳定的经费支持，严重制约了学科发展。我们建议设立软物质物理学的专门学科代码，以更好地凝练和扩大学科队伍、加速学科建设；同时将软物质科学纳入国家各级科技计划，如国家科技重大专项、国家重点研发计划、技术创新引导专项、基地和人才专项等。

（二）建立和支持有影响力的研究基地

建立并保持一支稳定的人才队伍，对软物质物理学这种新兴的交叉学科来说尤其重要。目前，在全国范围内的高校和研究机构中，真正开展软物质物理学研究的还比较少，在人才引进等方面也受到各种制约。建议国家层面或机构层面出台支持政策，在全国范围内建立并重点支持若干有特色的研究基地，吸引、培养和稳定一批高水平人才，逐渐扩大研究队伍，建成在国内外具有影响力的研究平台。

（三）加强产学研合作

软物质物理与我们的生产生活密切相关。缩小基础科研与实际应用之间的距离是我国科研界的大势所趋，是当前创新型社会的必然要求。建议以上述研究基地为依托，将促进学界和产业界的沟通及合作作为基地的重要使命，实现学界和产业界在人员、项目、数据、知识产权等多个层面的融合和共享。

（四）加强基础教育和普及教育

软物质不仅是新兴交叉学科，更是物理学重要的组成部分，但相关核心知识（如熵致有序）在国内物理学基础教育中未能得到体现。尤为严峻的是，作为软物质物理核心课程的统计物理学在各个高校被明显弱化，师资、教学量和教学水准难有保证。与之相反，相对论、量子物理等知识却越来越广泛地渗透到物理学各课程中，甚至市面上也多充斥着这类科普书籍。这些都对学生造成了一定程度的误导，导致学生严重缺乏对软物质物理学的认知，这对整个物理学的发展也是极为不健康的。我们建议组织人员，通过编写教材和出版科普读物，在高校本科教育中加强软物质基本概念的教学和普及，为培养具有全面素质的物理学后备人才构筑良好的学术氛围。

黎明（中国科学院大学），帅建伟、刘向阳（厦门大学），
欧阳钟灿（中国科学院理论物理研究所）

Abstract

Ⅰ. Scientific Relevancy and Strategic Importance of Soft Matter Physics Research

Soft matter refers to complex condensed matter with a state between that of a solid and an ideal liquid that generally shows weak interactions (at the ambient thermal energy level) among the basic building blocks and is only subtly modified by small temperature variations. Soft matter shows a significant entropic effect and easily acquires an ordered structure. Systems include colloids, liquid crystals, macromolecules and supramolecules, foams, emulsions, gels, granular materials, glass, biosystems, etc. Soft matter not only widely occurs in nature but can also be found in artificial products such as LCDs, plastic, rubber, detergent and skin care products. Since being first recognized by scientists in the middle and late 19th century, soft matter has developed into a highly multidisciplinary research field. Soft matter physics has become an important area of research in physics. Thus, physics research will on one hand inevitably evolve from a simple to a more complex system, and on the other hand will be driven by huge social demands.

1. Advancement of Soft Matter due to the Self-development of Physics

Since the term "soft matter" was formally coined by P. G. de Gennes in 1991, soft matter physics has experienced rapid advancement. Such development can be identified not only by greatly expanding the physical research subjects but also by imposing new challenges to basic physics, in particular, non-equilibrium physics (e.g., life phenomena). In 2005, "Science" proposed 125 world frontier scientific issues on its 125th anniversary. Among these issues, 13 are directly related to soft matter interdisciplinary research. In 2007, the CMMP 2010 Committee of the APS released a report entitled "Condensed Matter and Materials Physics: Science around Us" . The report listed six major subjects in physics to be examined in the next ten years, four of which are closely related to soft matter. In 2011, in the Materials Genome Initiatives announced by the US White House, with a particular emphasis on the computational science of materials science and technology, the significant relevancy of soft matter in materials science research was also listed. In 2013, in a white paper led by Prof. John Hemminger and sent to the State Department of Energy, it was written by the committee that the complexity and importance of soft matter at the mesoscopic scale will turn out to be a main focus. Indeed, the mesoscale is the scale where classic physics and quantum physics meet. This focus will have a far-reaching influence on physics research in the next several decades, with the structural characteristics of soft matter embodied at this scale.

2. Social Demands and Economical Requirements

Soft matter plays an increasingly important role in daily life. For instance, as a mainstream panel display technology that has achieved a global market value of over 100 billion USD, liquid crystal displays continue to renew innovative technologies, i.e., blue phase liquid crystal and flexible liquid crystal technologies. Furthermore, surfactants are one of the most widely applied types of soft matter and are applied in almost all economic and technical sectors, i.e., daily care chemicals, petroleum mining, agriculture, sanitation and environmental protection. To date, surfactants have acquired a share of the global market of several tens of billion USD. Moreover, macromolecular materials, with a turnover reaching 2 billion tons at the end of the 20th century, are widely employed in science and technology, national security and the domestic economy. These materials, together with metals and ceramics, are listed among the top three most important materials and have three main trends of high performance, multi-functionality and smartness. Note that smart soft materials are a major focus in current materials science research. For example, magnetorheological fluids are applied in intelligent vibration dampers for automobiles and equipment, while electrorheological fluids are considered to be one of the most promising materials for building vibration dampers applied in high speed railways and in national security and defence industries. Evidently, the identification and fabrication of new materials and devices with high performance, environmentally friendly materials, intelligent building materials, and innovations in biomedicine-related fields are the key elements driving soft matter research. As another example of emerging soft matter research, granular matter can play an important role in many practical issues of engineering, i.e., soil instability and the loss of riverbeds and river courses and the structural stability of large water conservation engineering.

In general, soft matter research has become one of the most important areas in contemporary materials science and engineering and has had a profound impact on the major issues faced by mankind in terms of materials, energy, the environment, medical treatment, etc. and is of great strategic value to our national welfare and the livelihood of the people.

Ⅱ. Characteristics of Soft Matter Physics

The term "colloid" , introduced by the British scientist Graham in 1861, marks the emergence of modern soft matter science. The Brownian motion theory of colloidal particles proposed by Einstein and Langevin, on the other hand, highlights the peak of soft matter physics research in the early 20th century. Due to their precise experimental confirmation of

Brownian motion, the French physicist Perrin and the Swedish physical chemist Svedberg were awarded the 1926 Nobel Prizes in Physics and Chemistry, respectively. Furthermore, the 1925 Nobel Prize in Chemistry was awarded to Zsigmondy for his remarkable work on the heterogeneous property of colloid solutions; the 1932 Nobel Prize in Chemistry, to Langmuir for his work on surface absorption; and the 1936 Nobel Prize in Chemistry, to Debye for his static shielding theory.

Another major discovery in soft matter within the same period was the identification of the "macromolecule" . This concept was formally coined by the German scientist Staudinger in 1920 and demonstrated in 1938. Staudinger was awarded the Nobel Prize in Chemistry in 1953, and his countrymen Ziegler and Natta were awarded the Nobel Prize in Chemistry in 1963 for their achievements on macromolecule catalytic synthesis. The American scientist Florey won the 1974 Nobel Prize in Physics for his systematic efforts in macromolecular polymerization mechanisms, macromolecular conformation statistics and rubber elasticity. The French physicist P.G. de Gennes exclusively won the 1991 Nobel Prize in Physics for his great theoretical achievements across the superconductor, liquid crystal and macromolecular fields. The efforts made by these scientists demonstrate the equal importance of basic and applied research in soft matter, pushing the entire field of soft matter research and application to a new level.

Liquid crystals are another class of soft matter of particular importance. The study of the optical and electrical properties of liquid crystals spurred the American scientist Heilmeier, and the German physicist Helfrich et al. to propose the tentative idea of using liquid crystals as display devices in the 1960s and 1970s, respectively. Due to these contributions, these scientists shared the Draper Prize, praised as the Nobel Prize in Engineering, in 2012. Helfrich also pioneered the theoretical study of liquid crystalline bio-films. On this basis, Chinese theoreticians, i.e., Ouyang Zhongcan et al., made great efforts to extend the impact and established a complete theory of liquid crystal films and biomembranes.

Since the 1960s, biological systems have gradually become one of the key subjects of soft matter research. Biological macromolecules and their ordered micro-nanostructures are most complex and diversified. Related studies have been accelerated due to the emergence of quite a number of new experimental and computational technologies, such as the multi-scale simulation technique (2013 Nobel Prize in chemistry) and ultrahigh resolution microscopic technique (2014 Nobel Prize in chemistry). The development of the single molecular technique in the 1990s brought biological study to the single-molecule level, which greatly extended the scope of soft matter research.

In the last few decades, another important soft matter, supramolecular materials, have emerged; these materials have been widely studied and applied in molecular machines, bio-sensors, nanoreactors, chemocatalysis, drug delivery, gene delivery, biological imaging and tissue engineering, which has led to great impact in these fields. It should be noted that wearable, implantable, bio-degradable/ absorbable and injectable flexible devices will exert a huge impact on human health and life in the future, and flexible materials, as novel functional materials, will play a key role in the conversion from materials to devices.

In general, since soft matter research was begun more than one hundred years ago, people have identified a variety of soft matter systems. These materials share some common characteristics, such as notable thermal fluctuations, multiple metastable states, mesoscopic multiscale self-assembled structures, entropy-driven order-disorder transitions, and macro flexibility. Briefly, these systems often show a large response to small stimulations and display strong nonlinearity. These characteristics are hard to explain based on their microstructures (either at the atomic or molecular level) but are more related to their mesoscopic self-assembled structures. Though people have gained some knowledge on these characteristics, a comprehensive understanding of soft matter physics has not yet been established. Establishing this understanding will become the main focus of future soft matter physics.

Ⅲ. Current Status and Trends

Due to the great relevance of soft matter physics in basic research and its strategic importance in economic and social developments, numerous research centres/initiatives have been set up in many countries to promote soft matter research.

In the past few decades, numerous physics departments and research institutions of most universities in developed countries, i.e., Harvard University, Princeton University, Los Alamos National Laboratory, Cambridge University, University of Oxford, Heidelberg University, Normale, Delft University of Technology, University of Queensland and Kyoto University, have established research directions for soft matter physics. It should be noted that Prof. Michael Cates of the University of Edinburgh was named the Lucasian Professor of Mathematics by Cambridge University in 2015 for his contributions to soft matter research. It is the first time that this honourary title with its long history and lofty status was awarded to a prominent scientist in the field of soft matter research, highlighting the importance of soft matter research in the international scientific community. Currently, ICAM, a soft matter-based international complex self-adaptive matter organization, has 72 member institutions. The development trend for soft matter research can also be seen from international academic publications. Since the 1990s, many well-known professional journals in the international physics society (e.g., *Phys. Rev. Lett., Adv. Mat., Adv. Func. Mat.*) have launched a soft matter subject or published special issues focused on soft matter. Accordingly, a large number of monographs or teaching materials for soft matter physics have emerged in the last decade.

China's research on soft matter physics dates back to the 1980s. The Institute of Theoretical Physics and the Institute of Biophysics, CAS, the University of Science and Technology, the Nanjing University, Xiamen University and the Beijing Normal University were among the first to carry out research on soft matter in China. The National Nature Science Foundation of China (NSFC) and the Chinese Academy of Sciences have devoted great efforts and substantial funding to research on soft matter physics and biophysics. *The National Program for Medium and Long-Term Scientific and Technological Development* (*2006–2020*) has listed "soft condensed matter" as one of the scientific frontier issues. NSFC started a "Controllable Self-Assembly Systems and Their Functionalization" key research project in 2010, and the Ministry of Science and Technology has also heavily sponsored soft matter and materials research, especially biological soft matter research, in the key nano research programme. In addition, in the last decade, a great number of domestic universities and research institutions have employed internationally well-known soft matter scientists and set up experimental or computational laboratories, such as the CAS Key Lab of Soft Matter Chemistry and the Key Lab of Soft Matter Physics, the Research Institute for Soft Matter and Biomimetics of Xiamen University, the Soft Matter Science & Engineering Center of Peking University, the Institute of Biophysics of Nanjing University, the BUAA International Research Center of Soft Matter and Application, the Soft Matter Research Center of Zhejiang University, the Center for Soft Condensed Physics & Interdisciplinary Research of Suzhou University, and the SJTU Soft Condensed Matter Physics Laboratory. In addition, several tens of workshops, lectures and summer schools for soft matter physics have been held, with more than ten types of relevant teaching materials and monographs compiled or translated and published since the early 1990s.

Currently, soft matter physics is one of the cutting edges in international physics research. China has built a solid foundation in this area and is now making rapid progress in funding and promoting soft matter research. This attention may not only lead to breakthroughs in basic research but also exert a major impact on industrial production, and, inevitably, on our lives.

Ⅳ. Key Issues in the Next Five to Ten years

Soft matter physics mainly includes three aspects: observing and measuring the complex phenomena of soft matter systems, establishing precise and quantitative models and theories for different soft matter systems, and uncovering the underlying universal laws governing soft matter systems. The key scientific issues and important research directions in the next few years are briefly described as follows.

1. Basic Theory of Soft Matter

Elasticity theory, phase translation theory, scaling theory, and self-consistent field theory will remain the mainstream in theoretical research on soft matter systems. Some new theories and computational models are expected to be developed for the dynamics of granular matter systems as well as viscous, diffusional and interfacial systems. For complex soft matter systems, new multiscale computational simulation methods are required.

2. Experimental Methods for Soft Matter

Soft matter experimental methods and techniques remain to be greatly improved in terms of both the signal-noise ratio and temporal and spatial resolution. it is hoped that he single-molecule manipulation technique will achieve a higher force resolution. Problems involving temporal resolution and system stability will soon be solved for the single-molecule fluorescence technique. The imaging of living cells at the single-molecule level needs the improvement of high-speed image scanning techniques. Many improvements can still be made in small-angle scattering, such as the establishment of more accurate structural factors and model-independent data analysis methods.

3. Soft Matter Mesoscopic Systems

3.1 Supramolecular materials

Supramolecular materials are a new system that has received wide attention in the last decade and has greatly influenced research in many related fields. Developing supramolecular gel materials that show biocompatibility, non-toxicity and degradability has become a main focus in recent years. In-depth study of the interaction between supramolecular materials and biosystems and the development of multifunctional supramolecular gels with better applicability is another important direction.

It should be noted that silk materials, as a special type of natural supramolecular material, show good biocompatibility, mechanical performance and optical/electrical properties, and have attracted particular attention. In addition to being applied in biomedicine, drug delivery and skin care, silk materials, especially functional silk materials, will play a key role in the development of biocompatible, resorbable, injectable, flexible photo-electronic devices, which are supposed to achieve a market value of trillions of USD in the future.

3.2 Polymers

Polymers have been proven to be the material with the highest practicability. Urgent problems in basic research in the near future include theories beyond the mean-field approximation and dynamic theory of non-equilibrium systems. Some other key issues include dephasing and crystallization, the glass transition, the non-equilibrium phase transition at various scales, the structure-property relationship of functional (optical, electrical and magnetic) materials and the assembly mechanism of confined polymer/nanoparticle composite systems. Intelligent polymer materials are one of the hottest research subjects in recent years, and it is hoped that in the future, novel systems with composited modification or multiple functions will be developed, that novel mechanisms for the memory driving and shape recovery of intelligent composite materials that respond to electrical, thermal, optical, and magnetic stimulation will be discovered, and that the response behaviour of these systems to coupled multi-physical fields will be studied.

3.3 Liquid crystals

For the theoretical research of liquid crystals, it is necessary to develop molecular models with higher precision than those currently available. The problems of confined liquid crystal systems, such as the self-consistent field simulation of liquid crystal structures on a curved surface and the orientation distribution of liquid crystal molecules in a confined system, will receive more attention. Liquid crystal displays such as three dimensional, blue-phase and flexible displays will remain a hot topic. As a new intelligent material, liquid crystalline elastomers will demonstrate a vast potential for applications such as dynamic keyboards, flexible devices, microlens and microfluidic devices.

3.4 Granular matter

In recent decades, granular matter has made important progress in terms of transport properties, phase transition mechanisms for granular flow, etc. However, researchers still have a very limited understanding of the properties and mechanisms of such systems. It is hoped to achieve great progress in the following four aspects: Granular gases: efforts should be made to reveal the speed distribution law similar to that for equilibrium statistical mechanics. Granular liquids: the focus should be on dynamic research; e.g., research on the shock wave structure in granular flow will facilitate our understanding of the dissipation characteristics and the phase transition in the flow. Granular solids: attention should mainly be devoted to the structure of the non-linear force-transmission chain structure, the constitutive relation of large-scale granular media, and the interaction between anisotropic particles and charged particles.

4. Low-Dimensional and Interfacial Systems of Soft Matter

4.1 Biological membrane

Membranes are quasi-2D systems, of which the biological membrane is the most typical. The spontaneous curvature model proposed by Helfrich in 1973 laid the foundation for the theory of biological membrane elasticity, and Chinese scholars have made tremendous efforts and influential achievements. Research on biological membrane elasticity theory will be continued in the near future, such as an analytic solution of the membrane equation, and the theory and application of a biphasic membrane. Another important topic is the interaction between membranes and proteins.

4.2 Surfaces

Super-wettable surfaces have great application potential and have long been studied; however, a number of problems remain in need urgent solutions, e.g., uncovering the biomimetic principles of the biological interface with special properties and studying the relationships among microstructure, composition and function; designing functional molecules and fine-tuning/characterizing their surface microstructure based on biomimetic principles; simulating special functions occurring in nature by controlling and optimizing the structure and composition of biomimetic intelligent multiscale interfacial materials; developing large-scale preparation technologies; and applying biomimetic intelligent multiscale interfacial materials in fields such as resource, energy, environmental studies and biomedicine.

4.3 Colloids

Colloids are widely used in daily life. Benefiting from the development of new technologies, the research on colloid physics and chemistry is going increasingly deeper, with a new trend emerging for utilizing colloid systems to study fundamental physical issues. In the future, it will be necessary to develop colloid systems with complex external shapes and complex interactions to study more complex phase transition behaviour, improve and develop various kinds of control means to obtain more complex crystal structures, set up self-driven colloid systems for the study of non-equilibrium self-organization behaviour, and study the phase transition and self-assembly behaviour of mixed systems with colloid particles and various types of organic molecules. In addition, composite colloid systems comprising surfactants and other soft matter are also a direction worthy of great effort.

4.4 Water at the microscopic scale

In recent years, people have gradually recognized that because of insufficient understanding of the extraordinary properties of water at the microscopic scale, it is hard to achieve breakthroughs for many water-related technologies. There are still unsolved problems, for example, the physical mechanism underlying the amazingly high flow (quasi-one dimensional water chain) in nanochannels, water flow behaviour in the presence of ions in carbon materials such as carbon nanotubes, structural changes when water transitions from high flow to intermediate flow; and water properties at the microscopic interface and the influence of these properties on wetting and flow. Since the concept of "biowater" was proposed, the impact of water on biomacromolecular structural stability has not yet been clarified, nor has the impact of the water environment on material structure and properties has not yet been fully clarified.

5. Biosystems

5.1 Biological soft matter physics

Biological soft matter physics is the most important interdisciplinary frontier between physics and biology, with its research scope covering molecular biophysics and cell biology. It utilizes and refers to the concepts, theories and technical means of condensed matter physics, statistical physics, computational physics, mathematics, information science and computer science. Current frontier research subjects include biomolecular interactions and structural and functional dynamics; non-coding sequences, non-coding genes and RNA; biomembrane-related structures and dynamics; cytoskeletal self-assembly, aggregation structure and dynamic behaviour, single-molecule biology, including various new techniques; the topology and dynamics of bio-networks; nerve systems; bacterial biophysics; and biology-inspired physical and engineering problems.

5.2 Interaction between nanoparticles and protein and cell membranes

The application of nanomaterials in biomedicine and health is the hotspot of nanoscience, with great worldwide efforts made to study the biological effects of nanomaterials. The current focus has gradually been transferred from the overall biological effects to the interactions between nanomaterials and cells and functional biomacromolecules. The frontier problems needing urgent solution include the following: What is the driving force for nanoparticles to enter cells? Do new trans-membrane processes and mechanisms exist for particles smaller than 100 nm? What is the molecular mechanism for the interaction between nanoparticles and biomacromolecules and are there any specificities? Can such specificities interpret the special biocompatibility and safety of different nanomaterials?

5.3 Bioinformatics and big data

Studies of intelligent analysis methods for understanding big data in biology are now at the cutting edge of bioinformatics. The research and development of advanced algorithms and methods for feature extraction from biological data is currently an important trend. Related subjects include the following: constructing new frameworks for massive bio-data-oriented dynamic learning; transferring learning algorithms for multi-source heterogeneous bio-data, especially transfer learning mechanisms and the corresponding intelligent modelling approaches for multi-type/multi-media bio-data or bio-data of different species and different levels; generating in-depth learning theories and methods for mining information hidden in hierarchical bio-data; and establishing a multi-source heterogeneous bio-data mining platform.

5.4 Soft matter biomedicine

The advances in cancer biophysics arising in the last decade have created a new direction for cancer research that focuses on the relationship between cancer cells and the microenvironment in the invasion and metastasis process, the route and mode of cell group invasion, and the collaboration and coevolution of cancer cells. In cancer invasion and metastasis research, soft matter physics may make groundbreaking contributions in the following aspects: quantitative research on biomechanical signals for intracellular signal pathway coupling and the construction of an accurate controllable 3-D tissue microenvironment platform. With respect to cancer-related theories and simulation, the future possible research scope may cover cell signalling transduction mechanisms; genetic expression regulatory mechanisms; and bionetwork structures, dynamics and functions.

6. Other Fields

6.1 Soft matter microrheology and microfluidic devices

Microfluidics is a technology for the accurate control and precise manipulation of fluids on the micrometer or sub-micrometer scale and has a huge application potential in biology, chemistry, medicine, etc. The rapidly evolving frontiers include microchannel design and manufacture, materials selection and preparation, microfluidic control system integration and encapsulation, and fluid transport control on the micrometre scale. In addition, many new applications are still in their infancy, such as the design and development of intelligent biomimetic materials, the simulation of porous media, the development of continuous flow organic synthesis, and the preparation of micro- or nano-particles. Microfluidic control research will bring great economic benefits to seawater desalination, oil-water separation, drug and biomolecular screening tests, degassing technology for complex fluids, etc.

6.2 Intelligent soft materials

Intelligent materials are fourth-generation materials, following natural materials, synthetic polymer materials and artificial materials. Intelligent materials possess seven functions, namely, sensing, feedback, information identification and accumulation, response, self-diagnosis, self-recovery and self-adaption, and are currently the hottest frontier in materials research. For example, giant electrorheological fluids are considered to be the most promising soft matter intelligent material that can be widely used in the high-speed rail, national defence and military industries. In addition to responding to electrical and magnetic stimulation, intelligent soft matter materials that can respond to thermal, optical and other stimulations will also become the key issues in this field. The relevant basic theoretical issues include the following: solid-liquid phase matching of electro- and magneto-rheological fluids and the diversity of the dispersion phases; physical mechanisms for improving electro- and magneto-rheological fluid performance by the surface modification of nanoparticles or the addition of surfactants; performance evaluation of solid nanoparticle precipitation and extreme applications; evaluations of the service and failure of intelligent materials under different practical conditions, etc.

Ⅴ. Suggestions on Discipline Development and Funding Policies

Soft matter physics is an extensive interdiscipline with greatly diversified and scattered research subjects, which makes it hard to rapidly form a clear theme. In China, the absolute number of physicists majoring in this field is small, and as the disciplinary attributes of this field are not very clear, it is hard to obtain sustainable support from fixed channels. This limitation not only hinders the consolidation of existing research teams and the promotion of academic exchange and cooperation but also impedes sustainable cultivation of young research talent. To promote the stable development of China's soft matter physics research, we propose the following suggestions.

1. Establishing Special Funds

At present, no independent channel exists for soft matter physicists to obtain financial support from national scientific research programmes such as those supported by the NSFC. As a striking contrast, traditional fields (e.g., particle physics and condensed matter physics) have specific application channels, while soft matter related fund applications have to be made via condensed matter physics and statistical physics channels in a dispersive manner. The innate inclination in policy has hindered soft matter researchers from obtaining steady financial support, thus seriously restricting disciplinary development. We suggest setting up a specific code for soft matter physics to better consolidate and expand the research community and to incorporate soft matter physics into higher-level national scientific research programmes such as the National Science and Technology Major Project, National Key Research and Development Plan, Targeted Research Project of Technology Innovation, and Base and Talent Project.

2. Establishing and Supporting Influential Research Bases

Establishing and keeping a scaled and stable research society is especially important to soft matter physics as an emerging interdiscipline. At present, among universities and research institutions in China, only a few have actually conducted soft matter physics research, with various restrictions also in talent introduction. We suggest that supportive policies at the national or institutional level should be released; a number of distinctive research bases should be set up and given key support; and high-level talent should be attracted, cultivated and stabilized to gradually expand the research team and build influential domestic and overseas research platforms.

3. Strengthening Industry-University-Research Cooperation

Soft matter physics is closely related to our daily lives. Shortening the distance between basic research and practical application is an irresistible trend in China's scientific research community and an inexorable requirement of the current innovative society. We suggest imposing obligations on the above bases by promoting communication and cooperation between the academic and industrial circles and by realizing multilevel infusion and the sharing of personnel, projects, data and IPR.

4. Enhancing Basic Education

Though soft matter physics is not only an emerging interdiscipline but also an important component of physics, its core knowledge (e.g., entropy-driven order) has not been embodied in domestic basic physics education. It is particularly concerning that statistical physics, which is a core course component of soft matter physics, has been obviously weakened in colleges, making it hard to guarantee the availability of qualified teachers and courses. In contrast, both the theory of relativity and quantum physics have increasingly and widely infiltrated into basic courses and even into popular scientific books, which can be seen everywhere on the market. This occurrence has misled students to a certain extent, making them seriously lack recognition of soft matter physics, which is extremely unhealthy to the overall development of physics. We suggest supporting the textbook writing, publishing popular scientific readings and enhancing the teaching and popularization of the basic concepts of soft matter physics in undergraduate education in order to create a good academic atmosphere for cultivating research talent in physics with comprehensive quality.

Li Ming (University of Chinese Academy of Sciences),
Shuai Jianwei, Liu Xiangyang (Xiamen University),
Ouyang Zhongcan (Institute of Theoretical Physics, CAS)

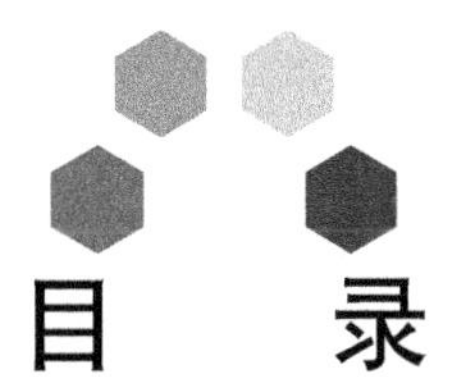

目 录

下

第七篇　软物质交叉领域

上

中

第八章　二维各向异性胶体热力学

第九章　活性胶体

第十章　纤维体软物质

第十一章　微观尺度下的水：从准一维、二维受限空间到生物以及材料表面

第六篇　软物质生物体系

第一章　DNA及DNA分子计算

第二章　核酸分子机器

第三章　蛋白质分子折叠、聚集和功能运动动力学

第四章　蛋白质结构预测

第五章 生物分子马达

第一节 生物分子马达简介

从物理学角度看，生物分子马达（以下简称分子马达）是实现化学反应（标量过程）与定向运动（矢量过程）之间相互转化的蛋白酶，其尺寸为几个或数十个纳米，是人类梦寐以求的天然纳米机器。通常分子马达可以催化ATP（有时为GTP）水解并利用其释放的能量，一些嵌膜分子机器还可利用跨膜离子的电化学势差，这些驱动过程都发生在马达的某些特殊位点上。由于马达自身结构“设计”得极为精巧，局部过程释放的化学能可进一步推动马达产生较大尺度的构象变化。一旦马达与轨道或膜结合，这种构象变化就会使得马达产生相应的相对运动，从而具备“运动性”（motility），实现将化学能转化为物理运动的目的[1]。

自20世纪70年代分子马达这一概念被明确提出之后，由于生物观测技术以及单分子生物物理学等领域的飞速发展，人们逐渐认识到很多已知的蛋白酶本质上也是分子马达（如DNA聚合酶），定向位移或对外发力这些运动特征是其实现生物功能所必需的。研究表明，在生命的所有基本过程（如DNA复制、基因转录、翻译、物质输运、细胞运动与分裂、ATP合成、肌肉收缩）中，分子马达均扮演了最关键的角色。与分子马达缺陷相关的疾病最近也陆续被发现，例如，扩张型或肥厚型心肌病、视网膜色素变性和失聪等与肌球蛋白突变有关；导致男性不育的原发性纤毛运动障碍和无脑回畸形症等与动力蛋白失活有关；心血管疾病等则与某种驱动蛋白基因的过度表达有关，而 ATP 合酶失活则直接导致代谢终止。可以说，分子马达是生命活动的真正推手。

粗略地看，生物体可视为“软物质 + ATP”的耦合非平衡系统，分子马达则是这类系统的微观原型，也是生物体宏观运动的根源。自分子马达发现后，物理学家就表现出浓厚的兴趣，这体现在两个层面。其一为结构层面，分子马达不是无结构高分子，而是具有精巧结构的蛋白酶，其结构如何形成、如何在工作循环中维持、如何响应环境变量（包括外界操控）？对于纤毛、鞭毛、肌纤维等以分子马达为主体的规则介观系统，其结构如何形成（是否仅依靠热力学自组装）、如何维系、什么情况下解体？这些问题与蛋白折叠、蛋白相互作用等问题密切相关，迄今仍是生物物理学中最困难的挑战。其二为动力学层面，分子马达一旦获得正确的结构，它是如何利用这个结构实现“力学-化学耦合”，从而展现出远比一般软物质系统精妙得多的时空有序行为？这不仅是实现多样化生命活动的关键，而且对物理学，尤其统计物理学的研究提出了重要课题。相比于结构层面的问题，动力学层面的问题在实验和理论上都更容易观测和分析，因此取得了不少进展，本章仅对这一方面进行介绍。

一、分子马达分类

从运动形式上，分子马达可分为转动马达和平动马达。转动马达包含 ATP 合酶（ATP synthase）、鞭毛马达（flagellar）和病毒 DNA 包装马达（DNA packaging motor）等；平动马达则包含驱动马达（kinesin）、肌球马达（myosin）和动力马达（dynein）、DNA 解旋酶（helicase）、DNA 和 RNA 聚合酶（polymerase）和核糖体（ribsome）等。令人惊叹的是，尽管这些马达本质上不过是单分子，但其晶体结构、电镜照片和实验分析显示，它们也具备了宏观机器的各种“组件”。转动马达结构上类似于人造电机，由“转子”和“定子”两部分组成，其中 ATP 合酶是可逆的，既可以是“电动机”也可以是“发电机”，“电”指的就是生命体能量货币 ATP。有的平动马达类似于人体的构造，具有“足”和“躯干”，可在轨道上步行或跳跃，如驱动蛋白、肌球蛋白；有的则类似于穿孔机，可以在轨道上滑动，如 DNA/RNA 聚合酶。平动马达可分为持续（processive）和非持续（nonprocessive）马达。所谓持续马达是指那些沿轨道做长距离定向运动而不脱轨的马达。如果定义马达在一个化学循环内吸附在轨道所占时间比为“占空比”(duty ratio)，则持续马达的“占空比”接近 100%，可在细胞内独立担当各种任务，如输运囊泡等；

而非持续马达的占空比＜2%，经常以集体形式发挥作用，如肌肉收缩就是肌球马达Ⅱ集体运动形式之一。ATP 合酶、驱动蛋白-1 和肌球蛋白Ⅱ等马达由于历史的原因受到更多关注，目前已成为该研究领域的模式马达，也是本章讨论和介绍的重点。

二、平动马达一例：驱动蛋白

驱动蛋白（kinesin-1）是通过水解 ATP 获取能量并沿微管定向运动的平动马达，自 30 年前被发现以来至今已累积了 14 个类别，它们的头部（马达域）序列是保守的[2-4]。驱动马达可以根据其马达域所处肽链的位置分为 N 端或 C 端马达；也可以根据肽链的数量分为单体、同型二聚体或异型二聚体等；还可以根据运动方向分为微管正向或反向马达。其中对传统的 kinesin-1 类（此后简称“驱动马达”）研究得最多，是分子马达研究的模式系统之一。

驱动马达是已知步行（hand-over-hand）马达中尺度最小的，是同型二聚体 N 端正向马达（图 6-5-1），每个单体独自构成四个功能域，分别是 N 端的球状头部、颈链（β9-β10）、茎区（coiled-coil stalk）和 C 端扇形尾部。尾部负责连接囊泡等“货柜”，两个单体通过茎区的 α 螺旋形成二聚体。每个头部有两个作用位点，其一结合 ATP 并催化其水解，为马达步行提供能量；其二则负责与微管结合[5]。马达运动具有两个主要特征，其一是持续性，它能沿微管长距离步进而不脱轨，原因是在两头交替步行时至少有一头保持与微管结合，当其中一个头（空态或结合 ATP）与微管强结合时，另一头（结合 ADP）则与微管弱结合，由此保证两头轮换。另一个特征是马达的力学与化学之间是紧耦合的，即马达步长严格为 8nm（微管极性周期），每前进一步消

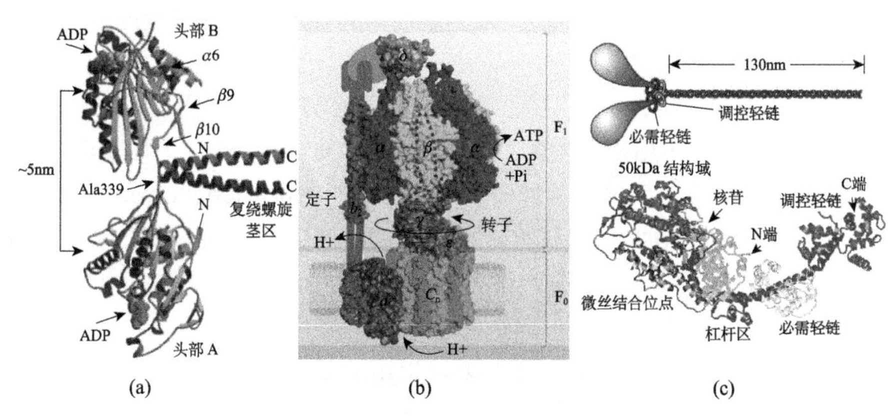

图 6-5-1　模式马达结构图

驱动蛋白（a）[5]、FoF$_1$-ATPase（b）[10]、肌球蛋白Ⅱ（c）[12]

耗一个 ATP。上述两个特征要求两个头部的核苷化学态及与微管的作用态需通过某个机构来协调统一，必须阻止两头同时脱离微管或同时水解 ATP。研究表明，担当此协调角色的正是颈链部分[6-8]，但关于其作用机制的理解却存在几种不同的模型。后文将对此进行详细讨论，此处不再赘述。

三、转动马达一例：ATP 合酶

ATP 合成酶（简称合酶，FoF_1-ATPase）是嵌膜复合蛋白，广泛存在于细菌、线粒体和叶绿体中。位于分泌泡上的合酶也称 V-ATPases，而古细菌中的合酶又称 A-ATPases。ATP 合酶在结构上可分为嵌膜的 Fo 和亲水的 F_1 两个转动马达（图 6-5-1）。三对 αβ 亚基环成 F_1 的“定子”，γ 则是其非轴对称“转子”（偏心轴）。膜内的 a 蛋白和质子通道 C_n 分别是 Fo 的“定子”和“转子”。两个马达的“定子”和“转子”分别通过 δb_2 蛋白组合和 ε 亚基连接成一体，耦合成 FoF_1-ATP 全酶[9,10]。

实验研究表明，Fo 部分在跨膜质子流驱动下发生转动。由于缺乏其结构的完整信息，该驱动机制迄今尚未阐明。可以肯定的是，Fo 部分将质子的跨膜电化学势转化为 F_1 的 γ 亚基的转动（图 6-5-1 中从下往上看的顺时针转动）。由于 γ 亚基与 $\alpha_3\beta_3$ 环直接接触，当其转动时可以推动三对 αβ 亚基陆续产生形变，这个形变导致 αβ 亚基将 ADP、P_i 合成为 ATP。F_1 是步进（stepping）马达，每步进一次转动 120°，因此，γ 亚基每转动一周可以合成 3 个 ATP。另外，如果 ATP 浓度足够高，则 ATP 会在 F_1 中水解，导致三对 αβ 亚基交替地发生构象变化，推动 γ 亚基逆时针转动，从而驱动 Fo 跨膜泵送质子[11]。因此，FoF_1-ATPase 是可逆马达。跨膜质子动力势与 ATP 浓度这两个标量是通过一个共同的物理矢量（转动）来实现高效转换的，物理矢量的方向取决于两个标量强度竞争的结果。

在体内，FoF_1-ATPase 往往是不可逆的。在富氧环境中，ATP 合酶利用细胞呼吸或光合作用积累的跨膜质子动力势 ΔP 驱动 Fo，进而带动 F_1 将 ADP 和 P_i 合成 ATP。某些细菌如酒石酸泥杆菌（Ilyobactertartaricus）利用跨膜 Na+ 动力势合成 ATP，其结构与质子驱动的合酶类似。在厌氧环境中，许多细菌通过 F_1 水解糖酵解过程产生的 ATP，从而驱动 Fo 跨膜泵送质子产生 ΔP，后者可被其他跨膜机器（如鞭毛马达等）利用实现细胞的一些基本功能，如化学趋化（chemotaxis）等。必须指出的是，在线粒体、叶绿体和某些细菌体内，F_1 的 ATP 水解功能是被禁止的。

四、非持续马达一例：肌球蛋白Ⅱ

肌球蛋白Ⅱ(myosin Ⅱ)是肌肉收缩的“始作俑者”，也是最早被确认的分子马达。肌球马达Ⅱ是双头马达，其结构类似于驱动马达，只是没有茎区的划分，两个单体通过尾部的α螺旋形成二聚体（图6-5-1）。每个单体（myosin Ⅰ）都有一条贯穿始终的重链（heavy chain）构成N端头部（马达域）、颈链和C端尾部。头部包括ATP水解和细丝结合两个位点，颈链则是一个较长的α螺旋结构，包含converter domain，并缠绕着两条轻链（ELC和RLC）。按照目前的理解，颈链是马达功能最关键的部件，它起到“杠杆”作用，能将马达头部微小的构象变化放大成马达整体的形变。肌球马达家族已经有18个成员，但其多样性的主要区别在尾部序列，头部序列则相当保守[12]。

与驱动蛋白不同，肌球蛋白Ⅱ双头同时结合在细肌丝（由肌动蛋白构成）上的概率极低。单分子实验表明其中任一头只能在细丝上向其一端做有偏“跳跃”（所需的分子形变由颈链介导，详见后），而无法沿细肌丝步行，因此它被称为非持续马达，并进化出另类的集体运动模式。例如，在肌肉收缩的发力单元肌节中，众多的肌球马达Ⅱ组成束状粗肌丝（thick filament）与细肌丝平行排列，突出的肌球马达Ⅱ类似横桥（cross-bridge）处在粗细肌丝之间，钙离子则调控马达与肌动蛋白发生作用并催化ATP水解。在马达的集体作用下，最终粗细肌丝之间产生相对滑移，导致肌肉收缩[13]。

分子马达尽管种类繁多并业已形成几大家族，但实验（尤其是单分子实验）显示它们似乎共享某些宏观人造机器的特征，如弹簧、杠杆、转轴和开关，各部件在ATP水解过程中以协调的方式推动马达整体做定向运动。同时，为了理解分子马达的运动机制也引入了描述人造机器的概念，如力、弹性、阻尼和功等。以上这些图像和概念对于宏观机器非常自然，但对于由热涨落主导的软物质分子机器，到底只是有效的比喻，还是确实存在对应的“组件”结构？这成为分子马达或分子机器研究的主题，与之密切相关的是分子马达研究中最核心的物理问题，即马达如何利用其结构实现力学-化学耦合。本章将以驱动蛋白、ATP合酶和肌球马达Ⅱ为例对此做重点介绍。另外，也适当涉及纳米机器的体外构建和实际应用。

第二节　研究进展及现状

分子马达最早可追溯到19世纪对肌肉收缩中肌球蛋白（myosin）的研究。1864年，Kühne及其同事首先从肌肉组织中分离出肌球蛋白和细肌丝的复合物[14]，但直到20世纪40年代，肌球蛋白才被单独分离出来[15,16]。在此期间，两位研究肌肉的先驱Meyerhof和Hill在1922的诺贝尔获奖感言中引入了“热机”的概念来阐述肌肉的收缩和能量转换机制。1953年，Hanson和Huxley系统地研究了独立的发力单元肌节[17]。为了描述肌节内粗肌丝与细肌丝之间的往复滑行运动，Huxley在1957年提出了肌节收缩的横桥模型（cross-bridges）[18]；同年，Huxley也提出了肌肉结构和收缩理论，认为马达所呈现的不同构型态与肌丝的相互作用和ATP水解循环有关[19]。其他如驱动精子纤毛拍打运动的动力马达（dynein）则在1963年获得了确认[20]。而沿微管输运囊泡的驱动马达（kinesin）直到1985年才得到纯化[2,3]。此后，得益于分子生物学和基因技术，马达的发现和分类获得了突飞猛进的发展。至于“分子马达”这一术语的起源，至少可追溯到Spencer于1976年发表的一篇文章，他在叙述鞭毛马达结构时明确使用了“molecular motor”这一词汇[21]。

一、分子马达的实验研究

分子马达研究尽管历史悠久，但长期处在生化层面，对其功能和构象的认知大都是基于一些结构信息的猜测，酶学性质的探索也停留在传统的系综实验。至于细致的力学化学耦合机制研究则要等到单分子技术的出现。20世纪80年代，Sheetz、Spudich和Vale三位先驱先后用小球和荧光技术实现了驱动马达沿微管定向运动、细丝和微管滑行的单分子实时示踪[22-24]，他们也因此获得了2012年度Albert Lasker基础医学研究奖。对分子马达的精细定量研究则始于光镊技术。

光镊是1970年由贝尔实验室的科学家Ashkin利用光散射产生梯度力的原理开发出来的单分子操纵技术之一，其操纵对象是直径为10～10^4nm的颗粒，而力的大小$f=k\Delta x$，此处k是弹性系数，Δx则是颗粒偏离光阱中心的距离，f属于皮（10^{-12}）牛量级[25]。1986年，美国能源部前部长朱棣文利用共振激光和磁场梯度陷阱（磁光阱）加载在光镊装置上，成功俘获并冷却了直径为

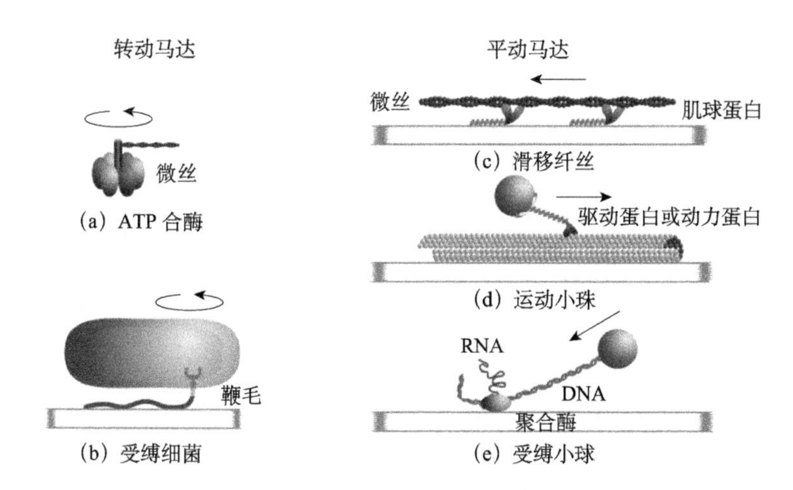

图 6-5-2 马达研究中的单分子技术示意图

(a) 转动马达 F_1 的单分子荧光实验揭示马达是 120°步进的，马达定子固定，转子粘上带荧光的细丝，显微镜可以探测 ATP 水解导致的荧光细丝转动。(b) 鞭毛马达驱动细菌胞体转动实验，鞭毛固定，胞身标记荧光，显微镜可以观察马达转动导致定子驱动细菌胞体的偏心转动。(c) 肌球马达驱动细丝滑行实验，肌球马达固定在底板，显微镜可观察到荧光细丝在集体马达作用下的滑行运动。(d) 驱动马达或动力马达受力作用的单分子实验，微管固定，马达尾部结合可以与光阱发生作用的微米直径的聚乙烯小珠，通过控制光阱中心与小珠中心的距离可以对马达施加定量的外力。(e)RNA 聚合酶受力单分子实验，聚合酶固定在底板，模板 DNA 一端结合小珠，光镊可以通过小珠对 DNA 施力。图片由 S.M.Block 惠赠

0.1nm 的中性原子，由此获得了 1997 年诺贝尔物理学奖。20 世纪 80 年代末，Ashkin 和 Dziedzic 首次利用光镊俘获了单个烟草花叶病毒和大肠杆菌，开启了光镊与生命科学研究相结合的新时代[26]。进入 20 世纪 90 年代，Spudich 和 Block 率先利用光镊在单分子水平上操纵分子马达，从而在纳米层次通过人为施力探测分子马达在涨落环境中的力学化学耦合机制。

（一）单分子实验揭示驱动马达为力学化学紧耦合的持续步行马达

尽管 Sheetz 和 Vale 在 1985 年用聚乙烯小珠固定在驱动马达尾部的方法实时观察到了马达沿微管的定向移动，但受制于仪器时空分辨率低的影响，实验结果得不到马达运动的高分辨轨迹，因此无法厘清下述两个问题：

（1）ATP 水解（化学过程）与马达运动（力学过程）是否有定量联系？

（2）马达前进的方式是步进的还是连续的？

从 1993 年开始，Block 等利用光镊针对上述问题展开了细致的研究。他们将微管固定在底板，马达尾部连上聚乙烯小珠。小珠可以被光阱中心俘

获，马达运动的轨迹可以由小珠轨迹表征，小珠和光阱中心轨迹可以同时测量。两者的距离 Δx 表示马达受力的情况［图 6-5-3(a)］，如果光镊位置固定，则马达受的外力随马达位移 Δx 而变，则光镊被称为位钳（position clamp）；反之，如果外围伺服驱动装置驱动光镊随时维持 Δx 不变，则马达受到恒定的外力，光镊也被提升为力钳（force clamp）[27]。位钳测得的马达轨迹［图 6-5-3(b)］表明双头驱动马达是步进（stepping）马达，步长是 8nm，对应于微管极性周期[28]。单分子随机度实验数据分析［图 6-5-3(c)］[29] 以及马达步进与 ATP 消耗速率之比数据分析［图 6-5-3(d)］[30] 表明，马达每步进 8nm 消耗一个 ATP，证明驱动马达的力学-化学是紧耦合的。光镊实验还测得马达的失速力（stall force）达 7pN[27,31,32]。

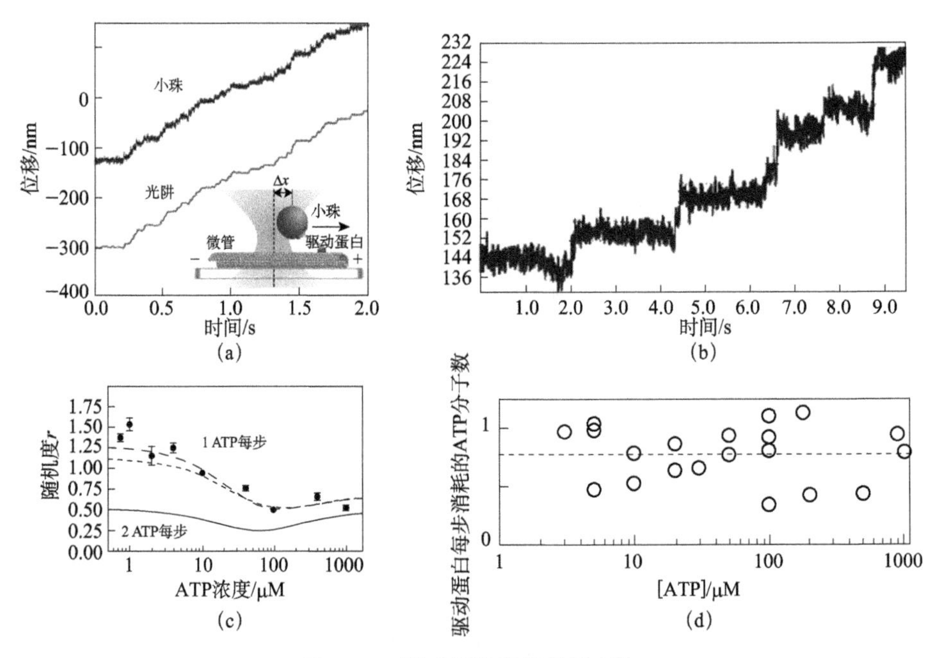

图 6-5-3　驱动马达的单分子实验

(a) 力钳实验装置示意图，微管固定在底板，马达尾部结合小珠，后者可以被光阱中心俘获，马达运动的轨迹可以由小珠轨迹表征，小珠轨迹和光阱中心轨迹可以同时测量。两者的距离 Δx 表示马达受力的情况，如果外围伺服驱动装置保持 Δx 恒定，则马达受到恒定的外力，光镊也被称为力钳[27]。(b) 驱动马达 8nm 步进轨迹[28]。(c) 随机度数据分析表明，驱动马达每步进 8nm 消耗一个 ATP[29]。(d) 驱动马达步进 8nm 的速率与 ATP 消耗速率之比接近 1[30]。(c) 和 (d) 结果表明驱动马达的力学化学之间是紧耦合的

小珠表征的驱动马达整体轨迹是步进的，但实现整体步进的马达双头采取怎样的策略呢？也就是说，是“步行”（hand-over-hand）还是“跛行”

（limping）或其他？ Selvin 等（图 6-5-4）利用荧光染料技术对驱动马达其中一头进行标记，然后用单分子技术观察荧光轨迹，发现马达其中一头的步长是 16.6nm（微管极性周期的两倍），表明驱动马达是步行马达[33]；2010 年，Ando 等利用自己开发的高速高分辨原子力显微镜实时获取了另一个大体积的双头持续马达肌球蛋白 V（myosin V）的步行影像[34]。这些结果似乎表明，双头持续马达的步行模式是自然进化所偏爱的模式。

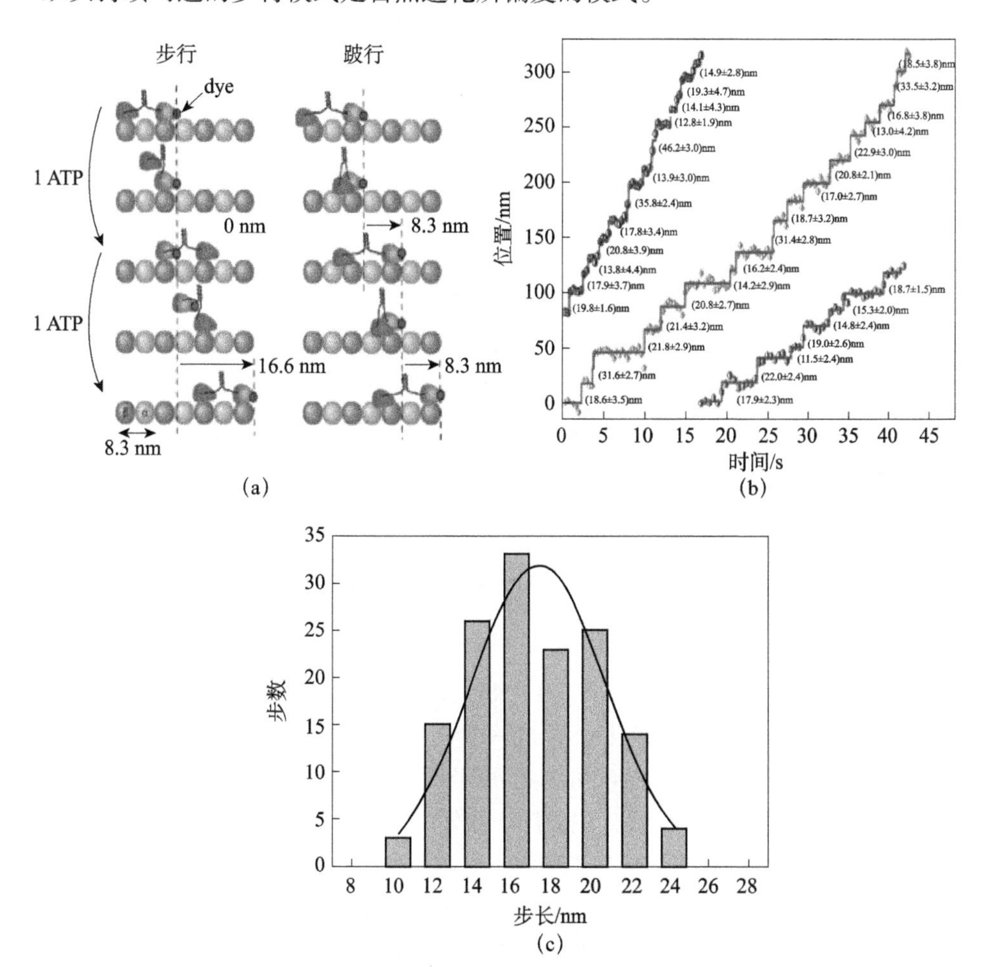

图 6-5-4 驱动马达的“步行”或“跛行”单分子实验

(a) 步行和跛行原理图，步行时每个头的行程是马达质心行程的 2 倍，而跛行时每个头的行程等同于马达质心行程；(b) 三个荧光轨迹展示驱动马达是步行马达；(c) 荧光轨迹步长（n=143）统计数据表明平均步长为（17.3 ± 3.3）nm，约等于微管极性周期或马达质心行程的 2 倍。图片摘自文献 [33]

另一个重要问题是马达两头的协同性机制，即马达的哪个部件负责协调双头交替步进且保证每跨一步水解一个 ATP？实验和动力学模拟研究表明，承担这个功能的是由 14 个氨基酸组成的颈链。颈链在此充当了弹簧、杠杆和开关的多重角色，对此存在若干不同理解。例如，较早提出的动力冲程（power stroke）模型[35]将之类比于直观的力传递。当两头都结合在微管上时，前导头为空态，后随头绑定 ATP，均为强结合态。此时，由于两段颈链伸直的总长度恰好约为 8nm，因此颈链会产生拉伸张力。当前导头也结合一个 ATP 时，与之连接的颈链向前倾斜并与头部对接（docking），这导致一个额外的张力通过整个颈链传递到后随头，后者水解所携带的 ATP 从微管上脱落并向前做有偏布朗运动，直至结合到微管上的前方位点，完成一次跨步。不过，最近人们综合大量单分子实验的结果，开始倾向于另一种“化学门控”（chemical gating）模型。该模型认为颈链的伸直状态本身即会阻碍前导头结合 ATP，进一步研究表明，前导头颈链后倾会阻碍头部与 ATP 的结合。换句话说，只有当后随头从微管上脱离，使得前导头颈链构象改变时，ATP 才能结合到前导头上，这样就造成两头之间的步调协同[36]。颈链上的拉伸张力对此协同性没有决定性贡献，但可以影响马达的发力以及能量转化效率[8]。至于最终哪个模型正确，或者两者都只是某个更完整的机制的极限情况，还有待于将来更细致的研究。

（二）单分子实验揭示 F_1 马达也是力学化学紧耦合的持续步进马达

在线粒体、叶绿体和细菌体内，绝大部分 ATP 都是由转动马达 ATP 合酶合成的，而且马达每步进 120°合成一个 ATP，即转一圈合成三个 ATP，但我们对图 6-5-1 所示的结构和功能的认知经历了很长的过程。

20 世纪四五十年代，科学家们已经发现细胞呼吸或植物光合作用时，线粒体或叶绿体内都有大量 ATP 聚集，从此引起了科学家的重视和探索。Racker 一直从事线粒体中有关 ATP 合成的酶的分离工作，1960 年他和同事发现剩余的线粒体膜片能够呼吸但不能合成 ATP，他们马上意识到将分离出来的组分放回去可以恢复氧化磷酸化功能，事实果然如此，因此他们将分离出的组分命名为因子 1（F_1：factor 1）[37]；1961 年 Mitchell 提出了“化学渗透偶联学说”（chemiosmotic hypothesis），首次建立了化学能（质子电化学势）与生物能（ATP）之间的转换理论[38]，但认为质子转移与 ATP 合成是直接耦合的；1965 年 Racker 和 Kagawa 发现了对寡霉素敏感的因子（Fo：factor

of oligomycin），由于寡霉素对 ATP 合成也很敏感，因此 F_1 与 Fo 之间必然存在某种内在的联系，同时也证明 ATP 合成（F_1）并不直接耦合到呼吸系统，而是间接耦合到跨膜质子梯度（通过 Fo）[39]，该发现也证实了 Mitchell 化学渗透偶联假说的正确性。

基于 Racker 提供的 Fo 和 F_1 分离组分，P.D.Boyer 进行了大量的生化实验，并于 1973 年提出了 ATP 合成的“结合变换机制”（binding change mechanism），该机制的核心是“定子”环 $\alpha_3\beta_3$ 的三个位点处在不同的构象：开放态（β_E：可以是空态，也可以是 ADP+P_i 态）、半开态（β_L：ADP 与 P_i 结合成 ADP.P_i，准备合成 ATP），以及闭合态（β_T：ATP 已经合成）。“转子”γ 亚基每转 120°，三催化位点的构象顺序轮换一次，即处于闭合态的位点转换到开放态并释放 ATP；处于半开态的位点切换到闭合态，合成 ATP；原先开放且已纳入 ADP、P_i 的位点切换到半开态。由于三对 αβ 亚基结构上是全同的，而同一时刻分别处于三种不同构型，实现这一目标的唯一可能性是中心转轴 γ 亚基结构上具有轴向非对称性 [40]。1994 年，Walker 解析了 F_1 的晶体结构，证明了 γ 亚基轴向非对称性及三个位点处在不同的状态 [41]。但所有这些研究都没有将 ATP 合酶上升到转动马达的层次，直到单分子技术的应用。

1997 年，日本 Yoshida 小组（图 6-5-5）首次创造性地将 F_1 的“定子”$\alpha_3\beta_3$ 环固定在底板上，“转子”γ 亚基顶端结合荧光细丝，在显微镜下观察到了 ATP 水解驱动的荧光细丝的定向转动，从而证明了 ATP 合酶是天然进化的分子马达 [42]，该工作直接促成了诺贝尔评奖委员会将当年的化学奖颁给了对 ATP 合酶的研究做出了开创性工作的两位健在的杰出代表 Boyer 和 Walker。

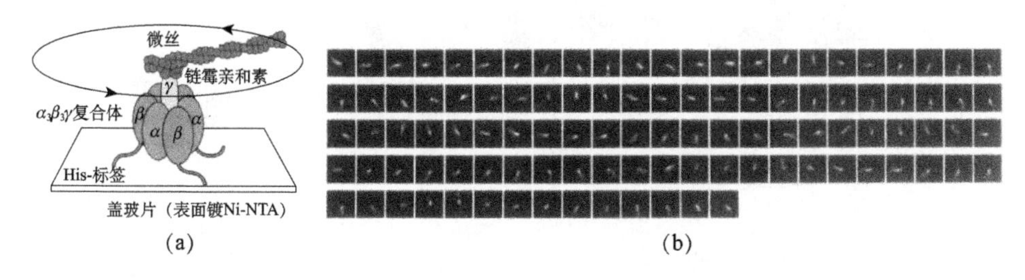

图 6-5-5 转动马达 F_1 的单分子实验

(a) 实验示意图，“定子”$\alpha_3\beta_3$ 环固定在底板上，“转子”γ 亚基顶端结合荧光细丝；(b) 荧光细丝转动时序图，图像时序间隔 133ms，细丝长 2.6μm，转速约 0.5rps。图片摘自文献 [42]

1998 年，Kinosita 小组（图 6-5-6）再接再厉，对单分子实验进行量化，发现 F_1 是步进马达、每步 120°，同时发现低浓度下马达每转一圈水解 3 个

ATP，且化学能转化为力学能的效率接近 100%[43]。2001 年，该小组用 40nm 直径的小球代替阻尼较大的荧光细丝并采用超高速摄影技术，发现 120°步进可进一步分成 90°和 30°两个子步，存在四个时序：与 ATP 浓度有关的结合等待、90°快速步进、决定最大速度的临时停留（interim dwell）以及 30°快速步进。同时在较大 ATP 浓度范围再次证明马达的力学化学是紧耦合的，即转一圈合成三个 ATP[44]。2003 年，他们进一步确认两个子步分别为 80°和 40°，且 ATP 水解发生在临时停留时段，耗时约 1ms[45]。

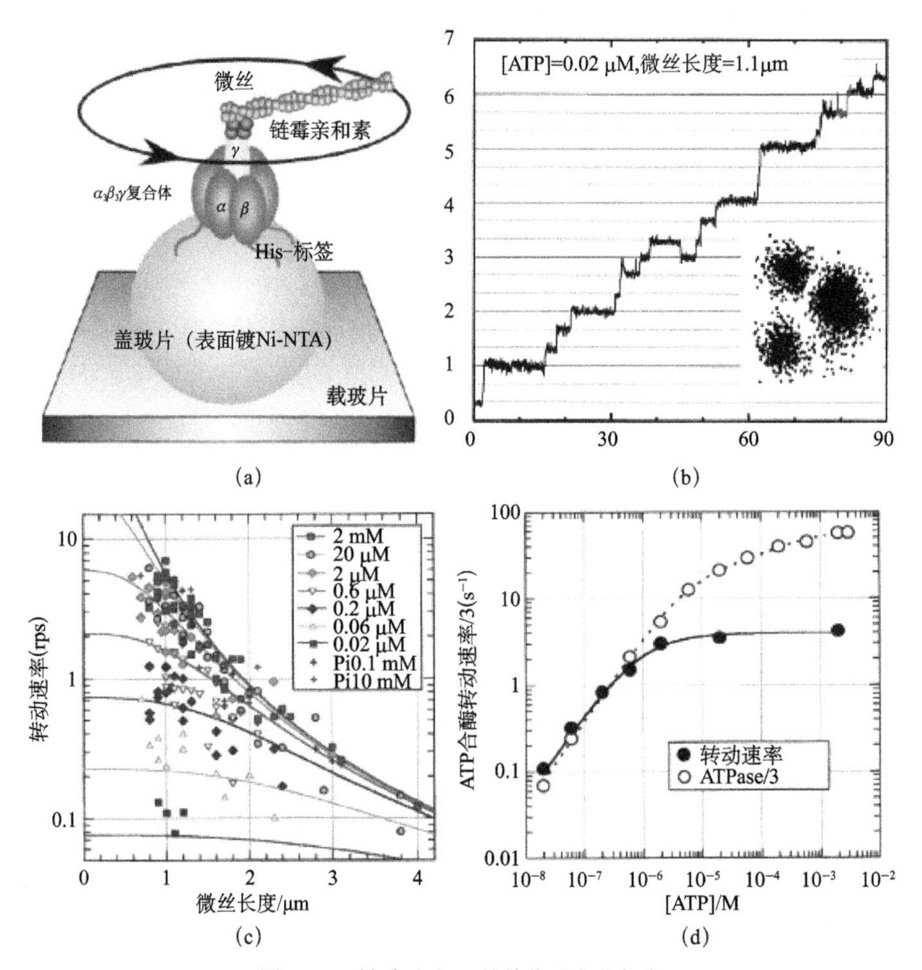

图 6-5-6 转动马达 F_1 的单分子步进实验

(a) 实验示意图，“定子”$\alpha_3\beta_3$ 环固定在大球表面，“转子”γ 亚基顶端结合荧光细丝；(b) 低 ATP 浓度时测得的马达转动轨迹，数据显示，马达每转一圈分三步，每步 120°，从而证实 ATP 合酶是步进马达；(c) 马达转速与 ATP 浓度和细丝长度之间的关系，流体动力学计算表明，马达将 ATP 水解能转化成细丝转动的机械能的效率达到 100%；(d) 在 μM 以下 ATP 浓度，马达每转一圈水解 3 个 ATP。图片摘自文献 [43]

2004 年，Itoh 小组（图 6-5-7）利用磁珠和旋转磁场的方法，在单分子水平上实现体外操控，并将物理矢量（扭矩）转化成化学标量（ATP 分子），即用生物的方法可控地合成 ATP[46]。2015 年，Toyabe 和 Muneyuki 使用相同的方法探测了 F_1 在 ATP 合成时的能量耗散，发现 ATP 合成即使远离平衡态，马达内部的耗散还是可以忽略的，从而证明 F_1 马达是一个高效的合成酶 [47]。

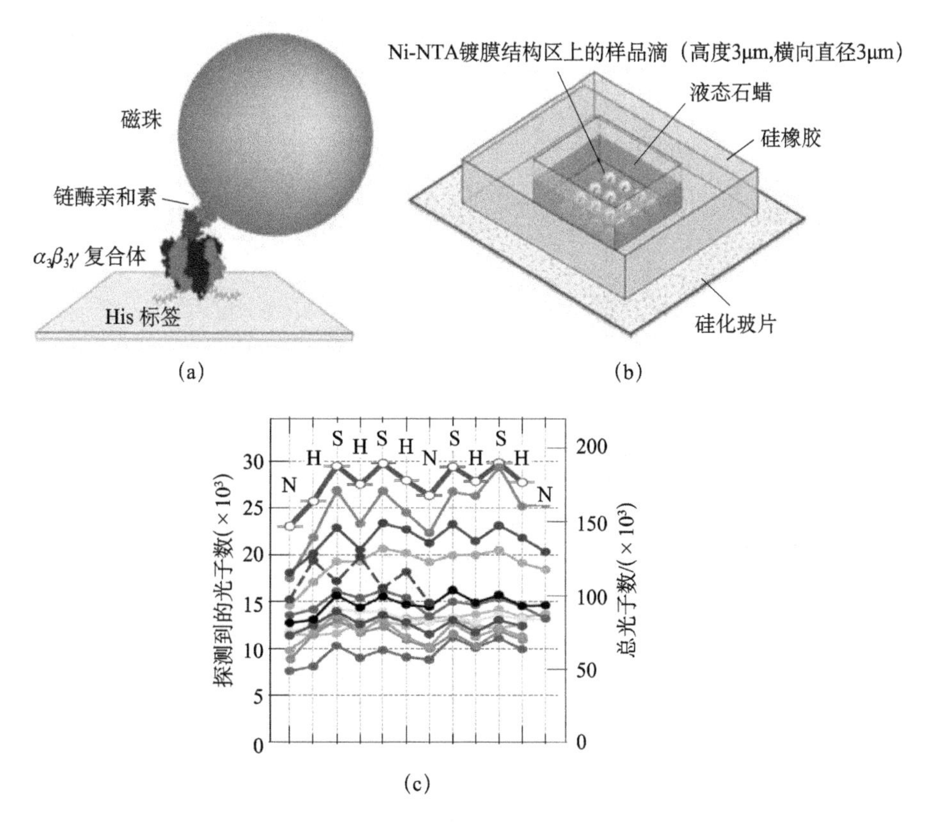

图 6-5-7 单分子操控 F_1 体外合成 ATP 实验

(a)“定子”$\alpha_3\beta_3$ 环固定在底板上，“转子”γ 亚基顶端结合磁珠; (b)16 个隔室内各置入一个马达，外置的旋转磁场可以操控马达合成或水解 ATP，同时利用荧光激素酶测量每个隔室的荧光强度变化表征 ATP 的合成或水解; (c) 12 个有效隔室的荧光强度变化（对应 12 条不同颜色）与马达转向的关系，N：不转; S：10Hz 合成转速; H：10Hz 水解转速。最高处品红粗线代表所有 12 个隔室的荧光总和，数据表明 ATP 总量的变化与马达转向正相关。图片摘自文献 [46]（文后附彩图）

（三）ATP 合酶全酶的运动学

2004 年，德国 Börsch 小组（图 6-5-8）利用 FRET 的方法对全酶 F_oF_1-

ATPase 的进行单分子研究。他们将供体结合在“转子”γ 亚基末端，受体固定在“定子”b_2 亚基上，γ 亚基的偏心转动导致供体与受体之间的距离发生变化，直接反映在能量转移的变化。实验结果表明，质子梯差驱动的全酶马达也是步进的，每步 120°，且合成与水解的转向相反[48]。跨膜质子梯差和 ATP 浓度两个化学标量之间的竞争通过打破物理矢量（马达正反转）之间的平衡表现出来。

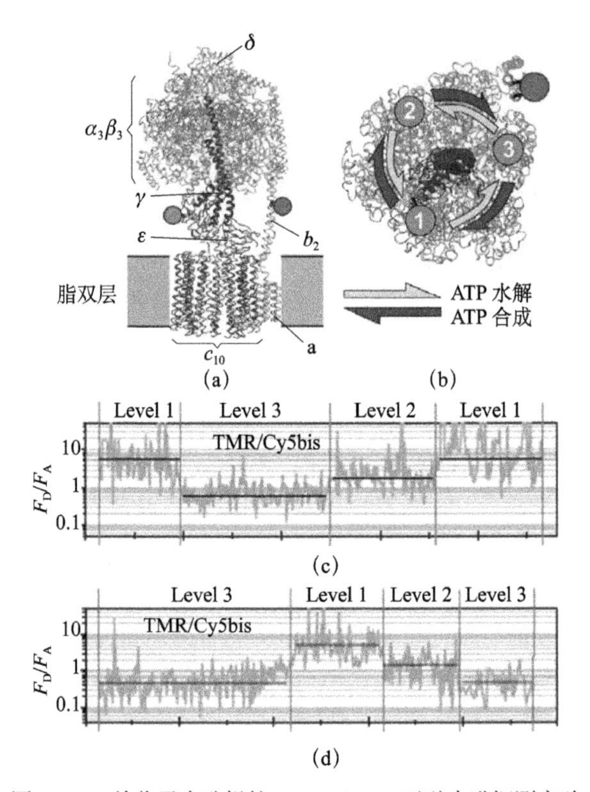

图 6-5-8 单分子全酶操控 FoF_1-ATPase 可逆步进探测实验

(a) FRET 实验示意图，Fo 嵌膜，受体固定在“定子”b_2 研究亚基，供体则固定在“转子”γ 亚基末端，跨膜质子梯差和 ATP 浓度两个化学标量之间的竞争 通过打破物理矢量（马达正反转）之间的平衡表现出来；(b) 供体所处三催化位点和受体的空间分布图，1-2-3 顺序对应距离为远－中－近，相应地从供体转移到受体的能量 F_A 为小－中－大，供体自身发光 F_D 则反之；(c) 水解转动顺序；(d) 合成旋转顺序。F_D/F_A 循环表明马达每转一圈分三步，每步 120°。图片摘自文献[48]

更深入的全酶研究需要直接观测 Fo 部分的运动，例如质子流如何驱动 C_n 转动。由于目前仍未获得 Fo 中 a 亚基的晶体结构，并且实验上也难以单独对 Fo 进行操控和观测，因此迄今还没有 Fo 质子驱动机制的准确模型。对此，人们只能根据 C_n 的结构信息给出一个大致猜想。如图 6-5-9 所示：a 蛋白的内侧（嵌于膜的内叶，即图中的上叶）、外侧（嵌于膜的外叶，即图中的下叶）各有一个相互错开的“┗”、“┏”拐形质子通道，其垂直于膜的部分分别通向膜的内、外两侧，水平部分则指向转盘 C_n 并与其两个残基接触。当跨膜质子动势（p.m.f）ΔP 足够大时，质子从膜外通过 a 蛋白外侧通道吸附到 C_n 并“中和”该处带负电的羧基，导致 C_n 发生转动（其中原因仍未确定）。当吸附质子的羧基转到 a 蛋白另一质子通道（内侧通道）时，被吸附的质子得以释放并通过该通道流向膜内，而羧基又恢复负电性，直到在 a 蛋白外侧通道处再次被另一质子“中和”[11]。如果质子动势较小、ATP 浓度足够高，则 ATP 会在 F_1 中水解，导致 $\alpha_3\beta_3$ 主动构象变化，并伴随 γ 亚基反向转动，从而驱动 Fo 跨膜（从上往下）泵送质子。

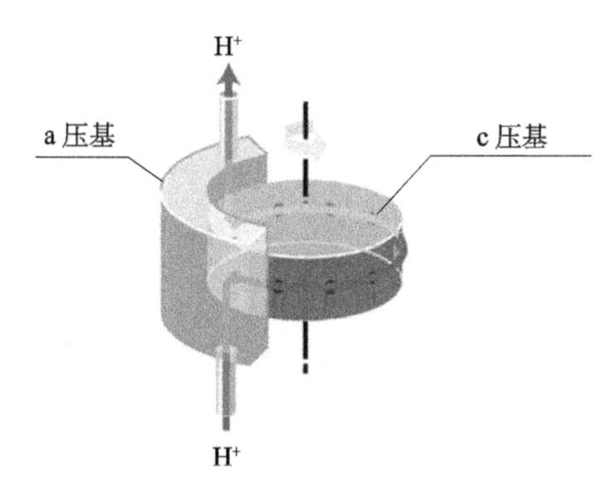

图 6-5-9　质子流驱动 Fo 马达转动的原理示意图 [11]

（四）单分子实验揭示 myosin Ⅱ为非持续马达

1994 年，Spudich 小组（图 6-5-10）利用双光镊操控两个小珠拉直一根细肌丝，小珠和细肌丝都带荧光，将肌球马达黏在大球表面，当细肌丝靠近大球时，可以测量两个小珠距离的变化及其涨落大小，并估算出平均位移和马达发力分别为 11nm 和 3～4pN。实验显示，单个 myosin Ⅱ没有持续的步进轨迹，因此是非持续马达 [49]。

目前，结合晶体数据和单分子实验数据，人们对 myosin Ⅱ单马达的工作机制已经有了大致了解。一般认为，myosin Ⅱ头部在空态下能与细肌丝牢固结合。当它再与 ATP 结合后会从细丝上脱离，而后将携带的 ATP 水解为 ADP 和 P_i，水解后马达蛋白（头部仍结合 ADP）的整体构象发生较大改变（即马达头部与茎秆之间的相对取向发生改变。将头部的微小构象变化放大的“杠杆”就是颈链）。当马达头部再次与细肌丝碰触并结合后，会释放

ADP 并回到原先的空态，导致马达整体构象也随之复原。这个构象复原的过程就是马达推动细肌丝运动的过程，即动力冲程（power stroke）。形象地看，myosin Ⅱ对细丝的发力过程非常类似于人类“划桨”的动作。

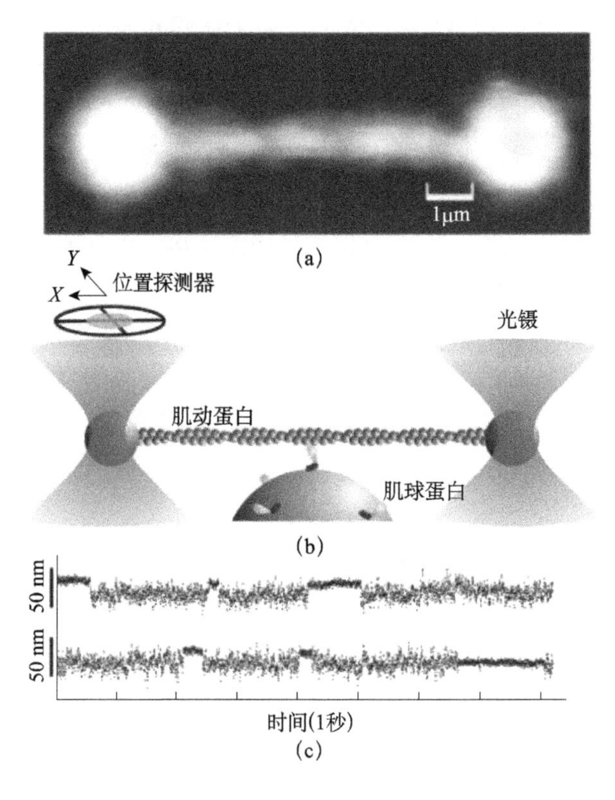

图 6-5-10　肌球马达的单分子实验

(a) 双光镊操控两个小珠拉直一根细肌丝。(b) 肌球马达的单分子力学实验，肌球马达粘在一个大球表面。(c) 细肌丝的轴向位移 - 时间曲线，注意：布朗运动减弱的间隙正是细肌丝与马达发生横桥作用的时段。图片摘自文献 [49]

由于肌球马达Ⅱ非持续的特征，它们通常不是单兵作战，而是以集团的方式协作运动。事实上，细胞内还存在多种具有重要生物学功能的集体马达系统，这些系统早就引起了科学家的兴趣和重视，肌球马达系统是其中最早被研究的例子。20 世纪 80 年代一些研究小组成功地在体外构建了肌球马达-细肌丝系统，实验观测到了一些有趣的性质。例如，Spudich 小组利用荧光技术观测到了细肌丝在集体肌球马达面上的定向滑移输运，并发现输运速度随 ATP 浓度上升而逐渐趋于饱和，输运速度和输运距离与马达密度密切相关 [24]。随后不少研究组利用类似系统研究了细肌丝-肌球马达的相互作用、运动学、酶学等。尽管与后来出现的单分子技术相比，这些研究难以揭示细肌丝-肌球马达相互作用的微观细节，但仍然提供了集体马达协同工作的极有价值的信息，例如，集体马达的“负载-速度”（*F-V*）曲线[50]是单个马达所不可能具备的，这就引出了集体马达研究中的重要问题，即马达之间如何协调工作，从而使得集体马达展现出远比单马达丰富的运动行为。这个话题在后面还要再次谈及。

二、分子马达的理论研究

在分子马达的研究中，理论是与实验互补的、不可或缺的研究手段。一方面，要对分子马达运动过程、工作机制的实验结果进行深入、定量的分析，通常需要先建立起明确的理论模型；另一方面，计算模拟作为在计算机上开展的虚拟实验，可提供当前实验技术尚无法观测到的细节信息。分子马达理论研究的对象包括单分子马达及多马达集成系统，主要运用分子动力学、随机动力学等方法，对分子马达的运动机制、结构设计、能量转化及物质输运规律等开展建模或模拟研究。

单分子马达的理论研究涉及微观及介观两个层面。微观层面主要是基于马达的分子结构，对其某些动态过程进行模拟，揭示实验尚无法观测的细节。最常用的方法是全原子或粗粒化分子动力学模拟，已经在一些典型分子马达的某个步骤或某个部分上取得了进展。这里仅以荣获 2013 年诺贝尔化学奖的三位分子动力学模拟科学家的工作为例。Levitt 在 2009 年用自己发展的变形方法（morphing methods）结合粗粒化模型成功模拟了 myosin Ⅱ的动力冲程（power-stroke）[51]。在 ATP 合酶方面，Karplus 小组在 2002 年就与另一位诺贝尔奖得主 Walker 小组合作，用分子动力学模拟的办法研究了 γ 亚基的转动如何导致 β 亚基的构象变化，发现位阻和静电相互作用都有贡献，且处于 γ 亚基突出位置的精氨酸和赖氨酸残基在 β 亚基的构象变化运动中起着主导作用[52]；Karplus 还发现 ATP 水解后 γ 的 40° 子步并没有马上发生，而是要滞后一段时间，原因是 P_i 从 β_E 位点释放需要时间，因此 P_i 结合在 β_E 位点就像门闩一样阻止了 40° 子步的 γ 亚基转动[53]，该结果与单分子实验观测一致[54]。2008 年，Karplus 小组又模拟了 kinesin 的发力机制，发现 ATP 结合会导致头部表面 9 个残基形成 *β* 结构，后者与颈链对接（docking）是马达发力和定向运动的分子基础[8]，该结果与实验观测[6]一致。Warshel 则在 2015 年拓展了有关 F_1 马达的粗粒化模型，模拟得到了基于结构的力学-化学紧耦合的自由能曲面，并强调静电相互作用在产生扭矩方面起主导作用[55]。笔者也使用全原子分子动力学模拟研究了驱动马达颈链的弹性，并结合介观模型研究了颈链弹性、长度与马达步行速度之间的优化关系[56]。然而，受计算机速度、容量等多方面的限制，微观方法还无法对大时空尺度上的马达运动提供描述和解释，这个目标只能诉诸介观方法。

介观方法忽略分子马达的结构细节，将其视为具有多个状态的布朗粒子，用随机动力学的方法描述粒子的大尺度运动，这是目前用于解释实验结果最主要的理论建模方法。这个方法的合理性在于两点。第一，分子马达本身具有异质、精巧的结构，无论是化学驱动还是力学驱动，只要驱动力适量，分子内的协同相互作用都能使得整个马达的结构不是连续变化，而是在一组确定的离散构象态之间跃迁，因此可视为多态粒子。第二，分子马达处于纳米尺度，受热涨落和黏性主导，因此可用过阻尼动力学描述。基于这样的考虑，德国马克斯-普朗克研究所（以下简称马普所）Jülicher 等于 1997 年提出了分子马达“力学-化学耦合”的非平衡统计物理理论“棘轮”模型，并用 Fokker-Planck 方程建立其数学描述，成为目前公认的分子马达运动学和热力学分析的理论框架之一[57]。

将马达整体比作布朗粒子，定义马达在时空 (x,t) 的概率密度为 $P(x,t)$，则马达运动状态可唯象地由 Fokker-Planck 方程描述：

$$\frac{\partial p(x,t)}{\partial t}=D\frac{\partial^2 p(x,t)}{\partial x^2}-\frac{\partial}{\partial x}\left[\frac{F(x)}{\gamma}p(x,t)\right]$$

概率密度满足归一化条件：

$$\int_{-\infty}^{+\infty}p(x,t)\mathrm{d}x=1$$

其中，D 和 γ 分别是马达运动的扩散和阻尼系数，满足爱因斯坦涨落耗散关系：

$$D=\frac{K_\mathrm{B}T}{\gamma}$$

此处 K_B 和 T 分别是玻尔兹曼常量和绝对温度。马达运动速度可以由概率流表征：

$$j(x,t)=-D\frac{\partial p(x,t)}{\partial t}+\frac{F(x)}{\gamma}p(x,t)$$

$F(x)$ 是作用在马达上的外力，比如棘轮势产生的力。只要棘轮势对称破缺导致一个定向的力，则上述理论可以定性解释马达的定向运动。

当考虑分子马达的物理扩散与化学反应（ATP 水解）耦合时，则可以用多态马达描述，反应-扩散方程为

$$\frac{\partial p_i(x_i,t)}{\partial t}=D\frac{\partial^2 p_i(x_i,t)}{\partial x^2}-\frac{\partial}{\partial x}\left[\frac{F(x_i)}{\gamma}p_i(x_i,t)\right]+\sum_j\left[k_{ji}(x_j)p_j(x_j,t)-k_{ij}(x_i)p_i(x_i,t)\right]$$

其中，k_{ij} 是态之间的转移速率。等号右边的前两项分别代表扩散、漂移（力学过程），它与化学过程（第三项）的耦合很明显：若每个化学步骤都处于细致平衡（求和项中每一项均为零）即平衡态时，分子马达无能量输入（驱动力为零），上述方程仅描述单纯的力学过程，马达无净漂移；当某个或某些化学步骤的细致平衡被打破（求和项中至少有一项都不为零）时，马达即受化学能驱动，产生定向运动。

对于不同的分子马达，可以通过对转移速率设定不同形式，从而反映出马达紧耦合或松耦合的特征。对于 myosin Ⅱ这样的非持续松耦合马达，使用上述方程描述其运动学和热力学是最自然的选择。但方程中很多参数难以获取，而且方程形式比较复杂，难以进行解析计算和分析，因此多用于原理性探讨，很难直接用于实验数据的分析。对于驱动马达和 ATP 合酶这样的紧耦合马达，一个工作循环内的每个步骤都伴随确定的空间位移（即确定的 power stroke），因此方程中扩散 + 漂移项可近似当成确定的“化学”步骤来处理，其总效应可间接体现在等效速率常数 $k_{ji}(x_j, D, F, \delta)$ 中，δ 就是该步骤所伴随的力学位移。于是上述偏微分方程可简化成常微分的随机化学动力学形式：

$$\frac{\partial p_i(x_i,t)}{\partial t}=\sum_j\left[k_{ji}(x_j,D,F,\delta)p_j(x_j,t)-k_{ij}(x_i,D,F,\delta)p_i(x_i,t)\right]$$

这个方程使得紧耦合马达的动力学研究大为简化。尤其在非平衡稳态情况下，由于每行一步只消耗一个 ATP，如果只关心关于马达运动的平均行为，则上述动力学还可进一步简化著名的米氏动力学方程：

$$V=\frac{V_{\max}[\mathrm{ATP}]}{[\mathrm{ATP}]+K_{\mathrm{M}}}$$

其中，V、K_{M} 和 [ATP] 依次代表马达速度、米氏常数和 ATP 浓度。

基于上述随机化学动力学的理论框架，Oster[58] 和 Karplus[59] 对 $\mathrm{F_1}$ 部分的水解动力学建立了详细的模型。笔者及合作者（图 6-5-11）则对全酶（包括 Fo、$\mathrm{F_1}$ 两部分）动力学进行了系统建模，成功地解释了生化实验及单分子实验的数据，并模拟了 ATP 合酶囊泡动力学 [60]，预测其最大合成转速约 14 rps（即每秒可以合成 40 个 ATP 分子），该预测值被江雷小组实验所证实 [61]。

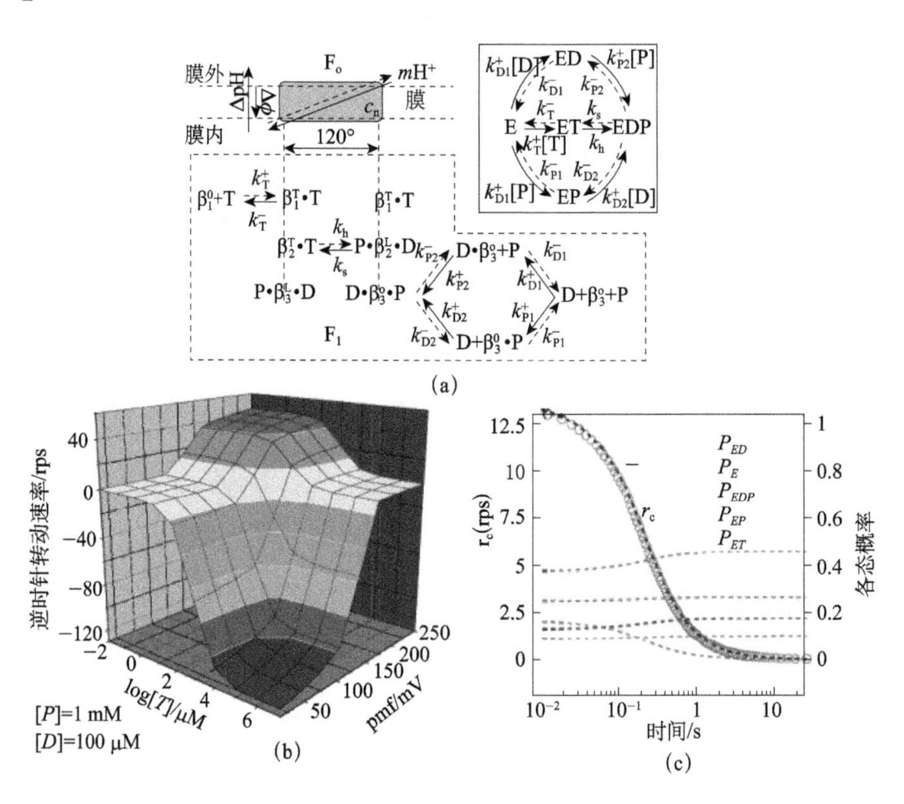

图 6-5-11 (a) F_oF_1-ATPase 全酶可逆动力学 5 态模型；(b) 马达转速与 ATP 浓度和跨膜化学势之间的定量关系；(c) 囊泡动力学模拟值，马达最大合成转速约 14 rps。图片摘自文献 [60]

目前看来，由于多态 Fokker-Planck 方程本身比较难以处理，因此它更适宜于简化描述的分子马达体系，尤其是多马达系统，例如 myosin Ⅱ这样的松耦合马达及其构成的多马达系统，使用该方程可对其运动机制进行粗粒化建模和定性讨论。而随机化学动力学方程比较容易处理，而且也更容易容纳分子结构等细节信息，因此更适宜对单马达的运动进行详细的描述和定量分析。事实上，对紧耦合单分子马达的实验数据的解释和拟合都是基于随机化学动力学的。

第三节 基础及应用研究中的前沿问题

早在 1959 年，物理学家费曼就在其讲演 *There's plenty of room at the bottom* 中畅想了纳米机器构成的世界。而生物分子马达作为纳米机器的典范，因其精巧的结构和丰富的功能，成为物理学、生物学、化学、纳米技术等多个学科的研究热点。对分子马达运动机制的定量研究可揭示其结构与运动特征之间的关系，从而理解其工作原理，这是目前分子马达基础研究的主要内容。在此基础上，如何利用分子马达的运动能力、甚至改造其结构以优化运动能力，是目前应用研究的主要内容。

一、基础研究的前沿问题

从基础研究的层面看，分子马达如何实现能量转化（力学-化学耦合机制），其效率如何，运动过程中马达之间如何协作，这三方面的问题是整个实验和理论研究中的核心共性问题。

（一）力学-化学耦合机制

力学化学耦合机制是分子马达实现化学标量与物理矢量相互转化的核心问题。从理论上可以证明，只要化学反应处于非平衡态（提供能源）、马达处于非对称势场中，则必然产生力学-化学耦合。事实上，目前已知的所有分子机器都满足这两个原则性条件。在这样的前提下，耦合的具体方式成为分子马达研究的主要内容。受实验技术和计算模拟技术所限，对马达力学-化学耦合机制的精细研究目前尚处于非常初级的阶段。

在定性研究方面，对于一些较简单的问题，例如马达如何产生特定方向的净运动（如 kinesin-1 只能向微管正极运动），人们已经找到了相应的分子部件并大致搞清了其工作原理。然而，对于力学化学耦合的更多结构细节，还有大量问题待解决。其中一个核心问题是力学化学耦合中的别构效应，即 ATP 结合、水解、ADP 释放等所导致的局域形变如何经过一系列“放大”装置最终转化为马达的整体形变或运动？这个问题本质上是马达内部各部分的协同性问题。

在前面对驱动蛋白、肌球蛋白的介绍中我们看到，人们通常引入“杠杆”等力学类比来定性理解这一过程。然而，杠杆这种“硬”机制与马达蛋白本身的“软”属性看似矛盾。对于不同分子马达，是否都存在着这类杠杆、哪些部分参与了其构成，这些都是需要明确回答的重要细节问题。Levitt 等关于 myosin Ⅱ 的分子动力学模拟[51]、Zheng 等关于 myosin Ⅱ 的弹簧网络模型和简正模分析[62]都显示存在显著的杠杆效应。对类似于驱动蛋白的步进马达 myosin Ⅴ，Kinosita 等的工作却表明，杠杆效应并不明显，决定其定向运动速度的主要因素是有偏布朗运动[63]。对于驱动马达，Karplus 等的模拟研究[8]也指出，像颈链这样的软连接物对马达两头之间的协同有决定性影响，但难以起到杠杆的传力作用，因此驱动马达的步进也是杠杆效应和有偏布朗运动的混合物。这些定性研究结果尚需进一步的模拟和实验工作确认。

ATP 合酶中也存在类似的“力学传动”图像。两个转子 γ 轴和 C_n 转盘之间连接物 ε 亚基，到底是承担“离合器”还是“万向节”的功能？前者意味着两个转子之间柔性耦合、可能存在打滑，后者意味着两者之间刚性咬合。由于 C_n 近于连续转动，而 γ 轴为步进运动，因此 ε 亚基很可能是柔性耦合。此外，在 ATP 合成过程中，F_1 中三组 $\alpha\beta$ 亚基之间是否真像 Boyer 设想的那样处于严格的异相同步状态？这种步调匹配要求它们通过与 γ 轴之间的刚性相互作用发生“通信”，使得各自保持不同构象。与上述杠杆效应一样，这种别构相互作用是否存在、如何发生，对此都还存在着争论。对于 Fo 部分也存在类似的问题，质子跨膜与 C_n 转动之间到底是紧耦合还是松耦合？如果是前者，则 C_n 每转动一周就会有 n 个质子跨膜，但 Gräber 等的脂质体（liposome）定量实验证明，在一定条件下，菠菜的 $CFoF_1$（n=15）和大肠杆菌的 $EFoF_1$（n=10）的 H+/ATP 都是 4.0[64]，即每转一周平均通过 12 个质子。这表明 C_n 的转动与质子跨膜并非完全紧耦合，C_n 可在热涨落下自发随机转动。在不同条件下这一耦合机制是否改变、其效率如何量化，这些问题对于精细理解 Fo 的设计原理与工作机制也非常重要。

此外，还有不少与分子内协同性密切相关的问题没有得到彻底解决。例如，在不同的力负载条件下，力学-化学耦合机制是否不同（例如，接近失速的高负载情况下是否存在着无效水解，从而从紧耦合变成松耦合）？如何对马达的关键部位进行修饰或改造从而改进马达的功能（如改变运动方向、运动速度、发力大小）[65,66]？等等。

在定量研究方面，外力或外力矩如何影响 ATP 水解循环（即外力如何影响转移速率 k_{ij}）仍然是实验研究中的难题，马达发力大小、运动速度快慢等问题对理论研究仍然构成巨大挑战。前面提及的 Karplus 关于 kinesin 的工作[8]，由模拟结果间接外推了马达产生的最大力，与实验测量的失速力并不精确吻合。Warshel 关于 ATP 合酶 F_1 的粗粒化模拟得到了马达水解 ATP 时产生的转矩[55]，虽与实验结果吻合，但其中能量面的构造有不少人为拼凑的因素。受模拟方法所限，这类工作仍无法严格、定量地研究马达的力学、运动学。这些问题对于全面理解马达的力学-化学耦合非常重要。然而，即使对于驱动马达、ATP 合酶等研究最多的模式系统，目前也还远没有建立起完整的图像。未来需要更精细的实验技术对某些步骤进行直接观测，同时需要将计算模拟技术延伸到更大时空尺度、发展更恰当的方法。

（二）能量转化效率

研究分子马达力学-化学耦合机制，一方面是为了阐明马达的运动学及其与马达结构之间的关系，另一方面也需要理解分子马达的另一个重要特征，即其能量转化的高效性。几种常见分子马达的能量效率都远高于人造机器，这也是人们对生物分子马达格外有兴趣的原因之一。那么，它们究竟是如何在热涨落显著的黏稠环境中实现能量的高效转换的？

按照宏观热力学的经验，高效率和高速度似乎是一对矛盾。如果工作步骤足够多且每个步骤接近于平衡态，则不难实现高的能量效率，但这将导致几近于零的运动速度。而有些分子马达（如 ATP 合酶）不仅有极高的能量效率，而且有很快的运动速度，似乎与上述经验相违背。如何更准确地理解并调和这一矛盾，不仅是基础研究的重要课题，对于分子马达的设计和应用也有重要意义。然而，无论从实验上还是理论模拟上看，对这一热力学问题的研究都远比运动学研究更为困难。

例如，位钳实验测得驱动蛋白马达的失速力达 7pN[31,32]，力钳实验测得其失速力为 5.5～7pN 且与 ATP 浓度有关[27]。以 7pN 计，在稍低于这个负载力的情况下马达仍可以步进但速度非常缓慢，此时 ATP 水解的化学能似乎将完全转化为对外做功。但粗略估算表明，此时外力功仅为 7×8=56pN·nm，只占到 ATP 水解能（约 80pN·nm）的 70%。换句话说，驱动马达对外做功的最大能量效率约为 70%。这个效率在分子马达中不算很高，这很可能与驱动马达的功能密切相关，毕竟它主要是长距离运输囊泡而不是对外做功。无论如何，从物理上探究剩余自由能究竟用于何处，这对了解驱动马达的设计原理是重要的。按照棘轮原理，马达运动中的任何一个步骤（无论是化学还是力学步骤），只要其细致平衡被打破，都必然导致耗散。要想在分子水平上直接测量这些效应，目前还没有可行的实验方法。在理论方面，虽然原则上可借助于前面提到的随机动力学来讨论马达的非平衡热力学，但要求知道所有基元步骤的正向、反向速率常数，这对于目前分子马达的单分子或生物化学实验都是巨大的困难。

与驱动蛋白等其他分子马达相比，ATP 合酶的能量转化问题远为重要，因为它本身的功能就是实现能量转化。2011 年，日本 Muneyuki 小组使用单分子技术探测了 F_1 的能量耦合效率，发现在较大范围的保守外力矩作用下，ATP 水解推动 F_1 马达转动的能量效率都接近 100%[67]，这与 1998 年 Kinosita

小组用黏滞力矩直接测得的结果[43]吻合。2015 年，他们又利用单分子技术结合非平衡统计热力学理论，对 F_1 在保守外力矩作用下合成和水解 ATP 的双向过程重新进行了更精细的分析，仍然发现在较大的力矩范围内，无论水解还是合成过程，其能量转化效率都接近 100%[47]。换句话说，在相当大的转动速度范围内，F_1 看似没有任何摩擦内阻，它将“高速度”和“高效率”这对表面上的矛盾完美统一了起来。如何理解？早在 1998 年，Oster 等结合分子结构数据提出的 64 态随机动力学模型[58]，就对 F_1 的高效率给出了直观解释，他们认为 ATP 结合时产生的能量可完全转化为分子内部的弹性能，而后又完全转化为偏心轴的转动。2015 年 Kawaguchi 等基于棘轮原理提出了一个非常简化的随机动力学模型[68]，某些定性结果可与实验数据比拟。但这类模型缺乏马达的结构及内部相互作用（如静电相互作用）的精确信息，因而还不能算是对 F_1 高效率的真正理解。更深入的研究需要将化学动力学与分子模拟巧妙结合起来。

至于 ATP 合酶的 Fo 部分，其能量效率至今无法阐明。这一方面是由于其“定子”蛋白 a 的结构至今没有解析，而该结构对于揭示跨膜质子流（化学）与 C_n 转动（力学）之间的耦合机制至关重要[69]。另一个问题更为现实，Fo 转动的驱动力即跨膜质子梯差 ΔpH，是指膜两边的体浓度差还是两侧的面浓度差？争论来自嗜碱的坚强芽孢杆菌（*Bacillus firmus*），其胞内 pH 接近 3（纯酸性），胞外为碱性（pH 大于 7），则 ΔpH>4，而一般情况下生物体跨膜电势 $\Delta\varphi \leqslant 200\,\mathrm{mV}$，于是 $\mathrm{pmf}=\Delta\varphi-\left(\frac{2.3k_{\mathrm{B}}T}{\mathrm{e}}\right)\Delta\mathrm{pH}\approx\Delta\varphi-59\Delta\mathrm{pH}<0$，不可能合成 ATP。因此，有观点认为 ΔpH 计算的是膜两侧的面浓度差，且膜两侧的质子是横向扩散的[70]。阐明 ΔpH 的准确意义，是深入理解 Fo 能量学的关键。

（三）集体马达系统的协同性问题

前文已经提及，在细胞内存在着多种集体分子马达系统。除了肌肉系统，细胞的整体运动和形变背后往往也是集体马达作用的结果，有丝分裂时纺锤体的形成、高尔基体膜管拉伸、甚至胞内膜泡沿微管的双向输运等过程也涉及数十个马达。这些集体马达系统不仅是分子马达基础研究的重要内容，而且与实际应用密切相关，因此成为当前研究的热点和前沿。实验上已经在体外观察到集体马达远比单马达复杂的行为，如双向性、自激振荡等[71,72]。如何理解这些集体行为？这涉及两方面的问题，其一，马达之间如何协同动作？其二，对于特定的集体运动，单马达的哪些性质是最关键的？

例如，Kinesin 家族成员 Ncd 的变异体 Nk11 本身并无定向性，但集体马达系统却展现出双向输运行为，即以某个速度输运微管一段时间，而后再以相近的速度反向输运微管[73]。Jülicher 等基于之前建立的棘轮理论[57]，提出了高度简化的对称二态棘轮模型，对此给出了定性解释[74]，其中的关键点是假设马达之间通过共享硬杆产生刚性耦合。对于肌肉收缩、纤毛拍动这样的系统，这种刚性耦合就是马达协同性的来源。在这样的假设下，笔者的研究还揭示了集体马达 ATP 消耗率与 ATP 浓度之间的非单调关系[75]。如果加入一些更实际的约束，如马达数目有限、马达之间不完全是刚性耦合（如肌球马达的横桥模型），仍可展示出双向运动等特征[76]。在上述研究中，马达本身的结构和动力学细节（如 F-V 曲线）并未明确考虑。它们对于集体马达的运动行为到底有多大影响，这是未来需要深入分析的问题。

有些集体马达系统不存在刚性耦合，例如在集体驱动马达从细胞中拉出“膜管”的过程中，由于膜无法固定马达，因此马达在运动过程中将逐渐富集在膜管前端。这种情况下马达之间的协同性如何体现？ Campàs 等利用刚性小球模型，通过模拟分析发现，如果考虑简单的在位体积排斥，需要三列马达一起作用才能保证膜管的形成[77]。笔者及合作者进一步考虑了驱动蛋白本身的弹性、具体结构、体积排斥等因素，模拟结果表明沿单根微管原丝上的驱动马达也能产生足够的拉力从膜泡中拉伸出膜管[78]。这说明这类集体马达的协同性比较复杂，针对不同问题可能需要考虑不同的单分子细节。

更有趣的一类集体马达是由不同种类的马达共同形成的系统，例如细胞内膜泡的双向输运系统，它是由分别朝着微管两极运动的驱动马达和动力马达构成的。直观地看，这两类马达独立工作、形成对抗，从而以“拔河”的方式来实现膜泡的双向输运。然而，体内实验发现这两类马达在细胞内似乎达成某种协同，使得当一种马达复活时另一种马达失活[79]。这个观察似乎不支持“拔河”的图像，而倾向于支持另一种生物化学的观点，即认为存在其他调控分子能控制这两类马达的活性切换。然而，Lipowski 等充分考虑单马达的动力学特征后（尤其是马达解离速率与力之间的非线性关系），在此基础上建立了集体马达运动的随机动力学模型，模拟分析表明“拔河”模型的确可以给出“一边倒”的正反馈机制，足以解释某些系统中的切换行为[80]。与前述硬杆或体积排斥导致的协同性不同，这种切换协同性并非源于结构，而是源于马达动力学参数之间的匹配，属于动力学失稳的范畴。不过，最近一些实验和理论又针对不同系统，从不同角度对“拔河”模型提出了质疑乃至否定，例如体内脂滴的双向输运数据似乎就难以用拔河模型进行解释[81]。针对这些新问题，一些新的机制和模型已经被提出（详情可参阅综述[82]）。这些进展表明，这类杂化马达系统的协同机制远比想象的复杂，无论是实验还是理论都还有大量的工作亟待开展。

二、分子马达的应用研究

近二十年来的生物物理研究（尤其是单分子生物物理学）揭示了各种分子马达的运动学或能量学特征，也深入研究了马达结构与这些特征之间的关系。这些基础研究成果极大地促成和推动了分子马达的应用研究，成为生物医学技术、纳米技术等多个学科新的研究方向。下面我们仅举几个近期出现的极具代表性的例子。

（一）基于驱动马达高持续性的免洗快速检测技术

传统的免疫酶联法需要多次清洗，将多余的标记物如金颗粒或荧光探针移除，不利于快速检测的执行［图 6-5-12(a)］。2009 年，佛罗里达大学的 Hess 等开发了基于驱动马达的智慧型主动传感器，通过微管与集体分子马达作用导致的高持续长距离定向移动，将酶联上的标记物快速移动到探测窗口，省却了清洗步骤，实现了快速定量探测［图 6-5-12(b)］[83]。

（二）基于 FoF_1-ATPase 高效能量转换和大扭矩输出的一体化纳米装置

利用磷脂膜、膜蛋白和膜被细胞器等构建具有特定功能的细胞一直是生物物理研究的热点之一。由于 F_1 能实现高效的能量转化并产生的大扭矩足以推动纳米机电系统（NEMS）装置，因此，F_1 成了受控纳米机器构建的新宠。最成功的案例是 F_1 消耗 ATP 驱动纳米螺旋桨[84]，F_1 马达将化学能近 100% 地转化成了物理转动，唯一的缺点是 ATP 需要外部供给。2005 年 Montemagno 等利用光照条件下视紫红质产生的质子梯度，驱动 ATP 合酶旋转合成 ATP，首次实现了双蛋白协同工作的纳米机器组装[85]。2008 年中国科学院化学研究所李峻柏等将自组装和层层组装技术相结合，制备出葡萄糖氧化酶和 ATP 合酶双蛋白复合纳米生物机器，此系统通过分解葡萄糖形成跨膜质子梯度，继而驱动 FoF_1-ATP 合酶旋转催化 ATP 合成[86]。2010 年，笔者和合作者借助超声将沼泽红假单胞菌（Rhodobacilluspalustris）实现胞翻，获得

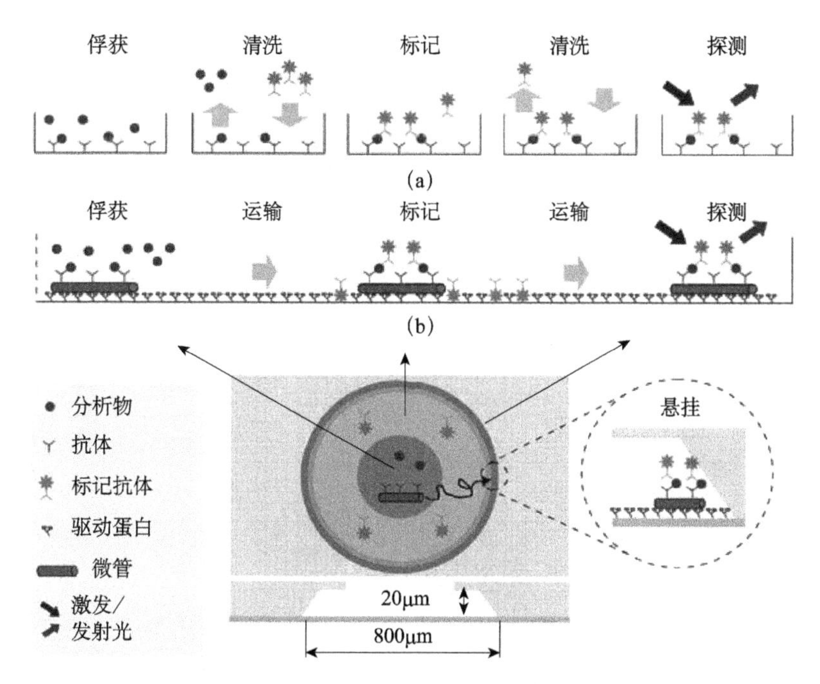

图 6-5-12 基于驱动马达的高持续主动输运的免疫快速检测原理图

(a) 传统双抗体三明治探测试验的（俘获—清洗—标记—清洗—探测）顺序图; (b) 智慧型探测器的俘获—输运—标记—输运—探测顺序。驱动马达驱动带有抗原的微管冲撞结合二抗的荧光分子，酶联荧光的微管输运到探测区进行定量探测。图片摘自文献 [83]

直径约 100nm 的带有一个 FoF_1-ATPase 的色素胞（chromatophore），并利用色素胞上的细菌叶绿素（Bacteriochlorophyll）实现光能和跨膜质子梯差的转换。在该结构中，色素胞起到马达“电池”的作用，可以通过光照对该“电池”实现非接触式充电。笔者和合作者还在马达定子上通过 β 亚基实现马达活性的外部调控。该装置可以用来高灵敏探测诸如病毒一类的大分子，有潜力发展成生物传感器 [87, 88]（图 6-5-13）。

（三）基于聚合马达高保真的 PCR 和实时单分子测序

分子马达应用开发方面最早也是最成功的例子无疑是 PCR(polymerase chain reaction) 中的 TagDNA 聚合酶。该酶由钱嘉韵等从美国黄石国家森林公园火山温泉的水生栖热菌（termus aquaticus）中分离获得 [89]，拥有很高的保真度，复制出错率可以低至 10^{-6}。1983 年，Mullis 充分利用 DNA 片段温度变性（单双链转换）特点，应用 Tag 酶实现了通过变温而达到片段倍增的目标 [90]，在基因工程尤其是特定序列的样品扩增中发挥了巨大作用（图 6-5-14）。

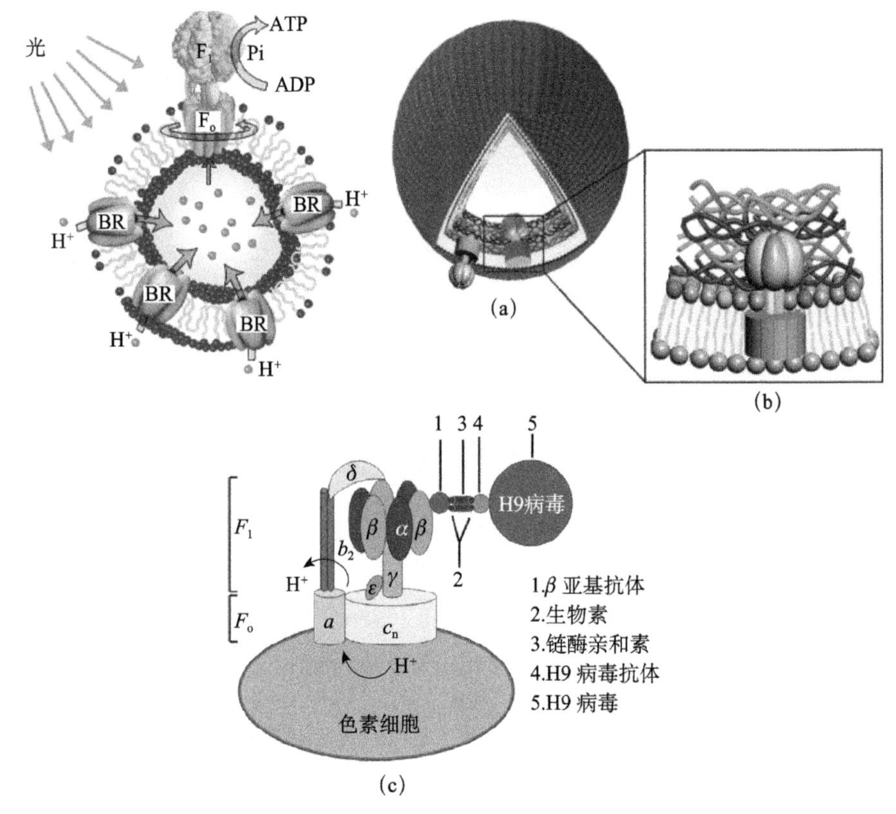

图 6-5-13　基于转动马达的纳米组织自组装

(a) 由视紫红质（BR）与 ATP 合酶构成的 ATP 合成系统，实现光能到化学能的转换；(b) 由葡萄糖氧化酶和 ATP 合酶构成的 ATP 合成系统；(c) 由超声胞翻获得的纳米生物传感器，可以通过非接触式光照激活。图片依次摘自文献 [85]～[87]

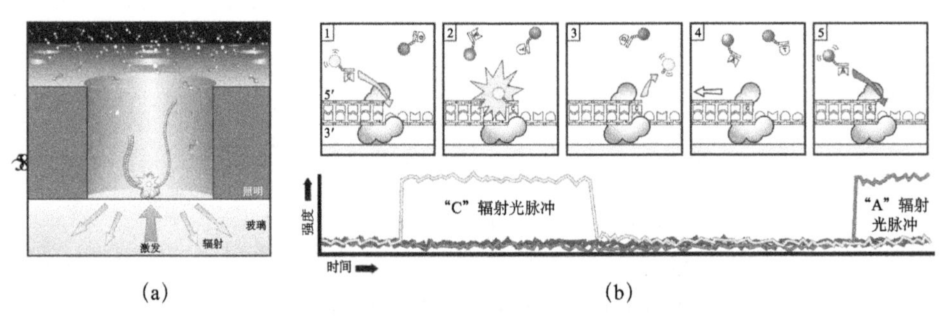

图 6-5-14　单分子实时测序原理图

(a)DNA 聚合酶被固定在零模波导管的底部，激发光只激发聚合酶周围仄升（10^{-21}L）区域的荧光分子，即只激发正在配对的核苷。(b) 测序原理图，核苷一旦配对聚合后，荧光分子将被剔除，因而双链 DNA 没有发光，保证不影响下一个核苷的聚合探测。通过阅读显示的时间色谱，就实时获得了与色谱配对的模板序列。图片摘自文献 [93]

2009年，太平洋生物科学公司的Korlach和Turner等利用康奈尔大学Webb等推出的零模波导技术[91]开发了实时单分子测序仪[92]。其中DNA聚合酶被固定，它利用模板链合成新链的同时也发挥其马达功能，拖曳DNA模板链从其孔道穿过，从而使新链合成反应得以持续。实验中将激发光能量压缩在聚合酶周边仄升（10^{-21}L）区域，再将参与聚合的四种核苷A、T、C、G分别标记蓝、绿、黄、红四色荧光分子，在模板链的引导下，参与聚合的数千个碱基可通过色谱一次性高保真实时测出。

第四节 未来5～10年重点发展方向

分子马达不仅是生物学、医学上重要的一类蛋白分子，也是当前非平衡物理学研究中一类重要的模型系统，对其结构、运动学、能量学的深入研究，是理解其设计原理、进而改造其结构和功能的基础。来自生物学、物理学、化学、纳米技术、计算科学等不同领域的科学家，对分子马达的不同方面进行了有效的探索，对不同分子马达的个性以及它们之间的某些共性已经积累了不少知识，但距离优化设计乃至实际应用的目标还非常遥远。未来的工作仍需要实验、理论、应用多方面的密切配合。以下大致列出从简单到复杂、从基础到应用的五个层次，层次之间相互依赖、相互促进。

第一，最基础的层面仍然是马达结构的研究。随着冷冻电镜及结晶技术的不断提高，有望在十年内对Fo的a蛋白和其他复杂机器（如动力蛋白等）在结构上有明确的解析。而对这些机器的自组装机制的研究则是更为困难的问题和更长远的目标。

第二，在结构信息的基础上，利用不断发展的单分子技术以及计算模拟技术，深入研究力学化学耦合机制的细节及能量转化效率等问题。这仍然是分子马达研究的主要内容，对理解乃至优化分子机器的结构设计和工作机制至关重要。

第三，在单马达运动机制研究的基础上，探索多马达体系的复杂动力学行为，尤其是精确揭示其中的协同效应，厘清不同分子马达在物质输运、细胞运动等复杂活动中扮演的角色。这部分涉及的问题比单分子马达更为复杂、烦琐，仍有大量简单问题无法回答，有待新的实验技术和数据分析技术的发展。

第四，建立分子马达与疾病之间的联系，从分子马达角度寻找对应的快速诊断和治疗方法。例如，扩张型或肥厚型心肌病是由肌球蛋白Ⅱ突变引起的，且大部分是遗传的；肌球蛋白ⅢA、Ⅵ、ⅦA、XV和ⅦA的突变会导致失聪；而肌球蛋白ⅦA的突变还会导致视网膜色素变性；动力蛋白Ⅱ的失活导致男性不育的原发性纤毛运动障碍；动力马达I重链的突变会导致发育疾病无脑回畸形症。驱动蛋白Ⅵ的Trp719Arg基因则可以是衡量心血管疾病风险的标志物。再如，Hausen证明HPV病毒是宫颈癌的唯一元凶[93]，因此获得了2008年诺贝尔奖，而HPV是DNA病毒，其环状双链DNA约8kb，即回转半径达200 nm，而病毒壳内径只有40nm[94]。病毒遗传物质的包装依赖于门马达，对此开展药物设计是一条可行之策。

最后，分子马达作为天然的纳米机器，除了可直接应用外，还能启发人类从头构建类似的高效纳米机器甚至纳米机器系统，实现费曼在其演讲中提出的纳米世界的梦想。化学家在这方面已经取得了一些进展，例如合成了可作为转子的小分子，此处不再详述。这方面的工作所面临的问题远比本章所述内容困难。构建纳米机器的首要难题是结构组装的问题（即各种部件如何自组装成完整的机器），其次才是其驱动机制。如同本章开头所谈及的，目前对天然分子马达的精细研究主要集中在后一层面，对前一个层面的理解还远远不够。相信在不久的将来，这一问题会随着计算模拟能力的提升和单分子实验技术的突破得以解决。

黎明（中国科学院大学物理学院），
舒咬根（中国科学院理论物理研究所）

参 考 文 献

[1] 舒咬根，欧阳钟灿．生物分子马达．物理，2007, 36: 735-741.

[2] Brady S T. A novel brain ATPase with properties expected for the fast axonal transport motor. Nature, 1985, 317: 73-75.

[3] Vale R D, Reese T S, Sheetz M P. Identification of a novel force-generating protein, kinesin, involved in microtubule-based motility. Cell, 1985, 42: 39-50.

[4] Lawrence C J, et al. A standardized kinesin nomenclature. J. Cell Biol., 2004, 167: 19-22.

[5] Kozielski F, et al. The crystal structure of dimeric kinesin and implications for microtubule-dependent motility. Cell, 1997, 91: 985-994.

[6] Rice S, et al. A structural change in the kinesin motor protein that drives motility. Nature, 1999, 402: 778-784.

[7] Rosenfeld S S, et al. Stepping and stretching: how kinesin uses internal strain to walk processively. J. Biol., Chem., 2003, 278: 18550-18556.

[8] Hwang W, Lang M J, Karplus M. Force generation in kinesin hinges on cover-neck bundle formation. Structure, 2008, 16: 62-71.

[9] 舒咬根，欧阳钟灿. 转动分子马达: ATP 合成酶 . 自然杂志，2007, 29: 249-254.

[10] Weber J. ATP synthase: Subunit-subunit interactions in the stator stalk. Biochim. Biophys. Acta., 2006, 1757: 1162-1170.

[11] Junge W. ATP synthase and other motor proteins. PNAS, 1999, 96: 4735-4737.

[12] Rayment I, et al. Three-dimensional structure of myosin subfragment-1: A molecular motor. Science, 1993, 261: 50-58.

[13] Geeves M A, Holmes K C. Structure mechanism of muscle contraction. Annu. Rev. Biochem., 1999, 68: 687-728.

[14] Kühne W, et al. Untersuchungenüber das Protoplasma und die Contractilität. W. Engelmann, Leipzig, 1864.

[15] Straub F B. Actin., Stud. Inst. Med. Chem. Univ. Szeged., 1941-1942, 2: 3-15.

[16] Szent-Györgyi A. Discussion. Stud. Inst. Med. Chem. Univ. Szeged., 1941-1942, 1: 67-71.

[17] Hanson J, Huxley H E. The structure basis of the cross-striations in muscle. Nature, 1953, 172: 530-532.

[18] Huxley H E. The double array of filaments in cross-striated in muscle. J. Biophys. Biochem. Cytol., 1957, 3: 631-648.

[19] Huxley A F. Muscle structure and theories of contraction. Prog. Biophys. Biophys. Chem., 1957, 7: 255-318.

[20] Gibbons I R. Studies on the protein components of cilia from Tetrahymemapyriformis. PNAS, 1963, 50: 1002-1010.

[21] Spencer M. Structure of bacterial flagella. Nature, 1976, 263: 370-371.

[22] Vale R D, Schnapp B J, Reese T S, et al. Movement of organelles along filaments dissociated from the axoplasm of the squid giant axon. Cell, 1985, 40: 449-454.

[23] Vale R D, Schnapp B J, Reese T S, et al. Organelle, bead, and microtubule translocations promoted by soluble factors from the squid giant axon. Cell, 1985, 40: 559-569.

[24] Kron S J, Spudich J A. Fluorescent actin filaments move on myosin fixed to a glass surface. PNAS, 1986, 83: 6272-6276.

[25] Ashkin A, Dziedzic J M, Bjorkholm J E, et al. Observation of a single-beam gradient force optical trap for dielectric particles. Optics Letters, 1986, 11: 288-290.

[26] Ashkin A, Dziedzic J M. Optical trapping and manipulation of viruses and bacteria. Science, 1987, 235: 1517-1520.

[27] Visscher K, Schnitzer M J, Block S M. Single kinesin molecules studied with a molecular force clamp. Nature, 1999, 400: 184-189.

[28] Svoboda K, Schmidt C F, Schnapp B J, et al. Direct observation of kinesin stepping by optical trapping interferometry. Nature, 1993, 365: 721-727.

[29] Schnitzer M J, Block S M. Kinesin hydrolyses one ATP per 8-nm step. Nature, 1997, 388: 386-390.

[30] Hua W, Young E C, Fleming M L, et al. Coupling of kinesin steps to ATP hydrolysis. Nature, 1997, 388: 390-393.

[31] Nishiyama M, Higuchi H, Yanagida T. Chemomechanical coupling of the forward and backward steps of single kinesin molecules. Nat. Cell Biol., 2002, 4: 790-797.

[32] Carter N J, Cross R A. Mechanics of the kinesin step. Nature, 2005, 435: 308-312.

[33] Yildiz A, Tomishige M, Vale R D, et al. Kinesin walks hand-over-hand. Science, 2004, 303: 676-678.

[34] Kodera N, Yamamoto D, Ishikawa R, et al. Video imaging of walking myosin V by high-speed atomic force microscopy. Nature, 2010, 468: 72-76.

[35] Block S M. Kinesin motor mechanics: binding, stepping, tracking, gating, and limping. Biophys. J., 2007, 92: 2986-2995.

[36] Dogan M Y, et al. Kinesin's front head is gated by the backward orientation of its neck linker. Cell Reports, 2015, 10: 1967-1973.

[37] Racker E, et al. Partial resolution of the enzymes catalyzing oxidative phosphorylation. J. Biol. Chem., 1960, 235: 3322-3336.

[38] Mitchell P D. Coupling of phosphorylation to electron and hydrogen transfer by a chemi-osmotic type of mechanism. Nature, 1961, 191: 144-148.

[39] Kagawa Y, Racker E. Partial resolution of the enzymes catalyzing oxidative phosphorylation. J. Biol. Chem., 1966, 241: 2466-2482.

[40] Boyer P D, Cross R L, Momsen W. A new concept for energy coupling in oxidative phosphorylation based on a molecular explanation of the oxygen exchange reactions. PNAS, 1973, 70: 2837-2839.

[41] Abrahams J P, Leslie A G W, Lutter R, et al. Structure at 2.8 Å resolution of F1-ATPase from bovine heart mitochondria. Nature, 1994, 370: 621-628.

[42] Noji H, Yasuda R, Yoshida M, et al. Direct observation of the rotation of F1-ATPase. Nature, 1997, 386: 299-302.

[43] Yasuda R, Noji H, Kinosita Jr K, et al. F_1-ATPase is a highly efficient molecular motor that rotates with discrete 120° steps. Cell, 1998, 93: 1117-1124.

[44] Yasuda R, et al. Resolution of distinct rotational substeps by submillisecond kinetic analysis

of F1-ATPase. Nature, 2001, 410: 898-904.

[45] Shimabukuro K, et al. Catalysis and rotation of F1 motor: Cleavage of ATP at the catalytic site occurs in 1 ms before 40° substep rotation. PNAS, 2003, 100: 14731-14736.

[46] Itoh H, et al. Mechanically driven ATP synthesis by F1-ATPase. Nature, 2004, 427: 465-468.

[47] Toyabe S, Muneyuki E. Single molecule thermodynamics of ATP synthesis by F1-ATPase. New J. Phys., 2015, 17: 015008.

[48] Diez M, et al. Proton-powered subunit rotation in single membrane-bound FoF1-ATP synthase. Nat. Struct. Mol. Biol., 2004, 11: 135-141.

[49] Finer J T, Simmons R M, Spudich J A. Single myosin molecule mechanics: piconewton forces and nanometer steps. Nature, 1994, 368: 113-118.

[50] Oiwa K, et al. Steady-state force-velocity relation in the ATP-dependent sliding movement of myosin-coated beads on actin cables in vitro studied with a centrifuge microscope. PNAS, 1990, 87: 7893-7897.

[51] Weiss D R, Levitt M. Can morphing methods predict intermediate structures? J. Mol. Biol., 2009, 385: 665-674.

[52] Ma J P, et al. A dynamic analysis of the rotation mechanism for conformational change in F1-ATPase. Structure, 2002, 10: 921-931.

[53] Nam K, Pu J Z, Karplus M. Trapping the ATP binding state leads to a detailed understanding of the F1-ATPase mechanism. PNAS, 2014, 111: 17851-17856.

[54] Masaike T, et al. Cooperative three-step motions in catalytic subunits of F1-ATPase correlate with 80° and 40° substep rotations. Nat. Struct. Mol. Biol., 2008, 15: 1326-1333.

[55] Mukherjee S, Warshel A. Dissecting the role of the γ-subunit in the rotary-chemical coupling and torque generation of F1-ATPase. PNAS, 2015, 112: 2746-2751.

[56] Shu Y G, Zhang X H, Ou-Yang Z C, et al. The neck linker of kinesin 1 seems optimally designed to approach the largest stepping velocity: a simulation study of an ideal model. J. Phys: Condens. Matter, 2011, 24: 035105.

[57] Jülicher F, Ajdari A, Prost J. Modeling molecular motors. Rev. Mod. Phys., 1997, 69: 1269-1281.

[58] Wang H Y, Oster G. Energy transduction in the F_1 motor of ATP synthase. Nature, 1998, 396: 279-282.

[59] Gao Y Q, Yang W, Marcus R A, et al. A model for the cooperative free energy transduction and kinetics of ATP hydrolysis by F_1-ATPase. PNAS, 2003, 100: 11339-11344.

[60] Shu Y G, Lai P Y. Systematic kinetics study of F_oF_1-ATPase: analytic results and comparison with experiments. J. Phys. Chem. B, 2008, 112: 13453-13459.

[61] Dong H, et al. Assembly of F_oF_1-ATPase into solid state nanoporous membrane. Cem.

Commun., 2011, 47: 3102-3104.

[62] Zheng W J, Doniach S. A comparative study of motor-protein motions by using a simple elastic-network model. PNAS, 2003, 100: 13253-13258.

[63] Shiroguchi K, Kinosita Jr K. Myosin V walks by lever action and brownian motion. Science, 2007, 316: 1208-1212.

[64] Steigmiller S, Turina P. Gräber P. The thermodynamic H^+-ATP ratios of the H^+-ATP synthases from chloroplasts and escherichia coli. PNAS, 2008, 105: 3745-3750.

[65] Yildiz A, Tomishige M, Gennerich A, et al. Intramolecular strain coordinates kinesin stepping behavior along microtubules. Cell, 2008, 134: 1030-1041.

[66] Clancy B E, et al. A universal pathway for kinesin stepping. Nat. Struct. Mol. Biol., 2011, 18: 1020-1028.

[67] Toyabe S, et al. Thermodynamic efficiency and mechanochemical coupling of F_1-ATPase. PNAS, 2011, 108: 17951-17956.

[68] Kawaguchi K, Sasa S I, Sagawa T. Nonequilibrium dissipation-free transport in F_1-ATPase and the thermodynamic role of asymmetric allosterism. Biophys. J., 2014, 106: 2450-2457.

[69] Walker J E. The ATP synthase: the understood, the uncertain and the unknown. Biochem. Soc. Trans., 2013, 41: 1-16.

[70] Cherepanov D A, et al. Low dielectric permittivity of water at the membrane interface: effect on the energy coupling mechanism in biological membranes. Biophys. J., 2003, 85: 1307-1316.

[71] Holzbaur E L F, Goldman Y E. Coordination of molecular motors: from in vitro assays to intracellular dynamics. Curr. Opin. Cell Biol., 2010, 22: 4-13.

[72] Guérin T, Prost J, Martin P, et al. Coordination and collective properties of molecular motors: theory. Curr. Opin. Cell Biol., 2010, 22: 14-20.

[73] Endow S A, Higuchi H. A mutant of the motor protein kinesin that moves in both directions on microtubules. Nature, 2000, 406: 913-916.

[74] Badoual M, Jülicher J, Prost J. Bidirectional cooperative motion of molecular motors. PNAS, 2002, 99: 6696-6701.

[75] Shu Y G, Shi H L. Cooperative effects on the kinetics of ATP hydrolysis in collective molecular motors. Phys. Rev. E, 2004, 69: 021912.

[76] Ma R, Li M, Ou-Yang Z C, Shu Y G. Master equation approach for a cross-bridge power-stroke model with a finite number of motors. Phys. Rev. E, 2013, 87: 052718.

[77] Campàs O, et al. Coordination of kinesin motors pulling on fluid membranes. Biophys. J. 2008, 94: 5009-5017.

[78] Wang Z Q, Shu Y G, Li M. A new cooperation mechanism of kinesin motors when extracting

membrane tube. Commun. Theor. Phys., 2013, 60: 753-760.

[79] Kural C, et al. Kinesin and dynein move a peroxisome in vivo: a tug-of-war or coordinated movement? Science, 2005, 308: 1469-1472.

[80] Müller M, Klumpp S, Lipowsky R. Tug-of-war as a cooperative mechanism for bidirectional cargo transport by molecular motors. PNAS, 2008, 105: 4609-4614.

[81] Kunwara A, et al. Mechanical stochastic tug-of-war models can not explain bidirectional lipid-droplet transport. PNAS, 2011, 108: 18960-18965.

[82] Hancock W O. Bidirectional cargo transport: moving beyond tug of war, Nat. Rev. Mol. Cell Biol., 2014, 15: 615-628.

[83] Fischer T, Agarwal A, Hess H. A smart dust biosensor powered by kinesin motors. Nat. Nanotech., 2009, 4: 162-166.

[84] Soong R K, et al. Powering an inorganic nanodevice with a biomolecular motor. Science, 2000, 290: 1555-1558.

[85] Choi H J, Montemagno C D. Artificial Organelle: ATP synthesis from cellular mimetic polymersomes. Nano. Lett., 2005, 5: 2538-2542.

[86] Qi W, et al. ATP synthesis catalyzed by motor protein $CFoF_1$ reconstituted in lipid-coated hemoglobin microcapsules. Adv. Mater., 2008, 20: 601-605.

[87] Cheng J, Zhang X A, Shu Y G, et al. FoF1-ATPase activity regulated by external links on β subunits. Biochem. Biophys. Res. Communi., 2010, 391: 182-186.

[88] Feniouk B A, et al. Chromatophore Vesicles of Rhodobactercapsulatus contain on average one F_oF_1-ATP synthase each. Biophys. J., 2002, 82: 1115-1122.

[89] Chien A, Edgar A B, Trela J M. Deoxyribonucleic acid polymerase from extreme thermophile thermusaquaticus. J. Bacteriology, 1976, 127: 1550-1557.

[90] Mullis K, et al. Specific enzymatic amplification of DNA in vitro: the polymerase chain reaction. Gold Spring Harbor Symposia on Quantitative Biology, 1986, L1: 263-273.

[91] Levene M J, et al. Zero-Mode waveguides for single-molecule analysis at high concentrations. Science, 2003, 299: 682-686.

[92] Eid J, et al. Real-Time DNA sequencing from single polymerase molecules. Science, 2009, 323: 133-138.

[93] Schwarz E, et al. Structure and transcription of human papillomavirus sequences in cervical carcinoma cells. Nature, 1985, 314: 111-114.

[94] Baker T S, et al. Structures of bovine and human papillomaviruses: analysis by cryoelectron microscopy and three-dimensional image reconstruction. Biophys. J., 1991, 60: 1445-1456.

第六章
细 胞 骨 架

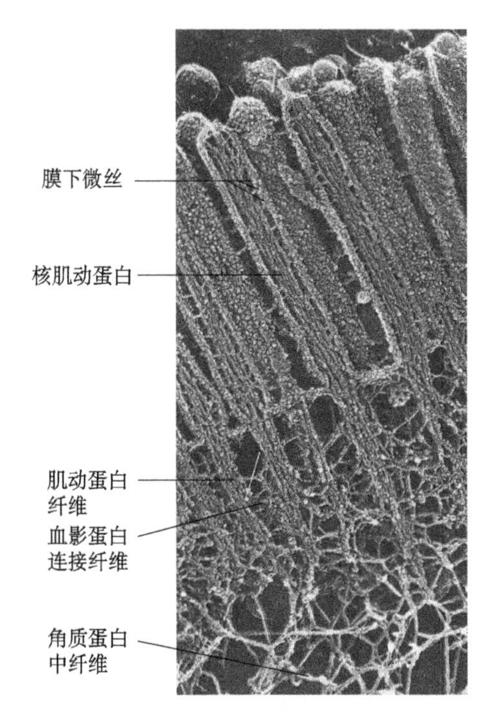

图 6-6-1 小肠上皮细胞顶部的电子显微镜图：小肠上皮细胞顶部表面覆盖了质膜的指尖部分微绒毛，每个微绒毛内部存在一束微丝，起到稳定结构的作用。微绒毛周围的质膜连接到微丝束的侧面。微丝束在根部和细胞相连。微绒毛的根部和血影蛋白连接纤维相连。在多种动物细胞内部质膜下面存在一条窄道，纤维肌动蛋白黏结蛋白主要分布在该道。根基主要和角质蛋白中纤维相连。这些蛋白构筑的网络结构主要为微绒毛提供支撑作用（该图片由 Hirokawa 惠赠）

细胞溶质是细胞新陈代谢的主要场所。其中，蛋白质在细胞溶质中的质量百分数高达 20%～30%，且细胞内的蛋白质总量中至少有 1/4 是分布在细胞溶质中的。据实验估算，蛋白质在细胞溶质中的浓度为 200～400 mg/mL。由于细胞溶质中蛋白质的浓度相当高，蛋白质又能以弱非共价键相连形成具有一定结构的络合物。与此同时，实验研究也表明，细胞溶质是高度有序的，其中可溶性蛋白质黏结到细胞骨架纤维或者位于特定蛋白结合点等待与其他蛋白相结合。图 6-6-1 为典型的动物细胞电子显微镜照片。从图中可以看出，细胞内部的蛋白质以一定的方式相互连接形成高度有序结构——细胞骨架。

从细胞骨架的分子构成来看，可以将细胞骨架分成三类：第一类是微管（microtubules），它是直径约为25nm、稍显弯曲的纤维。如果从持久长度的大小范围来划分，它属于刚性聚合物。图6-6-2给出了微管不同结构层次的构造示意图。从图中可以看出，微管是由8nm的微管蛋白亚基以首尾连接的方式组装形成的。该微管蛋白亚基重复单元是由α和β微管蛋白亚基形成的二聚体。微管是中空的圆柱体，其截面的一圈含13个微管蛋白亚基。沿着微管侧面看，每个亚基与其相邻有两类蛋白质-蛋白质相互作用：沿着环向，蛋白亚基之间有横向作用；而沿着长度方向，蛋白亚基之间存在头尾接触，使得所有亚基指向相同的方向。此外，微管具有极性，它的两端具有不对称性，分别将两端定义为正端（plus end）和负端（minus end）。

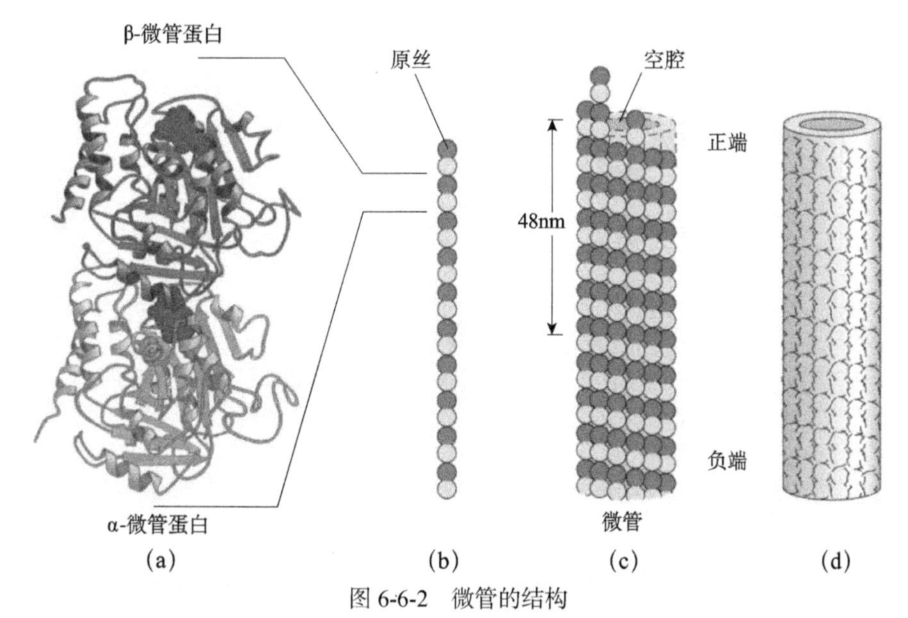

图6-6-2 微管的结构

(a) 微管蛋白亚基重复单元的结构图，它是由α和β微管蛋白亚基形成的二聚体。图中的红色块体为它们黏结的GTP分子所在的位置。(b) α和β微管蛋白亚基以首尾相连的方式形成单根原丝。(c) 微管是由13根原丝排列形成的中空圆柱体。微管蛋白亚基重复单元本身具有极性，从而由它组装形成的微管也具有极性，其中β微管蛋白永远存在于微管的正端，而α微管蛋白存在于微管的负端。(d) 微管通常被认为是一个直径为25 nm的空心圆柱体（摘自Alberts编写的Molecular Biology of the Cell，第六版，Garland Science，2014）（文后附彩图）

第二类是肌动蛋白纤维（actin filament），它是由尺寸约5nm的球状肌动蛋白亚基形成的螺旋组装体。该螺旋组装体具有2根原丝，直径约为8nm，见图6-6-3。该球状肌动蛋白亚基是375个氨基酸组成的肽链形成，它能黏结

ATP或者ADP分子。肌动蛋白亚基的两端本身就具有不对称性，而它们沿相同的方向组装，形成的肌动蛋白纤维也具有极性，两端具有明显不对称性：快速增长端（正端）和慢速增长端（负端）。它的正端也称为倒钩端（barbed end），负端也称为尖端（pointed end）。它的持久长度大小属于蠕虫状链的范畴，从而它也可称为半刚性链。

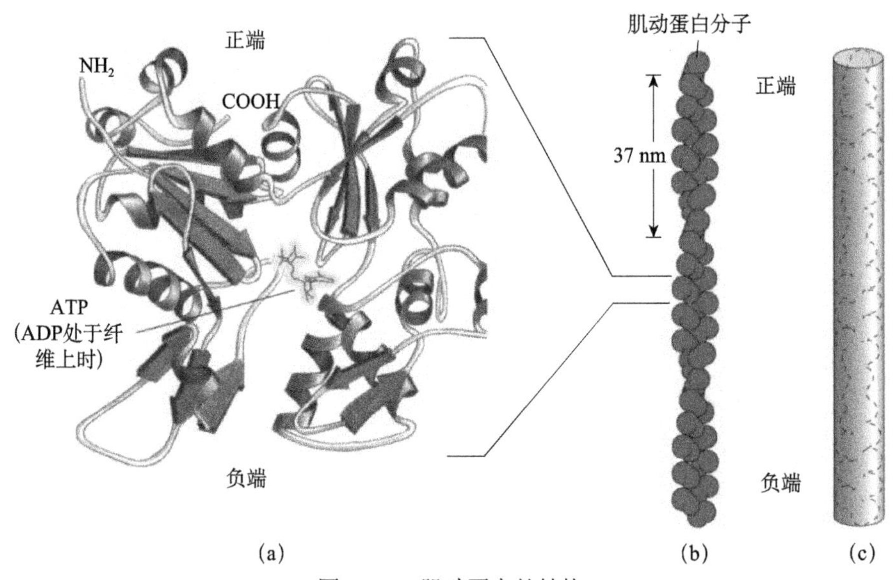

图 6-6-3　肌动蛋白的结构

(a) 单个肌动蛋白分子的结构图，它是直径为5.4nm的球状蛋白。图中黄色部分为它们黏结的ATP(ADP)分子所在位置。(b) 肌动蛋白纤维是由2根原丝组成的，它们形成的螺旋结构的螺距为37nm。所有的肌动蛋白亚基指向相同的方向。每新增一个肌动蛋白单体，纤维的长度增加2.7nm。(c) 肌动蛋白纤维可近似看成直径为8nm的圆柱体（摘自Alberts编写的Molecular Biology of the Cell，第六版，Garland Science，2014）（文后附彩图）

第三类是中间纤维（interfilament），它的直径大小约为10nm，介于肌动蛋白纤维和微管之间，而从持久长度的大小来看，它可近似看成柔性链。中间纤维的亚基与肌动蛋白和微管蛋白的亚基单元具有很大的差异，它们是长条形的蛋白质而不是球形蛋白，见图6-6-4。此外，最突出的地方在于它不同于肌动蛋白纤维和微管，中间纤维不能黏结核苷酸小分子（ATP或者GTP）。中间纤维的组装过程尚未清楚，但是目前认为它涉及亚基横向成束并扭转成卷曲的线团结构。首先形成卷曲线团的二聚体，该二聚体的中心有一个α螺旋的核，核两端连接N和C-球状蛋白，见图6-6-4 (a)。而该二聚体连接在一起形成反平行的四聚体，见图6-6-4 (b)，四聚体缠绕在一起形成绳子状的结构以提供中间纤维的力学强度，见图6-6-4 (c)。

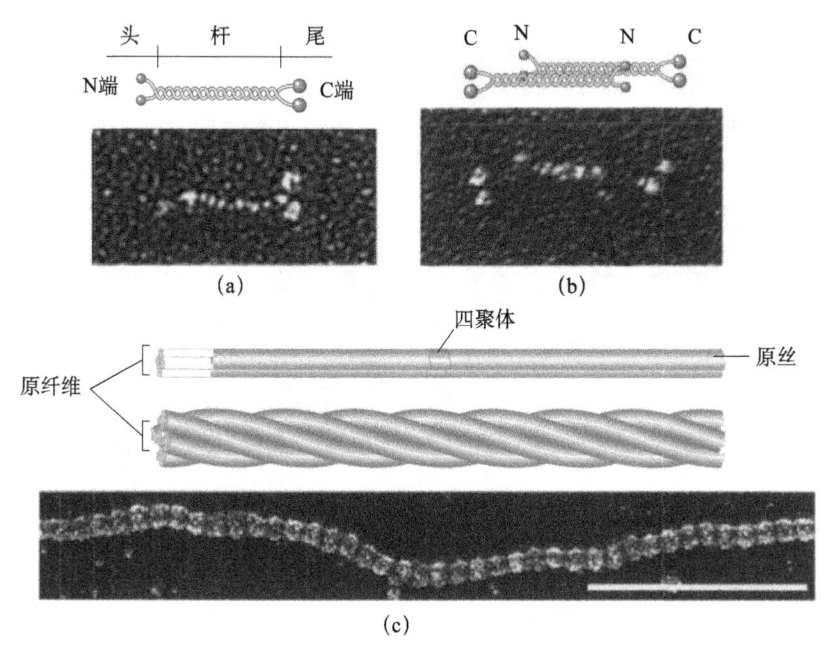

图 6-6-4 中间纤维的电子显微镜图

(a) 中间微丝是由卷曲线团的二聚体在一起形成反平行的四聚体，四聚体缠绕在一起形成绳子状，图中是卷曲线团的二聚体的结构; (b) 由卷曲线团的二聚体在一起形成反平行的四聚体结构; (c) 四聚体缠绕在一起形成绳子状原纤维（摘自 N. Geister et. al., J. Mol. Biol., 1998, 282: 601）

真核细胞中的细胞骨架通常都是聚合物，它们是由小分子蛋白亚基通过非共价键相连接而形成的。这些细胞骨架进一步聚集形成一定的组织结构，如束状、凝胶态结构以及网状结构等。这些细胞骨架构建出不同的聚集态结构，且表现出相应的力学行为，从而实现不同的生物功能，见图 6-6-5。肌动蛋白纤维主要为细胞质膜提供一定的支撑作用，且对细胞的表面形态起决定性作用。肌动蛋白纤维还为细胞运动提供一定作用力，也为细胞一分为二的分裂生物过程提供驱动力。微管决定了有机体囊泡的位置，提供细胞内分子的运输通道。微管还能在细胞分裂过程中形成有丝分裂纺锤体，帮助染色体实现分离。中间纤维主要起力学支撑的作用。这些细胞骨架纤维会和成千上万的小分子蛋白分子相互作用，从而将细胞骨架纤维和细胞其他组成部分相互连接起来。这些辅助小分子蛋白还能在特定位置控制细胞骨架纤维的组装和去组装，如分子马达蛋白分子，是一种分子机器。这种分子机器能通过 ATP 水解或者 GTP 水解所释放的能量来产生相应的作用力，而这些作用力能驱动有机体在细胞骨架纤维上面的运输或者驱动纤维自身的运动等。这些细胞骨架纤维具有一些共同的特征，例如，除了中间纤维，其他两种细胞骨架纤维都具有极性，但是这三种细胞骨架纤维形成的聚集态结构都具有高度的动态性质。此外，伴随着周围环境的变化，细胞能发出信号随时调整细胞骨架的结构和性能。

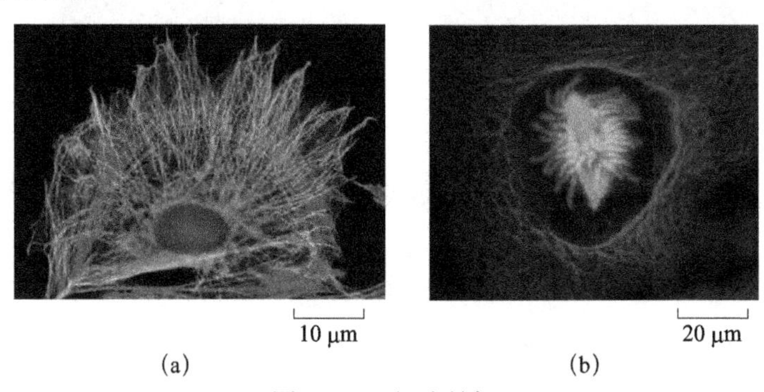

图 6-6-5　细胞骨架

(a) 细胞内细胞质阵列：微管（绿色）和肌动蛋白纤维（红色）。(b) 分裂细胞内组成纺锤体的微管（绿色）和周围的中间纤维（红色）在所有的细胞中 DNA 标记为蓝色（Albert Tousson, Conly Riender 惠赠）（文后附彩图）

通常，真核细胞比原核细胞更高级和复杂，典型的细菌细胞只有几个微米，形状也相对简单，以棒状为主。生物学家曾经认为细菌与真核生物的显著区别在于细菌类的简单原核细胞中没有细胞骨架。但是，现有的研究表明，将小蛋白亚基组装成细胞骨架纤维是细胞组装的普遍特性。在细菌中，也存在着真核细胞中。

三种细胞骨架纤维的同源体，而且相比于真核细胞，细菌中的肌动蛋白亚基和微管蛋白亚基更为多样化。几乎所有的细菌和古核生物中微管蛋白亚基的同源体被称为 FtsZ，而肌动蛋白亚基的同源体被称为 MreB 和 ParM。细菌中的 FtsZ 蛋白亚基会聚合形成纤维，进一步在特定位置组装形成 Z-环，从而在细胞分裂的过程中形成隔膜。Z-环的寿命通常有几分钟，而单根纤维具有高度动态性，它的半衰期只有 30s。随着细胞分裂的进行，Z-环变得越来越小，直到最终去组装。Z-环上的 FtsZ 纤维对细胞壁产生收缩作用力，而造成细胞分裂。此外，Z-环也为细胞分裂过程中隔膜的形成提供酶的结合位点。细菌中肌动蛋白亚基的同源物，MreB 和 Mbl 蛋白亚基，主要分布于棒状或螺旋状细胞中。它们在细胞的径向方向组装形成动态的结构，这些蛋白亚基与细胞壁的合成和基因组的定位等有关。此外，FtsZ、MreB 和 Mbl 纤维都是高度动态的，也能耦合核苷酸的水解过程。肌动蛋白亚基的同源物 ParM，其最有趣的地方在于在某种程度上，被认为是纺锤体的简化版本。

图 6-6-6 展示了 ParM 蛋白亚基在原核细胞内如何执行类似纺锤体的功能。ParM 纤维能自发地生长和收缩。首先，ParM 纤维一直生长到它两端能碰到抗生素抗性质粒的拷贝之一。当两个质粒分别结合纤维的两端时，纤维持续聚合而不再收缩，有效地将两个质粒推进到细胞的两极。

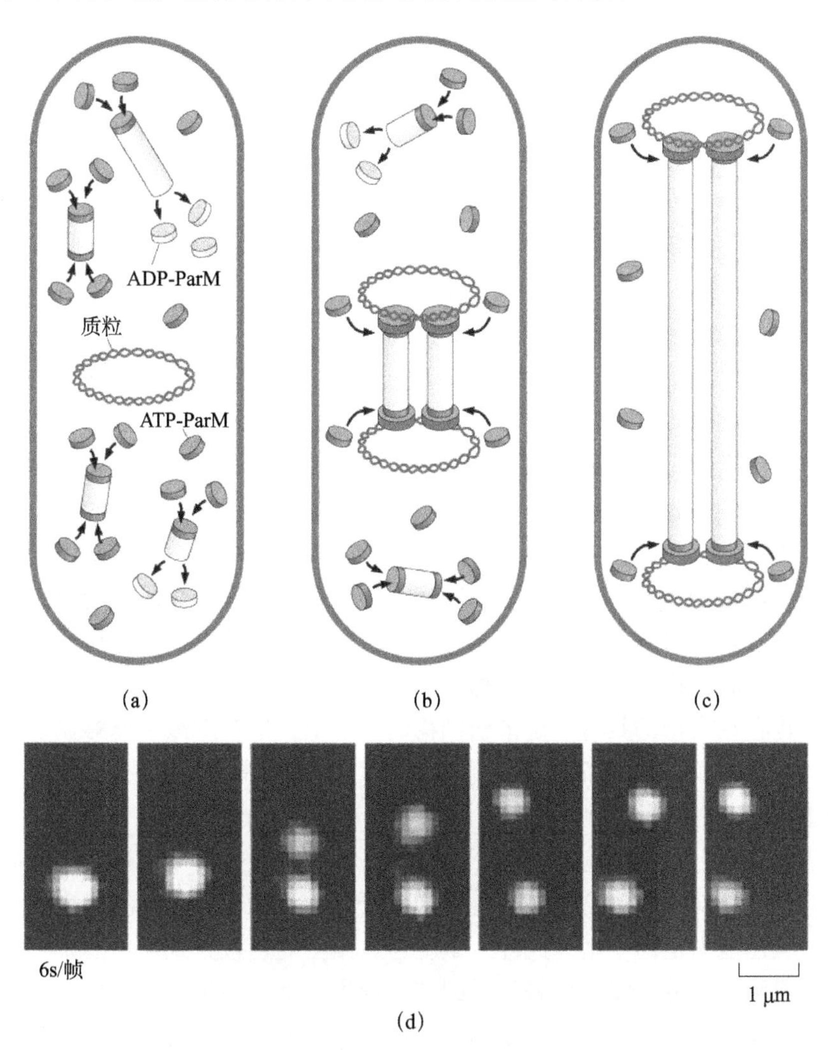

图 6-6-6　细菌内肌动蛋白同系物 ParM 驱动质粒分离的示意图

(a) 细胞内存在少量抗生素抗性质粒。为了保证抗生素抗性质粒的浓度水平，这里质粒也有着类似于染色体分离的生物过程。R1 质粒是由 ParM 组成的，而 ParM 纤维非常不稳定。(b) 在质粒被复制后，当不稳定的 ParM 纤维两端接触到质粒后，ParM 纤维的稳定性变强了。DNA 持续黏结，ParM 纤维不断增长。(c) 由于 ParM 纤维的持续增长，它们将两个质粒推到细胞的两极。(d) 活体大肠杆菌细胞中两个荧光染色的质粒的时间动力学过程 [(a)～(c) 摘自 E. C. Garner et al., Science, 306: 1021, 2004；(d) 摘自 C. S. Campbell and R. D. Mullins, J. Cell. Biol., 179: 1059, 2007.]

第一节　学科发展背景和现状

一、细胞骨架的动态稳定结构

细胞骨架系综不仅具有高度动态性，而且具有一定适应性。它们能组装成复杂而具有一定规律连接的网络结构。该大规模网络结构对外界的响应时间可根据环境的需要从几秒钟调整到长达细胞的寿命。但是，组成该大规模的网络结构中的单根细胞骨架纤维以一定的速度在不断变动中，而且其细胞内结构的重组也会伴随着体系能量的变化。

真核细胞通过调控细胞骨架纤维的动态行为和自组装能力，能构建出很多复杂的结构，如图 6-6-7 所示。肌动蛋白纤维主要分布于动物细胞的质膜内部，为薄的脂质体双层膜提供形态和力学上的支撑作用。它们在细胞表面形成具有一定形态的结构，而这些形态结构是动态的，会发生变化，如板状伪足（lamellipodia）和丝状伪足（filopodia）。细胞利用这些形态的动态变化能向四周移动，从而占领更多的势力范围。此外，肌动蛋白纤维的稳定阵列结构不但能给细胞提供力学支撑的作用，还能有助于产生作用力造成肌肉收缩。此外，耳朵内部毛细胞表层的硬纤毛也是由肌动蛋白纤维连接形成的具有规律的束状结构组成的。该结构对声音具有一定的响应能力。肠上皮细胞的表面也存在大量的具有一定束状结构的微绒毛，这些微绒毛能够显著增加细胞的局部表面积，从而提高细胞吸收所需营养的能力。在植物细胞内部，肌动蛋白纤维也能为细胞内细胞质的快速流动提供一定驱动力。

微管主要分布在细胞质内部，它能够在细胞分裂过程中通过重组形成双极的有丝分裂纺锤体。这些微管也能形成纤毛，而纤毛可作为细胞表面的传感器，且还能做定向摆动。此外，微管通过紧密的连接方式形成的束状结构可作为囊泡、养分、线粒体等物质的传输通道。微管在植物细胞中也会按一定的规律排列成微管阵列并且承担一定的生命活动。总而言之，在细胞的生长和凋亡周期过程中，不同时期的植物细胞内部微管会构建形成不同的微管阵列，从而各个微管阵列能执行相应于所在时期的独特功能。

中间纤维在细胞中围绕细胞核分布，成束成网，并扩展到细胞质膜并且与质膜相连，为细胞的 DNA 提供一个保护网。中间纤维从核纤层通过细胞质延伸，它不仅对细胞刚性有支持作用和对产生运动的结构有协调作用，而且与细胞分化、细胞内信息传递、核内基因传递、核内基因表达等生命活动有关。几乎所有的动物细胞中都能观察到中间纤维，但是在植物细胞和真菌细胞中几乎没有中间纤维。而且，实验发现中间纤维对细胞运动无关。

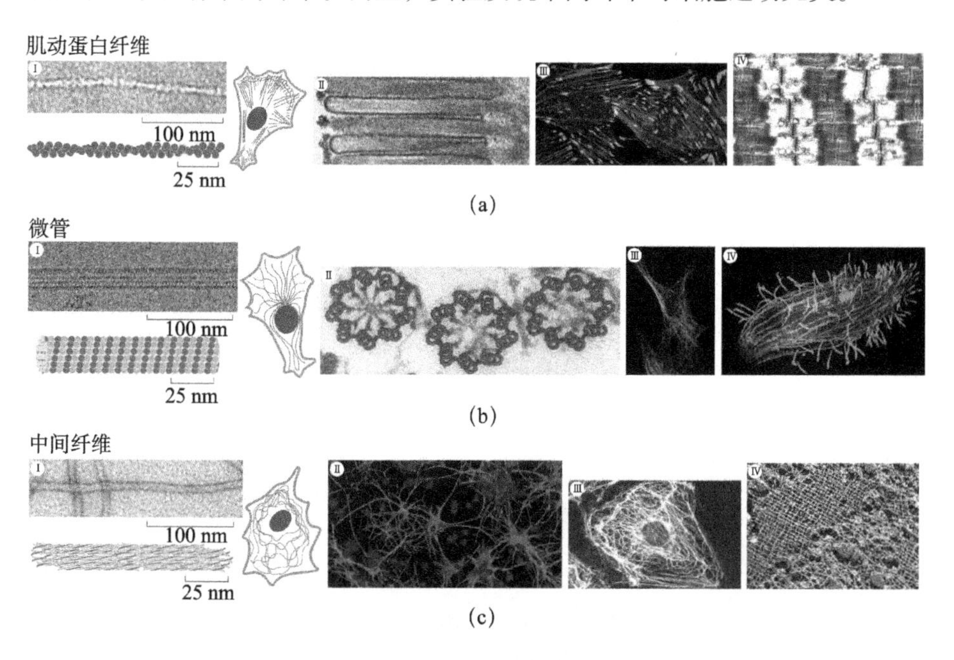

图 6-6-7　细胞骨架纤维

(a) 肌动蛋白纤维:它是由肌动蛋白亚基组成的螺旋状微丝。它的直径约为 8nm 且具有柔性结构，它能组装形成束状结构、二维的网状结构和三维的凝胶结构。肌动蛋白纤维分布于细胞内，主要集中分布于细胞的顶端和质膜下方。(Ⅰ) 单根肌动蛋白纤维，(Ⅱ) 微绒毛，(Ⅲ) 细胞局部黏结时形成的应力纤维，(Ⅳ) 肌肉内部纤维。(b) 微管：它是直径为 25 nm 的中空圆柱体。它比肌动蛋白纤维的持久长度要高。微管的一端通常黏结于微管组织中心（MTOC，中心体）。(Ⅰ) 单根微管，(Ⅱ) 纤毛内部微管三聚体的截面，(Ⅲ) 中间相微管阵列（绿色）和有机体（红色），(Ⅳ) 纤毛原生动物。(c) 中间纤维：它是直径为 10 nm 的绳子状纤维。一种中间纤维形成网状结构主要分布在细胞核内核纤层，另外一种中间纤维主要分布在细胞质内提供力学强度。在上皮细胞组织，它们能连接细胞以提供上皮组织的力学强度。(Ⅰ) 单根中间纤维，(Ⅱ) 神经元内的中间纤维（蓝色），(Ⅲ) 上皮细胞，(Ⅳ) 核纤层（R. Craig, P. T. Matsudaria 惠赠）(文后附彩图)

图 6-6-8 给出在细胞分裂过程中细胞骨架结构重组的过程图。运动的成纤维细胞是一种具有极性的、动态的肌动蛋白纤维结构。首先，染色体在细胞中间复制成功后，细胞质中的微管阵列将扩展重构形成双极的有丝分裂纺锤体，这些纺锤体将染色体分离到两个子核。而在这个过程中，肌动蛋白纤维也通过重构使成纤维细胞在爬行过程中从盘状的形态转变成球形的形态。

肌动蛋白和它的相关马达蛋白在细胞中间形成收缩环，而对细胞表面产生收缩作用力致使细胞分裂。当细胞分裂完成后，两个成纤维细胞中的细胞骨架将进一步重构使细胞的形态变回到母细胞的扁平爬行形态。

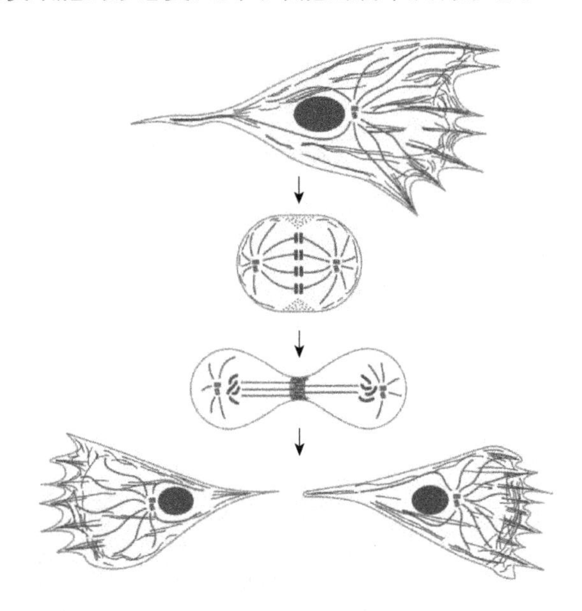

图 6-6-8　细胞分裂过程中细胞骨架的重构示意图

爬行的成纤维细胞具有极性、动态的肌动蛋白细胞骨架（红色）。在板状伪足和成纤维细胞内形成的肌动蛋白细胞骨架推动板状伪足和成纤维细胞向前移动。微管细胞骨架（绿色）对这些肌动蛋白细胞骨架提供支持作用。这些微管细胞骨架在细胞核内以一个中心点发散形成微管细胞骨架。当细胞分裂时，微管细胞骨架首先重组形成双极性纺锤体。纺锤体主要为染色体（棕色）分离提供作用力。在染色体分离后，肌动蛋白细胞骨架在细胞中间形成动态收缩环，它提供限制作用力将细胞一分为二。在细胞完全分裂后，子细胞将继续重组微管和肌动蛋白细胞骨架，从而子细胞继续爬行等待下一次分裂（摘自 Alberts 编写的 Molecular Biology of the Cell，第五版，Garland Science，2008）（文后附彩图）

二、蛋白质单元组装成细胞骨架纤维

通常，细胞骨架纤维的长度分布在几十到几百微米之间。而它的结构单元，组成细胞骨架纤维的小蛋白质亚基只有几个纳米。细胞骨架纤维就是由一系列的蛋白质亚基以一定的方式连接而形成的。蛋白质亚基在细胞质中的扩散速度很快，而组装形成的细胞骨架纤维几乎不能扩散。因此，细胞骨架纤维非常容易实现结构重组、在某些特定位置进行纤维的去组装，以及在特定位置实现再组装。

真核细胞中的三种细胞骨架都是由蛋白质亚基以螺旋结构的方式组装而成的。蛋白质亚基通过弱非共价键相互连接，或首尾相连，或者侧向相互作用连接而形成纤维。不同类型的细胞骨架纤维由于蛋白质亚基的结构不同，相互连接的作用力大小也不同。这些差异会造成组装成的纤维具有不同的热稳定性和力学性能。

肌动蛋白纤维和微管的蛋白亚基都是非对称的，而且蛋白亚基之间都是首尾相连的，且指向相同的方向。这些结构单元的极性赋予纤维两端能表现出不同的行为，从而具有极性。肌动蛋白纤维和微管的蛋白亚基都是酶能够分别黏结核苷酸——ATP 和 GTP 小分子，并且能够催化核苷酸水解。而核苷酸水解产生的能量将驱动细胞骨架纤维的重构。通过控制肌动蛋白纤维和微管组装的时间和地点，细胞能够利用细胞骨架纤维的极性和动态行为在特定方向产生作用力，驱动细胞前端的运动，如细胞分裂过程中染色体的分离。相比之下，中间纤维是对称的，因而它不能形成极性的纤维。中间纤维也不能催化核苷酸的水解。但是，中间纤维在特定条件下能够快速去组装，因此它的快速去组装能力赋予了中间纤维特定的生物功能。

细胞骨架纤维通常不能简单地按照单个蛋白亚基以非共价键相连的方式组装成单根纤维。如果成千上万个微管蛋白亚基按照简单的方式相连形成单根纤维，则该单根纤维在热涨落的情况下就极易去组装，除非蛋白亚基之间具有非常强的相互键接作用力。但是，强的键接作用力将阻碍细胞骨架纤维的去组装能力，从而造成细胞骨架处于静态的状态。为了提高微管的力学强度和适应性，微管是由 13 根原纤维纵向排列形成中空圆柱体的。每根原微丝的末端都能通过形成和断裂一个非共价键组装和去组装一个蛋白亚基。如果从微管中间去组装一个蛋白亚基，则需要同时断裂至少两个以上的非共价键。因此，断裂一个由多根原纤维组装形成的微管要比断裂单根纤维需要更多的能量。这也说明，以该种方式组装形成的微管完全能够抵抗热涨落的断裂，而微管的两端也能够快速地组装和去组装。具有螺旋结构的肌动蛋白纤维半径要比微管小很多，因此较小的能量就能够断裂肌动蛋白纤维，见示意图 6-6-9。但是，细胞内肌动蛋白纤维通过辅助蛋白连接形成束状聚集体后，就能够具有很好的力学强度，也保留了肌动蛋白纤维的动态行为。

为了调控细胞骨架纤维的长度、大小和空间稳定性，细胞需要控制细胞骨架纤维的连接方式和蛋白亚基的组成，以至于细胞骨架能够形成具有高度有序的动态结构。在细胞的整个蛋白调控网络中，上百种蛋白参与控制细胞骨架纤维的长度和动态行为、调控信号通道等生命活动。其中，最重要的蛋白有马达蛋白分子，下面的章节将详细阐述分子马达蛋白。

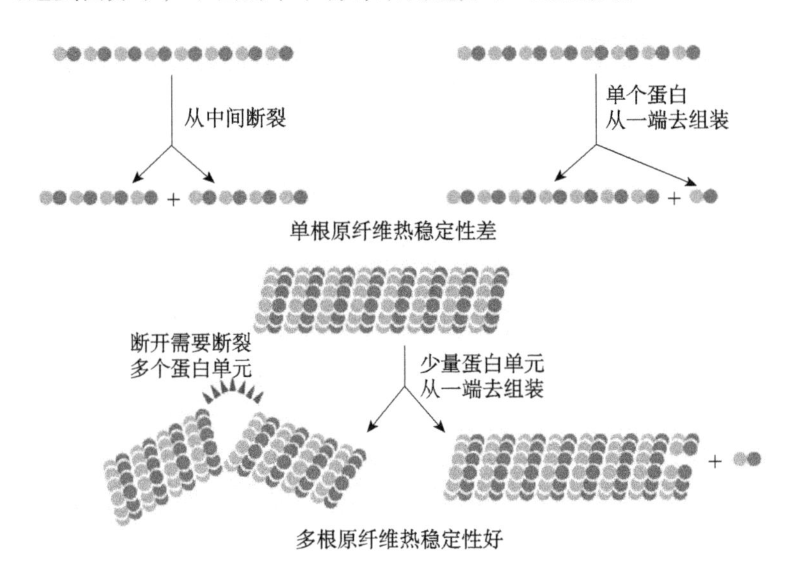

图 6-6-9　动态细胞骨架纤维的热稳定示意图

由单个蛋白亚基组成的单根原丝具有热力学不稳定性，因为断裂该单根原丝只需要蛋白连接的单键。相比之下，由多根原丝形成的细胞骨架纤维依然能赋予细胞骨架纤维的动态行为，还能抵抗热涨落。例如，在微管末端脱落一个微管蛋白亚基至少需要断裂 3 个非共价键，而从微管中间断裂至少需要断裂 13 个非共价键（摘自 Alberts 编写的 Molecular Biology of the Cell，第五版，Garland Science，2008）

三、细胞骨架的速率方程

大量的细胞过程都涉及蛋白亚基在生长的细胞骨架纤维末端的聚合和解聚，如转录和翻译、细胞骨架的组装及烟草花叶病毒等丝状病毒的聚合等。关于聚合反应最简单的描述是：假设单体以一定的聚合速率黏结到生长的长链末端，以及以一定的解聚速率从链末端脱落。以肌动蛋白亚基为例，图 6-6-10 所示的复杂性增加的层级结构模型。在最简单的情况下，就是在不考虑蛋白亚基结合核苷酸的状态不同，即忽略 ATP 酶（或 GTP 酶）水解的情况，由于肌动蛋白亚基本身的结构具有不对称性，肌动蛋白纤维两端的聚合和解聚速率不同。如果考虑蛋白亚基结合核苷酸状态的不同，在细胞骨架纤维的生长动力学中，水解反应起着核心作用。由于水解反应的存在，蛋白亚基结合的核苷酸状态通常被认为有三种：首先是蛋白亚基结合 ATP 酶（或者 GTP 酶）；其次，经过水解反应（或者说分裂过程）后，蛋白亚基结合的酶状态转变成蛋白亚基结合 ADP.Pi 酶（或者 GDP.Pi 酶）的状态；最后是无机磷酸释放后，蛋白亚基变成结合 ADP 酶（或 GDP 酶）的状态。由于肌动蛋白纤维上蛋白亚基结合核苷酸的状态不同，纤维两端的聚合速率和解聚速率差异非常大。图 6-6-11 给出了当考虑水解反应和两端不对称性的影响时纤维增长动力学的示意图。

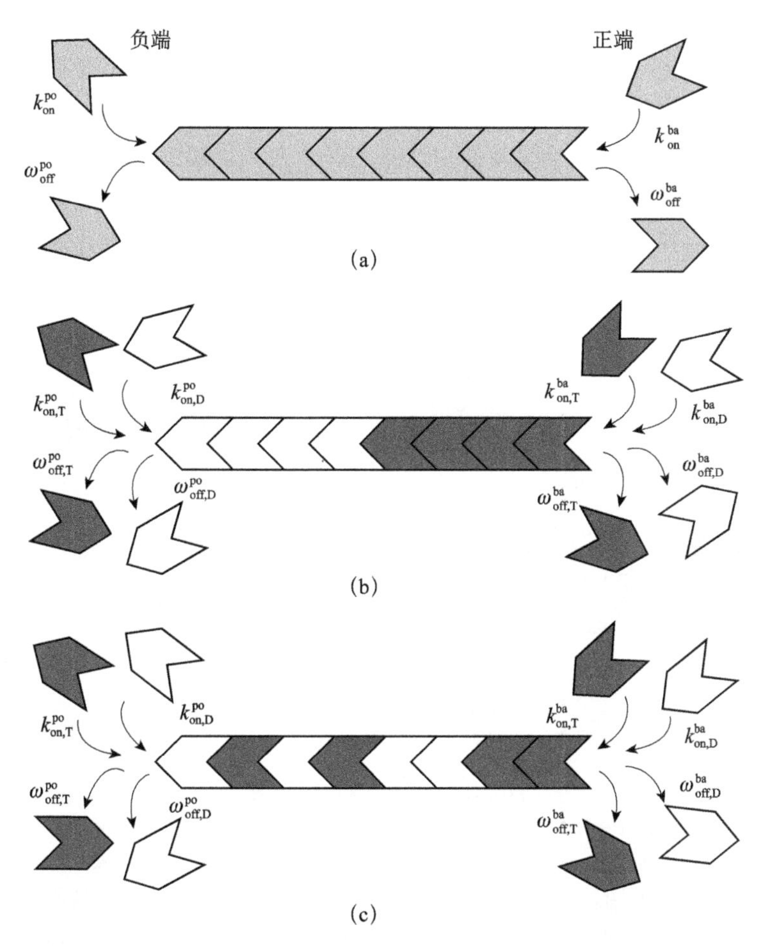

图 6-6-10　细胞骨架聚合的复杂性增加层次示意图，其中包括不对称速率和水解

(a) 由于肌动蛋白纤维的结构不对称性，它的两端具有不同的聚合和解聚速率，分别称为正端和负端，对于肌动蛋白纤维又称为倒钩端和尖端。(b) 矢量水解下，纤维两端以不同的速率聚合和解聚。(c) 随机水解下，纤维两端以不同的速率聚合和解聚。其中，红色表示 ATP- 肌动蛋白单体；白色表示 ADP- 肌动蛋白单体（文后附彩图）

为了处理细胞骨架动力学的复杂性，体外实验中先考虑图 6-6-11 中的简单情况。从图中可看出，肌动蛋白亚基在溶液中会自发地聚合形成纤维。在实验中，有一种简单的方法能够近似测量肌动蛋白纤维的合成速率，把荧光小分子结合到肌动蛋白亚基上。当肌动蛋白亚基处于单体状态时，它的荧光很弱，一旦聚合形成肌动蛋白纤维，荧光强度会突然增加。通过监测荧光信号强度随时间的变化，就能够测量肌动蛋白纤维成核与生长的速率，见图 6-6-11(c)。如图 6-6-11(a) 所示，肌动蛋白单体置于溶液中，经过初始滞后期，它们就成核并生长，直至达到平衡，即聚合速率等于解聚速率。如果溶液中存在预成核的“种子”，则单体会马上聚合成纤维，见图 6-6-11(b)，但生长速率与图 6-6-11(a) 中情况类似。这些实验让人们了解细胞如何通过调节肌动蛋白纤维的成核来控制纤维形成的时间和地点。活体细胞内含有数以百计的与肌动蛋白亚基相关的蛋白，它们能够调控肌动蛋白纤维动力学的各个方面，包括成核速率、延伸速率、解聚速率、ATP 交换速率甚至大尺度构造等。

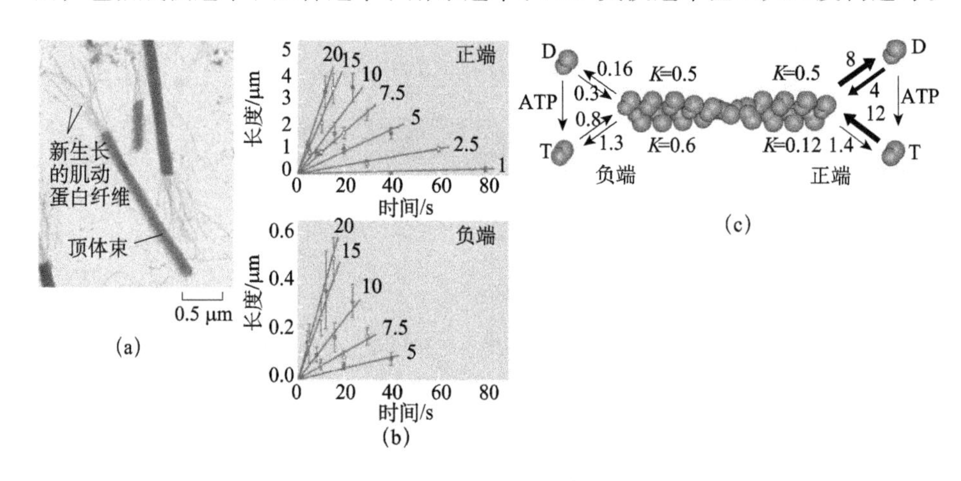

图 6-6-11　肌动蛋白聚合速率

(a) 从顶体束生长的肌动蛋白纤维电镜图。顶体束是由肌动蛋白纤维组装形成的平行矩阵。肌动蛋白单体会聚合到顶体束的两端，也能从顶体束的两端解聚。(b) 从顶体束两端生长的肌动蛋白纤维长度随时间的演化图。图中标注的是肌动蛋白单体浓度（以 μm 为单位）。从长度的标尺可看出，正端的聚合速率是负端聚合速率的 10 倍。(c) 肌动蛋白聚合的示意图。图中的聚合速率常数单位为 $\mu m^{-1} \cdot s^{-1}$，解聚速率的单位为 s^{-1}（摘自 T. D. Pollard, J. Cell. Biol., 103: 2747, 1986）

（一）蛋白亚基结合核苷酸状态为单态模型的速率方程

细胞骨架的聚合过程包括图 6-6-12 所示的三个不同阶段：成核、生长、稳态。如果初始时刻溶液中只有单体，那么这些单体首先要结合形成寡聚物凝结核，因此会有一个滞后期。成核期后，一旦成核种子形成，则成长期开始；细胞骨架分子就开始迅速生长，直到稳态期。在稳态期，溶液中的单体和纤维两端的原聚体存在交换，但纤维的长度不再有明显变化，而处于涨落中。

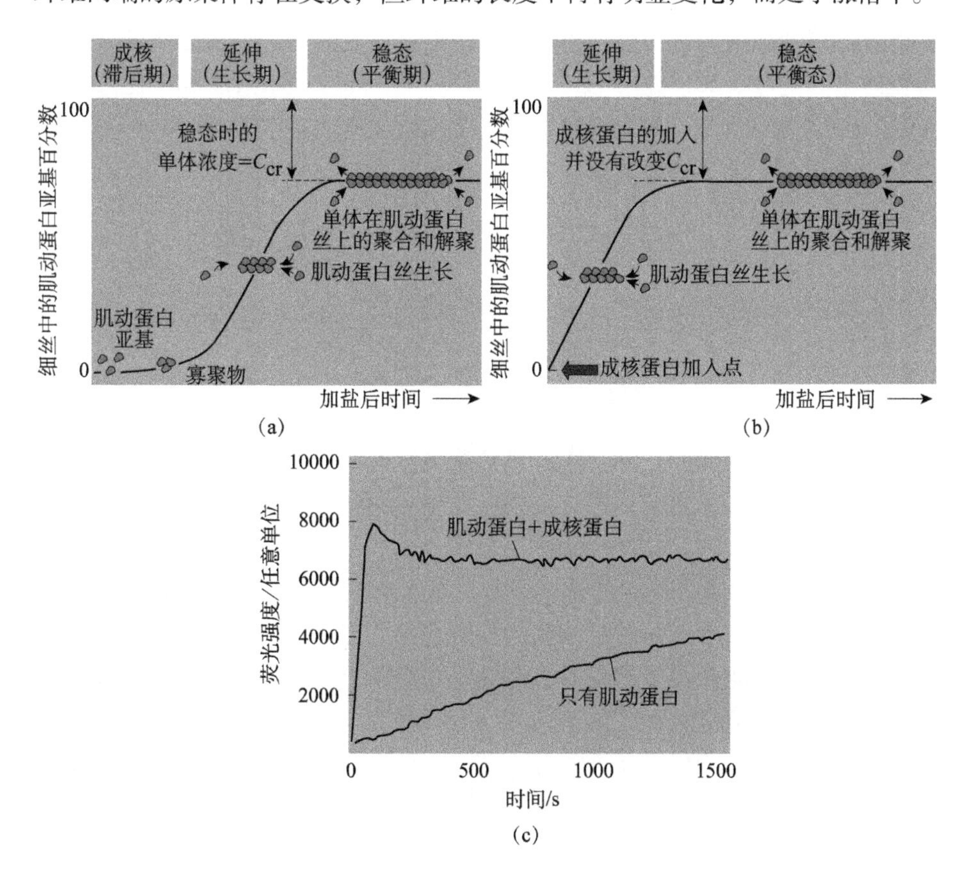

图 6-6-12 肌动蛋白聚合动力学的示意图

(a) 肌动蛋白单体从溶液中开始组装生长，首先，至少 3 个肌动蛋白亚基碰撞成核形成寡聚物，然后迅速生长。成核期比较慢，而生长期相对速度比较快。当聚合完成后，溶液中剩余少量的肌动蛋白单体，该浓度为临界浓度，c_{cr}。(b) 在溶液中加入预先形成的核，肌动蛋白单体迅速从溶液中聚合到核直到稳定。在该过程中没有滞后期。加入预先形成的核，并没有改变体系的临界浓度。(c) 实验测定的只有肌动蛋白和加入成核剂的增长（摘自 B. Alberts et al. Molecular biology of the cell，第 5 版，Garland Science, 2008, M. D. Welch et al. Science 281: 105, 1988）

首先，忽略 ATP 水解（GTP 水解），用简单的速率方程描述纤维的生长情况。考虑体系有 M 个核的最简单情况，我们要计算“平均”每根纤维两端结合单体数目 n 随时间变化曲线。速率方程并不考虑长度的涨落，单体数目 n 由以下方程给出：

$$\frac{\mathrm{d}n}{\mathrm{d}t}=\left(k_{\mathrm{on}}^{\mathrm{ba}}+k_{\mathrm{on}}^{\mathrm{po}}\right)\left(c_0-\frac{Mn(t)}{V}\right)-\left(\omega_{\mathrm{off}}^{\mathrm{ba}}+\omega_{\mathrm{off}}^{\mathrm{po}}\right) \tag{6-6-1}$$

式中，c_0 是溶液中初始单体浓度；V 是总体积；M 是核的数目；$k_{\mathrm{on}}^{\mathrm{ba}}$ 和 $k_{\mathrm{on}}^{\mathrm{po}}$ 分别表示单体黏结到纤维两端的聚合速率常数，而且在上述方程中假定单体在纤维两端的聚合速率，即 $\omega_{\mathrm{on}}^{\mathrm{ba}}$ 和 $\omega_{\mathrm{on}}^{\mathrm{po}}$，与溶液中自由蛋白单体浓度成正比；$\omega_{\mathrm{off}}^{\mathrm{ba}}$ 和 $\omega_{\mathrm{off}}^{\mathrm{po}}$ 是纤维两末端原聚体的解聚速率;上标 ba 表示纤维的正端，而 po 表示负端。式（6-6-1）是一个一阶线性微分方程，可将其改写为

$$\frac{\mathrm{d}n}{\mathrm{d}t}+\left(k_{\mathrm{on}}^{\mathrm{ba}}+k_{\mathrm{on}}^{\mathrm{po}}\right)\frac{Mn(t)}{V}=\left(k_{\mathrm{on}}^{\mathrm{ba}}+k_{\mathrm{on}}^{\mathrm{po}}\right)c_0-\left(\omega_{\mathrm{off}}^{\mathrm{ba}}+\omega_{\mathrm{off}}^{\mathrm{po}}\right) \tag{6-6-2}$$

可以猜测它有一个特解：

$$n_{\mathrm{particular}}(t)=\frac{V}{M\left(k_{\mathrm{on}}^{\mathrm{ba}}+k_{\mathrm{on}}^{\mathrm{po}}\right)}\left[\left(k_{\mathrm{on}}^{\mathrm{ba}}+k_{\mathrm{on}}^{\mathrm{po}}\right)c_0-\left(\omega_{\mathrm{off}}^{\mathrm{ba}}+\omega_{\mathrm{off}}^{\mathrm{po}}\right)\right] \tag{6-6-3}$$

而式（6-6-2）所对应的齐次方程（即令等号右边为 0）的通解为

$$n_{\mathrm{homogeneous}}(t)=A\mathrm{e}^{-\frac{M\left(k_{\mathrm{on}}^{\mathrm{ba}}+k_{\mathrm{on}}^{\mathrm{po}}\right)t}{V}} \tag{6-6-4}$$

式中，A 是由初始条件确定的待定常数。整个方程的通解则是上述特解加上齐次方程的通解。考虑初始时刻核的长度为零（当然，这与存在有限尺寸的核时聚合反应才能开始是矛盾的，但是它在数学上是方便的，并且能够得到与有限尺寸的核相同的关键特征）。令 $n(0)=0$，常数 A 为

$$A=-\frac{V}{M\left(k_{\mathrm{on}}^{\mathrm{ba}}+k_{\mathrm{on}}^{\mathrm{po}}\right)}\left[\left(k_{\mathrm{on}}^{\mathrm{ba}}+k_{\mathrm{on}}^{\mathrm{po}}\right)c_0-\left(\omega_{\mathrm{off}}^{\mathrm{ba}}+\omega_{\mathrm{off}}^{\mathrm{po}}\right)\right] \tag{6-6-5}$$

因此，方程的解可表示为

$$n(t)=\frac{V}{M\left(k_{\mathrm{on}}^{\mathrm{ba}}+k_{\mathrm{on}}^{\mathrm{po}}\right)}\left[\left(k_{\mathrm{on}}^{\mathrm{ba}}+k_{\mathrm{on}}^{\mathrm{po}}\right)c_0-\left(\omega_{\mathrm{off}}^{\mathrm{ba}}+\omega_{\mathrm{off}}^{\mathrm{po}}\right)\right]\left[1-\mathrm{e}^{-\frac{M\left(k_{\mathrm{on}}^{\mathrm{ba}}+k_{\mathrm{on}}^{\mathrm{po}}\right)t}{V}}\right] \tag{6-6-6}$$

肌动蛋白纤维的长度可以表示为

$$L(t)=L_0+an(t) \tag{6-6-7}$$

式中，a 为每个单体的长度；L_0 是初始时刻纤维的长度，相应的函数图像列于图 6-6-13 中。图 6-6-13 中有两个特征区域，反应的初始阶段，即当反应时间小于

$$\tau=\frac{V}{M(k_{\mathrm{on}}^{\mathrm{ba}}+k_{\mathrm{on}}^{\mathrm{po}})} \tag{6-6-8}$$

时纤维长度随时间呈线性增长。而当反应进行很长时间后，生长趋于饱和，纤维长度基本保持不变。

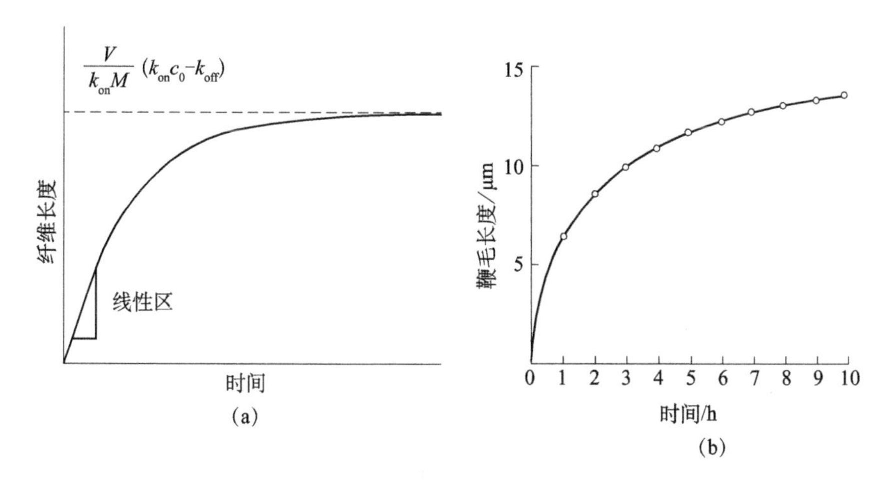

图 6-6-13　肌动蛋白纤维长度随时间的演化曲线

(a) 肌动蛋白纤维长度随时间的演化动力学，假定溶液初始具有 M 个核。在增长的初始阶段，增长速率是呈线性的，而反应进行一段时间后，体系将达到稳定，纤维的长度基本保持不变。(b) 细菌鞭毛组装过程中，纤维长度随时间的变化（摘自 T. Lino et al. J. Supramol. Struct. 2: 372, 1974）

其次，假定溶液中单体浓度固定为 c 的情况下肌动蛋白纤维增长的情况。纤维两端单体数目的平均增长速率分别为

$$\frac{\mathrm{d}\langle n^{\mathrm{ba}}\rangle}{\mathrm{d}t}=k_{\mathrm{on}}^{\mathrm{ba}}\,c-\omega_{\mathrm{off}}^{\mathrm{ba}} \tag{6-6-9}$$

和

$$\frac{\mathrm{d}\langle n^{\mathrm{po}}\rangle}{\mathrm{d}t}=k_{\mathrm{on}}^{\mathrm{po}}\,c-\omega_{\mathrm{off}}^{\mathrm{po}} \tag{6-6-10}$$

因此，纤维的增长速率，即 $J_{\mathrm{g}}(c)$，定义为纤维的长度平均对时间求导，简记为

$$J_{\mathrm{g}}(c)=\frac{\mathrm{d}\langle L\rangle}{\mathrm{d}t}=\frac{\mathrm{d}\langle n^{\mathrm{ba}}+n^{\mathrm{po}}\rangle}{\mathrm{d}t}=(k_{\mathrm{on}}^{\mathrm{ba}}+k_{\mathrm{on}}^{\mathrm{po}})c-(\omega_{\mathrm{off}}^{\mathrm{ba}}+\omega_{\mathrm{off}}^{\mathrm{po}}) \tag{6-6-11}$$

由于正端（倒钩端）的平均增长速率远大于负端（尖端），正端和负端相对于纤维的初始中心位置呈现出明显的不对称性。此外，当溶液中单体浓度为常数时，方程（6-6-11）的右端是一个常数，纤维中原聚体的数目随着反应时间的变化是线性增加或者减小，主要取决于溶液中浓度 c 是高于或低于临界浓度 c_{cr}，而纤维两端的临界浓度可表示为 $c_{\mathrm{cr}}=\omega_{\mathrm{off}}^{\mathrm{po}}/k_{\mathrm{on}}^{\mathrm{po}}=\omega_{\mathrm{off}}^{\mathrm{ba}}/k_{\mathrm{on}}^{\mathrm{ba}}$。图 6-6-14 中分别给出两个不同实验小组测得的纤维增长速率和浓度的依赖关系。从图中可以看出，它们具有不同的临界浓度，这主要是由于溶液的离子浓度不一样。而 Lipowsky 小组通过构建的粗粒化模拟模型也得到纤维增长速率和溶液中单体浓度的依赖关系，而且他们得到的模拟结果和实验结果非常一致。

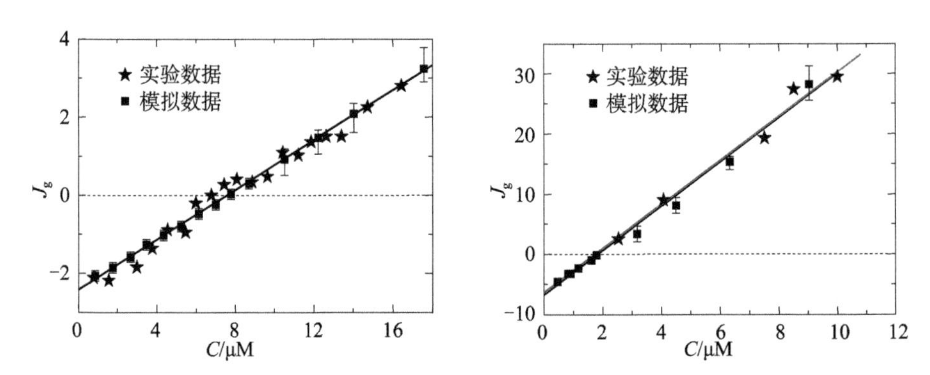

图 6-6-14　肌动蛋白纤维增长速率 $J_g(c)$ 随体系中溶液浓度 c 的演化曲线

(a) 和 (b) 中的实验数据都摘自 Pollard T D. J. Cell Biol., 1984, 99: 769; 1986, 103: 2747。黑色的点是模拟得到的数据结果

纤维的增长速率还可以通过单体黏结到纤维两端的频率来确定。纤维两端连续成功聚合单体的时间间隔或者等待时间的分布表现为随机的指数分布。如果从它们的指数分布图中提取出连续两次成功聚合的平均等待时间 τ_{on}，而且发现平均等待时间的倒数 $1/\tau_{on}$ 和溶液中单体的浓度 c 成正比。此外，纤维两端的聚合速率满足关系式 $\omega_{on}=k_{on}c=1/\tau_{on}$。同理，纤维的增长速率也可以通过平均等待时间来描述，即

$$J_g(c)=\frac{1}{\tau_{on}^{ba}}+\frac{1}{\tau_{on}^{po}}-\frac{1}{\tau_{off}^{ba}}-\frac{1}{\tau_{off}^{po}}=\omega_{on}^{ba}+\omega_{on}^{po}-\omega_{off}^{ba}-\omega_{off}^{po} \tag{6-6-12}$$

考虑肌动蛋白纤维的长度分布函数 $P_n(t)$ 随时间变化的规律，$P_n(t)$ 表示时刻 t 找到一个含有 n 个单体的纤维概率，它可以表示为

$$P_n(t)=\frac{N_n}{\sum_{n=1}^{\infty}N_n} \tag{6-6-13}$$

式中，N_n 为包含 n 个单体纤维的数目，即 n 聚体纤维的数目。图 6-6-15 描述了实验上如何通过直方图获得纤维长度的概率分布。

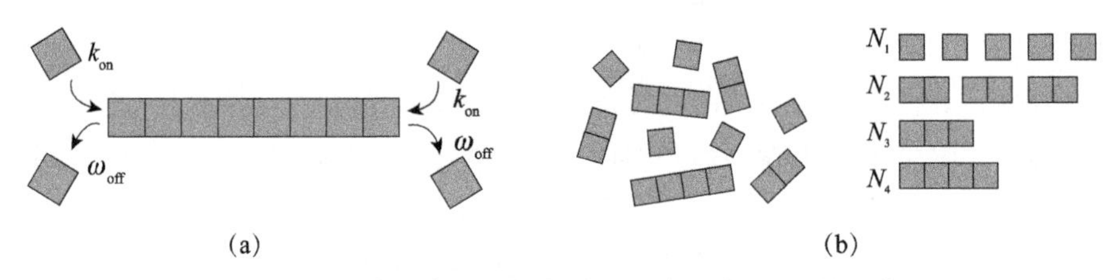

图 6-6-15　肌动蛋白纤维长度分布概率函数 $P_n(t)$ 的示意图

(a) 表示四个不同的过程；(b) 不同纤维长度的分布，以及其中的统计方法

描述 n 聚体长度变化的速率方程可以示意地写为

$$\frac{\mathrm{d}P_n}{\mathrm{d}t}=n\text{ 聚体的单体聚合速率}-\text{单体的解聚速率} \tag{6-6-14}$$

上式中做了简化，即假设 n 聚体长度的改变只能通过黏结或脱落单个单体实现。体系中排除了两个不同长度的纤维相互之间的黏结以及纤维从中间断开的两种情况。在这些假设下，反应可以近似写为

$$P_n+P_1 \underset{\omega_{\mathrm{off}}^{\mathrm{ba}}\text{或}\omega_{\mathrm{off}}^{\mathrm{po}}}{\overset{k_{\mathrm{on}}^{\mathrm{ba}}\text{或}k_{\mathrm{on}}^{\mathrm{po}}}{\rightleftarrows}} P_{n+1} \tag{6-6-15}$$

从上式我们需要注意的是这里不只是一个反应，而是彼此是耦合的反应。这也就是说 P_n 的演化依赖于 P_{n-1} 和 P_{n+1}。为求出 P_n 的演化，我们还要考虑“毗邻”的反应

$$P_{n-1}+P_1 \underset{\omega_{\mathrm{off}}^{\mathrm{ba}}\text{或}\omega_{\mathrm{off}}^{\mathrm{po}}}{\overset{k_{\mathrm{on}}^{\mathrm{ba}}\text{或}k_{\mathrm{on}}^{\mathrm{po}}}{\rightleftarrows}} P_n \tag{6-6-16}$$

即描述 n 聚体“脱落”一个单体成为 $n-1$ 聚体的反应。$P_n(t)$ 的演化由四种不同的过程决定的，而每个过程都可以从纤维的两端发生，但是纤维两端发生任何一过程的速率不同，数学上可写为

$$\begin{aligned}\frac{\mathrm{d}P_n}{\mathrm{d}t}=&\left(k_{\mathrm{on}}^{\mathrm{ba}}+k_{\mathrm{on}}^{\mathrm{po}}\right)P_{n-1}P_1+\left(\omega_{\mathrm{off}}^{\mathrm{ba}}+\omega_{\mathrm{off}}^{\mathrm{po}}\right)P_{n+1}\\&-\left(k_{\mathrm{on}}^{\mathrm{ba}}+k_{\mathrm{on}}^{\mathrm{po}}\right)P_nP_1-\left(\omega_{\mathrm{off}}^{\mathrm{ba}}+\omega_{\mathrm{off}}^{\mathrm{po}}\right)P_n\end{aligned} \tag{6-6-17}$$

从上式所描述的纤维的平均长度随时间的演化规律，可以将纤维的平均长度写成

$$\langle L\rangle=\sum_{n=1}^{n=\infty}naP_n \tag{6-6-18}$$

对此式求导，得到

$$\frac{\mathrm{d}\langle L\rangle}{\mathrm{d}t}=\sum_{n=1}^{n=\infty}na\frac{\mathrm{d}P_n}{\mathrm{d}t} \tag{6-6-19}$$

将式（6-6-17）代入上式，得到

$$\begin{aligned}\frac{\mathrm{d}\langle L\rangle}{\mathrm{d}t}=\sum_{n=1}^{n=\infty}na&\left(k_{\mathrm{on}}^{\mathrm{ba}}+k_{\mathrm{on}}^{\mathrm{po}}\right)P_{n-1}P_1+\left(\omega_{\mathrm{off}}^{\mathrm{ba}}+\omega_{\mathrm{off}}^{\mathrm{po}}\right)P_{n+1}\\&-\left(k_{\mathrm{on}}^{\mathrm{ba}}+k_{\mathrm{on}}^{\mathrm{po}}\right)P_nP_1-\left(\omega_{\mathrm{off}}^{\mathrm{ba}}+\omega_{\mathrm{off}}^{\mathrm{po}}\right)P_n\end{aligned} \tag{6-6-20}$$

上式右端重新整理得

$$\frac{\mathrm{d}\langle L\rangle}{\mathrm{d}t}=\sum_{n=1}^{n=\infty}na\left(k_{\mathrm{on}}^{\mathrm{ba}}+k_{\mathrm{on}}^{\mathrm{po}}\right)\left(P_{n-1}-P_{n}\right)P_{1}+\sum_{n=1}^{n=\infty}na\left(\omega_{\mathrm{off}}^{\mathrm{ba}}+\omega_{\mathrm{off}}^{\mathrm{po}}\right)\left(P_{n+1}-P_{n}\right) \quad (6\text{-}6\text{-}21)$$

接下来利用等式

$$\sum_{n=1}^{n=\infty}nP_{n-1}=\sum_{n=1}^{n=\infty}(n+1)P_{n}\pm \quad (6\text{-}6\text{-}22)$$

原式右端就可以化为

$$\begin{aligned}&aP_{1}\left(k_{\mathrm{on}}^{\mathrm{ba}}+k_{\mathrm{on}}^{\mathrm{po}}\right)\sum_{n=1}^{n=\infty}n\left(P_{n-1}-P_{n}\right)\\&=aP_{1}\left(k_{\mathrm{on}}^{\mathrm{ba}}+k_{\mathrm{on}}^{\mathrm{po}}\right)\sum_{n=1}^{n=\infty}\left[(n+1)-n\right]P_{n}\\&=aP_{1}\left(k_{\mathrm{on}}^{\mathrm{ba}}+k_{\mathrm{on}}^{\mathrm{po}}\right)\sum_{n=1}^{n=\infty}P_{n}=aP_{1}\left(k_{\mathrm{on}}^{\mathrm{ba}}+k_{\mathrm{on}}^{\mathrm{po}}\right)\end{aligned} \quad (6\text{-}6\text{-}23)$$

因此，可以通过这些代数式简化式（6-6-21）得到

$$J_{\mathrm{g}}(c)=\frac{\mathrm{d}\langle L\rangle}{\mathrm{d}t}=a\left[\left(k_{\mathrm{on}}^{\mathrm{ba}}+k_{\mathrm{on}}^{\mathrm{po}}\right)P_{1}-\left(\omega_{\mathrm{off}}^{\mathrm{ba}}+\omega_{\mathrm{off}}^{\mathrm{po}}\right)\right] \quad (6\text{-}6\text{-}24)$$

从上式可以看出，该结果和式（6-6-11）得到的结果相同。而从纤维的生长速率随单体变化的曲线可以看出，当浓度小于临界浓度 c_{cr} 时，纤维收缩；反之，当浓度大于临界浓度 c_{cr} 时，纤维将生长；当浓度等于 c_{cr} 时，平均长度保持不变，$\mathrm{d}\langle L\rangle/\mathrm{d}t=0$。

此外，我们还讨论一下纤维两端的聚合和解聚速率各不相同。前面已经提到过，纤维两端的非对称结构导致了反应速率的差别，正端（倒钩端）的聚合和解聚速率都要比负端（尖端）快。但是，这一模型中最有趣的是处于两端的聚合速率与解聚速率之比必须相等。在布朗动力学模拟中充分利用该原理校准两端的速率常数和临界浓度，而理解该原理比较简单的方式就是两端分子接触界面是一样的，因此与单体接触相关的自由能也是相同的。尽管单体分子本身是不对称的，但是分子之间形成的键接方式是相同的。因此，发生该反应的吉布斯自由能是相同的，两端的平衡常数是相等的，临界浓度也随之相同。数学上可以表述为

$$\frac{\omega_{\mathrm{off}}^{\mathrm{ba}}}{k_{\mathrm{on}}^{\mathrm{ba}}}=\frac{\omega_{\mathrm{off}}^{\mathrm{po}}}{k_{\mathrm{on}}^{\mathrm{po}}}=\frac{1}{V}\mathrm{e}^{\Delta G/k_{\mathrm{B}}T} \quad (6\text{-}6\text{-}25)$$

但是，当该蛋白单体耦合不同核苷酸时，它们形成的键接的自由能会发

生变化，从而也导致两端的临界浓度存在差别。它们有可能在一端生长而在另一端收缩，称为踏车现象，在后文会予以讨论。

（二）蛋白亚基结合核苷酸状态为单态模型的速率方程

本小节讨论蛋白亚基结合核苷酸的状态为两种状态的模型，即蛋白亚基耦合 ATP 分子和耦合 ADP 分子。因此，忽略磷酸释放过程，将水解过程直接简化为 ATP 直接水解成 ADP 分子。依然考虑体系有 M 个核的最简单情况，我们要计算“平均”每根纤维两端结合单体数目 n 随时间变化曲线。速率方程并不考虑长度的涨落，单体数目 n 由以下方程给出：

$$\frac{\mathrm{d}n}{\mathrm{d}t}=\left(k_{\mathrm{on}}^{\mathrm{ba}}+k_{\mathrm{on}}^{\mathrm{po}}\right)\left(c_{\mathrm{T}}-\frac{Mn(t)}{V}\right)-\left(\begin{matrix}\omega_{\mathrm{off,T}}^{\mathrm{ba}}P_{\mathrm{T}}^{\mathrm{ba}}+\omega_{\mathrm{off,D}}^{\mathrm{ba}}P_{\mathrm{D}}^{\mathrm{ba}}\\+\omega_{\mathrm{off,T}}^{\mathrm{po}}P_{\mathrm{T}}^{\mathrm{po}}+\omega_{\mathrm{off,D}}^{\mathrm{po}}P_{\mathrm{D}}^{\mathrm{po}}\end{matrix}\right)\quad(6\text{-}6\text{-}26)$$

式中，c_{T} 是溶液中初始单体耦合 ATP 分子的浓度，但是细胞内初始单体被近似认为全部耦合 ATP 分子；V 是总体积；M 是核的数目；$\omega_{\mathrm{off,T}}^{\mathrm{ba}}$和 $\omega_{\mathrm{off,T}}^{\mathrm{po}}$是纤维末端结合 ATP 分子原聚体的解聚速率；$\omega_{\mathrm{off,D}}^{\mathrm{ba}}$和 $\omega_{\mathrm{off,D}}^{\mathrm{po}}$是纤维末端结合 ADP 分子原聚体的解聚速率；$P_{\mathrm{T}}^{\mathrm{ba}}$，$P_{\mathrm{D}}^{\mathrm{ba}}$，$P_{\mathrm{T}}^{\mathrm{po}}$ 和 $P_{\mathrm{D}}^{\mathrm{po}}$ 是纤维末端结合 ATP 或者 ADP 分子原聚体的概率，其中纤维两端分别满足关系式 $P_{\mathrm{T}}^{\mathrm{ba}}+P_{\mathrm{D}}^{\mathrm{ba}}=1$ 和 $P_{\mathrm{T}}^{\mathrm{po}}+P_{\mathrm{D}}^{\mathrm{po}}=1$。

同上文，将式（6-6-26）改写为

$$\frac{\mathrm{d}n}{\mathrm{d}t}+\left(k_{\mathrm{on}}^{\mathrm{ba}}+k_{\mathrm{on}}^{\mathrm{po}}\right)\frac{Mn(t)}{V}=\left(k_{\mathrm{on}}^{\mathrm{ba}}+k_{\mathrm{on}}^{\mathrm{po}}\right)c_{\mathrm{T}}-\left(\begin{matrix}\omega_{\mathrm{off,T}}^{\mathrm{ba}}P_{\mathrm{T}}^{\mathrm{ba}}+\omega_{\mathrm{off,D}}^{\mathrm{ba}}P_{\mathrm{D}}^{\mathrm{ba}}\\+\omega_{\mathrm{off,T}}^{\mathrm{po}}P_{\mathrm{T}}^{\mathrm{po}}+\omega_{\mathrm{off,D}}^{\mathrm{po}}P_{\mathrm{D}}^{\mathrm{po}}\end{matrix}\right)\quad(6\text{-}6\text{-}27)$$

它有一个特解：

$$n_{\mathrm{particular}}(t)=\frac{V}{M\left(k_{\mathrm{on}}^{\mathrm{ba}}+k_{\mathrm{on}}^{\mathrm{po}}\right)}\left[\left(k_{\mathrm{on}}^{\mathrm{ba}}+k_{\mathrm{on}}^{\mathrm{po}}\right)c_{\mathrm{T}}-\left(\begin{matrix}\omega_{\mathrm{off,T}}^{\mathrm{ba}}P_{\mathrm{T}}^{\mathrm{ba}}+\omega_{\mathrm{off,D}}^{\mathrm{ba}}P_{\mathrm{D}}^{\mathrm{ba}}\\+\omega_{\mathrm{off,T}}^{\mathrm{po}}P_{\mathrm{T}}^{\mathrm{po}}+\omega_{\mathrm{off,D}}^{\mathrm{po}}P_{\mathrm{D}}^{\mathrm{po}}\end{matrix}\right)\right]\quad(6\text{-}6\text{-}28)$$

而它的通解为

$$n_{\mathrm{hom\,ogeneous}}(t)=A\mathrm{e}^{-\frac{M\left(k_{\mathrm{on}}^{\mathrm{ba}}+k_{\mathrm{on}}^{\mathrm{po}}\right)t}{V}}\quad(6\text{-}6\text{-}29)$$

式中，常数 A 为

$$A=-\frac{V}{M\left(k_{\mathrm{on}}^{\mathrm{ba}}+k_{\mathrm{on}}^{\mathrm{po}}\right)}\left[\left(k_{\mathrm{on}}^{\mathrm{ba}}+k_{\mathrm{on}}^{\mathrm{po}}\right)c_{\mathrm{T}}-\left(\begin{matrix}\omega_{\mathrm{off,T}}^{\mathrm{ba}}P_{\mathrm{T}}^{\mathrm{ba}}+\omega_{\mathrm{off,D}}^{\mathrm{ba}}P_{\mathrm{D}}^{\mathrm{ba}}\\+\omega_{\mathrm{off,T}}^{\mathrm{po}}P_{\mathrm{T}}^{\mathrm{po}}+\omega_{\mathrm{off,D}}^{\mathrm{po}}P_{\mathrm{D}}^{\mathrm{po}}\end{matrix}\right)\right]\quad(6\text{-}6\text{-}30)$$

因此，方程的解可表示为

$$n(t)=\frac{V}{M\left(k_{\rm on}^{\rm ba}+k_{\rm on}^{\rm po}\right)}\left[\begin{array}{l}\left(k_{\rm on}^{\rm ba}+k_{\rm on}^{\rm po}\right)c_{\rm T}\\-\left(\begin{array}{l}\omega_{\rm off,T}^{\rm ba}P_{\rm T}^{\rm ba}+\omega_{\rm off,D}^{\rm ba}P_{\rm D}^{\rm ba}\\+\omega_{\rm off,T}^{\rm po}P_{\rm T}^{\rm po}+\omega_{\rm off,D}^{\rm po}P_{\rm D}^{\rm po}\end{array}\right)\end{array}\right]\left[1-{\rm e}^{-\frac{M\left(k_{\rm on}^{\rm ba}+k_{\rm on}^{\rm po}\right)t}{V}}\right] \quad (6\text{-}6\text{-}31)$$

假定溶液中结合 ATP 分子单体浓度固定的情况下纤维增长的情况。纤维的增长速率，$J_{\rm g}(c_{\rm T})$ 为

$$J_{\rm g}\left(c_{\rm T}\right)=\frac{{\rm d}\langle L\rangle}{{\rm d}t}=\left(k_{\rm on}^{\rm ba}+k_{\rm on}^{\rm po}\right)c_{\rm T}-\left(\begin{array}{l}\omega_{\rm off,T}^{\rm ba}P_{\rm T}^{\rm ba}+\omega_{\rm off,D}^{\rm ba}P_{\rm D}^{\rm ba}\\+\omega_{\rm off,T}^{\rm po}P_{\rm T}^{\rm po}+\omega_{\rm off,D}^{\rm po}P_{\rm D}^{\rm po}\end{array}\right) \quad (6\text{-}6\text{-}32)$$

所有纤维末端耦合 ATP 或 ADP 分子原聚体都能够解聚。表 6-6-1 列出了实验上测得的解聚速率和聚合速率。从这些解聚速率参数可以看出，耦合 ADP 的末端原聚体从纤维两端脱落的速率要远大于耦合 ATP 的末端原聚体。当溶液的浓度较大时，正端（倒钩端）和负端（尖端）的原聚体几乎没有结合 ADP 分子，从而原聚体结合 ATP 分子在正端和负端的概率分别满足关系式 $P_{\rm T}^{\rm ba}=1-P_{\rm D}^{\rm ba}\approx 1$ 和 $P_{\rm T}^{\rm po}=1-P_{\rm D}^{\rm po}\approx 1$。图 6-6-16 给出了布朗动力学模拟中纤维两端原聚体结合核苷酸不同状态的分布概率和溶液中单体浓度的关系。从图中也可以看出，溶液中浓度较高时，纤维两端结合 ATP 分子的原聚体的概率接近 1。因此，在溶液中浓度较高时，式（6-6-32）可以改写为

$$J_{\rm g}\left(c_{\rm T}\gg c_{\rm cr,T}\right)\approx\left(k_{\rm on}^{\rm ba}+k_{\rm on}^{\rm po}\right)c_{\rm T}-\left(\omega_{\rm off,T}^{\rm ba}+\omega_{\rm off,T}^{\rm po}\right) \quad (6\text{-}6\text{-}33)$$

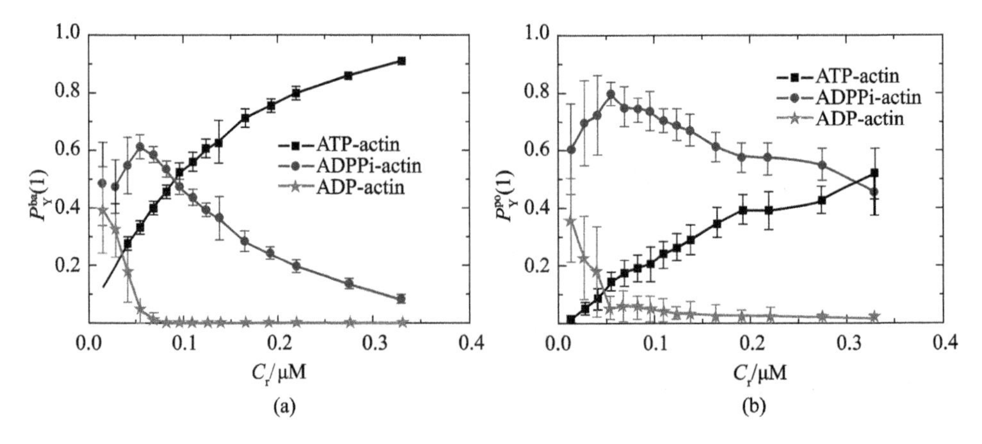

图 6-6-16　肌动蛋白纤维

末端 (a) 正端 (b) 负端黏结核苷酸三种状态的概率 $P_{\rm Y}^{\rm ba}$ 或者 $P_{\rm Y}^{\rm po}$ 对体系中自由肌动蛋白单体浓度 $c_{\rm T}$ 的依赖关系，其中 Y=T,Θ,D。所有的速率参数见表 6-6-1

当 $J_g=0$，即 $J_g^{ba}+J_g^{po}=0$ 时，溶液中单体的浓度定义为纤维的临界浓度，近似认为 $c_{cr,T}\approx(\omega_{off,T}^{ba}+\omega_{off,T}^{po})/(k_{on}^{po}+k_{on}^{ba})$。这是因为接近临界浓度时，纤维末端原聚体的状态不一定结合 ATP 分子。此时的临界浓度值应该介于纤维两端临界浓度值之间，即 $c_{cr,T}^{ba}<c_{cr,T}<c_{cr,T}^{po}$。

表 6-6-1 布朗动力学模拟中所用速率参数值

$k_{on,T}^{ba}$ (μMs^{-1})	11.6±0.8
$k_{on,T}^{po}$ (μMs^{-1})	1.3±0.2
$\omega_{off,T}^{ba}$ /s^{-1}	1.4
$\omega_{off,T}^{po}$ /s^{-1}	0.16
$\omega_{off,\Theta}^{ba}$ /s^{-1}	0.2
$\omega_{off,\Theta}^{po}$ /s^{-1}	0.02
$\omega_{off,D}^{ba}$ /s^{-1}	5.4
$\omega_{off,D}^{po}$ /s^{-1}	0.25
ω_c/s^{-1}	0.3
ω_r/s^{-1}	0.003

k_{on} 是 ATP- 肌动蛋白的聚合速率常数；ω_{off} 是 T、Θ 和 D 原聚体的解聚速率；ω_c 是 ATP 分裂速率；ω_r 是无机磷酸 Pi 释放速率

图 6-6-17 给出了模拟得到的纤维增长速率和溶液中单体浓度的依赖关系。当单体浓度大于 $c_{cr,T}$ 时，纤维的增长速率 $J_g>0$，而单体浓度小于 $c_{cr,T}$ 时，纤

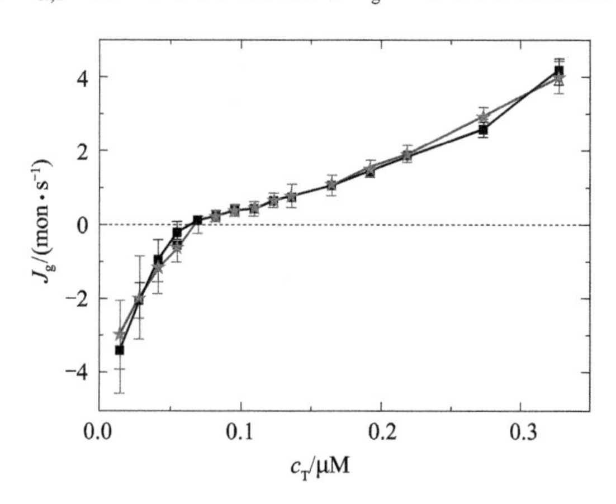

图 6-6-17 肌动蛋白纤维总增长速率和体系中肌动蛋白单体浓度 c_T 的依赖关系。其中，方块的数据是从肌动蛋白纤维平均长度对时间的依赖关系中得到的，见式（6-6-32），而星形的数据是从聚合和解聚时间的平均等待时间 τ 中得到的，见式（6-6-34）

维的增长速率 $J_g<0$。从图中还可以看出，从纤维的增长速率和浓度的依赖关系可以明显看出曲线分为两个不同区域，纤维增长速率和浓度之间的斜率明显不同。当体系中浓度远低于临界浓度时，纤维上的原聚体几乎全部结合ADP分子。纤维两端的聚合速率接近零，而解聚速率 $\omega_{\text{off}}\approx\omega_{\text{off,D}}^{\text{ba}}+\omega_{\text{off,D}}^{\text{po}}$。

纤维的增长速率也可以通过单体黏结到纤维两端的频率来确定。同上文描述

$$J_g(c_T)=\frac{1}{\tau_{\text{on}}^{\text{ba}}}+\frac{1}{\tau_{\text{on}}^{\text{po}}}-\frac{1}{\tau_{\text{off}}^{\text{ba}}}-\frac{1}{\tau_{\text{off}}^{\text{po}}}=\omega_{\text{on}}^{\text{ba}}+\omega_{\text{on}}^{\text{po}}-\omega_{\text{off}}^{\text{ba}}-\omega_{\text{off}}^{\text{po}} \tag{6-6-34}$$

根据上文对纤维长度分布函数的定义，同理可将 $P_n(t)$ 的演化由四种不同的过程决定，而每个过程都可以从纤维的两端发生，但是纤维两端发生任何一过程的速率不同，数学上可写为

$$\begin{aligned}\frac{\mathrm{d}P_n}{\mathrm{d}t}=&\left(k_{\text{on}}^{\text{ba}}+k_{\text{on}}^{\text{po}}\right)P_{n-1}P_1+\left(\omega_{\text{off,T}}^{\text{ba}}P_{\text{T}}^{\text{ba}}+\omega_{\text{off,D}}^{\text{ba}}P_{\text{D}}^{\text{ba}}+\omega_{\text{off,T}}^{\text{po}}P_{\text{T}}^{\text{po}}+\omega_{\text{off,D}}^{\text{po}}P_{\text{D}}^{\text{po}}\right)P_{n+1}\\&-\left(k_{\text{on}}^{\text{ba}}+k_{\text{on}}^{\text{po}}\right)P_nP_1-\left(\omega_{\text{off,T}}^{\text{ba}}P_{\text{T}}^{\text{ba}}+\omega_{\text{off,D}}^{\text{ba}}P_{\text{D}}^{\text{ba}}+\omega_{\text{off,T}}^{\text{po}}P_{\text{T}}^{\text{po}}+\omega_{\text{off,D}}^{\text{po}}P_{\text{D}}^{\text{po}}\right)P_n\end{aligned} \tag{6-6-35}$$

从上式所描述的纤维的平均长度随时间的演化规律，可以将纤维的平均长度写成

$$\langle L\rangle=\sum_{n=1}^{n=\infty}naP_n \tag{6-6-36}$$

对此式求导，得到

$$\frac{\mathrm{d}\langle L\rangle}{\mathrm{d}t}=\sum_{n=1}^{n=\infty}na\frac{\mathrm{d}P_n}{\mathrm{d}t} \tag{6-6-37}$$

将式（6-6-17）代入上式，得到

$$\begin{aligned}\frac{\mathrm{d}\langle L\rangle}{\mathrm{d}t}=&\sum_{n=1}^{n=\infty}na\left(k_{\text{on}}^{\text{ba}}+k_{\text{on}}^{\text{po}}\right)P_{n-1}P_1\\&+\left(\omega_{\text{off,T}}^{\text{ba}}P_{\text{T}}^{\text{ba}}+\omega_{\text{off,D}}^{\text{ba}}P_{\text{D}}^{\text{ba}}+\omega_{\text{off,T}}^{\text{po}}P_{\text{T}}^{\text{po}}+\omega_{\text{off,D}}^{\text{po}}P_{\text{D}}^{\text{po}}\right)P_{n+1}\\&-\left(k_{\text{on}}^{\text{ba}}+k_{\text{on}}^{\text{po}}\right)P_nP_1-\left(\omega_{\text{off,T}}^{\text{ba}}P_{\text{T}}^{\text{ba}}+\omega_{\text{off,D}}^{\text{ba}}P_{\text{D}}^{\text{ba}}+\omega_{\text{off,T}}^{\text{po}}P_{\text{T}}^{\text{po}}+\omega_{\text{off,D}}^{\text{po}}P_{\text{D}}^{\text{po}}\right)P_n\end{aligned} \tag{6-6-38}$$

对上式右端重新整理得

$$\begin{aligned}\frac{\mathrm{d}\langle L\rangle}{\mathrm{d}t}=&\sum_{n=1}^{n=\infty}na\left(k_{\text{on}}^{\text{ba}}+k_{\text{on}}^{\text{po}}\right)\left(P_{n-1}-P_n\right)P_1\\&+\sum_{n=1}^{n=\infty}na\left(\omega_{\text{off,T}}^{\text{ba}}P_{\text{T}}^{\text{ba}}+\omega_{\text{off,D}}^{\text{ba}}P_{\text{D}}^{\text{ba}}+\omega_{\text{off,T}}^{\text{po}}P_{\text{T}}^{\text{po}}+\omega_{\text{off,D}}^{\text{po}}P_{\text{D}}^{\text{po}}\right)\left(P_{n+1}-P_n\right)\end{aligned} \tag{6-6-39}$$

接下来利用等式

$$\sum_{n=1}^{n=\infty} nP_{n-1} = \sum_{n=1}^{n=\infty} (n+1) P_n \tag{6-6-40}$$

原式右端就可以化为

$$\begin{aligned} aP_1\left(k_{\text{on}}^{\text{ba}} + k_{\text{on}}^{\text{po}}\right)\sum_{n=1}^{n=\infty} n\left(P_{n-1} - P_n\right) &= aP_1\left(k_{\text{on}}^{\text{ba}} + k_{\text{on}}^{\text{po}}\right)\sum_{n=1}^{n=\infty}\left[(n+1)-n\right]P_n \\ &= aP_1\left(k_{\text{on}}^{\text{ba}} + k_{\text{on}}^{\text{po}}\right)\sum_{n=1}^{n=\infty} P_n = aP_1\left(k_{\text{on}}^{\text{ba}} + k_{\text{on}}^{\text{po}}\right) \end{aligned} \tag{6-6-41}$$

因此，可以通过这些代数式简化式（6-6-21）得到

$$J_{\text{g}}(c_{\text{T}}) = \frac{\mathrm{d}\langle L\rangle}{\mathrm{d}t} = a\left[\left(k_{\text{on}}^{\text{ba}} + k_{\text{on}}^{\text{po}}\right)P_1 - \left(\begin{aligned} &\omega_{\text{off,T}}^{\text{ba}}P_{\text{T}}^{\text{ba}} + \omega_{\text{off,D}}^{\text{ba}}P_{\text{D}}^{\text{ba}} \\ &+\omega_{\text{off,T}}^{\text{po}}P_{\text{T}}^{\text{po}} + \omega_{\text{off,D}}^{\text{po}}P_{\text{D}}^{\text{po}} \end{aligned}\right)\right] \tag{6-6-42}$$

从上式可以看出，该结果和式（6-6-32）得到的结果相同。

关于蛋白亚基结合核苷酸状态为三态的完整模型，依次类推，在此不再一一阐述了。

第二节　细胞骨架的前沿问题

一、细胞骨架伴随着核苷酸水解

细胞骨架的奇妙和复杂之处在于蛋白亚基与核苷酸之间有相互作用，尤其是细胞骨架上面的原聚体可以催化核苷酸水解。例如，肌动蛋白亚基既可以结合 ATP 分子，也可以结合 ADP 分子；而微管蛋白亚基则结合 GTP 或 GDP 分子。这些细胞骨架蛋白亚基通过结合 ATP 酶（或 GTP 酶），并伴随着这些酶的水解过程，蛋白亚基分子的构象会发生变化，从而改变相邻两个蛋白亚基分子之间的接触界面，进而导致不同的聚合和解聚速率。下面以肌动蛋白纤维为例，ATP 结合肌动蛋白亚基的水解过程可以分为两个过程，首先是 ATP 分裂过程，即肌动蛋白亚基结合的 ATP 酶状态转变成结合 ADP.Pi 酶（此时的磷酸分子通过非共价键和 ADP 酶结合）的过程，随后伴随着无机磷酸释放的过程，即肌动蛋白亚基变成结合 ADP 酶的状态。由此可见，肌动蛋白亚基可以结合 ATP 酶、ADP.Pi 酶和 ADP 酶，因此它存在三种不同的构象具有不同的聚合和解聚速率。同时，这三种不同的状态在纤维的两端均具有

不同的临界浓度。

关于肌动蛋白纤维上肌动蛋白亚基的水解机制，目前认为主要有三种：第一种是“随机水解”，该机制认为纤维上的任何一个结合 ATP 的肌动蛋白亚基以一种独立于相邻蛋白亚基结合的核苷酸状态且以随机的方式发生水解，肌动蛋白纤维上的水解速率和未发生水解单体的浓度有关；第二种是“矢量水解”，结合 ATP 肌动蛋白亚基的水解过程只能发生在纤维两端的 ATP 帽子和 ADP 结合肌动蛋白亚基的界面上，肌动蛋白纤维上只有一个 ATP 结合的蛋白亚基能发生水解；第三种是“协同水解”，该机制认为肌动蛋白纤维上结合 ATP 的肌动蛋白亚基的水解速率依赖于和它相邻蛋白亚基的核苷酸状态，因此，肌动蛋白纤维上结合 ATP 肌动蛋白亚基的水解速率各不相同。图 6-6-18 分别描述了这三种水解机制的示意图。

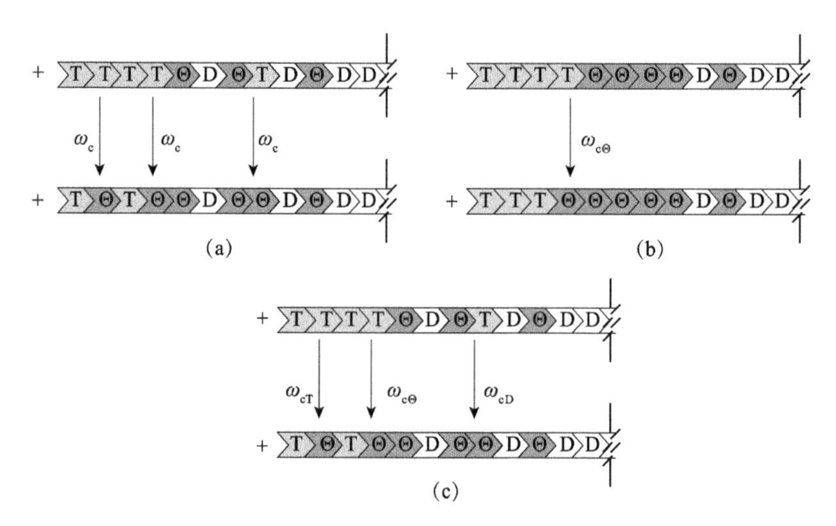

图 6-6-18 (a)ATP- 肌动蛋白原聚体（T，红色）以分裂速度 ω_c 随机分裂成 ADP/Pi- 肌动蛋白原聚体（Θ，蓝色），该过程不依赖于邻近原聚体的核苷酸状态；(b)ATP- 肌动蛋白原聚体（T，红色）以分裂速度 $\omega_{c\Theta}$ 矢量分裂成 ADP/Pi- 肌动蛋白原聚体，该分裂只局限于 TΘ 或者 TD 的边界处；(c) 协同分裂，该分裂速率依赖于邻近单体的核苷酸状态，具有三个分裂速率 ω_{cT}、$\omega_{c\Theta}$ 和 ω_{cD}。关于随机、矢量和协同无机磷酸 Pi 释放过程也有类似的定义（文后附彩图）

（一）随机水解

为了深入理解肌动蛋白亚基结合 ATP 分子发生随机水解时，肌动蛋白纤维上的增长动力学以及肌动蛋白纤维上核苷酸状态的分布，Lipowsky 教授小组首次采用布朗动力学模拟单根肌动蛋白纤维的增长动力学。当肌动蛋白纤维长度基本不变时，肌动蛋白纤维的增长速率接近于零，即 $J_g \cong 0$。以纤维正端位置开始记起，统计了纤维上面各个位置出现肌动蛋白结合不同核苷酸状态的概率，见图 6-6-19。从图中可以看出，纤维在正端有一个非常短的 ATP 帽子和结合 ADP.Pi 酶的帽子，纤维的负端呈现零星的结合 ADP.Pi 酶的状态，纤维上靠近负端的大部分单体都是处于结合 ADP 酶的状态。从不同位置出现蛋白亚基结合不同酶状态的概率，可以得到纤维上结合 ATP 酶和结合 ADP.Pi 酶的肌动蛋白单体个数之和分别为

$$\langle N_{\mathrm{T}}\rangle=\sum_{x=1}^{\infty}P_{\mathrm{T}}(x),\quad \langle N_{\Theta}\rangle=\sum_{x=1}^{\infty}P_{\Theta}(x) \tag{6-6-43}$$

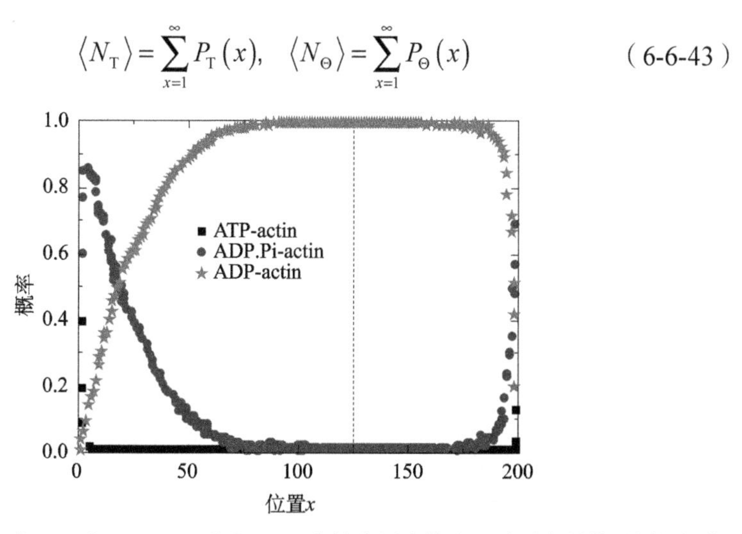

图 6-6-19　当稳态时 ATP 酶、ADP.Pi 酶和 ADP 酶结合原聚体在肌动蛋白纤维不同空间位置出现的概率：其中定义正端（倒钩端）的位置为 1，而负端（尖端）的位置为 200

从模拟的结果大约得到 $\langle N_{\mathrm{T}}\rangle \cong 0.9$，而 $\langle N_{\Theta}\rangle \cong 28.01$。对于随机 ATP 分裂过程和无机磷酸随机释放的过程，纤维上任何位置 ATP 分裂速率和磷酸速率可以分别表示为 $j_{\mathrm{c}}(x)=P_{\mathrm{T}}(x)\omega_{\mathrm{c}}$ 和 $j_{\mathrm{r}}(x)=P_{\Theta}(x)\omega_{\mathrm{r}}$，从而它们的总速率为 $J_{\mathrm{c}}=\sum_{x=1}^{\infty}j_{\mathrm{c}}(x)$ 和 $J_{\mathrm{r}}=\sum_{x=1}^{\infty}j_{\mathrm{r}}(x)$。因此，可以得到 $J_{\mathrm{c}}=\langle N_{\mathrm{T}}\rangle\omega_{\mathrm{c}}$ 和 $J_{\mathrm{r}}=\langle N_{\Theta}\rangle\omega_{\mathrm{r}}$。图 6-6-20 中给出了布朗动力学模拟中纤维上 ATP 帽子长度 $\langle N_{\mathrm{T}}\rangle$ 和体系中肌动蛋白单体浓度的依赖关系。当体系中肌动蛋白单体浓度进一步增大时，纤维上 ATP 帽子的长度 $\langle N_{\mathrm{T}}\rangle$ 趋于基本不变。而且从图中可以看出，模拟得到的结果和理论分析的结果非常一致。在纤维的增长速率接近于零，即处于稳态时，ATP 分裂的速率 J_{c} 和无机磷酸释放速率 J_{r} 满足下列关系式：

$$J_{\mathrm{c}}^{\mathrm{ba}}=J_{\mathrm{r}}^{\mathrm{ba}}+P_{\Theta}^{\mathrm{ba}}(1)\omega_{\mathrm{off},\Theta}^{\mathrm{ba}} \tag{6-6-44}$$

和

$$J_{c}^{po}=J_{r}^{po}+P_{\Theta}^{ba}(1)\omega_{off,\Theta}^{po} \tag{6-6-45}$$

因此，纤维上结合 ATP 和 ADP.Pi 酶的肌动蛋白亚基的总数目也满足下列关系：

$$\langle N_{T}^{ba}\rangle\omega_{c}=\langle N_{\Theta}^{ba}\rangle\omega_{r}+P_{\Theta}^{ba}(1)\omega_{off,\Theta}^{ba} \tag{6-6-46}$$

和

$$\langle N_{T}^{po}\rangle\omega_{c}=\langle N_{\Theta}^{po}\rangle\omega_{r}+P_{\Theta}^{po}(1)\omega_{off,\Theta}^{po} \tag{6-6-47}$$

从图 6-6-16 中给出纤维两端核苷酸不同状态的概率分布函数也可以验证上述两式。

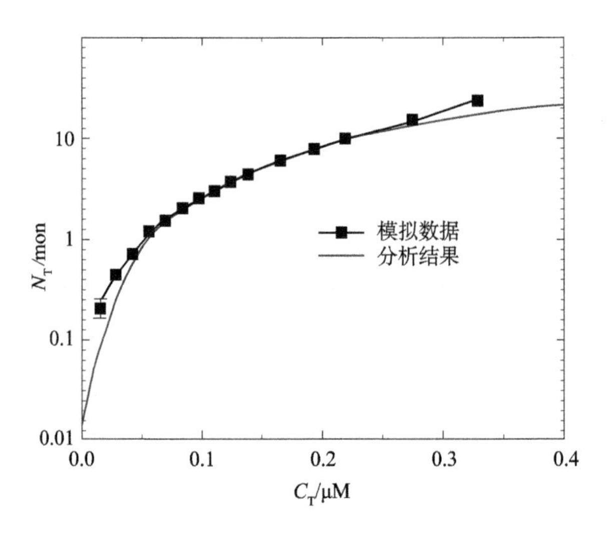

图 6-6-20　纤维上 ATP 酶结合原聚体数目和肌动蛋白单体浓度 c_T 的依赖关系。方块为布朗动力学模拟的结果，实线是理论分析结果

（二）协同水解

不同的水解机制直接影响纤维上核苷酸状态的分布，而肌动蛋白纤维上结合不同的核苷酸状态会导致纤维的结构和力学性能存在差异。最近，生物物理学家采用荧光显微镜观察单根肌动蛋白纤维的解聚动力学，为肌动蛋白纤维上面结合不同核苷酸状态的分布情况提供了最直接的证据，并且实验结果充分描述了纤维的结构塑性。通过分析纤维上面结合的不同核苷酸的状态分布，可以进一步探究出纤维上 ATP 的水解机制。图 6-6-21 中给出了两根肌

动蛋白纤维的解聚过程，其中包括实验结果（空心圆点）、协同分裂和协同磷酸释放的水解机制（SS）、随机分裂和随机磷酸释放（RR）与矢量分裂和随机磷酸释放（VR）三种机制得到的理论解聚过程。从图中可以看出，协同分裂和协同磷酸释放的水解机制（SS）和实验结果非常吻合。

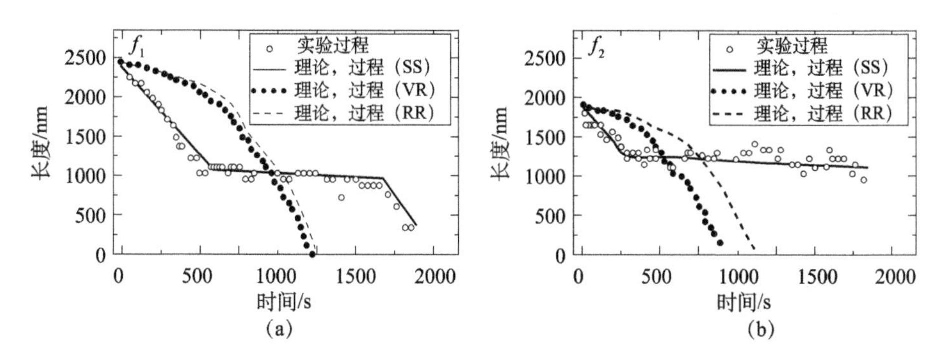

图 6-6-21　两根肌动蛋白纤维的解聚动力学过程：纤维长度 (a) f_1 和 (b) f_2 随时间的变化过程。实验的结果和水解过程（SS）的结果非常一致，其中水解过程（SS）表示强烈的协同 ATP 分裂过程和协同无机磷酸 Pi 释放过程；水解过程（RR）为随机 ATP 分裂过程和随机无机磷酸 Pi 释放过程；水解过程（VR）为矢量 ATP 分裂过程和随机无机磷酸 Pi 释放过程

根据邻近蛋白质亚基的核苷酸状态，协同分裂和协同磷酸释放速率可分别用 ω_{cT}，$\omega_{c\Theta}$，ω_{cD}，ω_{rT}，$\omega_{r\Theta}$，ω_{rD} 等参数来表示。假定 $\omega_{c\Theta}=\omega_{cD}=\omega_c$ 和 $\omega_{cT}=\rho_c\omega_c$、$\omega_{r\Theta}=\omega_{rT}=\rho_r\omega_r$ 和 $\omega_{rD}=\omega_r$，其中无量纲参数 ρ_c 和 ρ_r 满足关系式 $0\leqslant\rho_c\leqslant1$ 和 $0\leqslant\rho_r\leqslant1$。当 $\rho_c=1$ 时表示随机分裂，$\rho_r=1$ 时表示随机磷酸释放，而 $\rho_c=0$ 时表示矢量分裂，$\rho_r=0$ 时表示矢量磷酸释放。通过实验测得的速率参数以及大量模拟计算的结果，并参考图 6-6-21 中给出的实验结果的进行对比，得到了模型中所有的速率参数，具体见表 6-6-2。

表 6-6-2　水解过程 RR, VR 和 SS 的转变速率参数

	k_{on}/ μMs^{-1}	$\omega_{off,T}/s^{-1}$	$\omega_{off,\Theta}/s^{-1}$	$\omega_{off,D}/s^{-1}$	ω_c/s^{-1}	ρ_c	ω_r/s^{-1}	ρ_r
RR	11.6	1.4	0.2	5.4	0.3	1	0.003	1
VR	1.7	5.1	0.2	5.0	13.6	0	0.003	1
SS	11.6	2.2	0.1	2.7	1.0	$3/10^6$	0.57	$2/10^6$

（三）稳态时的速率平衡关系

由于负端几乎所有蛋白亚基都结合 ADP 酶，而且负端的聚合解聚速率都比较小，下面的理论描述都基于正端。在稳态时，肌动蛋白纤维的增长速率

可以用末端单体各核苷酸状态的概率函数表示，即

$$J_g=k_{on}c_T-P_T(1)\omega_{off,T}-P_\Theta(1)\omega_{off,\Theta}-P_D(1)\omega_{off,D} \tag{6-6-48}$$

同时，ATP 分裂的总速率为 $J_c=k_{on}c_T-P_T(1)\omega_{off,T}$，而无机磷酸释放的总速率为 $J_r=k_{on}c_T-P_T(1)\omega_{off,T}-P_\Theta(1)\omega_{off,\Theta}$。根据这些关系式，可以得到下列关系：

$$J_r=J_c-P_\Theta(1)\omega_{off,\Theta} \tag{6-6-49}$$

当体系中肌动蛋白单体的浓度非常高，或者非常低时，$P_\Theta(1)\simeq 0$，因此 $J_c\simeq J_r$。事实上，ATP 分裂速率和磷酸释放速率与浓度的依赖关系非常相似，具体见图 6-6-22。

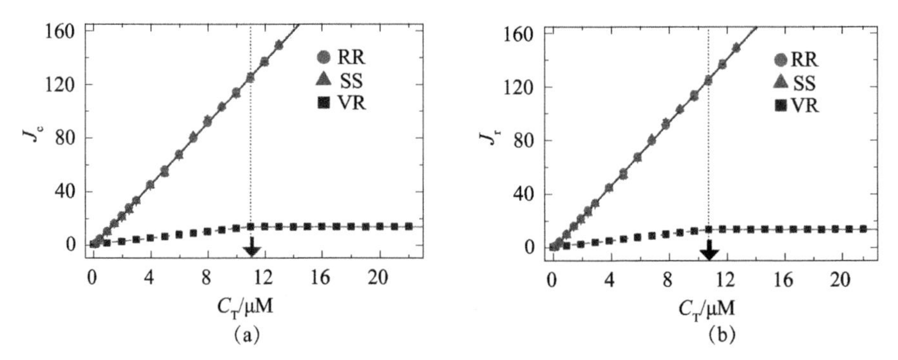

图 6-6-22　ATP 总分裂速率 J_c 和无机磷酸 Pi、总释放速率 J_r（单位：原聚体 /s）和肌动蛋白单体浓度 c_T 的依赖关系。圆、正三角和方块分别表示水解过程 RR、SS 和 VR。黑色的箭头表示矢量分裂的临界浓度，$c_{c,T}$=11.0μM（文后附彩图）

当体系中肌动蛋白单体浓度非常高时，肌动蛋白纤维末端单体总是结合 ATP 酶，即 $P_T(1)\cong 1P_T(1)\simeq 1$，$P_\Theta(1)\cong 0P_\Theta(1)\simeq 0$ 和 $P_D(1)\cong 0P_D(1)\simeq 0$，此时肌动蛋白纤维的增长总速率为 $J_g=k_{on}c_T-\omega_{off,T}$。

当体系中肌动蛋白单体浓度非常低时，纤维的增长速率近似为

$$J_g\approx-\omega_{off,D}+\frac{\omega_c(\omega_r+\omega_{off,D})/(\omega_r+\omega_{off,\Theta})}{\omega_c+\omega_{off,T}}k_{on}c_T \tag{6-6-50}$$

（四）矢量分裂和矢量磷酸释放

在矢量分裂中，肌动蛋白纤维上最多只有 1 个 TΘ 和 TD 边界。因此，体系中总的 ATP 分裂速率 J_c 不能超过 ω_c。在稳态时，纤维上 ATP 结合蛋白单体的增长速率为

$$J_{+T}=k_{on}c_T-P_T(1)\omega_{off,T} \tag{6-6-51}$$

当体系中肌动蛋白单体浓度非常低时，刚聚合到纤维的 ATP 酶结合的

蛋白单体还没有等到下一个单体聚合，就已经发生分裂。此时，纤维上没有ATP酶结合的蛋白单体，从而也没有TΘ和TD边界。此时，体系中总的分裂速率$J_c<\omega_c$，而且$J_{+T}=J_c$。

当体系中肌动蛋白单体浓度非常高时，体系中ATP结合蛋白亚基增长的速率J_{+T}远大于ATP分裂速率ω_c。纤维上面只存在1个TΘ或TD边界，因此，$J_c=\omega_c$。

此时，纤维正端的核苷酸总是结合ATP酶，即$P_T(1)=1$，因此结合ADP.Pi和ADP酶的概率为零：$P_\Theta(1)=P_D(1)=0$。体系的临界浓度$c_{c,T}$介于高浓度和低浓度之间，此时体系ATP结合蛋白亚基的增长速率为

$$J_{+T}=k_{on}c_{c,T}-P_T(1)\omega_{off,T}=\omega_c \tag{6-6-52}$$

其中，$P_T(1)=1$。因此可以推出，$c_{c,T}=(\omega_c+\omega_{off,T})/k_{on}$。

矢量磷酸释放即只有当纤维上存在ΘD边界时，边界上ADP.Pi酶结合的蛋白亚基才发生磷酸释放，因此，纤维上总磷酸释放速率J_r不可能超过单个蛋白单体的磷酸释放速率$\omega_r=\omega_{rD}$。在稳态时，纤维上ADP.Pi酶结合的蛋白亚基的增长速率为

$$J_{+\Theta}=J_c-P_\Theta(1)\omega_{off,\Theta}$$

或者

$$J_{+\Theta}=k_{on}c_T-P_T(1)\omega_{off,T}-P_\Theta(1)\omega_{off,\Theta} \tag{6-6-53}$$

当体系中肌动蛋白单体浓度非常低时，纤维上可能不存在ADP.Pi酶结合的蛋白亚基，因此不存在ΘD边界。此时，$J_r<\omega_r$且$J_{+\Theta}=J_r$。当体系中肌动蛋白浓度非常高时，$J_{+\Theta}$远大于ω_r直至体系达到稳定时，纤维有一个持续增长的ATP帽子或ADP.Pi帽子，此时$J_r=\omega_r$。在两种状态中间也有一个分界浓度$c_{r,T}$，它满足下列关系式：

$$J_{+\Theta}=k_{on}c_{r,T}-P_T(1)\omega_{off,T}-P_\Theta(1)\omega_{off,\Theta}=\omega_r \tag{6-6-54}$$

和

$$P_T(1)+P_\Theta(1)=1$$

（五）协同ATP分裂

当协同分裂参数$\rho_c>0$时，纤维可以达到一个稳定的状态，具有稳定的原聚体核苷酸状态分布和长度分布等。肌动蛋白纤维上ATP酶结合单体的总数目包括ATP酶结合蛋白单体邻近蛋白单体结合核苷酸可能三种状态，即$\langle N_T\rangle=\langle N_{TT}\rangle+\langle N_{T\Theta}\rangle+\langle N_{TD}\rangle$。因此，可将纤维的总ATP分裂速率改写为

$$J_c=\langle N_{TT}\rangle\rho_c\omega_c+\left(\langle N_{T\Theta}\rangle+\langle N_{TD}\rangle\right)\omega_c \tag{6-6-55}$$

在稳态时，依据上面提到的速率关系，$J_{+T}=J_c$。因此，稳态时体系满足

$$\langle N_{TT}\rangle\rho_c\omega_c+\left(\langle N_{T\Theta}\rangle+\langle N_{TD}\rangle\right)\omega_c=J_{+T} \tag{6-6-56}$$

其中，$\langle N_{TT}\rangle\approx\sqrt{c_T}J_{+T}/\omega_c\sqrt{\rho_c}$。当协同分裂参数$\rho_c$非常小时，体系中ATP酶结合蛋白亚基邻近的蛋白大多是也是结合ATP酶，此时$\langle N_T\rangle\approx\sqrt{\pi}k_{on}c_T/\sqrt{2\omega_{cD}\rho_c\omega_c}$。

（六）ATP酶结合原聚体的长度分布函数

忽略蛋白质亚基结合中间态ADP.Pi酶，肌动蛋白纤维可近似看成一端是结合ATP酶的状态，而另一端则是结合ADP酶的状态。但是，在纤维结合ATP酶端也存在少量的结合ADP酶的状态，把纤维结合ATP酶分割成不同的小部分。因此，可以进一步研究纤维正端在协同水解情况下蛋白结合ATP酶的长度分布。蛋白结合ATP酶长度k（k=0,1,2⋯）的概率分布函数p_k随时间演化方程为

$$\begin{aligned}\partial_t p_k = J_T\left(p_{k-1}-p_k\right)-\omega_c\left[1+\rho_c(k-1)\right]p_k\left(1-\delta_{k,0}\right)\\ +\omega_c p_{k+1}+\omega_c\rho_c\sum_{s\geqslant k+2}p_s\end{aligned} \tag{6-6-57}$$

边界条件p_{-1}=0。方程（6-6-57）右端的第一项表示通过聚合过程造成ATP原聚体长度k的增加和减少；第二项表示正端ATP长度k>0时，ATP分裂过程造成的ATP原聚体长度k的减少；最后两项表示ATP分裂过程造成的ATP原聚体长度k的增加，其中可以是ATP原聚体长度大于或等于k+2通过协调分裂形成新的ATP原聚体长度为k，也可以是TD界面处ATP分裂形成新的长度为k的ATP原聚体。在稳态，即$\partial_t p_k$=0时，定义函数$p_k=\sum_{l\geqslant k}p_l$，其中$p_0$=1和$p_1=p_{1,T}$。稳态时，$k\geqslant1$，这些函数满足

$$P_{k+1}-P_k=\frac{J_T}{\omega_c(1-\rho_c)}(P_k-P_{k-1})+\frac{\rho_c}{1-\rho_c}kP_k \tag{6-6-58}$$

矢量水解时，ρ_c=0且$J_T<\omega_c$时，$p_k\sim\left(J_T/\omega_c\right)^k$；而当$J_T>\omega_c$时，正端蛋白结合ATP酶的长度持续增加，没有静态解。

一般情况下的通解为

$$P_k=\exp(-ak/2)\frac{Ai(b^{-2/3}a^2/4+b^{1/3}k)}{Ai(b^{-2/3}a^2/4)} \tag{6-6-59}$$

其中，$a\equiv2(1-\rho_c-J_T/\omega_c)/(1-\rho_c+J_T/\omega_c)$和$b\equiv2\rho_c/\left(1-\rho_c+J_T/\omega_c\right)$，

$Ai(x)$ 是艾林函数。因此，蛋白结合 ATP 酶的长度分布函数可以通过 $p_k=P_k-P_{k+1}$ 得到。当 $J_T/\omega_c \gg 1$ 时，纤维快速增长阶段，$a\approx-2$，而 $b\approx 2\rho_c\omega_c/J_T \ll 1$。当 $x\gg 1$ 时，艾林函数也近似为 $Ai(x)\sim e^{-2x^{3/2}/3}$。因此，$P_k \approx e^{-bk^2/2|a|}$，而 $p_k \approx -\partial_k P_k(k)$，从而

$$p_k \approx \left(kb/|a|\right)e^{-bk^2/2|a|} \approx k\left(\omega_c\rho_c/J_T\right)e^{-k^2\omega_c\rho_c/2J_T} \tag{6-6-60}$$

蛋白质结合 ATP 酶的总长度可定义为 $\langle N_T\rangle = \sum_{k\geq 1} kp_k$，从式（6-6-60）可得到蛋白质结合 ATP 酶的总长度：

$$\langle N_T\rangle \approx \sqrt{\pi/2}\left(|a|/b\right)^{1/2} \approx \sqrt{\pi/2}\left(J_T/\omega_c\rho_c\right)^{1/2} \tag{6-6-61}$$

图 6-6-23 给出了 p_k 和 k 的依赖关系。从图中可以看出，理论结构和蒙特卡罗模拟得到的结果非常吻合。

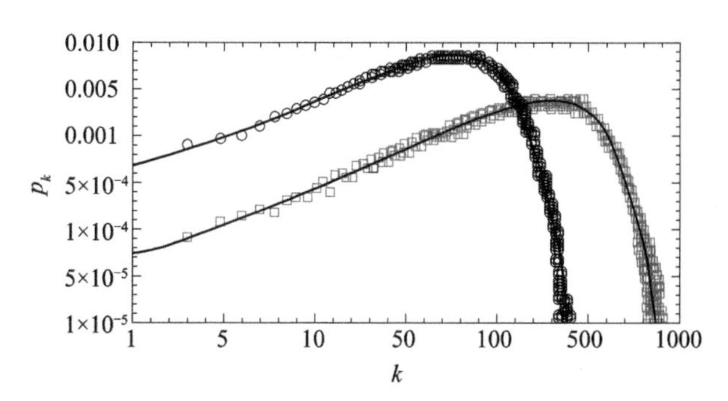

图 6-6-23 当肌动蛋白单体浓度为 c_T=1μM 时，ATP 帽端纤维长度分布函数和 ATP 数目的双对数函数图，其中 $\rho_c=10^{-2}$（空心圆），$\rho_c=10^{-3}$（空心方块），其他参数见表 6-6-2。实线分别是式（6-6-59）得到的分析结果，点表示随机模拟的结果

（七）蛋白质结合 ATP 酶的生成图谱

为了探究纤维的结构塑性，不仅要考究蛋白质结合 ATP 酶在纤维上的长度分布函数，还要研究纤维上 ATP 酶长度为 k 岛屿的平均数目 I_k，它满足下列时间演化方程：

$$\begin{aligned}\partial_t I_k = {} & J_T\left(p_{k-1}-p_k\right)-\omega_c\left[1+(k-1)\rho_c\right]I_k \\ & +\omega_c\left(1+\rho_c\right)I_{k+1}+2\omega_c\rho_c\sum_{s\geqslant k+2} I_s\end{aligned} \tag{6-6-62}$$

其中 $k\geqslant 1$，方程右边的第一项表示聚合和解聚造成 ATP 岛屿数目的变化；第

二项是 ATP 岛屿上 $k-1$ 位置的水解以及 TD 界面上面的水解；最后两项是来源于 ATP 岛屿内部的水解。第三项是长度为 $k+1$ 的 ATP 岛屿在 TD 界面上面的水解而形成长度 k 的 ATP 岛屿。第四项是长度为 $s\geqslant k+2$ 的 ATP 岛屿通过水解分裂成一个长度为 k 的 ATP 岛屿和另外一个长度的 ATP 岛屿。

为了方便和实验对照，定义了其他物理量：ATP 岛屿的平均数目 $I\equiv\sum_{k\geqslant1}I_k$ 和 ATP 原聚体的平均数目 $\langle N_{\mathrm{T}}\rangle\equiv\sum_{k\geqslant1}kI_k$。根据定义和式（6-6-62），可以分别得到这两个物理量的速率方程：

$$\partial_t I=J_{\mathrm{T}}p_0+\omega_{\mathrm{c}}\rho_{\mathrm{c}}\left(\langle N_{\mathrm{T}}\rangle-2I\right)-\omega_{\mathrm{c}}\left(1-\rho_{\mathrm{c}}\right)I_1 \tag{6-6-63}$$

和

$$\partial_t\langle N_{\mathrm{T}}\rangle=J_{\mathrm{T}}-\omega_{\mathrm{c}}\rho_{\mathrm{c}}\left(\langle N_{\mathrm{T}}\rangle-I\right)-\omega_{\mathrm{c}}I \tag{6-6-64}$$

而总的 ATP 分裂速率为 $J_{\mathrm{c}}=\omega_{\mathrm{c}}\rho_{\mathrm{c}}\left(\langle N_{\mathrm{T}}\rangle-I\right)+\omega_{\mathrm{c}}I$，且式（6-6-64）可改写为 $\partial_t\langle N_{\mathrm{T}}\rangle=J_{\mathrm{T}}-J_{\mathrm{c}}$。在稳态时，ATP 单体的聚合速率和分裂速率保持平衡，即 $J_{\mathrm{T}}=J_{\mathrm{c}}$。

当体系达到稳态时，通过推导得到

$$I_1\approx\sqrt{\frac{\pi}{2}}\frac{J_{\mathrm{T}}\sqrt{\rho_{\mathrm{c}}}}{\omega_{\mathrm{c}}},\ I\approx\frac{J_{\mathrm{T}}}{\omega_{\mathrm{c}}},\ \langle N_{\mathrm{T}}\rangle\approx\sqrt{\frac{\pi}{2}}\frac{J_{\mathrm{T}}}{\omega_{\mathrm{c}}\sqrt{\rho_{\mathrm{c}}}} \tag{6-6-65}$$

图 6-6-24 给出了蒙特卡罗随机模拟和理论分析得到的结果。从图可发现，两者非常吻合。

（八）ADP.Pi 酶结合原聚体的平均数目

在随机 ATP 分裂和随机磷酸释放中，体系处于高肌动蛋白浓度时，肌动蛋白纤维上 ADP.Pi 酶结合原聚体的平均数目为 $\langle N_{\Theta}\rangle\approx\langle N_{\mathrm{T}}\rangle\omega_{\mathrm{c}}/\omega_{\mathrm{r}}$。在矢量 ATP 分裂和随机磷酸释放的模型中，稳态时纤维上 ADP.Pi 酶结合原聚体数目 $\langle N_{\Theta}\rangle=\omega_{\mathrm{c}}/\omega_{\mathrm{r}}$，当 $c_{\mathrm{T}}>c_{\mathrm{c,T}}$。在协同 ATP 分裂和协同磷酸释放的模型中，当 $c_{\mathrm{T}}>c_{\mathrm{r,T}}$ 时，

$$\langle N_{\Theta}\rangle\approx\langle N_{\mathrm{T}}\rangle\frac{\omega_{\mathrm{c}}\sqrt{\rho_{\mathrm{c}}}}{\omega_{\mathrm{r}}\sqrt{\rho_{\mathrm{r}}}}=\langle N_{\mathrm{T}}\rangle\frac{\sqrt{\omega_{\mathrm{cD}}\omega_{\mathrm{cT}}}}{\sqrt{\omega_{\mathrm{rD}}\omega_{\mathrm{rT}}}} \tag{6-6-66}$$

图 6-6-25 中给出了肌动蛋白纤维上 ATP 酶结合原聚体和 ADP.Pi 原聚体结合的平均数目随肌动蛋白单体浓度的依赖关系。

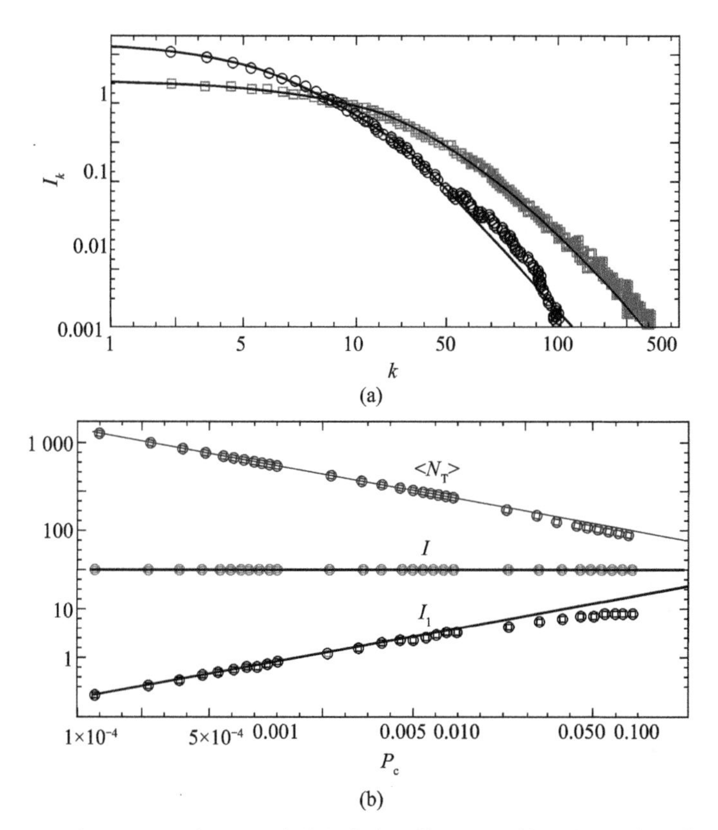

图 6-6-24 (a) 当 c_T=1μM 时，ATP 岛屿的分布函数和 ATP 数目的双对数函数图，其中 $\rho_c=10^{-2}$（空心圆），$\rho_c=10^{-3}$(空心方块)，其他参数见表 6-6-2。实线分别是式（6-6-65）得到的分析结果，点表示随机模拟的结果。(b) 当 c_T=1μM，稳态时，$\langle N_T\rangle$、I 和 I_1 与分裂参数 ρ_c 的依赖关系，其中实线分别是式（6-6-65）得到的分析结果，点表示随机模拟的结果

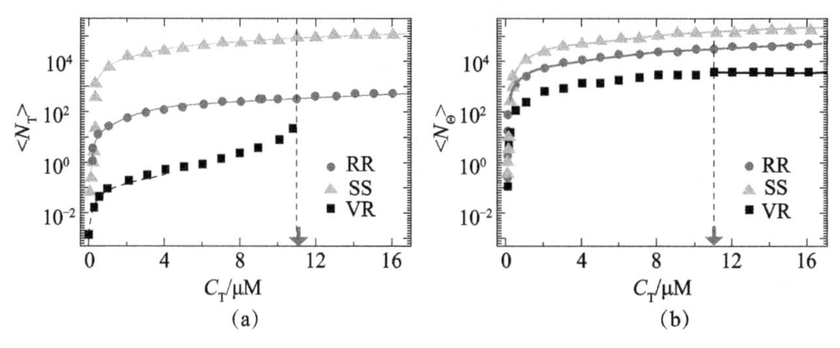

图 6-6-25 肌动蛋白纤维在三种不同水解机制过程中，ATP 酶结合原聚体$\langle N_T\rangle$和 ADP.Pi 酶结合原聚体$\langle N_\Theta\rangle$与肌动蛋白单体浓度 c_T 的依赖关系，具体参数见表 6-6-2。图中箭头表示的是 $c_{T,c}$=11μM

二、动态不稳定性

动态不稳定性的基本现象是：即使化学环境基本保持不变，微管随反应体系中游离的二聚体的浓度变化而呈现出生长和收缩转变的循环。在体外实验中，把许多微管放在同一试管内观察，发现在任何时刻，总是有些微管在生长，而有些在收缩。测量某个微管长度随时间的变化函数，发现该曲线呈锯齿状，见图 6-6-26。因此可见，微管生长动力学不时被塌缩事件所打断，从而突然收缩，其后又重新恢复生长。动态不稳定性也存在于细菌的肌动蛋白类似物 ParM 中。

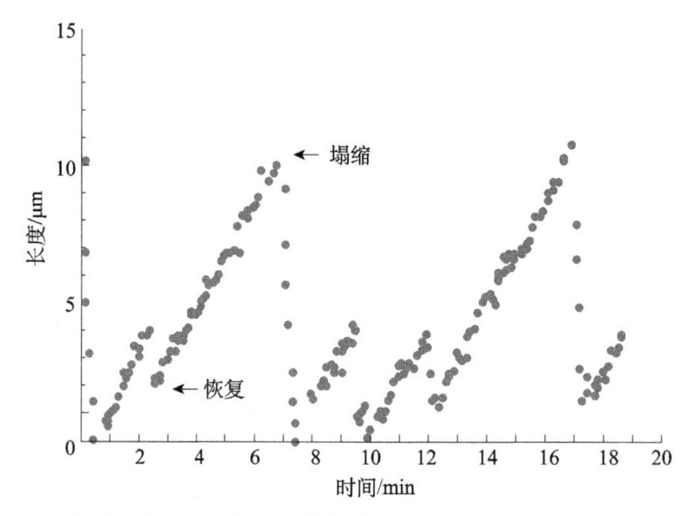

图 6-6-26　单根微管长度 L 随时间的变化规律。观察一根 10μm 长的微管长度随时间的变化，每分钟观察 10 次，并记录微管长度变化。实验发现，微管会在很短的时间内塌缩，长度迅速降为零，也会继续恢复生长过程，微管持续增长（摘自 D. K. Fygenson et al. Phys. Rev. e50: 579, 1994）

为了更深入地理解微管的动态不稳定性，首先需要建立合适的理论模型，了解塌缩动力学的第一步就是考虑一个不发生水解的帽体的简单玩具模型。前面已经提到过，生长微管的一端存在 GTP 酶结合的帽体，而且帽体的长度取决于聚合速率与微管内水解速率之间的竞争。通常情况，塌缩发生在 GTP 帽体长度收缩至零时。关于水解的问题，前面已经提到过，通常认为微管或者肌动蛋白纤维上 GTP/ATP 酶结合的蛋白亚基存在三种水解机制。在讨论塌缩动力学时，先假设体系中只存在矢量水解，即只有位于帽体与微管剩余部分交界处的蛋白质亚基结合的 GTP 分子才发生核苷酸水解，而且这个单体的核苷酸水解速率可以用时间寿命的物理量 τ_T 表示。这个假定意味着交界面以一定的速率沿微管移动。另外，负端的速率常数通常假设由 ADP、GDP 结合的单体决定。塌缩是在水解面移动的速率追上了正端的生长点才发生。可以导致该效应的一个物理机制是溶液中有效单体逐渐耗尽而降低生长速率。通常，正端以速度 v_{tip} 向前移动，水解以速率 $1/\tau_T$ 发生。发生塌缩的临界条件是生长端前进的速率刚好等于水解界面（交界处）前进的速率，即 $dx_{tip}/dt=a/\tau_T$。这个模型中隐含了生长端的帽体只有一个单体长度的假设。该临界条件实际上设定了边界，即只有当水解面移动速率比生长点前进快时，才能保证前者追上后者，从而发生塌缩现象。

塌缩的物理图像则是一旦水解面追上生长点时，塌缩就发生了，因为在这个时间点，GTP（或者 ATP）结合蛋白的增长速率被同伴 GDP(或 ADP) 结合蛋白的增长速率所取代。GDP 单体主导解聚速率，而 GTP 蛋白主导聚合速率。该模型预言了一个非常特殊的关系，即塌缩速率与有效单体浓度的依赖关系。

顶端（GTP 帽体）的生长速率可以用下式描述：

$$\frac{\mathrm{d}x_{tip}}{\mathrm{d}t}=a\frac{\mathrm{d}n}{\mathrm{d}t}=a\left[k_{on}\left(c_0-\frac{x_{tip}M}{Va}\right)-\omega_{off}\right] \tag{6-6-67}$$

式中，M 为微丝的数目；c_0 是体系中初始溶液中蛋白质浓度；x_{tip}/a 是每根微丝帽体端所含单体的数目，可以得到该方程的解

$$x_{tip}(t)=\frac{aV}{Mk_{on}}\left[k_{on}c_0-\omega_{off}\right]\left(1-\mathrm{e}^{-Mk_{on}t/V}\right) \tag{6-6-68}$$

只有当单体聚合的速率低于水解点前进速度时塌缩才发生。塌缩发生在某个临界帽体长度 x_{tip}^{crit}，在该临界点，由于溶液中单体数量逐渐减少，生长端速率会降到水解速率以下。临界帽体长度 x_{tip}^{crit} 可以通过求解方程

$$a\left[k_{on}\left(c_0-\frac{x_{tip}M}{Va}\right)-\omega_{off}\right]=\frac{a}{\tau_T} \tag{6-6-69}$$

而得到。具体表达式为

$$x_{tip}^{cri}=\frac{aV}{Mk_{on}}\left[k_{on}c_0-\omega_{off}-\frac{1}{\tau_T}\right] \tag{6-6-70}$$

除临界长度以外，塌缩现象发生的频率为多少，这也是人们非常关心的问

题。为求解停留时间 t_{crit}，令式（6-6-59）的右端等于临界长度，即式（6-6-61）的右式，可以解得

$$e^{Mk_{on}t_{crit}/V}=\tau_T\left(k_{on}c_0-\omega_{off}\right) \tag{6-6-71}$$

因此，$t_{crit}\approx V\ln(\tau_T k_{on}c_0)/Mk_{on}$，其中假定 $\omega_{off}\ll k_{on}c_0$，即帽体的生长速率远大于脱落速率。

三、踏车现象

肌动蛋白纤维和微管都存在“踏车现象”。踏车现象的出现归根结底是由于它们具有极性，在组装时候，正端就是聚合速率和解聚速率快的一端，而负端为聚合和解聚速率慢的另一端。细胞在运动或者形态改变时，肌动蛋白纤维的生长始终发生在正端。在一定条件下，即体系中单体的浓度处于正端和负端临界浓度之间时，聚合到正端的肌动蛋白分子的速率正好等于从负端脱落单体的速率。此时，微丝的总长度几乎保持不变，单体从正端聚合到纤维上，随着聚合的不断进行，单体在纤维的位置向负端移动直至最终从纤维上脱落。由此可见，整个纤维处于动态平衡的状态，该过程为肌动蛋白的“踏车现象”，又称“轮回现象”。

随着实验技术水平的不断发展，最近生物物理实验学家在单分子水平上深入研究了肌动蛋白纤维的增长动力学行为。对单根肌动蛋白纤维的组装动力学的研究中发现，当肌动蛋白纤维的增长速率接近于零时，肌动蛋白纤维达到稳态，但是同时也观察到踏车现象。踏车现象是一个非平衡的过程，肌动蛋白单体从正端聚合而表现出来从另一端脱落，从而整个肌动蛋白纤维朝正端向前移动。通过控制溶液中不同蛋白的浓度以及离子浓度，实验上观察到踏车速率从几秒达到几天。关于踏车现象，当体系中浓度处于正端和负端临界浓度之间时，即 $c_{cr,T}^{ba}<c_T<c_{cr,T}^{po}$，肌动蛋白纤维正端的增长速率大于零，而负端的增长速率小于零，表现为收缩。肌动蛋白纤维的踏车速率满足该关系式: $J_{tm}\equiv\frac{1}{2}(J_g^{ba}-J_g^{po})$。图 6-6-27 为从布朗动力学模拟得到的单根肌动蛋白纤维处于踏车现象时，纤维总长度随时间的变化函数，以及纤维两端的空间位置随时间的变化函数。从图中可以明显看出，纤维的总长度随着时间呈现出明显的涨落，而纤维的正端（倒钩端）朝一个方向以一定的速度移动，可得到肌动蛋白纤维的踏车速度约为 0.07 mon/s，和实验结果基本吻合。

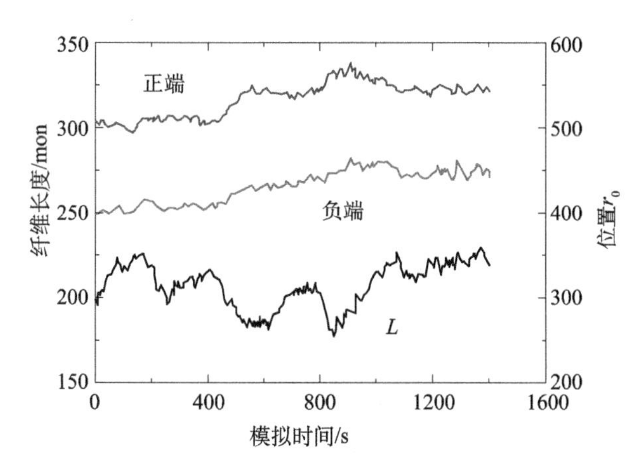

图 6-6-27　肌动蛋白纤维总长度 L 和肌动蛋白纤维两端（正端和负端或者倒钩端和尖端）空间位置与布朗动力学模拟时间的依赖关系，肌动蛋白纤维的长度以肌动蛋白单体数目为单位，而肌动蛋白纤维两端的空间位置以模拟的基本长度量纲 r_0

四、肌动蛋白纤维的长度扩散

在肌动蛋白纤维聚合动力学的单分子实验研究中发现，稳态时纤维表现出来的长度涨落（30～45mon^2/s）远大于通过肌动蛋白单体聚合和解聚速率得到的理论长度扩散系数值（1 mon^2/s）。

为了研究肌动蛋白纤维的长度扩散问题，首先选择最简单的理论模型，即忽略 ATP 水解过程，也就是不考虑核苷酸的状态变化。肌动蛋白纤维的总长度可以表示为 $L=L(t)$，或采用肌动蛋白纤维两端增加的蛋白亚基数目来表示 $n(t)=n^{\mathrm{ba}}(t)+n^{\mathrm{po}}(t)$，从而肌动蛋白纤维的长度扩散系数为

$$\left\langle\left[L-\langle L\rangle\right]^2\right\rangle=\left\langle\left[n-\langle n\rangle\right]^2\right\rangle\approx\left\langle\left[n^{\mathrm{ba}}-\langle n^{\mathrm{ba}}\rangle\right]^2\right\rangle+\left\langle\left[n^{\mathrm{po}}-\langle n^{\mathrm{po}}\rangle\right]^2\right\rangle \tag{6-6-72}$$

肌动蛋白纤维两端增加的蛋白亚基数目 n^{ba} 和 n^{po} 满足偏倚无规行走模型。纤维的长度分布函数 $P_n(t)$ 由四个过程决定，可记为

$$\begin{aligned}\frac{\mathrm{d}P_n}{\mathrm{d}t}=&\left(k_{\mathrm{on}}^{\mathrm{ba}}c+k_{\mathrm{on}}^{\mathrm{po}}c\right)P_{n-1}P_1+\left(\omega_{\mathrm{off}}^{\mathrm{ba}}+\omega_{\mathrm{off}}^{\mathrm{po}}\right)P_{n+1}\\&-\left(k_{\mathrm{on}}^{\mathrm{ba}}c+k_{\mathrm{on}}^{\mathrm{po}}c\right)P_nP_1-\left(\omega_{\mathrm{off}}^{\mathrm{ba}}+\omega_{\mathrm{off}}^{\mathrm{po}}\right)P_n\end{aligned} \tag{6-6-73}$$

其中，初始条件为 $P_1(0)=1$，其他为零。定义辅函数 $Q(z,t)=\sum\limits_{n=-\infty}^{\infty}z^nP_n(t)$ 和 $Q_k(t)\equiv\left.\dfrac{\partial^k}{\partial z^k}Q(z,t)\right|_{z=1}$，因此，长度的平均值为 $\langle n(t)\rangle=Q_1(t)$，方差为

$\left\langle\left[n(t)-\langle n(t)\rangle\right]^2\right\rangle=Q_2+Q_1-Q_1^2$。通过前面提到的方法求解式（6-6-73）可以得到 $Q(z,t)=\mathrm{e}^{g(z)t}$，其中 $g(z)\equiv\left(k_{\text{on}}^{\text{ba}}+k_{\text{on}}^{\text{po}}\right)cz+\left(\omega_{\text{off}}^{\text{ba}}+\omega_{\text{off}}^{\text{po}}\right)/z-\begin{bmatrix}\left(k_{\text{on}}^{\text{ba}}+k_{\text{on}}^{\text{po}}\right)c\\+\left(\omega_{\text{off}}^{\text{ba}}+\omega_{\text{off}}^{\text{po}}\right)\end{bmatrix}$。因此，纤维的平均长度和长度的方差分别为

$$\langle n(t)\rangle=\left[\left(k_{\text{on}}^{\text{ba}}+k_{\text{on}}^{\text{po}}\right)c-\left(\omega_{\text{off}}^{\text{ba}}+\omega_{\text{off}}^{\text{po}}\right)\right]t \tag{6-6-74}$$

和

$$\left\langle\left[n(t)-\langle n(t)\rangle\right]^2\right\rangle=\left[\left(k_{\text{on}}^{\text{ba}}+k_{\text{on}}^{\text{po}}\right)c+\left(\omega_{\text{off}}^{\text{ba}}+\omega_{\text{off}}^{\text{po}}\right)\right]t \tag{6-6-75}$$

而纤维的长度扩散系数 D_{L} 满足 $\left\langle\left[n(t)-\langle n(t)\rangle\right]^2\right\rangle\approx 2D_{\text{L}}t$。因此，纤维的长度扩散系数为

$$D_{\text{L}}=\frac{1}{2}\left[\left(k_{\text{on}}^{\text{ba}}+k_{\text{on}}^{\text{po}}\right)c+\left(\omega_{\text{off}}^{\text{ba}}+\omega_{\text{off}}^{\text{po}}\right)\right] \tag{6-6-76}$$

图 6-6-28 中给出了忽略核苷酸状态模型中，肌动蛋白纤维的长度扩散系数模拟和理论的结果。从图中可以看出，纤维的长度扩散系数和肌动蛋白单态的浓度呈明显的线性关系。当肌动蛋白浓度非常低时，$D_{\text{L}}\simeq\left(\omega_{\text{off}}^{\text{ba}}+\omega_{\text{off}}^{\text{po}}\right)/2$。

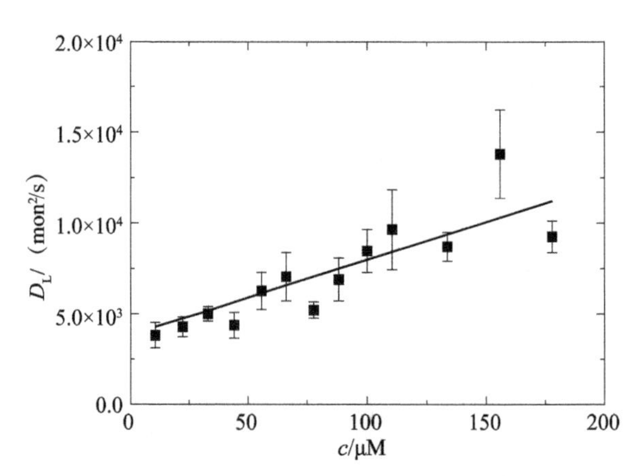

图 6-6-28　纤维长度扩散系数与肌动蛋白单体浓度 c 的依赖关系：实心方块表示布朗动力学模拟的结果；实线表示理论分析的结果，见方程（6-6-76）

在考虑 ATP 矢量水解的情况下，将肌动蛋白结合核苷酸的状态简化为两

种状态，即 ATP 结合肌动蛋白和 ADP 结合肌动蛋白两种状态。为了方便得到纤维的扩散系数，首先定义几个物理量：稳态时非零 ATP 帽子的概率 q，它满足关系式 $q=(k_{\text{on}}^{\text{ba}}+k_{\text{on}}^{\text{po}})c_{\text{T}}/(\omega_{\text{off,T}}^{\text{ba}}+\omega_{\text{R}})$，其中 ω_{R} 为纤维上矢量水解时的磷酸释放的总速率。当 q=1 时，体系中最小肌动蛋白浓度为 $c_{\text{T,tr}}$。当 $c_{\text{T}}\leqslant c_{\text{T,tr}}$ 和 $q\leqslant 1$ 时，肌动蛋白纤维的增长速率和扩散速率分别为

$$J_{\text{g}}=(k_{\text{on}}^{\text{ba}}+k_{\text{on}}^{\text{po}})c_{\text{T}}-\omega_{\text{off,T}}^{\text{ba}}q-\omega_{\text{off,D}}^{\text{ba}}(1-q)-\omega_{\text{off,D}}^{\text{po}} \tag{6-6-77}$$

和

$$D_{\text{L}}=\frac{1}{2}\left[\begin{array}{l}(k_{\text{on}}^{\text{ba}}+k_{\text{on}}^{\text{po}})c_{\text{T}}+\omega_{\text{off,T}}^{\text{ba}}q+\omega_{\text{off,D}}^{\text{ba}}(1-q)+\omega_{\text{off,D}}^{\text{po}}\\ +2\dfrac{(\omega_{\text{off,D}}^{\text{ba}}-\omega_{\text{off,T}}^{\text{ba}})\left[(k_{\text{on}}^{\text{ba}}+k_{\text{on}}^{\text{po}})c_{\text{T}}+\omega_{\text{off,D}}^{\text{ba}}q\right]}{\omega_{\text{off,T}}^{\text{ba}}+\omega_{\text{R}}}\end{array}\right] \tag{6-6-78}$$

当 $c_{\text{T}}\geqslant c_{\text{T,tr}}$ 时，肌动蛋白纤维的增长速率和扩散速率分别为

$$J_{\text{g}}=(k_{\text{on}}^{\text{ba}}+k_{\text{on}}^{\text{po}})c_{\text{T}}-P_{\text{T}}^{\text{ba}}(1)\omega_{\text{off,T}}^{\text{ba}}-P_{\text{D}}^{\text{ba}}(1)\omega_{\text{off,D}}^{\text{ba}}-P_{\text{T}}^{\text{po}}(1)\omega_{\text{off,T}}^{\text{po}}-P_{\text{D}}^{\text{po}}(1)\omega_{\text{off,D}}^{\text{po}} \tag{6-6-79}$$

和

$$D_{\text{L}}=\frac{1}{2}\left[\left(k_{\text{on}}^{\text{ba}}+k_{\text{on}}^{\text{po}}\right)c_{\text{T}}+\left(\omega_{\text{off,T}}^{\text{ba}}+\omega_{\text{off,T}}^{\text{po}}\right)\right] \tag{6-6-80}$$

图 6-6-29 中给出了在考虑核苷酸状态变化下，布朗动力学模拟中肌动蛋白纤维长度扩散系数和肌动蛋白单体浓度的依赖关系。从图可以看出，当体系

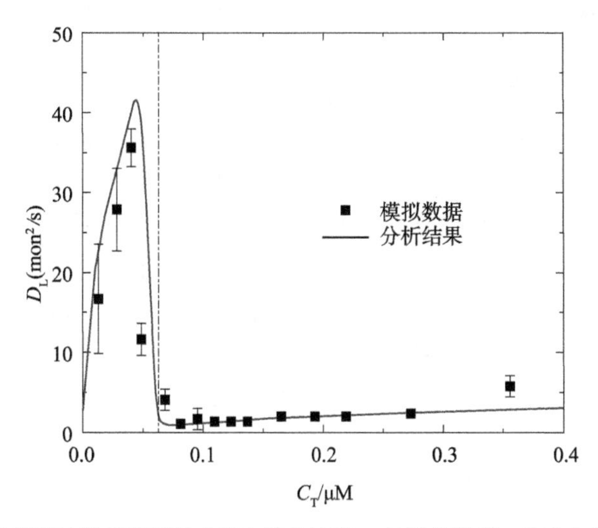

图 6-6-29 纤维长度扩散系数随肌动蛋白单体浓度 c_{T} 的依赖关系：实心方块表示布朗动力学模拟的结果；实线表示理论分析的结果，见方程（6-6-78）；垂直的虚线是 $c_{\text{T,tr}}$

的浓度高于 $c_{T,tr}$ 时，纤维的扩散系数比较小，$D_L \simeq 2.5\ mon^2/s$，而且和肌动蛋白单体的浓度呈线性关系；而当体系的浓度低于临界浓度时，纤维的扩散系数急剧地增加到 30～40 mon^2/s，然后随着体系中肌动蛋白单体的浓度进一步降低时，纤维的扩散系数随着单体浓度的降低而降低，并且也呈线性关系。从图中还可以发现，上述解析得到的扩散系数和布朗动力学模拟的结果非常吻合。

五、细胞骨架的力学性质

前面已经提到过，细胞骨架纤维决定细胞的组织结构和力学性质，它们可以提供力学支撑以反抗外力并克服诸如膜的弹性、结合反应、表面张力等不同机制导致的力的竞争。要想知道细胞骨架结构单元在这些生命过程中如何起作用，首先要研究它们作为单独的力学单元的响应特征。

对于细胞骨架单元，三种不同因素决定了它们在细胞内的等效刚度：第一种因素是纤维本身的柔软性；第二种因素是纤维结合蛋白的出现或缺失如何调节纤维的刚度；第三种因素是纤维大尺度组装成纤维束、网络或凝胶。现在已经发展了多种技术手段来测量这些不同尺度的刚度。对于单根纤维，热运动足以导致纤维弯曲，而且通过光学显微镜可以直接测量。因此，最直接而又最少侵扰的测量纤维刚度的方式是直接观察荧光标记的纤维或者用增强相差显微镜观察无荧光标记的微管。其他的测量是对纤维进行机械操纵从而观察它们在外力下的变形情况。这类测量可用标定的显微操作针或光控制的珠子将已知的力传递到所俘获的纤维的一端。图 6-6-30 为光镊子将肌动蛋白纤维打结的生动例子。

细胞内细胞骨架通常能构筑形成一定的结构，而这些结构比单根细胞骨架纤维具有更强的承受弯曲的单元。因此，研究细胞骨架组合结构的大尺度力学性质也是非常重要的。成束的肌动蛋白纤维可形成人类肠上皮细胞膜上的突起物，从而极大地增加表面积以利于营养吸收。从另外的角度看，膜会对肌动蛋白纤维束产生作用力。而从肠上皮细胞表面持续存在束结构，这表明膜产生的作用力不足以使纤维束弯曲或屈曲。因此，纤维成束的生物功能是提高复合纤维的刚度，使其大于单根纤维的刚度。

细胞骨架纤维在聚合过程中可以产生作用力。一个特别有趣的过程是肌动蛋白纤维聚合的前端使得细胞膜变形，如运动细胞的丝足。尤为重要的问题是纤维发生屈曲变形之前能够承受多大的力。在体外实验中，微管的聚合从一个表面开始向一个沟槽生长，沟槽阻止微管生长并导致屈曲，见图 6-6-31。

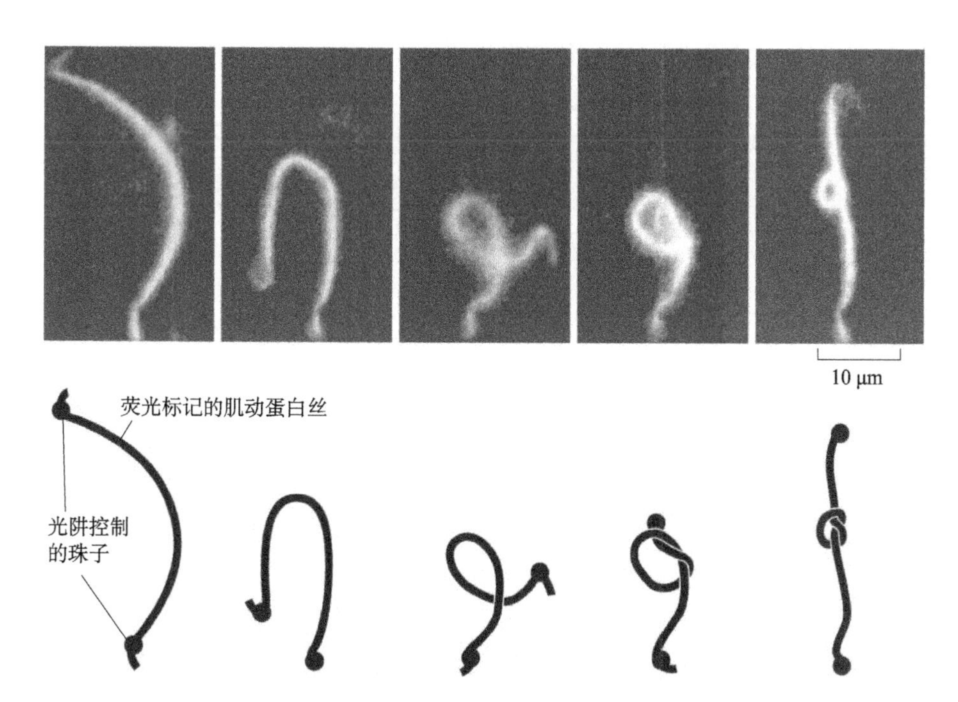

图 6-6-30 肌动蛋白打结的过程图。将两个珠子固定肌动蛋白纤维的两端，通过光阱控制两个珠子致使肌动蛋白纤维打结（摘自 Y. Arai et al. Nature 399: 446, 1999）

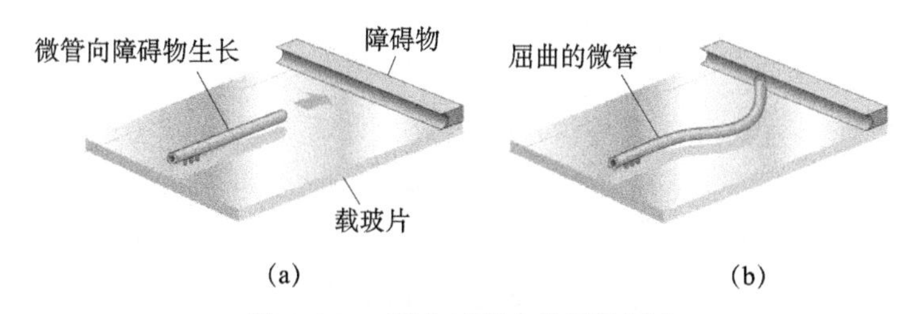

图 6-6-31 聚合过程中微管的屈曲

(a) 实验开始时，将微管紧密结合在载玻片上，朝向微管的正端放置一障碍物。随着微管蛋白亚基聚合到微管，微管持续生长直至碰到障碍物; (b) 为了在生长的微管正端留有足够的空间让微管蛋白亚基持续生长到微管上，微管蛋白亚基将对微管施加作用力，而造成微管屈曲（摘自 M. Dogterom, B. Yurke, Science 278: 856, 1997）

为了理解微管的弯曲作用力问题，考虑微管两端受到压力 F，当外力增加到一定程度时，微管会发生屈曲。为了估算屈曲临界作用力，采用圆弧的简单几何模型，见图 6-6-32。假定微管的形状是一个半径为 R、长为 L 的圆弧，其中 $\theta=L/R$ 是圆弧对应的圆心角。首先确立 $\theta\neq0$ 对应能量极小值的临界力。在微管屈曲的过程中，随着微管的屈曲，外力贡献的能量是减小的，而弯曲能是增加的。如果弯曲能的增加占主导地位，那么微管保持为直的

（θ=0）时候，外力最小。当外力非常大时，微管将屈曲。因此，利用弹性屈曲的圆弧几何构型，估算临界力 F_{crit}。

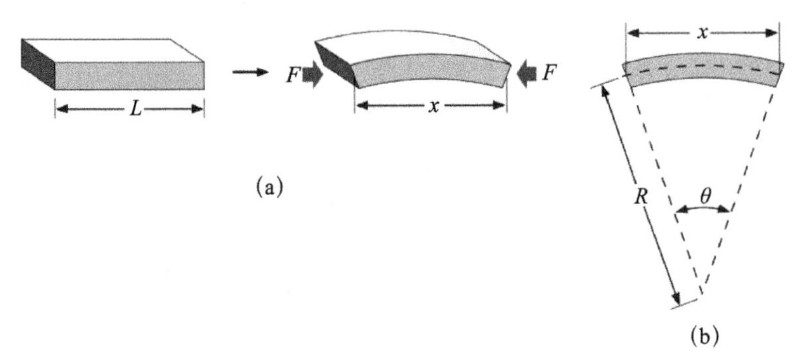

图 6-6-32　屈曲过程的示意图

(a) 长度为 L 的悬梁两端实际压缩作用力 F，悬梁屈曲；(b) 屈曲过程的参数

体系的总能量可包括微管的弯曲能和外力的贡献，即

$$E_{tot} = \frac{\xi_p k_B T}{2}\frac{L}{R^2} - F(L-x) \tag{6-6-81}$$

其中，$x=2R\sin(\theta/2)$ 是微管两端的距离。根据角度 θ，屈曲微管的总能量写为

$$\frac{E_{tot}}{k_B T} = \frac{\xi_p \theta^2}{2L} - \frac{FL}{k_B T}\left(1 - \frac{2}{\theta}\sin\frac{\theta}{2}\right) \tag{6-6-82}$$

要获得临界力，首先要搜寻屈曲失稳，即看足够大外力时最低能量态对应的 θ 是否非零。图 6-6-33 给出了两种情形时，即 $F<F_{crit}$ 和 $F>F_{crit}$ 时，作用力 F 和角度 θ 的依赖关系。当 $F<F_{crit}$ 时，体系总能量的极小值对应于 $\theta=0$，不会发生屈曲，微管永远是直的；当 $F>F_{crit}$ 时，能量的极小值发生在 $\theta\neq0$ 时，微管倾向于屈曲。

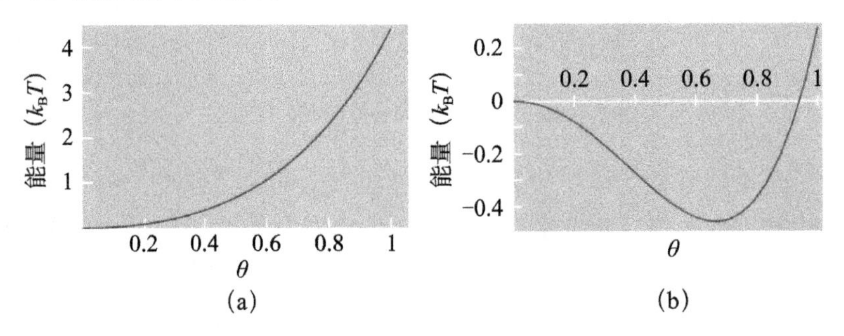

图 6-6-33　悬梁屈曲过程中能量和角度 θ 的依赖关系

(a)$F < F_{crit}$ 和 (b)$F > F_{crit}$。图中的参数为 L=20μm 和 $\xi_p k_B T$=30pN·μm^2

将式（6-6-82）对小角度 θ 展开，可得到

$$\frac{E_{\text{tot}}}{k_{\text{B}}T}=\frac{\xi_{\text{p}}\theta^2}{2L}-\frac{FL}{k_{\text{B}}T}\frac{\theta^2}{24} \tag{6-6-83}$$

由此可以得到，临界力为

$$F_{\text{crit}}=12\frac{k_{\text{B}}T\xi_{\text{p}}}{L^2} \tag{6-6-84}$$

图 6-6-31 所示的实验中，微管长度约为 20μm，特征刚度为 $\xi_{\text{p}}k_{\text{B}}T$= 30pN·μm^2 时，相应的临界力为 0.9pN。

迄今，已经发展了几种技术都能够直接测量细胞骨架聚合时所产生的作用力大小。首次直接测量微管聚合时产生的作用力是利用了微管本身的弯曲弹性特性，通过测量弯曲程度间接测力。在上小节中讨论了这个实验，但是其中只计算了弯曲一根微管所需的力。该实验并不是测量弯曲力，而是通过测量微管的弯曲情况来推算微管与障碍物的界面产生的聚合力。该经典实验证明单根微管在这种受限的几何环境中聚合时可以产生数皮牛的推力，如图 6-6-34 所示。该数值接近于单个驱动马达或动力马达产生的力。

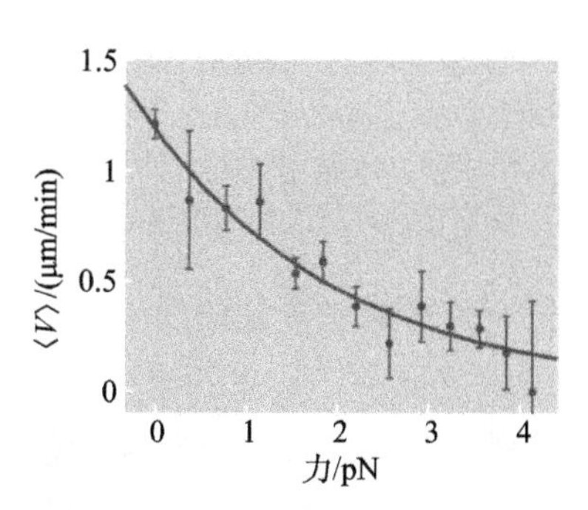

图 6-6-34　在对障碍物施加不同外力 F 时，推动障碍物的生长微管的平均聚合速度（M. Dogterom and B. Yurke, Science 278: 856, 1997）

微管本身作为一个装置来测量发力大小衍生了一些问题。尤其是微管的刚度随着长度的变化而变化，会导致实验不能准确地推算聚合力，因此需要采用其他方法来验证测得的聚合力。图 6-6-35 所示的光镊可以实现该目的。光镊在纳米尺度的位移内很适合测量皮牛量级的力。光镊的测量结果与微管弯曲的测量结果非常一致，表明微管生长影响其弹性的观点不足以明显改变测得的聚合力。

光镊的方法在测量肌动蛋白纤维网络所产生的巨大作用力时具有一定的

困难，但是基于悬臂的显微镜能测量大作用力。图 6-6-36 给出了悬臂实验，肌动蛋白纤维网络的生长是在非常接近玻璃板的悬臂表面时触发的。悬臂的偏斜反映了组织生长，可以用其来测量力和生长速度。该实验得到了肌动蛋白阵列并不是以简单的累加模式一起发力的结果。而且最为突出的结果是，在较大范围的变力作用下，肌动蛋白纤维组织的生长几乎是恒速的，这个与单个分子马达表现的单调下降的力-速度曲线形成鲜明的对照。

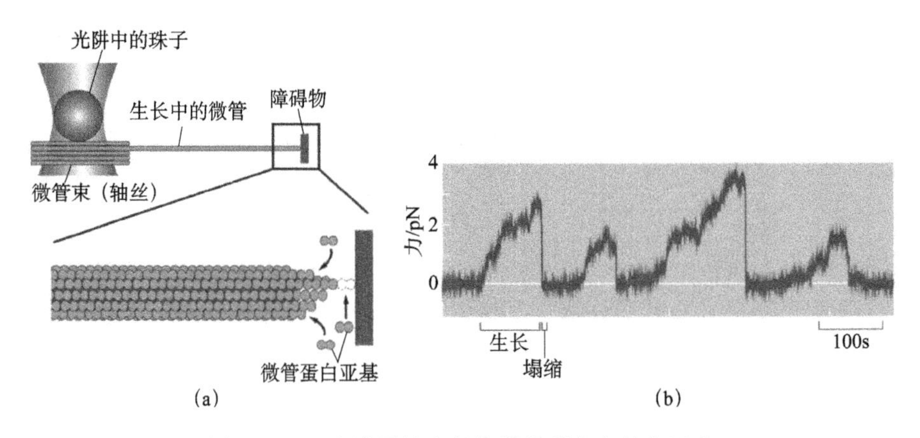

图 6-6-35　光阱测量生长的微管所产生的作用力

(a) 光阱中的珠子固定一束微管后，微管持续生长直至接触到障碍物。当微管进一步聚合微管蛋白亚基时，微管将推动珠子移动；(b) 微管对珠子产生的作用力随时间的演化图将进一步描述微管的生长和塌缩过程

细胞可以利用纤维聚合力来做非常复杂的工作。特别有趣的是：在巨大的细胞体内，细胞骨架组织就像一个宏大的坐标系统可以精确定位细胞器。如果用荧光标记细胞内的微管，可以看见他们从中心体向外散射，呈星状分布。中心体是一个直径小于 0.5μm 的微小实体，却能定位在细胞中心数十微米的区域附近。中心体是如何发现细胞中心的呢？如果设计一个实验，将一个分离出来的中心体放置纳米结构制备孔，孔的尺度与细胞相当，中心体将漫无目的地来回扩散。将微管蛋白加入，则微管将会从中心体生长出来，中心体立刻定位于孔的几何中心，且与孔的形状无关。如果用光镊抓住中心体使其偏离其中心位置，它将缓慢而坚定地恢复到中心位置。该机制与微管端部接触孔壁时的推力有关。只有当中心体处于区域的几何中心时，所有的合力才会消失。在裂殖酵母菌的活细胞内，也可以直接观察到，利用微管两端的推力作用，细胞核定位于杆状细胞的中部。

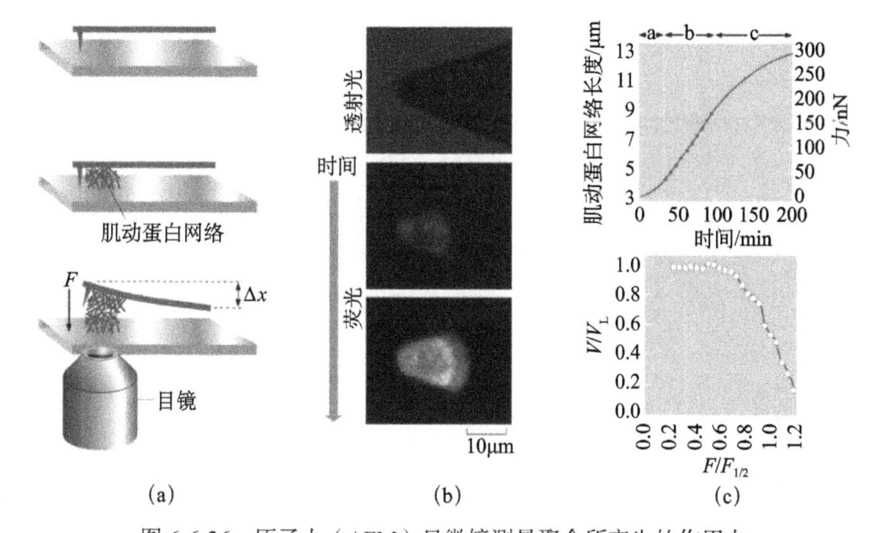

图 6-6-36　原子力（AFM）显微镜测量聚合所产生的作用力

(a) 实验的示意图：首先将原子探针上面黏结肌动蛋白网络，不断增长的肌动蛋白网络将会上移原子探针；(b) 随着实验的进行，荧光标记的肌动蛋白网络的信号强度随时间的变化；(c) 肌动蛋白网络长度随时间的演化曲线。随着肌动蛋白网络的生长，原子探针将不断弯曲，聚合所产生的作用力持续增加。(c) 中下图表示网络生长速度和作用力的依赖关系，图中的平台表示网络生长速度保持不变（摘自 S. H. Parekh et al. Nat. Cell Biol., 7: 129, 2005）

第三节　未来 5～10 年重点发展方向

细胞骨架是一个典型的生命软物质体系，其中涉及细胞骨架的自组装和聚集态行为，例如，生物分子（如能量驱动蛋白等）的构型转变和折叠，细胞骨架之间以及细胞骨架在其他蛋白分子协同作用下的聚集态结构和动态行为等。此外，细胞膜也是典型的生命软物质体系，细胞骨架的聚集态结构和动态行为将与细胞膜上的特殊结构（如黏附班、细胞间连接等）相互联系，并与细胞核的支架系统相互作用等。细胞骨架性质的变化与细胞内部和外部环境的物质、能量和信息的交换密切相关。研究这些变化的内部物理和化学规律，将是未来细胞骨架的重点发展方向。

在生物体内机体中各组织细胞的结构和功能是密切相关的。实验发现，在恶性病变的细胞中，将伴随着细胞骨架结构的破坏、组装和分布的异常、微管的解聚等现象。肿瘤细胞的主要特征之一是失去控制细胞的生长和增殖。科学家们研究了胃癌、食管癌等肿瘤细胞，发现肿瘤细胞质内免疫荧光染色的微管数量明显减少甚至缺失。因此，微管数量的减少是细胞恶性转化的一个重要标志。而进一步的研究表明，细胞内存在特定结合蛋白会改变微管蛋白的结构，例如在有丝分裂的中后期，秋水仙碱和长春花碱等蛋白可与纺锤体微管蛋白或微管结合而抑制细胞增殖。因此，科学家们可以通过设计一些特异性蛋白（抗肿瘤药物）改变微管的动力学行为，致使机体内存在稳定的微管和细胞结构。此外，中间微丝的形态虽然相似，但其生化成分差异很大，具有严格的组织特异性。通常，人们根据中间纤维的种类来鉴别、区分不同组织来源的肿瘤细胞及各种肿瘤细胞的亚型，为肿瘤的诊断和治疗提供了决定性的依据。同时，细胞骨架结构的异常表达与许多神经系统疾病也有关，如阿尔茨海默病（alzheimer disease）病、亨廷顿症、帕金森病等[10-12]。阿尔茨海默病最主要的细胞病理变化是细胞内细胞骨架异常交联而形成的不溶性神经纤维缠结（蛋白的聚集）等现象。神经纤维缠结主要归功于高磷酸化状态的tau蛋白（一种微管结合蛋白）和微管蛋白相结合导致微管蛋白聚集态结构发生演变的结果。因此，从细胞骨架结构的角度认识肿瘤细胞、神经退行性疾病等发病的分子机制，全面认识这类疾病的生物学机制也是未来的重点发展方向。

第四节　结　　语

细胞中细胞质主要是由蛋白质纤维——细胞骨架网络构筑而成的。该细胞骨架网络包含三种细胞骨架纤维：肌动蛋白纤维、微管和中间纤维。这三种细胞骨架纤维都是蛋白亚基以首尾相连或者侧面相互作用等形式组合而成的螺旋结构。蛋白亚基结构的差别和组装方式的不同造就了这三种细胞骨架纤维的力学性能存在差异。蛋白亚基的组装和去组装能力能重构这三种细胞骨架纤维。肌动蛋白亚基和微管蛋白亚基能够黏结和催化核苷酸（ATP和GTP分子），而且它们能够以首尾形式组装形成极性细胞骨架纤维而产生作用力。活体细胞中，辅助蛋白能调控细胞骨架纤维的动力学和组织结构，能积极参与细胞分裂、细胞迁移以及在细胞内部形成极性的组织结构等生命活动。细菌类细胞也存在肌动蛋白、微管蛋白、中间纤维的同源物，它们也能形成动态结构并有助于控制细胞的形状和细胞分裂等。

郭坤琨（湖南大学材料科学与工程学院）

参考文献

[1] Bray D. Cell movements: from molecules to motility. New York: Garland Publishing, 2001.

[2] Alberts B. Molecular Biology of the Cell. New York: Garland Publishing, 2014.

[3] Lodish H. Molecular Cell Biology. San Francisco: Freeman, 2012.

[4] Phillips R. Physical Biology of the Cell. New York: Garland Publishing, 2012.

[5] Geisler N, et al. Assembly and architecture of invertebrate cytoplasmic intermediate filaments reconcile features of vertebrate cytoplasmic and nuclear lamin-type intermediate filaments. J. Mol. Biol., 1998, 282: 601-617.

[6] Bretscher A. Regulation of cortical structure by the ezrin-radixin-moesin protein family. Curr. Opin. Cell Biol., 1999, 11: 109-116.

[7] Sheterline P J, Clayton J, Sparrow J C. Protein profile. Actin., 1995, 2: 1-103.

[8] Howard J. The mechanics of Motor Proteins and the Cytoskeleton. Sinauer, 2001.

[9] Kreis T, Vale R. Guidebook to the Cytoskeletal and Motor Proteins. Oxford: Oxford Univ. Press, 1999.

[10] Mullins R D, Pollard T D. The interaction of Arp2/3 complex with actin: nucleation, high affinity pointed end capping, and formation of branching networks of filaments. Proc. Natl. Acad. Sci. USA, 1998, 95: 6181-6186.

[11] Fehon R G, Bretscher A. Organizing the cell cortex: the role of ERM proteins. Nat. Rev. Mol. Cell Biol., 2010, 11: 276-287.

[12] Garner E C, Bernard R, Wang W, et al. Coupled, circumferential motions of the cell wall synthesis machinery and MreB filaments in B. subtilis. Science, 2011, 333: 222-225.

[13] Garner E C, Campbell C S, Mullins R D. Dynamic instability in a DNA-segregating prokaryotic actin homolog. Science, 2004, 306: 1021-1025.

[14] Hill T L, Kirschner M W. Bioenergetics and kinetics of microtubule and actin filament assembly-disassembly. Int. Rev. Cytol. 1982, 78: 1-125.

[15] Jones L J, Carballido-Loepez R, Errington J. Control of cell shape in bacteria: helical, actin-like filaments in Bacillus subtilis. Cell, 2001, 104: 913-922.

[16] Oosawa F, Asakura S. Thermodynamics of the Polymerization of Protein. Academic Press, 1975.

[17] Osawa M, Anderson D E, Erichson H P. Curved FtsZ protofilaments generate bending forces on liposome membranes. EMBO J., 2009, 28: 3476-3484.

[18] Theriot J A. Why are bacteria different from eukaryotes? BMC Biol., 1977(2013), 11: 119.

[19] Aldaz H, Rice L M, Stearns T, et al. Insights into microtubules nucleation from the crystal

structure of human gamma-tubulin. Nature, 2005, 435: 523-527.

[20] Dogterom M, Yurke B. Measurement of the force-velocity relation for growing microtubules. Science, 1997, 278: 856-860.

[21] Doxsey S, McCollum D, Theurkauf W. Centrosomes in cellular regulation. Annu. Rev. Cell Dev. Biol., 2005, 21: 411-434.

[22] Galjart N. Plus end tracking proteins and their interactions at microtubule ends. Curr. Biol., 2010, 20: 528-537.

[23] Hotani H, Horio T. Dynamics of microtubules visualized by darkfield microscopy: treadmilling and dynamic instability. Cell Motil. Cytoskeleton, 1988, 10: 229-236.

[24] Kerssemaker J W, Munteanu E L. Assembly dynamics of microtubules at molecular resolution. Nature, 2006, 442: 709-712.

[25] Mitchison T, Kirschner M. Dynamic instability of microtubule growth. Nature, 1984, 312: 237-242.

[26] Singla V, Reiter J F. The primary cilium as the cell's antenna: signaling at a sensory organelle. Science, 2006, 313: 629-633.

[27] Verhey K J, Kaul N, Soppina V. Kinesin assembly and movement in cells. Annu. Rev. Biophys., 2011, 40: 267-288.

[28] Helfand B T, Chang L, Goldman R D. The dynamic and motile properties of intermediate filaments. Annu. Rev. Cell Dev. Biol., 2003, 19: 445-467.

[29] Campbell C S, Mullins R D. In vivo visualization of type II plasmid segregation: bacterial actin filaments pushing plasmids. J. Cell Biol., 2007, 179: 1059-1066.

[30] Carlier M F, Pantaloni D, Korn E D. Evidence for an ATP cap at the ends of actin filaments and its regulation of the F-actin steady state. J. Biol., Chem., 1984, 259: 9983.

[31] Dogterom M, Leibler S. Physical aspects of the growth and regulation of microtubule structures. Phys. Rev. Lett., 1993, 70: 1347-50.

[32] Fygenson D K, Braun E. Phase diagram of microtubules, Phys. Rev. E, 1994, 50: 1579-88.

[33] Lino T. Assembly of Salmonella flagellin in vitro and in vivo. J. Supramol. Struct., 1974, 2: 372-84.

[34] Ouellete L, Morales M F. Molecular kinetics of muscle adenosine triphosphatase. Arch. Biochem. Biophys., 1952, 39: 37-50.

[35] Pollard T D. Rate constants for the reactions of ATP- and ADP-actin with the ends of actin filaments. J. Cell Biol., 1986, 103: 2747-54.

[36] Pollard T D. Polymerization of ADP-actin. J. Cell Biol., 1984, 99: 769-77.

[37] Welch M D, Rosenblatt J, Mitchison T J. Interaction of human Arp2/3 complex and the Listeria monocytogenes ActA protein in actin filament nucleation. Science, 1998, 281:

105-8.

[38] Wu J Q, Pollard T D. Counting cytokinesis proteins globally and locally in fission yeast. Science, 2005, 310: 310-314.

[39] Guo K K, Shillcock C J, Lipowsky R. Self-assembly of actin monomers into long filaments: Brownian dynamics simulations. J. Chem. Phys., 2009, 131: 015102(1-11).

[40] Guo K K, Shillcock C J, Lipowsky R. Treadmilling of actin filaments via Brownian dynamics simulations. J. Chem. Phys., 2010, 133: 155105(1-9).

[41] Guo K K, Qiu D. Polymerization of actin filaments coupled with adenosine triphosphate hydrolysis: Brownian dynamics and theoretical analysis. J. Chem. Phys., 2011, 135: 105101(1-6).

[42] Fujiwara I, Takahashi S, Ishiwata. Microscopic analysis of polymerization dynamics with individual actin filaments. Nat. Cell Biol., 2002, 4: 666-673.

[43] Kueh H Y, Mitchison T J. Dynamic stabilization of actin filaments. Proc. Natl. Acad. Sci. USA, 2008, 105: 16531-16536.

[44] Li X, Kierfeld J, Lipowsky R. Actin polymerization and depolymerization coupled to cooperative hydrolysis. Phys. Rev. Lett., 2009, 103: 048102-048105.

[45] Li X, Kierfeld J, Lipowsky R. Coupling of actin hydrolysis and polymerization: Reduced description with two nucleotide states. EPL, 2010, 89: 38010-38016.

[46] Holy T E, Dogterom M, Yurke M, et al. Assembly and positioning of microtubule asters in microfabricated chambers. Proc. Natl Acad. Sci. USA, 1997, 94: 6228-6231.

[47] Parekh S H, Theriot J A, Fletcher D A. Loading history determines the velocity of actin-network growth. Nat. Cell Biol., 2005, 7: 1219-1223.

第七章 细胞软物质力学及其在生理病理机制中的作用

黏附细胞可以说是自然界中最为复杂的软体材料：其内部由上百种蛋白质构成，不同蛋白质之间存在相互作用，蛋白质分子聚合成链并团聚成致密的网络状细胞骨架结构。细胞的化学与物理结构如此复杂，对其力学行为的研究和认识具有巨大的挑战性，并且需要生物学、化学、物理学等多学科的共同努力。近年来，关于细胞软物质力学的研究已经成了新的热点，并逐步揭示了一些重要的细胞软物质力学行为。例如，细胞骨架的热力学非平衡态行为，细胞力学行为与其内部预应力的显著关系，细胞具有无特征松弛时间尺度的流变行为和在短暂拉伸后发生迅速流态化等。这些行为很可能对于细胞的许多功能（如黏附、迁移、分化等）发挥重要的调控甚至决定作用，进而在相关的生理病理机制中发挥重要的作用。因此，深入研究细胞软物质力学，不仅对于正确认识细胞生命活动的基本规律具有重大意义，而且为揭示疾病发生发展的内在机制并探索相应的防治途径提供新的视角和思路。

第一节 引　　言

作为生命基本构件的细胞，不仅其构造十分复杂，其功能更是十分奇妙并存在大量令人费解之谜。过去很长时期，细胞生物学的研究中还原主义方法（reductionist approach）一直占据主要地位。这种方法对于解析细胞的物质结构，包括细胞膜与细胞内骨架系统的物质基础，细胞各结构间的化学信号的传导和调控等有着不可替代的作用。但是，近年来的研究也逐渐表明仅仅采用传统还原主义方法是不能完全揭示细胞功能的所有谜团的，许多重要的问题可能需要引入解决现代物理学问题的思路才能解决 [1]。

以黏附细胞为例。我们知道绝大部分黏附细胞内部存在由肌动蛋白纤丝、中间纤丝和微管共同构成的聚合物网络——细胞骨架。细胞骨架使得细胞在自发和 / 或外力作用下运动与变形时依然能够保持其形状和结构的稳定性 [2-7]。许多重要疾病的病理特征，如哮喘病的气道过度收缩、动脉血管硬化、癌细胞扩散等都与细胞骨架在收缩，变形和响应外应力等物理过程中的异常行为有潜在的联系 [8-12]。

从还原论的角度来看，细胞骨架是由各种骨架蛋白聚合长链及其捆绑蛋白、运动蛋白等构成的具有主动性的半柔性纤维网络。这种主动的纤维网络进而被组装进柔性的细胞膜内，并通过跨膜蛋白与细胞外环境连接感知外环境的变化，如图 6-7-1(a)、(b) 所示 [13]。关于主动的半柔性聚合物纤维网络的动力学行为特征和规律都已在实验和理论模型方面得到了较好的描述和解释 [14,15]。但是，这些研究的结果对于解释真实细胞的力学行为是极为有限的。因为上述研究中的聚合物纤维网络都被假定为稀疏的物理系统，这与真实细胞中的情况有很大不同。真实细胞内部是一个十分拥挤的物理和化学空间，蛋白质分子含量高达 40% 以上，分子之间仅有几纳米的物理空间距离，如图 6-7-1(c) 所示 [16]。

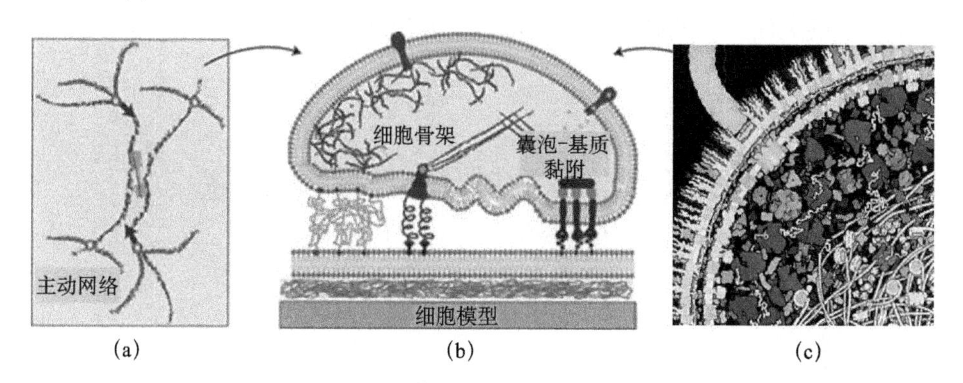

图 6-7-1 内部高度致密和拥挤的细胞软物质结构模型（改编自 Phillips, 2008, Smith, 2010[13,16]）
(a) 肌动蛋白纤维（长链）、交联蛋白（圆点）和运动蛋白束（齿状条）构成的主动网络。(b) 主动网络被组装进双极脂质细胞膜形成细胞骨架结构，并通过跨膜蛋白与细胞外环境连接和感受来自细胞外部的物理信号，从而实现对细胞的主动控制。(c) 细胞内蛋白质分子间的自由空间非常小（＜ 10 nm），实际上形成高度致密和拥挤的空间结构

即使在无生命的软体材料世界中，如此拥挤致密的胶体物质的力学行为已经是极为复杂的。作为有生命的细胞的力学性质更有其特殊性，因为它与细胞的生物功能联系在一起[16]。作为细胞的主要结构，细胞骨架同时也起到非常关键的生物化学作用[17]。例如，细胞骨架蛋白纤维交联的解折叠既是一个物理过程，同时也会触发相应的生物化学信号[18,19]。细胞内蛋白质的致密限制了分子的扩散以及细胞质内的各种细胞器，从而影响细胞内化学反应的速率[20,21]。因此，理解细胞力学不仅需要理解细胞如何变形和重构，而且需要掌握外部应力如何触发生物信号并下行调节细胞功能。构建一个如此复杂综合的框架来理解细胞力学是一个十分艰巨的任务，不仅需要传统的生物学、化学、物理学等多学科的融会、交叉，更可能需要在实验技术和方法以及理论框架方面的创新。

尽管细胞具有结构上和力学行为上的复杂性，过去二十年间随着观测技术的发展和理论上的探索，人们逐渐发现了细胞的一些具有普遍性的软物质力学行为，其中得到较为广泛认可的有：①细胞骨架是一个处于预应力状态的张力整体（tensegrity）；②细胞骨架呈现无尺度流变学行为（scale-free rheology）；③细胞骨架内的扩散行为异常（anomalous diffusion）和涨落耗散定理（fluctuation dissipation theorem）失效；④细胞骨架在短暂拉伸后发生快速短暂的流态化（fluidization）。这些行为与细胞类型和细胞状态无关，具有相当的普遍性，而且相互之间存在密切的关联，很可能受同样的内在机制的调控。这些发现无论在细胞生命科学领域还是在凝聚态软物质科学领域都引起了广泛的关注。与此同时，细胞软物质力学这一前沿领域正在以前所未有的速度向前发展。

第二节　学科发展现状及前沿问题

一、细胞软物质力学的表征和测量方法

由于细胞的硬度低，尺寸小，因此机械力和形变也很小，分别在皮牛顿和纳米范围内。近年来随着微流变技术的发展，对于单个活细胞的精准定量力学测量变得可能，这些技术包括磁镊（magnetic tweezers）、光镊（optical tweezers）、原子力显微技术（atomic force microscopy）、细胞穿刺（cell poking）、微孔反应板（microplates）和细胞伸张器（cell stretcher）。1922 年，

Heilbronn 利用早期的磁镊技术，首次报道了细胞质的微流变研究。在 20 世纪 40 年代后期，Francis Crick 和 Arthur Hughes 对该技术进行了改进，虽然实验手段仍有局限性，但他们观察到了细胞的全域黏弹性行为，包括塑性（plasticity）和剪切稀化（shear thinning），并报道了细胞内部的硬度范围为 2～50Pa。

为什么细胞如此柔软？单独看待细胞骨架的丝状网络，它们实际是由硬度和强度相当高的材料构成的。例如，中间纤丝角蛋白（keratin），同时也是羊毛的主要成分，具有 1～5GPa 的硬度和 0.2GPa 的抗张强度（tensile strength）；细胞骨架肌动蛋白（actin）作为细胞内最丰富的蛋白，其杨氏模量为 1～2GPa。然而，细胞整体反而很软，其杨氏模量比肌动蛋白的要低大约 6 个数量级，这跟羊毛衫触摸起来很柔软的原因一样：因为蛋白纤维的体积分数（volume fraction）很低，而且单个纤维受到的是弯曲而不是拉伸形变。

但是，通过细胞硬度的测量值无法估算出体积分数、纤丝弯曲度、交联拉伸和动态重构机制等因素对细胞硬度的相对贡献。细胞硬度作为一个单一的数值时，忽略了或者说平均了任何不均匀性（inhomogeneity）、间隔尺寸（granularity）、各向异性（anisotropy）、非线性（nonlinearity）和时间波动（time fluctuations）。此外，这种平均的特质取决于所用的探针（probe），如它的形状、尺寸、表面修饰、与细胞的接触时间、与细胞体的相对位置。这种平均进一步取决于所施加的力或者形变的量级和测量过程的时间或频率范围。因此，不同的测量方法实际测量细胞力学的不同方面，每种方法都有自己的特点、优势和局限性。

此外，不管是通过接合（ligation）或者是力学传导（mechanotransduction），探针（如原子力显微尖端，配体包被的微珠，或者微孔板）与细胞的相互作用不可避免地导致局部重构，从而引起探测区域的结构改变[22-27]。这种局部重构是所有采用外部探针的方法的通病。对于采用微珠接合的方法，其局限性表现为微珠在细胞表面的位置不受控制[28]，细胞与微珠之间的几何交互也不受控制。

除了两点质粒追踪（two-point particle tracking）微流变学[29,30]，其余所有测量细胞力学的方法都需要调用相应的长度尺度来将原始的偏移量数据转化成细胞应变，进而得出一个恰当的弹性模量。细胞内的应变分布可以通过分析推导得出[31]，可以通过细胞应变的有限元模型得出[32]，也可以简单地通过空间参数得出[33]。但以上方法都需要假设细胞是一个均质的各向同性的线弹性材料。就算在这样一个简化的情况下，细胞内与探针接近部位的应力和应变场是极其复杂的[34]，但探针对细胞的影响会随着远离探针的距离而迅速衰减。

考虑到以上所有因素，进行准确的细胞力学定量测量并非易事，采用不同方法测量得到的细胞弹性模量可能相差一个数量级以上[4]。不过目前已有一些实验测量揭示了一套可以被重复，可以用不同方法和不同细胞类型比较细胞力学属性的方法[35]。这种通用的技术方法是从物理学角度理解复杂的细胞力学的基础。

二、细胞的主要软物质力学行为及其特征

（一）细胞是处于预应力状态的张力整体

Harris 和同事最早发现，与多数无生命的惰性软物质不同，细胞处于内部存在预应力的状态[39]。随后科学家们不断完善精确测量细胞内张力的技术[40,41]。运用这些技术，世界上许多不同的实验室分别独立地观察和证明了细胞骨架是处于预应力状态的结构[42,43]。细胞骨架预应力主要由肌球蛋白马达产生。大部分预应力通过肌动蛋白细胞骨架，即微丝结构（MFs）上的张力平衡，而张力则通过微丝与跨膜蛋白的连接进而在细胞-细胞黏着间和细胞-细胞外基质黏着间传递。小部分预应力被微管结构中的压缩力所平衡[图 6-7-2(a)]。

细胞骨架中的预应力不仅维持细胞形状和力学稳定性，而且对于维持细胞的生物功能也很重要。例如，绑定到细胞骨架的离子通道的打开或者隐秘多肽序列的暴露都依赖于细胞骨架中存在一定的预应力；细胞内的生物活性分布与细胞骨架预应力的分布表现出很强的相关性；细胞骨架预应力的存在使得细胞能够感知到其外部环境的物理特性，如细胞外基底的硬度等，进而对细胞的迁移、增殖、分化等重要的生物学行为实施调控[44,45]。

根据细胞骨架预应力的存在及其力学性质，哈佛大学的 Ingber 受张力整体（tensegrity）几何概念[图 6-7-2(b)]的启发提出了著名的细胞骨架张力整体模型[37,38]。此后大量的研究成果证明了该模型的重要价值。但是，细胞骨架张力整体模型是一个宏观的模型，其局限性在于只能较好地解释细胞骨架在稳态条件下的力学行为，但却不能很好地预测和解释细胞骨架的动力学行为。

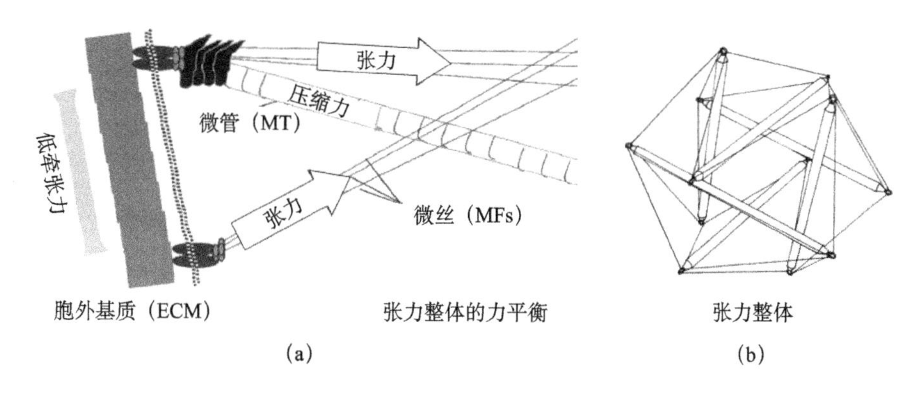

图 6-7-2 细胞骨架的力平衡与张力整体模型

(a) 黏附细胞内存在预应力。细胞内张力通过粘着斑和微丝（microfilaments，MFs）支撑，而压缩力通过微管（microtubule，MT）支撑。(b) 张力整体模型（引自文献[36]～[38]）

（二）细胞的无尺度流变行为

流变学研究是认识软物质，包括细胞和组织，动力学特性的重要途径。然而，过去的细胞流变学实验往往只能在很窄的时间尺度范围内进行，因此对于细胞骨架的动力学行为难以进行综合全面的观察和诠释。最近，许多新技术的发展克服了此局限性。例如，哈佛大学 Fredberg 实验室发明的磁微粒扭转细胞测量技术可以同时直接测量大量细胞在很宽频率带宽内（0.1～1000 Hz）的流变学行为，并采用该技术测量了一系列黏附细胞的细胞骨架的存储模量（G'）和损耗模量（G''）[46,47]。有趣的是，尽管细胞的类型不同，但它们的 G' 与频率 f 之间始终满足幂函数关系 $G'(f)\sim f^{\alpha}$。Deng 等进一步发现细胞骨架在快慢不同的时间尺度下至少表现出两个显著不同的动力学行为[48]。在长时间尺度下（低频），细胞骨架动力学行为遵守可变幂率规律，幂指数 α 的取值范围为 0.1～0.2。在短时间尺度下（高频）细胞骨架动力学则遵守固定幂律规律（α=3/4）。后者与被动的蛋白聚合物纤维网络动力学特征相同，间接地证明了磁微粒扭转细胞测量技术确实反映了细胞骨架的动力学[49]。大量其他研究也证明了幂律行为是细胞骨架动力学的一个普遍内在特质，如图 6-7-3 所示。

幂律流变学对于细胞骨架动力学有两个重要含义。第一，细胞骨架动力学不是由离散的特征弛豫时间决定的，而是由以幂函数形式分布的连续的弛豫时间谱决定的，这种行为被称为无尺度（scale-free）流变学。第二，细胞骨架内的耗散是与弹性应力联系在一起，而不是黏性应力，这种行为被称为

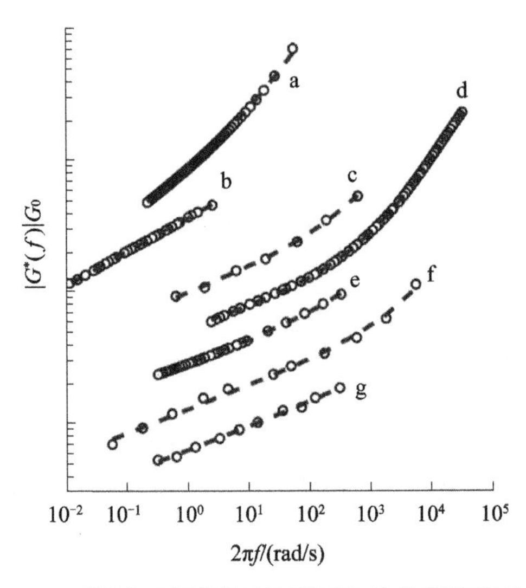

图 6-7-3　不同实验测得的不同类型细胞的幂律流变响应

(a) 内皮细胞，磁牵拉法; (b) 成肌细胞，单轴流变仪; (c) 上皮细胞，原子力显微镜; (d) COS-7 细胞，激光追踪微流变学; (e) 气道平滑肌细胞，磁微粒扭转细胞技术; (f) 气道平滑肌细胞，磁振荡流变仪; (g) 成肌细胞，光镊（引自文献 [50]）

结构性减振（structural damping）。细胞运动和变形时，其摩擦和弹性来自于相同的细胞骨架结构。因此，传统流变学研究中将弹性与摩擦分别归于由少量弹簧和阻尼器构成的黏弹性模型中的特定组成部分，并不能用来描述这种特别的细胞流变学行为。

细胞流变学与尺度无关并不是指时间尺度没有起作用。事实上，特征时间点已经被观察到，但这些时间点并不是对应某个分子弛豫时间或者时间常数，而是它们区分了不同区域行为的过渡。上述高频区域被认为源自单个纤丝的动力学 [48,49]，低频区域的行为虽然也具有普遍性，但对其物理机制目前还存在争议。最近还发现在更低频区域可能存在频率响应更强的区域 [49]。另外，当体外对肌动蛋白与细丝蛋白的交联网络施加预应力时，交联网络表现出细胞骨架相似的流变学行为 [51,52]。

（三）细胞骨架内的扩散行为异常，涨落耗散定理失效

一般采用观察包埋或者绑定到细胞骨架的微粒的自发运动来研究细胞骨架的波动和重构。只有当细胞骨架重组时，牢固锚定在细胞骨架上的微粒才会运动［图 6-7-4(a) 下部］。因此，微粒在时间段 Δt 的均方位移（mean square displacement，MSD）对细胞骨架重构过程提供了有效的测量。如果

均方位移可以被记为 $MSD=D^{*}\Delta t^{\beta}$，那么任何微粒在理论上都可能有三种不同的运动行为。当指数 $\beta=1$ 时，微粒做纯粹的布朗运动，称为正常扩散；当指数 $\beta>1$ 时，微粒表现出持续性（persistence），增量位移，称为超扩散（superdiffusive）；当指数 $\beta<1$ 时，微粒运动表现出反持续性（antipersistence），称为亚扩散（subdiffusive）。

Bursac 等研究发现，绑定在细胞骨架上的微粒的运动行为时断时续，有短时的停顿紧跟着短时的跳跃［图 6-7-4(a) 上部］。分析表明，微粒并不始终遵循某一种运动行为，而是随观察时间段的长短表现不同的运动行为。在短时间段 Δt 内，微粒运动状态近乎停顿，均方位移表现为亚扩散（$\beta\sim0.2$）；在长时间 Δt 内，微粒看似处于大幅度的跳跃运动状态，均方位移表现为超扩散（$\beta\sim1.6$）［图 6-7-4(b)］。从亚扩散到超扩散的过渡时间约为 1s，并取决于热力学温度[53,54]。这种细胞骨架内的反常扩散在许多不同种类的细胞中都被观察到，具有相当的普遍性。

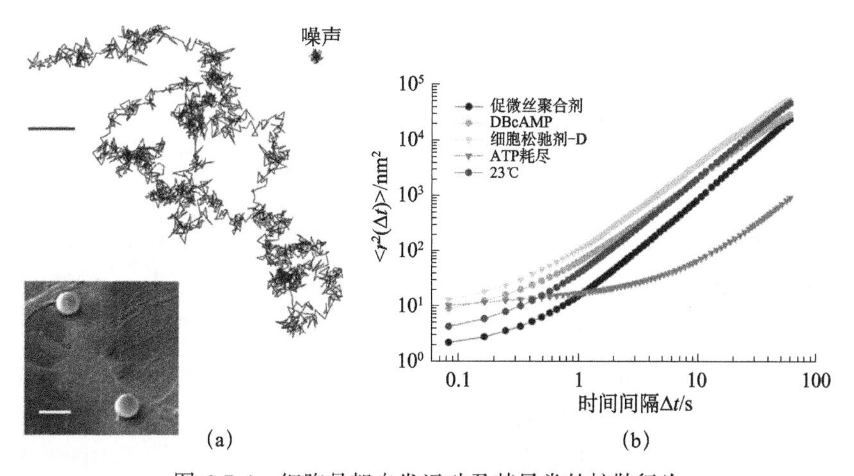

图 6-7-4　细胞骨架自发运动及其异常的扩散行为

(a) 绑定到细胞骨架的微粒及其自发运动轨迹；(b) 绑定到细胞骨架的微粒的均方位移（纵坐标）与观察时间段（横坐标）的关系，在短时间段内为亚扩散行为，在长时间段内为超扩散行为（引自文献 [53]，[54]）

根据 Bursac 等的分析，上述异常扩散行为可能归结到细胞骨架运动的作用力的来源本质。他们认为，如果微粒的自发运动仅仅来源于细胞骨架系统热运动对微粒作用力的波动，那么通过涨落耗散定理（fluctuation dissipation theorem）推导得到的细胞骨架剪切复模量（$G^{*}(f)=G+iG''$）应该与测量得到的均方位移值很接近。但实际测量结果只有在 Δt 很小时才与预测比较接近，说明在这个条件下系统接近热力学平衡状态；当 Δt 较大时，实验测量值与基于热运动推导的预测值的差异逐渐增大，表明细胞骨架系统逐渐远离热力学平衡状态，基于平衡态热运动的涨落耗散定理不再适用。而细胞骨架之所以处于一种远离平衡态的热力学状态，是由于细胞内除热运动能源外还有来源于生物代谢、外界施加的物理作用力等其他能量，这些能量的波动幅度可能远大于热运动的能量波动幅度，从而推动细胞骨架系统发生较大程度的重构现象[53]。

（四）细胞受短暂拉伸后发生快速短暂的流态化

细胞另一个有趣的力学行为是其对外界施加其上的拉伸作用的响应。许多重要的生理器官在生命活动周期中都在经受不断的拉伸应变，这种拉伸可能是持续反复的，也可能是快速短暂的，拉伸应变的幅度也可以从很小到超过 20%[55]。细胞受拉伸后的响应问题从 19 世纪末就已受到关注，但早期的研究中，有的发现拉伸导致细胞硬化，有的则发现拉伸使细胞软化[56]。长期以来，人们将这些研究结果的矛盾简单地归因于实验技术和细胞类型的不同，除此之外对其内在机制莫衷一是。

直到最近，人们才逐渐认识到细胞骨架对于外界施加的拉伸作用的方式是非常敏感的。不同的拉伸方式对同一个细胞可能导致不同甚至相反的响应。细胞骨架这一奇妙力学特性极有可能是其对拉伸作用具有主动和被动两种不同但同时存在的响应机制。对于研究最多的持续拉伸作用，细胞随着持续拉伸逐渐变硬和趋近固态[58-60]。这种响应一部分源自于细胞骨架网络的被动非线性弹性，另一部分源自于拉伸作用触发和/或激活一系列生物化学信号通路的活性，从而导致细胞在数秒内或者数分钟内变硬[14,61,62]。这种硬化行为被认为是维持细胞和组织结构完整性的一种保护机制[63]。相映成趣的是细胞骨架对于短暂拉伸的力学响应。短暂拉伸虽然在生物学上有重要的意义，如深吸气对于正常人和哮喘病人的呼吸功能就有着相反的作用，但过去对其研究较少。Trepat 等[56]、Chen 等[57]和 Krishnan 等[64]分别采用不同的方法对不同的细胞施加短暂的拉伸应变（应变率 =2.5%～10%），并实时地观测了拉伸前后细胞的力学性质及其随时间的变化规律。结果表明，当细胞受到短暂拉伸作用时，细胞骨架发生迅速的流态化反应，即拉伸后瞬间，细胞骨架的硬度迅速下降、摩擦黏度迅速增高。此后，细胞骨架逐渐缓慢地再固化，如图 6-7-5(a) 所示[56]。更重要的是，对细胞骨架的特定信号通路实施干扰并不能改变其对短暂拉伸作用的响应特征，表明细胞对短暂拉伸的响应在很大程度上不是通过特异性的信号转导实现的［图 6-7-5(b)］[57]。此外，短暂拉伸作用还使细胞骨架重构的速率在拉伸结束后迅速加快，进一步向超扩散行为偏离[56]。Chen 等[57]的研究结果表明短暂拉伸导致细胞骨架流态化的内在机制在很大程度上可以归因于肌动蛋白聚合体纤维在受到拉伸时的单纯物理解聚现象，而并不涉及生物化学信号通路的激活。

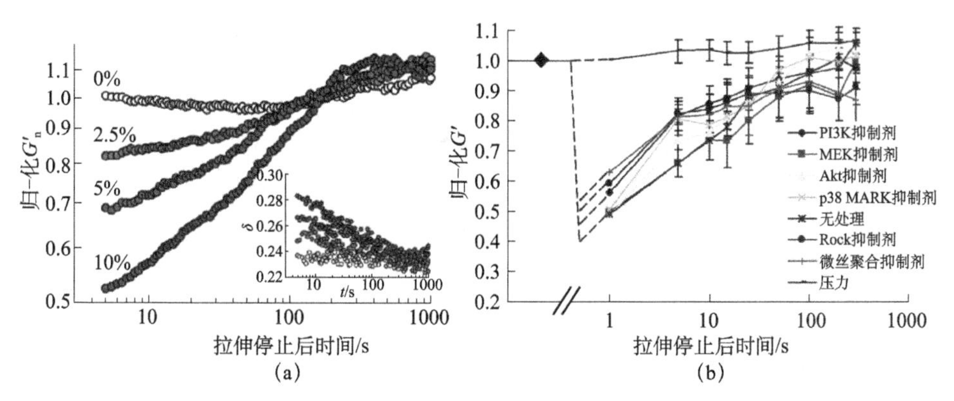

图 6-7-5 短暂的拉伸作用导致黏附细胞流态化：(a) 气道平滑肌细胞在受到持续时间为 4s 的不同幅度的拉伸后，硬度迅速减小然后缓慢回升（引自文献 [56]）；(b) 膀胱平滑肌细胞在受到 10% 的短暂拉伸后，以及在不同药物作用下，硬度迅速减小然后缓慢回升（引自文献 [57]）

三、细胞软物质行为的理论模型

（一）线性黏弹性模型

根据所用的实验方法和探测的时间尺度，细胞展现出类似固体的弹性和类似液体的黏性。这种行为被称为黏弹性。在工程学中，黏弹性材料通常可以被等同描述为胡克弹性弹簧（spring）和牛顿黏性阻尼器（dashpot）。任意的线性黏弹性行为都可以通过利用弹簧和阻尼器的串联或并联来模拟。

弹簧-阻尼器方法的目的是利用力学等同的器件来描述不同的结构元素。在细胞中，不同的弹性和黏性元素可以定义细胞膜、肌动蛋白皮层、细胞骨架等元素。由于早期实验得出的细胞中黏弹性蠕变和应力松弛结果精度有限，包含一个时间常量的简单弹簧-阻尼器模型都足以拟合实验数据[65-67]。随着更精确复杂的实验技术的发展，能够覆盖的时域和频域范围更广，数据精度更高，因此需要引入比弹簧-阻尼器更复杂的细胞力学模型[68]。

（二）幂律流变模型

随着微流变学的时间和频率分辨率的提高，研究者们开始记录更广阔的黏弹性频谱，找寻精确的时间常量来描述细胞力学。在跨多数量级的时间和频率范围内，细胞的黏弹性行为可以通过单一指数的幂次定律来描述。

早在19世纪，关于生物样本的幂律应力松弛就有报道。随着线性黏弹性的发展，用幂律描述应力应变的方法被力学上更直观的具有指数弛豫行为（relaxation behavior）的弹簧-阻尼器模型取缔。但是100多年过后，Hildebrandt发现了幂次定律在描述组织生物力学上的简洁性和准确性[69]。

幂律行为可以最好地通过蠕变实验来解释。在时间 t=0 时，材料受到恒力 F 作用，材料形变 d 随着时间的变化被记录。形变 $d(t)$ 与力 F 的比值就定义了蠕变函数 $J(t)$：

$$J(t)=d(t)/F=j_0^*(t/\tau_0)^\beta \quad (6\text{-}7\text{-}1)$$

在这个公式中，前因子 j_0 代表材料的柔软度（softness）或者屈服度（compliance），即材料硬度的倒数。时间刻度 τ_0 对时间进行标准化，τ_0 的变化不会引起幂律指数 β 的改变。多数时候，材料不是受到一个恒力，而是受到正弦变化的力。在这种情况下，傅里叶变换后的位移与力的比值定义了材料的复模量（complex modulus）$G(\omega)=d(\omega)/F(\omega)$，其中 ω 是角频率。公式（6-7-1）即转换为

$$G(\omega)=1/j_0^*(i\omega\tau_0)^\beta\Gamma(1-\beta)$$

其中 Γ 为伽马函数，$i^2=-1$。

幂律指数非常便捷地描述了材料的黏弹性行为。如果 β 接近0，公式（6-7-1）则简化为遵循胡克定律的弹性形变，所有的形变能量被储存在材料中，当外力撤销后，材料弹簧恢复到受力前的形状。如果 β 接近1，公式（6-7-1）则简化为遵循牛顿定律的黏性形变，材料不能弹性储存形变能量，能量以热能的形式耗散，当外力撤销后，材料仍然保持形变后的状态。当 β 介于0和1之间时，弹性和耗散机制共存。较高的 β 值表示更为耗散的行为，而较低的 β 值表示更为弹性的行为。对于细胞来说，其 β 值一般介于0.1到0.5之间。

幂律流变学和公式（6-7-1）的运用已经通过不同的实验手段在许多类型的细胞中得到了证明[70]。与尺度无关（scale-free）的幂律流变学看似黏附细胞的普遍性规律，甚至在细胞受到作用于细胞骨架的药物刺激后仍然成立。考虑到这种普遍性，公式（6-7-1）可以说是极其简洁明了的经验关系，它通过单一的幂律指数 β 就抓住了数据的本质。

这种行为可能看起来或多或少令人失望，因为如果没有明显的特征时间刻度，就无法鉴别一个主要的弛豫过程。如果所有的药物介导或者说不同细胞间的差异都可以通过单一的幂律指数 β 来表征，那么 β 肯定无法解释特定的分子机制。那么 β 可以解释什么呢？一个可能的答案来自于软玻璃物质（soft glassy materials）理论。软玻璃物质广泛包括了泡沫（foam）、糨糊（paste）、胶体（colloid）、乳剂（emulsion）、料浆（slurry）等。这些物质非常柔软（帕到千帕范围之间），根据公式（6-7-1），它们的幂律指数在 0.1 这个数量级。因为这些材料的多样化，所以它们共同的流变学特征不太可能是反映某个特定的分子机制，而是这一大类物质的结构组成特点。所有的软玻璃物质都是由大量聚集的离散的元素组成的，这些元素之间的相互作用很弱。此外，软玻璃物质远未达到热力学平衡（thermodynamic equilibrium），其内部是无序和亚稳态的。活细胞满足软玻璃物质的所有条件特征，因此可以被归为软玻璃材料。

软玻璃流变学理论认为基质里的每单个元素处于包含不同能量深度的陷阱中。对于活细胞来说，这些陷阱可能是相邻细胞骨架蛋白的亲水作用绑定能量，电荷效应，或者是简单的空间约束。软玻璃物质内的元素无法通过热扰动跳出这些陷阱，它们需要一些其他的能量来源。在细胞里，ATP 被认为能提供能量来完成这种非热扰动[53]。扰动的量级是由一个有效温度或者说是噪声级 x 来表示的。

一个软玻璃物质进行弹性形变，需要它内部的元素保持在能量陷阱内；进行流动，需要元素跳出能量陷阱。一旦元素跳出能量陷阱，元素储存的形变能量就以热能的形式耗散。因此，耗散是跟弹性应力而不是跟黏性应力关联。

随着有效温度 x 的升高，元素更频繁地跳出陷阱再落入更深的陷阱。这种跳跃过程可以被认为是基础的分子重构过程，以及类似液化行为的起源。x 的升高导致系统的融化（melt），x 的降低导致系统的凝固（freeze）。在呈指数分布的能量陷阱中，系统的蠕变响应遵从公式（6-7-1）。公式中的参数定义如下：幂律指数 β 定义为噪声温度 $x-1$；前因子 j_0 定义为系统最低的屈服度（最高的硬度），这个值发生在 $x=1$ 时；时间尺度是一个元素跳出陷阱需要的最短时间。前因子和时间尺度取决于系统的结构常量，而幂律指数 β 反映系统的动力学。

如果 j_0 和 τ_0 是常量，其对细胞行为的影响是显著的。细胞无法独立地改变弹性和耗散属性，只能通过一个特殊的轨迹和通过改变 β 实现。细胞必须变得更类似于固体来增加其硬度；变得更类似于液体来降低其硬度。细胞蠕变响应的实验证实了这一行为。蠕变曲线可能差别很大，但是它们在时间 τ_0 时相交于点 j_0。这一现象已经通过不同的实验方法在很多细胞中观察到。这个行为仍然很神秘，但是它暗示了幂律指数 β 在将所有数据拟合到叠合曲线中起到的重要作用[47]。

（三）细胞骨架预应力模型

细胞软物质力学的第二条定律就是黏附细胞的硬度随着收缩张力的增加而增加[73]。这可以通过一个简单的实验来说明，当用二头肌提重物时，肌细胞收缩得越厉害就会变得越硬。这个行为在除肌细胞以外的其他细胞也能看到。弹性模量 K' 与收缩基调的关系为线性，加上一个很小的偏移 K_0'。

$$K'=K_0'+a\sigma_{\mathrm{p}} \tag{6-7-2}$$

这里收缩基调（tone）表达为细胞的内部应力 σ_{p}[42]。这种内部应力在细胞未受到外部应力作用时仍然存在，因此被称为预应力（prestress）。K' 即为在固定预应力下细胞对小力或者小形变的响应。

软玻璃物质力能并不能解释力的生成，预应力，或者收缩硬化（contractile stiffening）。与胶体、泥浆等不同，细胞并不是一个惰性材料。相反，细胞非常活跃，并能通过肌动蛋白纤丝与肌球蛋白马达的相互作用产生较强的收缩应力。这种相互作用发生在分子水平上肌动-肌球蛋白桥的形成。由 Huxley 提出的 sliding-filament 理论捕捉到了这种行为[70]。Huxley 的理论与 Sollich 的软玻璃物质理论在数学结构上很类似。软玻璃物质的元素可以被看成肌球蛋白马达，能量陷阱可以被看成肌球蛋白与肌动蛋白的绑定能量。两种理论重要的区别是，在 Huxley 的模型中，自由元素不会在平衡位置落入陷阱，因此这些元素会产生一个力。布朗棘轮（Brownian ratchet）的原理与之相同，而且可以运用到 Sollich 的模型中，将惰性的软玻璃物质转变成活跃的力生成材料。Huxley 的模型和活跃软玻璃模型对力与硬度之间的严格线性关系给出了有力的解释。

对于收缩硬化的另一种解释是收缩性的肌动蛋白和肌球蛋白网络的空间排列。当预应力的纤维网络受到横向形变时，它们的表观刚度（apparent stiffness）与应力成严格的正比，这好比拨动的琴弦和支紧帐篷的绳子。细胞流变学的张拉整体模型（tensegrity model）也表达了同样的观点。

细胞力学的两个通用定律，一个证明细胞硬度是幂律指数的函数，另外一个证明细胞硬度是收缩预应力的函数。要保证两个定律同时成立，以下关于预应力、细胞硬度与幂律指数的关系必须成立[71]：

$$\beta=\ln[j_0(K_0'+a\sigma_p)]/\ln\tau_0 \quad (6\text{-}7\text{-}3)$$

根据公式（6-7-3），软玻璃物质流变学中的指数 β 反映了决定细胞骨架预应力的噪声温度，进一步将公式（6-7-1）和（6-7-2）中的通用关系关联起来。对于细胞和交联了肌球蛋白马达的肌动蛋白凝胶，ATP 消耗过程设定了收缩预应力，这同时也是扰动能量和噪声温度 x 的根源。

四、细胞内吞及其力学行为

细胞软物质行为还表现在细胞内及细胞间的相互作用，特别是细胞内吞作用（endocytosis）及其相关力学行为。内吞作用，作为细胞重要软物质行为之一，通过质膜变形运动将细胞外物质转运入细胞内。如图 6-7-6 所示，由于进入细胞物质及进入细胞机制的不同，将采用不同路径实现内吞。一般微米尺寸的生物粒子等通过吞噬作用[72]（phagocytosis）或巨胞饮作用[73]（macropinocytosis）进入细胞。吞噬作用［图 6-7-6(a) 中 A］主要通过形成杯状突出吞噬凋亡细胞、病原体及细胞残骸；巨胞饮作用是肌动蛋白调控过程，通过细胞膜褶皱将大量流体及生物粒子内吞至细胞［图 6-7-6(a) 中 B］，一旦闭合，就形成胞饮泡。肌动蛋白运动在吞噬作用或巨胞饮作用过程中起到显著作用[74]。网格蛋白（clathrin）调控的内吞过程［图 6-7-6(a) 中 D］，受体-配体结合导致质膜在细胞质一侧出现包被网格蛋白的凹陷[75]，然后聚集为有助于内吞的多边笼状结构，同时网格蛋白也参与内吞后期囊泡颈部形成及夹断过程，网格蛋白调控的内吞过程对于病毒粒子比较多见[76]。对于小窝蛋白（caveolin）参与的内吞［图 6-7-6(a) 中 C］，发卡结构的小窝蛋白使质膜在细胞质一侧形成直径为 50～80nm 的长瓶状质膜微囊[77]。同时，网格蛋白、小窝蛋白调控的内吞过程涉及复杂的化学信号转导级联反应[78]。对于不依赖于网格蛋白、小窝蛋白的内吞过程[79]，受体-配体结合提供了内吞驱动力［图 6-7-6(a) 中 E］，一般修饰共轭聚合物的生物粒子通过非特异相互作用实现内吞［图 6-7-6(a) 中 F］，直接跨膜穿透［图 6-7-6(a) 中 G］主要针对较大粒子，一般需要提供较大作用力，但会对细胞膜造成损伤[80]。

伴随纳米技术发展，内吞作用及相关机制将指导纳米粒子输运及设计。通常，在携带造影剂或相关治疗药物的纳米粒子表面修饰相应配体，保证纳米药物粒子与具有特定受体的靶细胞结合，将药物输运至病灶区域，提高药物吸收，改善治疗效果或者提高造影质量。如图 6-7-6(b) 所示，目前纳米粒子内吞主要取决于纳米粒子尺寸[81]、形状[82]、表面化学修饰及硬度[83]、带电情况[84]及所处微环境等因素。

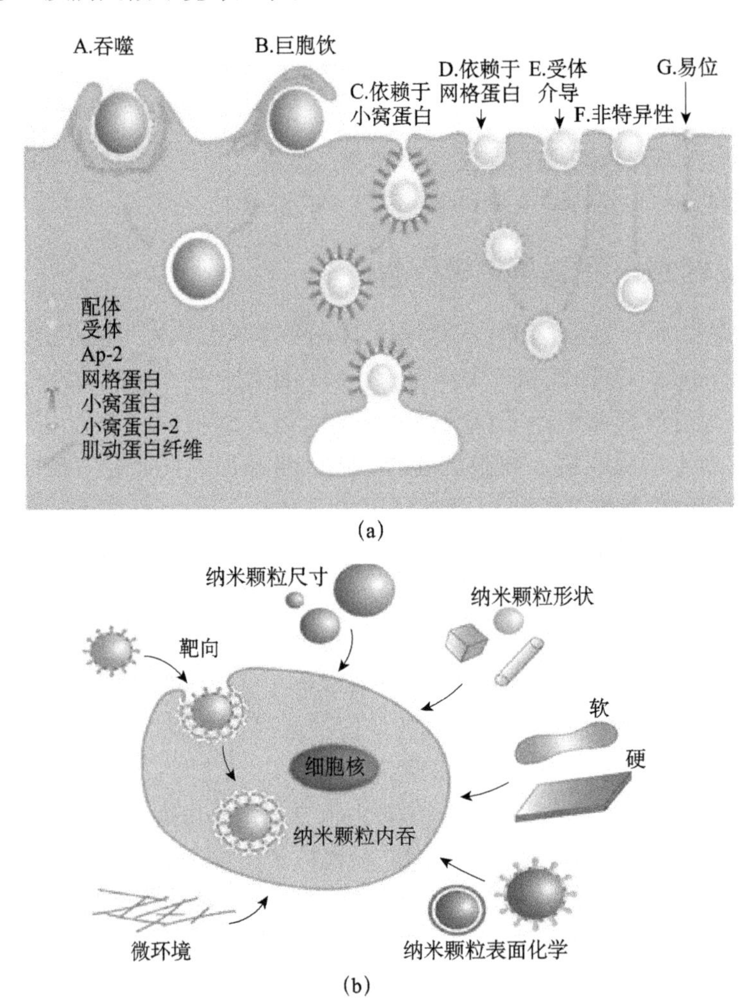

图 6-7-6 内吞作用示意图（a）及影响生物粒子内吞的主要因素 (b)

（一）纳米粒子尺寸与形状等对细胞内吞的影响

纳米粒子尺寸、形状等因素对纳米粒子胞吞效率有显著影响。实验发现，球形纳米粒子的胞吞效率依赖于纳米粒子的尺寸[85]。实验利用 transferrin 蛋白修饰球形纳米粒子，形成“蛋白冠”，纳米粒子直径分别为 14nm、30nm、50nm、74nm 及 100nm，相对于其他直径，直径为 50nm 的纳米粒子胞吞效率最高［图 6-7-7（a）、（b）］。同样，实验发现纳米粒子的胞吞效率依赖于纳米粒子形状。柱形纳米粒子（中间为圆柱，两端为两个半球，同样利用

transferrin 蛋白修饰）胞吞效率低于球形纳米粒子胞吞效率，并依赖于柱形纳米粒子的长度-直径比。如图 6-7-7（c）所示，直方图对应的长度-直径比分别为 1∶1、1∶3、1∶5，长度-直径比为 1∶1 时，为球形纳米粒子，故前两个直方图分别代表直径为 14nm、74nm 的球形纳米粒子，后两个直方图代表直径为 14nm，长度分别为 40nm、74nm 的柱形纳米粒子，显然，随着长度-直径比增加，柱形纳米粒子的胞吞效率依次降低。

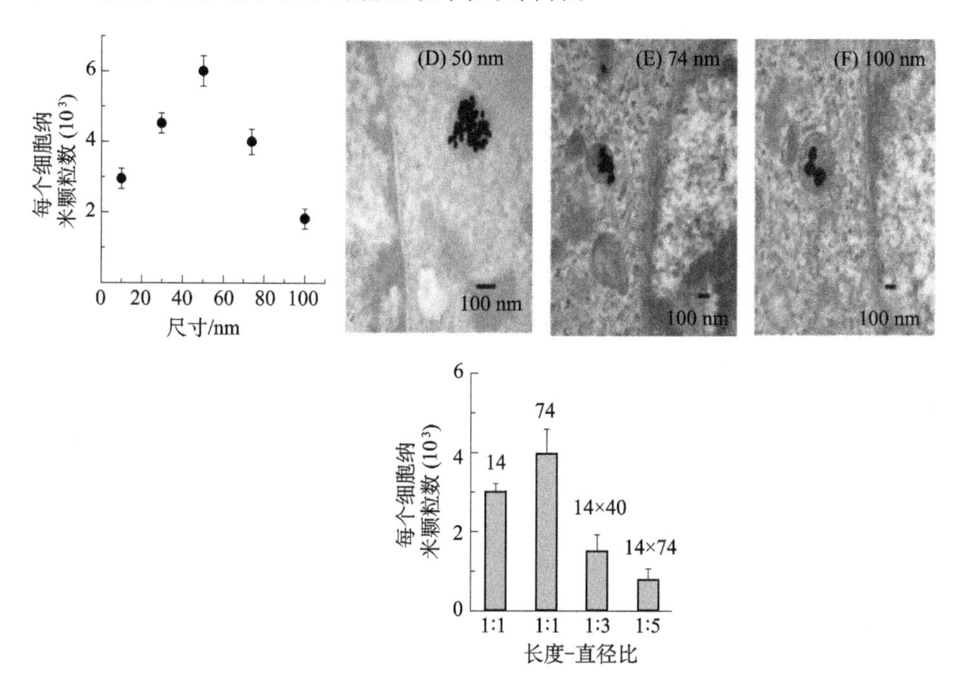

图 6-7-7 球形纳米粒子胞吞实验：胞吞纳米药物粒子数目与纳米粒子直径、形状的关系（图片取自 Chrithrani B D, Ghazani A A, Chan W C）

（二）纳米粒子表面化学对细胞内吞的影响

纳米粒子是否能够被内吞入靶细胞取决于纳米粒子表面的修饰配体，称为“蛋白冠”[86]。“蛋白冠”一般有两种形成方式，一类是纳米粒子通过消化道等进入血液，并吸附血液中的血清蛋白，形成“蛋白冠”[87]；另一类是通过体外化学修饰，使配体蛋白吸附于纳米粒子表面形成“蛋白冠”[88]。无论采用哪种方式，“蛋白冠”均会改变纳米粒子尺寸及其表面组成。纳米粒子利用表面“蛋白冠”进入靶细胞与病毒粒子胞吞过程一致。如图 6-7-8 所示，首先是纳米粒子的扩散与识别，其次是识别后的胞吞过程。无论是识别过程还是胞吞过程，纳米粒子“蛋白冠”中的配体蛋白与细胞膜表面的受体蛋白发生相互作用。因此，配体-受体结合是实现纳米粒子胞吞过程的主要途径[89, 90]。

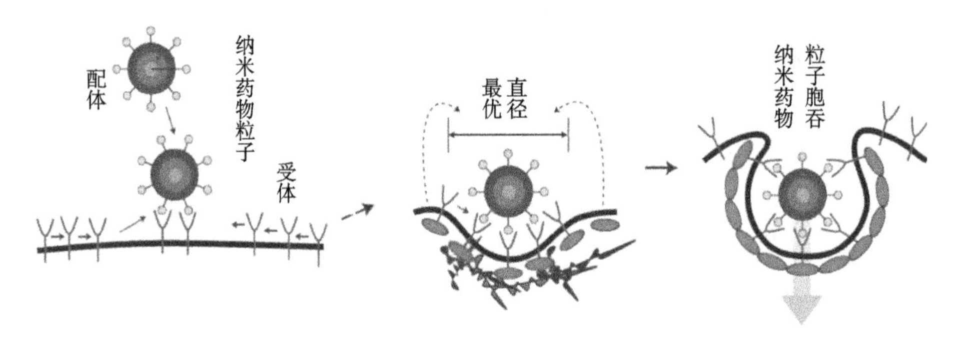

图 6-7-8　纳米药物粒子胞吞过程示意图

（三）纳米粒子硬度变化对细胞内吞的影响

纳米粒子硬度变化影响细胞对其内吞的效率。如图 6-7-9 所示，在 HIV（human immunodeficiency virus）入侵细胞的过程中，病毒粒子硬度发生显著变化，病毒粒子未成熟阶段的硬度约为成熟阶段硬度的 14 倍，而且未成熟病毒粒子稳定性依赖于病毒粒子的两个功能域 Evn 和 CT。在二者分别缺失的情况下，未成熟病毒粒子硬度显著降低。从未成熟到成熟阶段，病毒调整其自身硬度以提高感染效率[91]。

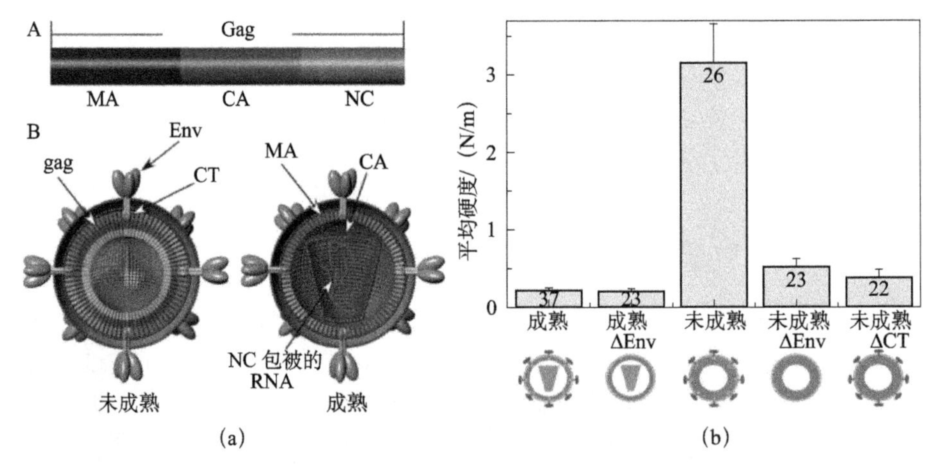

图 6-7-9　(a) HIV 未成熟/成熟阶段结构示意图；(b) HIV 不同阶段硬度（参见文献 [91]）

以上为主要影响内吞的纳米粒子方面因素，除此以外，生物膜表面性质也会影响纳米粒子内吞。生物膜与纳米粒子相互作用可以归纳为有利或不利于内吞的两类作用力（表 6-7-1）。

表 6-7-1 纳米粒子内吞过程涉略的相互作用 [92]

有利于内吞的作用力	不利于内吞的作用力
特异性结合：受体-配体相互作用	细胞膜的拉伸及弹性
非特异相互作用：粒子表面特性	细胞膜的热涨落
接触位点自由能降低	极性表面疏水相互作用
最优粒子尺寸及形状	受体向黏附区域扩散
能量相关的膜及细胞骨架组分，原动力	受体-配体复合体拉伸及弹性

（四）内吞过程所处的微环境对内吞效率影响

基于荧光标记和量子点技术的细胞实验 [93] 指出，配体-受体复合体尺寸影响生物膜间的相互作用并决定生物膜间的生物粒子尺寸（图 6-7-10）。在生物膜间的荧光粒子（黑色）直径大于配体-受体复合体长度的情况下，导致生物膜变形，相邻配体-受体复合体之间彼此排斥，导致荧光粒子及配体-受体复合体重新分布（图 6-7-10，Ⅰ区）。若两生物膜间荧光粒子直径小于配体-受体复合体长度，则无以上情况（图 6-7-10，Ⅱ区）。明显发现配体-受体尺寸对两生物膜间的荧光粒子有选择性，即两生物膜间只可以容纳小于配体-受体复合体长度的粒子，这种选择性实际体现的是生物膜间的排空效应（depletion effects）。

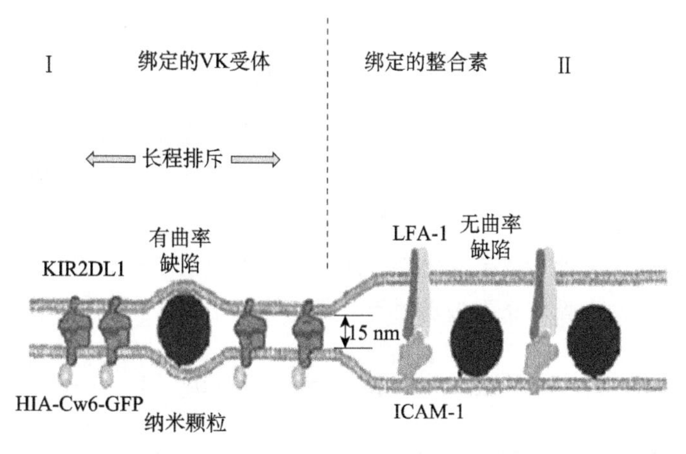

图 6-7-10 排空效应的量子点实验示意图

由此可以认识到，纳米粒子胞吞进入细胞的过程，通过纳米粒子表面

“蛋白冠”的配体与细胞膜表面受体结合实现，而配体-受体复合体尺寸可与微环境中生物粒子尺寸相比拟，对细胞膜间的粒子具有选择性。实验表明，不同种类配体-受体复合体的长度处于 13～42nm 范围内[94]。例如，在 T 细胞黏着斑区域，几类非常重要的配体-受体复合体长度：TCR-MHCp 复合体长度为 13nm，LFA1-ICAM1 复合体长度为 42nm。因此，有必要考虑配体-受体尺寸在纳米药物粒子胞吞过程中的效应及排空效应与受体-配体尺寸之间竞争性对内吞过程的影响。

（五）纳米粒子内吞的理论模型

目前，关于内吞过程的主要理论模型可以分为两类。第一类模型（图6-7-11）着眼于内吞过程的绑定区域[95]，主要考虑特异性相互作用（受体-配体结合）作为主导因素克服内吞过程中纳米粒子导致生物膜的形变能（包括弯曲、拉伸）及压缩能（来源于细胞骨架形变）。

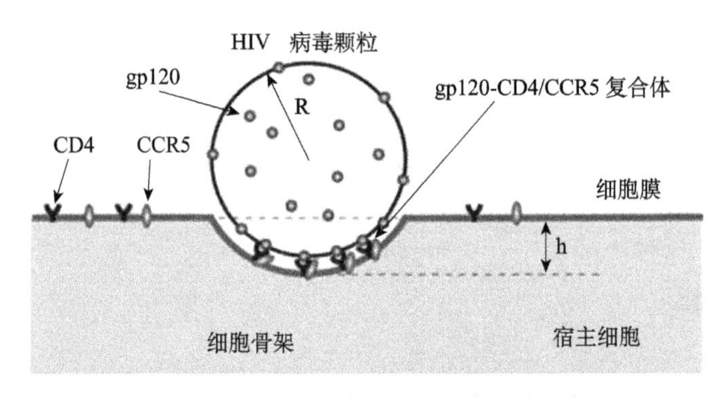

图 6-7-11　HIV 粒子内吞过程黏附区域示意图

这类工作目前对形变能及压缩能在内吞过程中哪个占优势存有争议，Sun 等[95]指出，内吞过程中压缩能占优势，而 Wang 在其工作[96]中指出，内吞过程中存在临界的内吞深度（h），当内吞深度小于临界深度时，形变能相对于压缩能占优势；相反则压缩能占优势。这一类工作忽略了细胞骨架的不连续性，将细胞骨架、生物膜体系看成弹性半空间，忽略细胞内吞过程中的非线性形变，即压缩能（E_3）采用 $E_3=\dfrac{2\sqrt{R}}{5\mu}h^{\frac{5}{2}}$ 表示，其中 $\mu=\dfrac{3}{4}\left(\dfrac{1-\sigma_1^2}{\varepsilon_1}+\dfrac{1-\sigma_2^2}{\varepsilon_2}\right)$，$\varepsilon_1$、$\varepsilon_2$ 分别为生物粒子和生物膜的杨氏模量，σ_1、σ_2 分别为生物粒子和生物膜的泊松比，对于弹性半空间，压缩能（E_3）一般只使用于小形变情况。

第二类模型充分考虑了受体-配体相互作用及内吞过程的一些重要特征[97]。首先，与非特异性相互作用不同，受体-配体结合将会导致时间延迟，即内吞过程中细胞膜上受体需要向黏附区域扩散，决定生物粒子内吞必然存在特征时间尺度；其次，受体-配体结合是不连续过程，由受体-配体结合而导致的化学能释放要求每个配体所占据的面积被生物膜包被。因此，在两个相邻绑定事件之间必然存在生物膜形变导致的势垒，包被在生物粒子表面的配体密度调谐每一步绑定配体所占面积及势垒。势垒将进一步延迟生物粒子内吞，一般来说，在配体密度足够大的情况下，该时间延迟相对于受体扩散导致的时间延迟是可以忽略的；最后，受体-配体结合不仅在黏附区域提供了黏附力，受体在向黏附区域扩散过程中提供了平移熵。因此，生物粒子内吞不是一个局部事件，其关乎整个细胞，导致细胞膜表面受体再分布。

前面提到 HIV 粒子在内吞过程中调整病毒的硬度，实际上是为了提高病毒粒子的内吞效率。如图 6-7-12(a) 所示，关于软生物粒子内吞的理论工作证实其内吞过程依赖相对硬度[98]，即内吞粒子硬度与生物膜硬度的比值，内吞过程主要有三种状态：部分内吞，完全内吞，不内吞。如图 6-7-12(b) 和 (c) 所示关于膜表面张力 $\bar{\alpha}$ 及黏附能 $2\alpha R^2/B$ 的相图，其中，α 为黏附能密度，B 为生物膜弯折刚度。随着软粒子硬度的降低，部分内吞与完全内吞之间的转变界限逐渐下移，表明软粒子完全内吞越来越困难；相反，硬粒子则容易完全内吞。软粒子之所以难于被内吞，如图 6-7-12(a) 所示，软粒子一旦锚定在生物膜上，则被生物膜浸润，在生物膜上完全铺展，而且浸润角很大，导致大曲率出现，若被内吞，生物膜则需要较大的弯折能克服较大势垒。

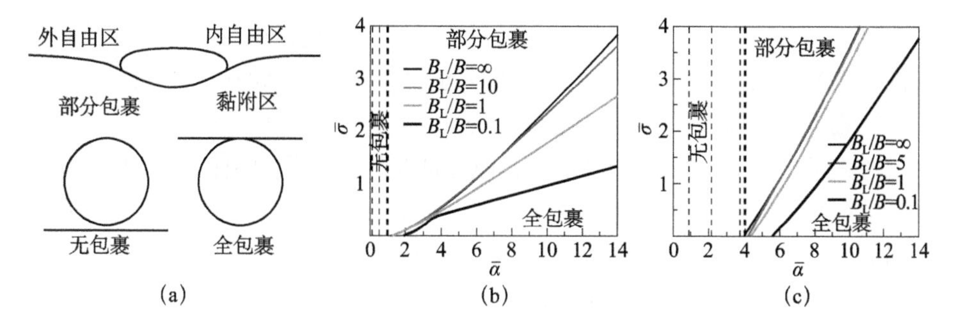

图 6-7-12　不同内吞状态示意图 (a) 及在相对弯折刚度（软粒子弯折刚度与生物膜弯折刚度比值 B_L/B）取不同值时利用黏附能和膜张力表征的内吞相图 (b)、(c)；(b) 二维情况；(c) 三维情况。虚线：部分内吞与没有内吞之间的边界；实线：部分内吞与完全内吞之间的边界，参见文献［98］

五、细胞软物质力学的学科前沿问题

（一）细胞的被动应力硬化

如果外加应力或者应变很小，细胞的黏弹性响应就可以认为是线性的，即不受应力或者应变大小的支配[99,100]。之前讨论的研究大部分是在线性区域进行的。在某些情况下，偏离线性的行为也被观察到，但这种偏离不足以重要到否定线性响应理论[31]。只有很少一部分研究报道了细胞硬度依赖于外部应力的行为[66,101]，但这些应变硬化后来被发现归因于基于微球的微流变学[32]。

Fenandez 对细胞的非线性黏弹性进行了全面研究，首次探索了生理范围内细胞骨架应力造成的细胞力学响应。相同的响应在体外构建包含有肌动蛋白和交联蛋白 filamin A 的细胞骨架网络也被观察到。这些结果说明细胞和细胞骨架聚合物是一种应力硬化材料，内部应力控制着材料硬度。

重构的细胞骨架网络可以通过调节 ATP 和交联剂浓度来形成收缩性材料。利用该系统，研究者发现，材料主动生成的应力和外部施加应力对影响材料微分刚度具有相同作用。大量活细胞的研究数据表明，不同程度的应力硬化可以拟合到一个主曲线，细胞骨架的总应力（预应力与外部应力之和）是调节微分刚度的唯一参数。这个主曲线是对公式（6-7-2）的概括：

$$K'(\sigma)=\delta\sigma/\sigma\gamma=K'_0+a\sigma \qquad (6\text{-}7\text{-}4)$$

其中，K' 是微分刚度（differential stiffness）；σ 是预应力 σ_p 与外部应力 σ_e 之和。公式（6-7-4）描述了活细胞和细胞骨架网络的通用属性。此外，公式（6-7-4）解释了为什么具有高收缩预应力的细胞与柔软细胞相比在相同外力下产生较低的应力硬化：因为高预应力的细胞只能产生较小比例的应力增加。

公式（6-7-4）的积分生成指数的应力-应变关系。大应变对应的指数级硬化对大部分普通材料是不寻常的，但是却是许多生物材料的普遍属性，特别是结缔组织。利用微孔板流变仪研究固定的细胞（防止细胞在大应变下发生屈服），观察到指数级硬化。这种普遍的应力或者应变硬化指向一种在多长度尺度适用的结构机制，而不是某种特定的分子机制。

当把细胞与体外网络的应力硬化进行比较时，需要考虑到细胞能展现主动的力感知响应。细胞能感受外力，并将力学信号转导成生物化学信号，引起细胞骨架的重构和强化（reinforcement），导致力学性质的显著改变。一些交联蛋白，例如 filamin A 在构建的细胞骨架网络中起到重要的被动响应作用，却在活细胞中表现出主动的力学感知和调节作用，这让两者的比较变得更复杂。对于主动的力化学响应，通常可以通过振荡载荷（oscillatory loading）或者内锁（lock-in）技术将其过滤，或者把测量时间缩到足够短来最小化主动响应的影响，而细胞的低频或者长时间被动行为却非常难以分析。

（二）拉伸导致的细胞流态化及其恢复

如果我们将预应力 σ_p 替换为整个细胞骨架网络的应力 $\sigma_p+\sigma_e$，细胞硬度与预应力的关系［式（6-7-2）］就可以从线性区域扩展到非线性区域。同样的方法对公式（6-7-3）不适用，因为这样推测出的结果是当细胞随着应力增加而硬化时会变得更类似固体。而实际的情况恰恰相反，细胞对大应力的响应表现为流态化。

这种情况的出现并不是完全出人意料。如果细胞如上面所讨论的，属于软玻璃类物质，那么机械拉伸则会注入能量，引起基质的扰动，升高噪声温度，从而导致内部元素更频繁地跳出能量陷阱，引发流态化，这已被实验证明[56]。但这里需要阐明的是，流态化发生在应力加载的过程中，被同时发生的硬化行为掩盖。而且重要的是，细胞的流态化程度直接取决于其在拉伸前的骨架预应力。硬的收缩性强的接近固体的细胞的流态化程度，高于软的收缩性弱的接近液体的细胞。

（三）细胞力学响应模型

为了研究细胞硬度和黏弹性的物理起源，近年来出现了一系列的细胞力学响应模型。当然，每一种模型的产生都是为了解释一系列特定的细胞力学的实验结果。鉴于不同实验方法之间缺乏统一性，寄希望于一个单一的模型能够解释所有的实验数据和结论是不切实际的。因此，不仅仅是从单一的实验着手，新的模型需要能够描述更广义的细胞力学统一性。目前面临的一个特殊挑战是，所创建的物理模型既要足够粗放，以能够描述相关的分子水平的网络或者细胞亚结构，又要足够详细，以具有力学上的意义。

目前主要研究的有以下几种模型。

1. 溶胶-凝胶模型（sol-gel hypothesis）

对于细胞的黏弹性，一个很富有吸引力的解释是细胞是经受溶胶-凝胶转换的动态系统[102,103]。比较短的细胞骨架聚合物或者浓度较低的交联蛋白没有彻底地机械连接起来，它们对应力的响应很像液体对应力的响应。这称为溶胶态（sol state）。随着纤丝长度增长和/或者交联蛋白浓度的增加，表现出弹性的凝胶态（gel state）最终形成。在细胞迁移中，这种模型表明，细胞通过解聚肌动蛋白纤丝来爬行，通过聚合肌动蛋白纤丝来停止运动。然而，与细胞不同[29,53]，溶胶和凝胶是平衡态的材料，而且只受到热扰动（thermal fluctuations）[104]。

在这个模型中，细胞的力学性质应该与纯化的细胞骨架成分的网络的力学性质对应。这一理论促进了大量关于纯化的生物聚合物系统的研究[105-107]，通过分析纯化系统的流变学，我们可以预测许多细胞力学量。

尽管生物聚合物，特别是肌动蛋白网络的流变学特性和细胞的流变学表现出了一些相似性，它们之间仍然存在很大的差异。与细胞相比，肌动蛋白聚合物[107]、波形蛋白（vimentin）[108]和微管[109]都在低频下显示出了一个弹性稳态（elastic plateau），$\left|G^*(\omega)\right|\sim\omega^0$。此外，肌动蛋白在中间频率区域显示出了在细胞中没有观察到过的尺度依赖性[110]，$\left|G^*(\omega)\right|\sim\omega^{1/2}$。然而，肌动蛋白和微管在高频下的流变学实验都显示$\left|G^*(\omega)\right|\sim\omega^{3/4}$，这与在细胞中观察到的一致[35,48]。此外，现已有广泛的理论研究来理解这些网络的力学的分子机制，但是由于在细胞内没有观察到纯弹性区域，并且$\left|G^*(\omega)\right|\sim\omega^{3/4}$是在生理上不相关的时间尺度上（<1ms）观察到的，在这些时间尺度下的信息可能包含细胞骨架微结构的信息和细胞内分子马达（molecular motors）的动力学。

交联的肌动蛋白凝胶的力学特性与细胞的力学响应有更接近的相似性。有趣的是，非结构化的交联，比如肌动蛋白与亲和素（avidin-biotin）[111]或者 scruin[14]的交联在低频显示出纯弹性响应，在高频的响应为$\left|G^*(\omega)\right|\sim\omega^{3/4}$。与之相比，生理的交联蛋白比如细丝蛋白（filamin）[52]，高浓度的辅肌动蛋白（actinin），或者两者的混合都显示出幂律流变$\left|G^*(\omega)\right|\sim\omega^{\beta}$，和在细胞中观察到的很相似，只是幂律指数的取值范围更低，$0.05<\beta<0.15$。这个弱幂律的生理起源并不清楚，可能与交联蛋白的动力学有关。

2. 张力整体模型（tensegrity model）

tensegrity 是由 R. Buckminster Fuller 提出的一个著名的建筑学概念。它的理论是采用承受拉伸力的材料设计的建筑比采用承受压缩力或者剪切力的材料设计的建筑更具有力学稳定性。对于 tensegrity 结构的一个定义为，如果其力学结构元素中没有预应力，那么该结构在承重时不能维持形状的稳定性[38]。

tensegrity 模型提出，整个细胞是预应力的结构[37]。应力张力可以通过肌动蛋白-球蛋白网络产生，通过细胞黏着斑产生，通过细胞与细胞的粘连以及细胞骨架组成部分的聚合产生。通常情况下，拉伸力由肌动蛋白[42]和中间纤丝网络[112]支撑，而压缩力由微管和黏着斑支撑。与这一论断一致，已有的研究发现，当受到压缩力作用时，细胞里的微管将会褶皱[36,113]；而当细胞受到拉伸力时，用激光切断肌动蛋白应力纤维，它们将会回缩[114]；在外力比较大时，细胞将会硬化。更进一步，tensegrity 对于细胞的预应力和细胞的总体硬度呈线性关系这一定量推断已经得到实验结果支持。

尽管细胞的力学性质很大程度上建立在预应力之上已经得到观察和证实[71]，但 tensegrity 是一个宏观的模型，它不能解释细胞水平上的热扰动，也不能预测细胞力学的频率相关性。此外，细胞核周围区域的力学性质被证实完全与肌动蛋白或者球蛋白无关，这也和 tensegrity 模型的大前提相矛盾。tensegrity 模型和相关支持该模型的实验结果并不能提供一个完整的细胞力学模型[115]。

3. 软玻璃态流变模型（soft glassy rheology model，SGR）

从理论上讲，许多软的材质，比如泡沫、乳剂、糨糊、泥浆等都具有非常相似的流变学特性。这些材料的一个共同特征是它们的微观组成结构不能足够自由地移动来达到热力学平衡，它们被称为软玻璃态物质（soft glassy materials）。这个模型最早是由 Sollich 在 1998 年提出来描述这些软物质的。2001 年，Fabry 等通过磁微粒扭转细胞实验发现 SGR 模型也可以用来描述细胞。将 SGR 模型运用到细胞最初始的原理是因为它可以解释在细胞内观察到的幂指数定律，并且可以将不同细胞的流变学性质归纳成同一条流变学曲线。从那以后，与软玻璃态物质相关的其他现象也陆续在细胞力学实验中得到观察证实，进一步巩固了 SGR 在细胞生物物理学领域的地位。

软玻璃态物质的应力松弛主要是通过它们内在的挤得很紧的质点或者泡沫的重排列实现的。在这样一个密集拥挤的微环境，其内部的重排列需要比自身热能高出几个数量级的能量 E，这里假设重排列所需的能量由其他能量源提供，并且由变量 x 调控。这里 x 是被该材料的平均陷阱能量深度 E_0 均一化的有效驱动能量。如果能量源衰减到小于某关键值（$x=1$），应力松弛将急剧减弱，而且该系统达到动态平衡所需时间函数像达到玻璃相变（glass transition）的过程方式一样发散。如果 x 的取值范围高于玻璃相变（$1<x<2$），动态平衡就会受到阻碍而很难达到，因此导致很长的松弛时间和相应的幂定律流变学，$\left|G^*(\omega)\right|\sim\omega^\beta$，其中 $x=\beta+1$。

在细胞里，x 的值被假定是由 ATP 相关的非热过程决定的，比如肌动蛋白聚合和肌球蛋白的运动。许多研究者已经发现，抑制或者促进肌动蛋白聚合或者肌球蛋白活性可以改变一些细胞类型的流变学指数。由于 SGR 缺乏细胞分子水平上的细节阐述，它很难将流变学性质和已有的生物化学领域的实验数据对应和匹配。

许多细胞流变学实验在抽空 ATP 后，并没有表现出流变特性的显著改变，这和肌球蛋白活性和流变学指数 β 或 x 的密切联系不一致。SGR 模型还主张应力松弛，这一点也和观察到的预应力增大而导致的细胞硬化矛盾。

4. 动态交联的主动胶体模型（active gels and dynamic cross-links）

最近，一个由纯化的肌动蛋白、细丝蛋白和外加张力的体外细胞骨架模型被创造出来，其产生与细胞非常相似的力学行为，包括弱幂律流变，应变硬化行为和细胞流变学，施加应力，流变的频率响应三者之间的联系。然而，实验结果显示这种模型不能完美再现细胞骨架的力学性质[116]。因此，借鉴这个模型系统，提出了一个基于外力导致的交联蛋白的解折叠（unfolding）的简单模型。在这个模型里，随着缓慢变形的施加，许多交联聚集在解折叠尖的张力处[117,118]，从而系统进入非常态。动力学蒙特卡罗模拟（Kinetic Monte Carlo simulations）显示观察到一些在基于交联蛋白解绑定（unbinding）的模型中无法观察到幂律动力学行为[119]。

在这个领域内一个持续的争论是细胞骨架交联蛋白的解折叠与解绑定，哪一种对细胞的力学性质影响更大。单分子测量显示，解折叠如辅肌动蛋白[120]或者细丝蛋白[121]这类蛋白质的结构域需要的力的大小为 50～200pN，而解绑定需要的力的大小为 1～40pN[122]。这导致许多作者认为交联蛋白的解绑定应该主宰细胞的力学性质[123,124]。然而，最近的一系列实验显示，当拉动率（pulling rate）匹配时（在 5～50Pn/s），解绑定和解折叠细丝蛋白和辅肌动蛋白所需力的大小差不多，两种现象都能在实验中观察到[125]，尽管在低拉动率下解绑定更容易发生。解折叠现象看似是在高拉动率时占主导，或者在低伸展率，绑定作用被加强时占主导。对钙结合蛋白 calponin，辅肌动蛋白和肌动蛋白的纯化系统的流变学测量发现，calponin 可以增强辅肌动蛋白与肌动蛋白的相互作用[126]。最近的一篇论文显示，低至 2pN 的力可以暴露出人踝蛋白（talin）的隐秘结合位点（binding sites），便于黏着斑蛋白（vinculin）与之结合[127]。这些数据表明细胞内交联蛋白的解折叠仍然是一种可能性很大的调控机制。

这个模型似乎将那些经典模型的关键方面结合起来了。像溶胶-凝胶模型一样，它是基于半柔性聚合物和交联蛋白的。幂律动力学仅仅在高应力/应变的情况下观察到，因此预应力是一个很关键的变量。在无序网络里，力的分布可能导致弛豫时间的广泛分布，从而产生幂律流变学，这种解释似乎也是很合理的。

包含有肌球蛋白马达的交联肌动蛋白凝胶也提供了一个有趣的模型系统。在没有交联蛋白的纯肌动蛋白网络里，分子马达加强了纤丝的滑动，从而导致更多的流化力学性质[128,129]。在交联非常强，基本上是不可逆的交联蛋白的肌动蛋白网络里，肌动运动导致应变硬化的力学性质[130]。这可能是因为聚合物纤丝的熵波动（entropic fluctuation）被引发了。这个效应类似于对肌动蛋白凝胶施加外部应力，也能观察到明显的应变硬化。在细胞骨架里，肌动运动（motor activity）可能会被可逆的交联抵制，但是到目前为止还没在实验中和理论上研究这一情形。不过，合乎逻辑的解释是动态的交联可以支撑小的应力，导致应变硬化，而大的应力加速了交联的解绑定或者解折叠，导致流化或者应变。

最近发表的有关应变脉冲影响细胞力学行为的文献的争论核心在于施加力和动态交联的关系。如果施加步进应力，可以在细胞中和聚合物网络中观察到应变硬化，这和细胞力学的 tensegrity 模型和 sol-gel 模型预测的一致。然而，当施加三角应力时，观察到的是与肌动蛋白-肌球蛋白凝胶性质相似的，与软玻璃流变一致的流化。此外，当用单轴流变仪（uniaxial rheometry）对细胞施加三角应变时，类似于应变硬化和流化的效应都有发生[131]。然而，流化效应［也被称为黏塑性（viscoplasticity）］被细胞固定抑制，因为细胞固定抑制交联蛋白动力学，而应变硬化未受到细胞固定的抑制。因此，应力软化和应变硬化的明显差异可能通过部分交联抵抗力，促进应变硬化，而其他交联解绑定/解折叠促进应力软化来解释。

虽然还没有完全发展，纯化的生物聚合物模型似乎能够再现细胞力学的许多现象，不过仍然还有大量的未完成的工作摆在我们面前。

5. 其他物理模型

最近又有两种新的细胞力学物理模型被提出来。这两种模型暂时还不能与细胞的低频效应关联，但是将来可能会在细胞力学领域得到相应的运用。

第一种模型受启发于高分子物理学中的一个常见模型，即普通蠕虫状链（worm-like chain）的非线性效应。明确地讲，热焓效应（enthalpic effect）导致化学键之间的非线性弹性，当整个链的硬度增加时，就会引起应变硬化，并且当施加应力时，整个链的流变学的频率相关曲线趋于平坦。类似的响应在细胞内也被发现。这个模型是热能驱动的，阐明了 tensegrity 模型中的预应力这个概念和聚合物动力学（polymer dynamics）之间的潜在联系。但是，该模型目前的表述并不能支持幂定律流变学。

第二种模型叫作软玻璃态蠕虫状链（glassy worm-like chain）模型[132]。它是基于生物聚合物本身是黏性的，并且在张力作用下，它们之间相互松弛滑动产生相互作用这样一个前提。和普通的蠕虫状链模型相比，软玻璃态蠕虫状链由于聚合物之间摩擦增大，而其松弛状态减缓，很接近类似细胞的行为，如反常扩散（anomalous diffusion）和低频幂定律流变学。该流变学的频率响应是由链内的相互摩擦的强度决定的。这个模型也是热能驱动的，并且主要用来解释由一些弱绑定的交联引起的细胞现象。

（四）细胞内吞的排空效应

最近有工作基于第一类细胞内吞理论模型，在引入受体-配体尺寸效应及排空效应（depletion effects）之间竞争关系的情况下，指出内吞球形生物粒子最优尺寸依赖于受体-配体复合体尺寸，对于直径为 50nm 的生物粒子，对应的受体-配体复合体长度为 40nm。如图 6-7-7 所示，表面修饰 transferrin 蛋白的纳米金颗粒，直径为 50nm 的纳米金颗粒内吞效率最高，而且在排除误差的情况下，动态光散射实验证实纳米金颗粒表面修饰的 transferrin 蛋白尺寸为 38.2nm，与理论值 40nm 完全相当。而之前的理论模型、分子动力学模拟均没有考虑受体-配体尺寸效应及排空效应之间的竞争关系，这对于生物粒子内吞过程是值得再考虑的问题。

理论建模中的第一类模型，将细胞膜（包括细胞骨架部分）看成弹性半空间，而第二类模型则利用生物膜替代细胞膜，通过调整其弯折刚度反映细胞骨架网络在内吞过程中的抵抗作用，显然两类模型均忽略了细胞骨架网络的不连续性。细胞骨架是由微管、中间丝、微丝组成的，其中由交联蛋白连接微丝构成的网络主要分布在细胞膜内侧，微丝形成的网络结构不规则，相邻交联蛋白之间的距离处在 100nm 到 150nm 之间，而且最近基于原子力显微镜研究细胞力学实验表明，细胞弹性模量表现显著的各向异性，这说明有必要在研究内吞的过程中仔细考量细胞骨架网络。只有在内吞生物粒子尺寸大于细胞骨架的特征尺度的情况下，以往的理论模型才适用，反之，则需要考量细胞骨架各向异性的问题。

第三节 未来5～10年重点发展方向

研究表明，机械力对组织和细胞都有很重要的影响作用。最明显的例子就是我们的听觉和触觉，不太为人熟悉的例子包括内皮细胞根据流体剪切应力排列，并且流体剪切应力的改变会改变内皮细胞的转录，以及细胞周围环境的力学性质能够影响细胞分化和癌细胞转移潜能（metastatic potential）等。

生物学证据表明，外力作用可以改变蛋白质的构象，从而改变信号蛋白质的绑定或者激发相关酶的活性，这些都是力学信号传导的可能原因。此外，单分子实验表明，许多与力学性能相关的分子，比如辅肌动蛋白、细丝蛋白、血影蛋白（spectrin）[133]和肌联蛋白（titin）[134]都包含能够被50～200pN的外力解折叠的区域。

未来重点研究的方向将包括，但不局限于：

（1）细胞力学对力学信号传导提供了哪些信息？正如前面所讨论的，蛋白质的解折叠或许为细胞骨架的应力松弛提供了一种机制。蛋白质的折叠状态在很多生物过程中得到感知，而从理论上讲细胞可以通过合计解折叠区域的数量来估算应力大小。与其他传导分子应力的传感机制不同，对于应力的感受和估测能够使细胞对大规模的力学变量进行响应，比如细胞的形状和细胞外基质的力学顺应性。此外，如果细胞牵张力在1000Pa这个数量级，那么大小为1Pa的外加应力是怎么被细胞感知到的呢？交联的肌动蛋白胶体中的构象改变引起的结构分子的重排似乎是一种理想的化学-力学传感器，因为外力的轻微增大也能引起构象改变。

（2）通过观测交联蛋白的结合配体来识别在力学信号传导路径中起作用的新元素。许多交联体都和很多信号蛋白绑定，包括heat shock protein，protein kinase C，Ral A，PIP2，PIP3，PI3-kinase，MEKK1 和 Rho GTPases调控蛋白[135,136]。在外力作用后，交联区域的蛋白质绑定或解绑定可以传导所受到的剪切应力或者张力。此外，多结构域交联蛋白和它们的异构体分布在细胞的不同区域，因此不同的细胞亚结构可以有自己独特的机械感应能力。

（3）结合多种细胞力学实验技术，以进一步认知细胞的生物进程及其在生理病理机制中的作用和临床应用，服务于精准医疗。例如，肌动蛋白斑显微测量技术（actin speckle microscopy）可以追踪亚微米的荧光标记的肌动蛋白组来测定流率和肌动蛋白细胞骨架的聚合动力学[137]。这些分子自身也受到热扰动，因此可以根据它们探测不受外力干扰时其内部的力学性质。通过激光对细胞进行挤压高通量地检测细胞在相关生物进程，包括细胞迁移[138]、细胞对剪切应力的响应[139]、细胞硬度的改变[140]、细胞周围物理环境的改变[141]、信号分子的表达[142]、细胞分化[143]、肿瘤细胞的转移等中的力学特性和行为变化，据此实现对病理（如癌症）细胞的早期检测。

（4）细胞内吞的相关理论及实验研究，包括：相关蛋白如何参与细胞内吞过程，配体-受体尺寸效应及排空效应的竞争关系对内吞过程的影响；细胞骨架网络不连续性在内吞过程中的考量等也是未来需要重点研究的问题。

邓林红（常州大学生物医学工程与健康科学研究院），
陈诚（第三军医大学西南医院关节外科），刘艳辉（贵州大学物理学院）

参考文献

[1] Zewail A H. Physical Biology: From Atoms to Medicine. Imperial College Press, 2008.

[2] Bao G, Suresh S. Cell and molecular mechanics of biological materials. Nat. Mater., 2003, 2(11): 715-25.

[3] Hoffman B D, Crocker J C. Cell mechanics: Dissecting the physical responses of cells to force. Annu. Rev. Biomed. Eng., 2009, 11: 259-88.

[4] Janmey P A, McCulloch C A. Cell mechanics: integrating cell responses to mechanical stimuli. Annu. Rev. Biomed. Eng., 2007, 9: 1-34.

[5] 龙勉．细胞-分子生物力学：力学-生物学，力学-化学耦合．医用生物力学，2007, 22(1): 1-3.

[6] 吕守芹，杨帆，龙勉．细胞-分子生物力学研究进展．医用生物力学，2009, 24(2): 79-84.

[7] 李良，陈槐卿．细胞核结构与力学生物学．医用生物力学，2009, 24(1): 1-5.

[8] 邓林红．气道平滑肌生物力学与哮喘病理机制的研究进展．医用生物力学，2009, 24(4): 237-245.

[9] An S S, et al. Do biophysical properties of the airway smooth muscle in culture predict airway hyperresponsiveness? American Journal of Respiratory Cell and Molecular Biology, 2006. 35(1): 55-64.

[10] Fredberg J J, Kamm R D. Stress transmission in the lung: pathways from organ to molecule. Annu. Rev. Physiol., 2006, 68: 507-541.

[11] Haga J H, Li Y S J, Chien S. Molecular basis of the effects of mechanical stretch on vascular smooth muscle cells. Journal of Biomechanics, 2007, 40(5): 947-960.

[12] Suresh S. Biomechanics and biophysics of cancer cells. Acta Materialia, 2007, 55(12): 3989-4014.

[13] Smith A S. Physics challenged by cells. Nature Physics, 2010, 6(10): 726-729.

[14] Gardel M, et al. Elastic behavior of cross-linked and bundled actin networks. Science, 2004, 304(5675): 1301-1305.

[15] Liverpool T B. Active gels: where polymer physics meets cytoskeletal dynamics. Philosophical Transactions of the Royal Society of London A: Mathematical, Physical and Engineering Sciences, 2006, 364(1849): 3335-3355.

[16] Fletcher D A, Mullins R D. Cell mechanics and the cytoskeleton. Nature, 2010, 463(7280): 485-492.

[17] Kim S, Coulombe P A. Emerging role for the cytoskeleton as an organizer and regulator of translation. Nature Reviews Molecular Cell Biology, 2010, 11(1): 75-81.

[18] Sawada Y, et al. Force sensing by mechanical extension of the src family kinase substrate p130Cas. Cell, 2006, 127(5): 1015-1026.

[19] Johnson C P, et al. Forced unfolding of proteins within cells. Science, 2007, 317(5838): 663-666.

[20] Ellis R J, Minton A P. Cell biology: Join the crowd. Nature, 2003, 425(6953): 27-28.

[21] Minton A P. How can biochemical reactions within cells differ from those in test tubes? Journal of Cell Science, 2006, 119(14): 2863-2869.

[22] Plopper G, Ingber D E. Rapid induction and isolation of focal adhesion complexes. Biochem. Biophys. Res. Commun., 1993, 193(2): 571-578.

[23] Schmidt C E, et al. Integrin-cytoskeletal interactions in migrating fibroblasts are dynamic, asymmetric, and regulated. J. Cell. Biol., 1993, 123(4): 977-991.

[24] Felsenfeld D P, Choquet D, Sheetz M P. Ligand binding regulates the directed movement of beta1 integrins on fibroblasts. Nature, 1996, 383(6599): 438-440.

[25] Choquet D, Felsenfeld D P, Sheetz M P. Extracellular matrix rigidity causes strengthening of integrin-cytoskeleton linkages. Cell, 1997, 88(1): 39-48.

[26] Deng L, et al. Localized mechanical stress induces time-dependent actin cytoskeletal remodeling and stiffening in cultured airway smooth muscle cells. Am. J. Physiol. Cell Physiol., 2004, 287(2): 440-448.

[27] Metzner C, et al. Fluctuations of cytoskeleton-bound microbeads—the effect of bead-receptor binding dynamics. J. Phys. Condens. Matter., 2010, 22(19): 194105.

[28] Park C Y, et al. Mapping the cytoskeletal prestress. Am. J. Physiol. Cell Physiol., 2010,

298(5): 1245-1252.

[29] Lau A W, et al. Microrheology, stress fluctuations, and active behavior of living cells. Phys. Rev. Lett., 2003, 91(19): 198101.

[30] Levine A J, Lubensky T C. One- and two-particle microrheology. Phys. Rev. Lett., 2000, 85(8): 1774-1777.

[31] Bausch A R, et al. Local measurements of viscoelastic parameters of adherent cell surfaces by magnetic bead microrheometry. Biophys. J., 1998, 75(4): 2038-2049.

[32] Mijailovich S M, et al. A finite element model of cell deformation during magnetic bead twisting. J. Appl. Physiol. (1985), 2002, 93(4): 1429-1436.

[33] Kasza K E, et al. Filamin A is essential for active cell stiffening but not passive stiffening under external force. Biophys. J., 2009, 96(10): 4326-4335.

[34] Hu S, et al. Intracellular stress tomography reveals stress focusing and structural anisotropy in cytoskeleton of living cells. Am. J. Physiol. Cell Physiol., 2003, 285(5): 1082-1090.

[35] Hoffman B D, et al. The consensus mechanics of cultured mammalian cells. Proc. Natl. Acad. Sci. USA, 2006, 103(27): 10259-10264.

[36] Wang N, et al. Mechanical behavior in living cells consistent with the tensegrity model. Proceedings of the National Academy of Sciences, 2001, 98(14): 7765-7770.

[37] Ingber D E. Cellular tensegrity: defining new rules of biological design that govern the cytoskeleton. Journal of Cell Science, 1993, 104: 613.

[38] Ingber D E, Tensegrity I. Cell structure and hierarchical systems biology. Journal of Cell Science, 2003, 116(7): 1157-1173.

[39] Harris A K, Wild P, Stopak D. Silicone rubber substrata: a new wrinkle in the study of cell locomotion. Science, 1980, 208(4440): 177-179.

[40] Butler J P, et al. Traction fields, moments, and strain energy that cells exert on their surroundings. American Journal of Physiology-Cell Physiology, 2002, 282(3): 595-605.

[41] Dembo M, Wang Y L. Stresses at the cell-to-substrate interface during locomotion of fibroblasts. Biophysical Journal, 1999, 76(4): 2307-2316.

[42] Wang N, et al. Cell prestress. I. Stiffness and prestress are closely associated in adherent contractile cells. American Journal of Physiology-Cell Physiology, 2002, 282(3): 606-616.

[43] Coughlin M F, Stamenović D. A prestressed cable network model of the adherent cell cytoskeleton. Biophysical journal, 2003, 84(2): 1328-1336.

[44] Ingber D E. Cellular mechanotransduction: putting all the pieces together again. The FASEB Journal, 2006, 20(7): 811-827.

[45] Engler A J, et al. Matrix elasticity directs stem cell lineage specification. Cell, 2006, 126(4): 677-689.

[46] Fabry B, et al. Scaling the microrheology of living cells. Physical Review Letters, 2001, 87(14): 148102.

[47] Fabry B, et al. Time scale and other invariants of integrative mechanical behavior in living cells. Physical Review E, 2003, 68(4): 041914.

[48] Deng L, et al. Fast and slow dynamics of the cytoskeleton. Nature materials, 2006, 5(8): 636-640.

[49] Stamenović D, et al. Rheological behavior of living cells is timescale-dependent. Biophysical Journal, 2007, 93(8): 39-41.

[50] Hoffman B D, et al. The consensus mechanics of cultured mammalian cells. Proceedings of the National Academy of Sciences, 2006, 103(27): 10259-10264.

[51] Gardel M, et al. Stress-dependent elasticity of composite actin networks as a model for cell behavior. Physical Review Letters, 2006, 96(8): 088102.

[52] Gardel M, et al. Prestressed F-actin networks cross-linked by hinged filamins replicate mechanical properties of cells. Proceedings of the National Academy of Sciences of the United States of America, 2006, 103(6): 1762-1767.

[53] Bursac P, et al. Cytoskeletal remodelling and slow dynamics in the living cell. Nature Materials, 2005, 4(7): 557-561.

[54] Lenormand G, et al. Directional memory and caged dynamics in cytoskeletal remodelling. Biochemical and Biophysical Research Communications, 2007, 360(4): 797-801.

[55] Tschumperlin D J, Margulies S S. Alveolar epithelial surface area-volume relationship in isolated rat lungs. Journal of Applied Physiology, 1999, 86(6): 2026-2033.

[56] Trepat X, et al. Universal physical responses to stretch in the living cell. Nature, 2007, 447(7144): 592-595.

[57] Chen C, et al. Fluidization and resolidification of the human bladder smooth muscle cell in response to transient stretch. PloS One, 2010, 5(8): e12035.

[58] Trepat X, et al. Viscoelasticity of human alveolar epithelial cells subjected to stretch. American Journal of Physiology-Lung Cellular and Molecular Physiology, 2004, 287(5): 1025-1034.

[59] Fernández P, Pullarkat P A, Ott A. A master relation defines the nonlinear viscoelasticity of single fibroblasts. Biophysical Journal, 2006, 90(10): 3796-3805.

[60] 张鹏，房兵，江凌勇．机械刺激对成骨细胞骨架的影响．医用生物力学，2011, 26(1): 87-91.

[61] Choquet D, Felsenfeld D P, Sheetz M P. Extracellular matrix rigidity causes strengthening of integrin-cytoskeleton linkages. Cell, 1997, 88(1): 39-48.

[62] Matthews B D, et al. Cellular adaptation to mechanical stress: role of integrins, Rho,

cytoskeletal tension and mechanosensitive ion channels. Journal of Cell Science, 2006, 119(3): 508-518.
[63] Weitz D, Janmey P. The soft framework of the cellular machine. Proceedings of the National Academy of Sciences, 2008, 105(4): 1105-1106.
[64] Krishnan R, et al. Reinforcement versus fluidization in cytoskeletal mechanoresponsiveness. PloS One, 2009, 4(5): e5486.
[65] Schmid-Schonbein G W, et al. Passive mechanical properties of human leukocytes. Biophys. J., 1981, 36(1): 243-256.
[66] Wang N, Ingber D E. Control of cytoskeletal mechanics by extracellular matrix, cell shape, and mechanical tension. Biophys. J., 1994, 66(6): 2181-2189.
[67] Thoumine O, Ott A. Time scale dependent viscoelastic and contractile regimes in fibroblasts probed by microplate manipulation. J. Cell. Sci., 1997, 110 (Pt 17): 2109-2116.
[68] Laurent V M, et al. Partitioning of cortical and deep cytoskeleton responses from transient magnetic bead twisting. Ann. Biomed. Eng., 2003, 31(10): 1263-1278.
[69] Hildebrandt J. Comparison of mathematical models for cat lung and viscoelastic balloon derived by Laplace transform methods from pressure-volume data. Bull. Math. Biophys., 1969, 31(4): 651-67.
[70] Huxley A F. Muscle structure and theories of contraction. Prog. Biophys. Chem., 1957, 7: 255-318.
[71] Stamenovic D, et al. Rheology of airway smooth muscle cells is associated with cytoskeletal contractile stress. J. Appl. Physiol. (1985), 2004, 96(5): 1600-1605.
[72] Swanson J A. Shaping Cups into Phagosomes and Macropinosomes. Nat. Rev. Mol. Cell. Biol., 2008, 9: 639-664.
[73] Lewis W H. Pinocytosis. Bull. Johns. Hopkins. Hosp, 1931, 49: 17-27.
[74] McNeil P L. Mechanisms of nutritive endocytosis. Ⅲ. a freeze-fracture study of phagocytosis by digestive cells of chlorohydra. Tissue Cell, 1984, 16: 519-533.
[75] Mooren O L, Galletta B J, Cooper J A. Roles for actin assembly in endocytosis. Annu. Rev. Biochem, 2012, 81: 661-686.
[76] Roth T F, Porter K R. Yolk protein uptake in the oocyte of the mosquito aedes aegypti. J. Cell Biol., 1964, 20: 313-332.
[77] Marsh M, Helenius A. Virus entry: open sesame. Cell, 2006, 124: 729-740.
[78] Mercer J, Schelhaas M, Helenius A. Virus entry by endocytosis. Annu. Rev. Biochem., 2010, 79: 803-833.
[79] Palade G E. An electron microscope study of the mitochondrial structure. J. Histochem. Cytochem., 1953, 1: 188-211.

[80] Yamada E. The fine structure of the gall bladder epithelium of the mouse. J. Cell. Biol., 1951, 1: 445-458.

[81] Cavalli V, Corti M, Gruenberg J. Endocytosis and signaling cascades: a close encounter. FEBS Lett., 2001, 498: 190-196.

[82] Lakadamyali M, Rust M J, Zhuang X. Endocytosis of influenza viruses. Microbes Infect, 2004, 6: 929-936.

[83] Sieczkarski S B, Whittaker G R. Influenza virus can enter and infect cells in the absence of clathrin-mediated endocytosis. J. Virol., 2002, 76: 10455-10464.

[84] Ding H M, Tian W D, Ma Y Q. Designing nanoparticle translocation through membranes by computer simulations. ACS Nano, 2012, 6: 1230-1238.

[85] Wong-Ekkabut J, Baoukina S, et al. Computer simulation study of fullerene translocation through lipid membranes. Nat. Nanotechnol., 2008, 3: 363-368.

[86] Yang K, Ma Y Q. Computer simulation of the translocation of nanoparticles with different shapes across a lipid bilayer. Nat. Nanotechnol., 2010, 5: 579-583.

[87] Zhang H, et al. Cooperative transmembrane penetration of nanoparticles. Sci. Rep., 2015, 5: 10525.

[88] Lin J, Alexander-Katz A. Cell membranes open doors for cationic nanoparticles/biomolecules: insights into uptake kinetics. ACS Nano, 2013, 7: 10799-10808.

[89] Chithrani B D, Ghazani A A, Chan W C. Determining the size and shape dependence of gold nanoparticle uptake into mammalian cells. Nano Lett., 2006, 6(4): 662-668.

[90] Chithrani B D, Chan W C. Elucidating the mechanism of cellular uptake and removal of protein-coated gold nanoparticles of different sizes and shapes. Nano Lett., 2007, 7(6): 1542-1550.

[91] Kol N, Shi Y, Tsritov M, et al. A stiffness switch in human immunodeficiency virus. Biophys J, 2007, 92: 1777-1783.

[92] Nel A E, et al. Understanding biophysicochemical interactions at the nano-bio interface. Nature, 2009, 8: 543-557.

[93] Alakoskela J M, et al. Mechanisms for size-dependent protein segregation at immune synapses assessed with molecular rulers. Biophysical Journal, 2011, 100: 2865-2874.

[94] Springer T A. Adhesion receptors of the immune system. Nature, 1990, 346: 425-434.

[95] Sun S X, Wirtz D. Mechanics of enveloped virus entry into host cells. Biophysical Journal, 2006, 90: 10-12.

[96] Li L, et al. On resistance to virus entry into host cells. Biophysical Journal, 2012, 102: 2230-2233.

[97] Zhang S, Gao H, Bao G. Physical principles of nanoparticle cellular endocytosis. ACS

Nano, 2015, 9: 8655-8671.

[98] Yi X, Shi X, Gao H J. Cellular uptake of elastic nanoparticles. Phys. Rev. Lett., 2011, 107.

[99] Lenormand G, et al. Linearity and time-scale invariance of the creep function in living cells. J. R. Soc. Interface, 2004, 1(1): 91-97.

[100] Desprat N, et al. Creep function of a single living cell. Biophys. J., 2005, 88(3): 2224-2233.

[101] Wang N, Butler J P, Ingber D E. Mechanotransduction across the cell surface and through the cytoskeleton. Science, 1993, 260(5111): 1124-1127.

[102] Janmey P A, et al. Resemblance of actin-binding protein/actin gels to covalently crosslinked networks. Nature, 1990, 345(6270): 89-92.

[103] Condeelis J S, Taylor D L. The contractile basis of amoeboid movement. V. The control of gelation, solation, and contraction in extracts from Dictyostelium discoideum. J. Cell. Biol., 1977, 74(3): 901-927.

[104] Tempel M, Isenberg G, Sackmann E. Temperature-induced sol-gel transition and microgel formation in alpha -actinin cross-linked actin networks: a rheological study. Phys. Rev. E Stat. Phys. Plasmas Fluids Relat. Interdiscip Topics, 1996, 54(2): 1802-1810.

[105] Crocker J C, et al. Two-point microrheology of inhomogeneous soft materials. Phys. Rev. Lett., 2000, 85(4): 888-891.

[106] Gardel M L, et al. Chapter 19: Mechanical response of cytoskeletal networks. Methods Cell Biol., 2008, 89: 487-519.

[107] Gardel M L, et al. Microrheology of entangled F-actin solutions. Phys. Rev. Lett., 2003, 91(15): 158302.

[108] Janmey P A, et al. Viscoelastic properties of vimentin compared with other filamentous biopolymer networks. J. Cell. Biol., 1991, 113(1): 155-160.

[109] Sato M, et al. Mechanical properties of brain tubulin and microtubules. J. Cell. Biol., 1988, 106(4): 1205-1211.

[110] Liu J, et al. Microrheology probes length scale dependent rheology. Phys. Rev. Lett., 2006, 96(11): 118104.

[111] Koenderink G H, et al. High-frequency stress relaxation in semiflexible polymer solutions and networks. Phys. Rev. Lett., 2006, 96(13): 138307.

[112] Wang N, Stamenovic D. Contribution of intermediate filaments to cell stiffness, stiffening, and growth. Am. J. Physiol. Cell. Physiol., 2000, 279(1): 188-194.

[113] Stamenovic D, et al. Cell prestress. II. Contribution of microtubules. Am. J. Physiol. Cell. Physiol., 2002, 282(3): 617-624.

[114] Kumar S, et al. Viscoelastic retraction of single living stress fibers and its impact on cell shape, cytoskeletal organization, and extracellular matrix mechanics. Biophys. J., 2006,

90(10): 3762-3773.

[115] Storm C, et al. Nonlinear elasticity in biological gels. Nature, 2005, 435(7039): 191-194.

[116] Coughlin M F, et al. Filamin-a and rheological properties of cultured melanoma cells. Biophys. J., 2006, 90(6): 2199-2205.

[117] DiDonna B A, Levine A J. Unfolding cross-linkers as rheology regulators in F-actin networks. Phys. Rev. E. Stat. Nonlin. Soft Matter Phys., 2007, 75(4 Pt 1): 041909.

[118] DiDonna B A, Levine A J. Filamin cross-linked semiflexible networks: fragility under strain. Phys. Rev. Lett., 2006, 97(6): 068104.

[119] Hoffman B D, Massiera G, Crocker J C. Fragility and mechanosensing in a thermalized cytoskeleton model with forced protein unfolding. Phys. Rev. E Stat. Nonlin. Soft Matter Phys., 2007, 76(5 Pt 1): 051906.

[120] Rief M, et al. Single molecule force spectroscopy of spectrin repeats: low unfolding forces in helix bundles. J. Mol. Biol., 1999, 286(2): 553-561.

[121] Furuike S, Ito T, Yamazaki M. Mechanical unfolding of single filamin A (ABP-280) molecules detected by atomic force microscopy. FEBS Lett., 2001, 498(1): 72-75.

[122] Miyata H, Yasuda R, Kinosita K. Strength and lifetime of the bond between actin and skeletal muscle alpha-actinin studied with an optical trapping technique. Biochim. Biophys. Acta., 1996, 1290(1): 83-88.

[123] Tharmann R, Claessens M M, Bausch A R. Viscoelasticity of isotropically cross-linked actin networks. Phys. Rev. Lett., 2007, 98(8): 088103.

[124] Wachsstock D H, Schwarz W H, Pollard T D. Cross-linker dynamics determine the mechanical properties of actin gels. Biophys. J., 1994, 66(3 Pt 1): 801-809.

[125] Ferrer J M, et al. Measuring molecular rupture forces between single actin filaments and actin-binding proteins. Proc. Natl. Acad. Sci. USA, 2008, 105(27): 9221-9226.

[126] Leinweber B, et al. Calponin interaction with alpha-actinin-actin: evidence for a structural role for calponin. Biophys. J., 1999, 77(6): 3208-3217.

[127] del Rio A, et al. Stretching single talin rod molecules activates vinculin binding. Science, 2009, 323(5914): 638-641.

[128] Humphrey D, et al. Active fluidization of polymer networks through molecular motors. Nature, 2002, 416(6879): 413-416.

[129] Liverpool T B, Maggs A C, Ajdari A. Viscoelasticity of solutions of motile polymers. Phys. Rev. Lett., 2001, 86(18): 4171-4174.

[130] Mizuno D, et al. Nonequilibrium mechanics of active cytoskeletal networks. Science, 2007, 315(5810): 370-373.

[131] Fernandez P, Ott A. Single cell mechanics: stress stiffening and kinematic hardening. Phys.

Rev. Lett., 2008, 100(23): 238102.

[132] Semmrich C, et al. Glass transition and rheological redundancy in F-actin solutions. Proc. Natl. Acad. Sci. USA, 2007, 104(51): 20199-20203.

[133] Law R, et al. Cooperativity in forced unfolding of tandem spectrin repeats. Biophys. J., 2003, 84(1): 533-544.

[134] Rief M, et al. Reversible unfolding of individual titin immunoglobulin domains by AFM. Science, 1997, 276(5315): 1109-1112.

[135] Ohta Y, Hartwig J H, Stossel T P. FilGAP, a Rho- and ROCK-regulated GAP for Rac binds filamin A to control actin remodelling. Nat. Cell. Biol., 2006, 8(8): 803-814.

[136] Mammoto A, Huang S, Ingber D E. Filamin links cell shape and cytoskeletal structure to Rho regulation by controlling accumulation of p190RhoGAP in lipid rafts. J. Cell. Sci., 2007, 120(Pt 3): 456-467.

[137] Ponti A, et al. Two distinct actin networks drive the protrusion of migrating cells. Science, 2004, 305(5691): 1782-1786.

[138] Kole T P, et al. Intracellular mechanics of migrating fibroblasts. Mol. Biol. Cell., 2005, 16(1): 328-338.

[139] Lee J S, et al. Ballistic intracellular nanorheology reveals ROCK-hard cytoplasmic stiffening response to fluid flow. J. Cell. Sci., 2006, 119(Pt 9): 1760-1768.

[140] Solon J, et al. Fibroblast adaptation and stiffness matching to soft elastic substrates. Biophys. J., 2007, 93(12): 4453-4461.

[141] Panorchan P, et al. Microrheology and ROCK signaling of human endothelial cells embedded in a 3D matrix. Biophys. J., 2006, 91(9): 3499-3507.

[142] Kole T P, et al. Rho kinase regulates the intracellular micromechanical response of adherent cells to rho activation. Mol. Biol. Cell., 2004, 15(7): 3475-3484.

[143] Collinsworth A M, et al. Apparent elastic modulus and hysteresis of skeletal muscle cells throughout differentiation. Am. J. Physiol. Cell. Physiol., 2002, 283(4): 1219-1227.

第八章 生物大分子的静电学

本章首先叙述了生物大分子的静电学的研究对象，体系的特点和现象，以及主要研究内容，简单介绍了学科各个研究方向的进展：生物大分子静电学的物理基础和计算方法；与蛋白质相关的静电相互作用的各个层次的模型；这些理论和方法在各类生物大分子中的应用。

第一节 引 言

静电相互作用是化学键的基石，其中也包括在生物大分子中起重要作用的氢键。生物大分子中，普遍存在大量的带电和极性基团。因此，静电相互作用在生物大分子的组成和生物物理过程中都扮演着重要的角色[1-4]。在溶液中，所有核酸和大部分蛋白质都携带电荷。比如，在生理条件下，球形蛋白中大约 20% 的氨基酸是电离带电的，而另外 25% 的氨基酸侧链是极性基团。核酸属于最强的天然聚电解质，比如双链 DNA 的线电荷密度大约为每 1.7Å 一个电子电荷。看静电相互作用是否对生物大分子的结构和功能起作用的一个判据是将静电作用能与热运动能 k_BT 相比较。我们可以引进一个特征长度 l_B，在这个距离两个单位电荷之间的库仑作用能等于其热运动能，称为 Bjerrum 长度。

$$l_B=\frac{e^2}{\varepsilon k_B T} \tag{6-8-1}$$

在室温下（T=300K）水溶液（介电常数 ε=80）中的 Bjerrum 长度大约

为 7Å。大部分生物膜都包含有少量的涉及生理功能的带电磷脂。在生理条件下，这些裸露在生物大分子表面的电荷被水溶液及含伴随离子和平衡离子的盐溶液所屏蔽。这个物理图像，初看起来简单，但实际上要复杂得多。生物大分子的行为还取决于其他因子，如电荷、尺寸、水合作用、浓度以及伴随离子和平衡离子的几何特性。在生物大分子发生表面-表面接触之前，静电相互作用在纳米尺度起主导作用，如自发组装和长程有序结构的构建。这些带电的生物聚合物的化学物理性质是一个丰富的领域，它经常有反直觉的现象，在理论、实验、计算方面需要广泛的研究。对于生物大分子的静电作用而言，主要有以下几个方面的课题。

（1）生物大分子与溶剂和离子的相互作用[5]。生物大分子在溶液中，在极性水分子的作用下，通常都会离解成本体的宏观离子和溶液中的平衡离子。在宏观离子附近平衡离子的分布对宏观离子的性质和功能有决定性的影响。溶液中通常有各种盐，在稀溶液的条件下，基于平均场的泊松-玻尔兹曼[1]理论可以给出与实验一致的平衡离子的分布。但是，在很多情况下，平均场理论无法用于生物大分子相关的静电学现象，原因是在这些体系中有较强的静电关联现象。强关联导致了两个重要的结果：电荷逆转（或者叫电荷过补偿）和同类电荷吸引。这两个现象与我们的物理直觉都是不一致的。电荷逆转是指宏观离子吸引了比其自身电荷还多的平衡离子，使得整体净电荷变号。同类电荷吸引指的是两个带相同电荷宏观离子在溶液中会互相吸引而聚集。这两个问题是生物大分子静电学的中心课题。

（2）DNA-DNA 相互作用[6]。DNA 是带电最强的生物大分子之一，同时具有很大的刚性。在生命物质中，DNA 通常都是被有序包装在很小的空间内。一个重要的问题就是它们是如何克服库仑排斥力和熵力聚集在一起的。这个问题也是同类电荷吸引的一个典型的例子。主要问题有 DNA 的凝聚和聚集以及包装[7,8]，平衡离子和包装蛋白的作用。

（3）DNA-蛋白质相互作用[9,10]。DNA 与蛋白质结合有重要的生物学功能，比如转录因子和限制性内切酶与 DNA 的结合。DNA 与蛋白质的结合分为特异性结合和非特异性结合。前者还涉及蛋白质对 DNA 特定序列的识别。通常 DNA 结合蛋白质表面有带电区块，以利于识别 DNA 序列并与之结合。带电区块的形状和电荷分布决定其与双链 DNA 的相互作用模式。

（4）蛋白质相关静电学问题[11,12]。静电作用对蛋白质有三个方面的作用：稳定蛋白质结构，分子识别功能和实现酶的催化功能。这些现象的研究通常涉及大量的数值计算。但目前对于与蛋白质相关的静电学问题还不能全部用全原子的方法进行分析和计算。半微观的连续模型方法是蛋白质静电势和其他电学性质计算的可行手段，这些计算都涉及蛋白质的介电常数。水分子的处理和蛋白质介电常数校准是计算中的关键问题。

（5）RNA 折叠相关的静电学问题[13]。RNA 的平衡离子可分成固定和流动两类。固定平衡离子是与化学键相关的，而流动离子分布与前面问题提到的一样是物理相关的。为了搞清 RNA 折叠的机制，有几个关键问题：静电力（包括关联）和非静电力（如熵力）的大小，离子的特定结合和非特定结合的贡献，镁离子的关联效应的大小以及溶剂效应。

（6）电泳及相关问题[14,15]。电泳是指胶体微粒在电场作用下在分散介质中向阴极或阳极做定向移动。电泳技术广泛用于蛋白质、核酸和其他生物大分子的分离和分析。电转动和介电泳是传统电泳在近年来的进一步发展，在癌症研究、生物组织分离和鉴别以及生物芯片设计等方面有广泛的应用。电转动和介电泳现象涉及多极电荷效应和交变电场效应等，理解其物理机制需要进一步的模型构建和数值模拟。

生物大分子的静电作用与生物制药及基因治疗也有密切联系。比如，研究 DNA 包装，RNA 折叠和细胞骨架自组装行为的静电作用机制和调控等，对于疾病治疗和开发新的药物有重要的启示。

第二节　学科发展背景和现状

一、生物大分子静电学的物理基础和计算方法

生物大分子静电学的出发点是离子溶液的 PB 方程[16-20]。PB 理论是一个平均场理论，依赖于以下几个假定：①带电体之间只有库仑相互作用；②不考虑固有的和诱导的偶极子-偶极子相互作用；③电荷按点电荷处理，忽略有限尺寸效应和任何短程非静电相互作用；④水溶液按连续介质模型处理，介电常数为 ε，对于水，介电常数 $\varepsilon=80$；⑤对于每个离子感受到的静电势 $\phi(r)$ 是一个依赖于所有其他离子的平均场的连续函数。电荷密度 $\rho(r)$ 也是仅依赖于位置 r 的平均场连续函数。

考虑含有两种离子的离子液体。其正电荷密度为 ρ_+，而负电荷密度为 ρ_-，则总电荷密度为 $\rho_\pm=\rho_++\rho_-$。设两种离子的电荷数密度为 $n_\pm$，则 $\rho_\pm(r)=ez_\pm n_\pm(r)$，其中 $z_+>0$ 是阳离子价数，而 $z_-<0$ 是阴离子价数。离子可以移动且处于热力

学平衡状态，则任意点的电势和电荷密度满足泊松方程：

$$\nabla^2\phi = -\frac{4\pi}{\varepsilon}\rho(r) = -\frac{4\pi e}{\varepsilon}\left[z_+ n_+(r) + z_- n_-(r)\right] \tag{6-8-2}$$

由于每种离子都处于热力学平衡状态，其对应的数密度满足玻尔兹曼分布：

$$n_\pm = n_\pm^0 \mathrm{e}^{-ez_\pm\phi/k_\mathrm{B}T} \tag{6-8-3}$$

其中，$n_\pm^0$为零电势的参考数密度。联合式（6-8-2）和式（6-8-3），可得电势的 PB 方程：

$$\nabla^2\phi = -\sum_{i=\pm}\frac{4\pi e n_i^0 z_i}{\varepsilon}\mathrm{e}^{-\frac{ez_i\phi(r)}{k_\mathrm{B}T}} \tag{6-8-4}$$

PB 方程是研究生物大分子和软物质静电学的出发点和基础，也是很多数值方法的非常有用的分析近似。由于 PB 方程是非线性的，只有在几种简单边界条件下有解析解。另外，如果做进一步的近似或借助数值方法，我们可以获得某些复杂结构的离子分布和自由能。作为一种近似理论，PB 方程也有局限性。比如，PB 理论可以很好地解释单价离子的行为。而对于高价离子，有些重要的特性无法用 PB 理论描述。

对于低静电势的情形，PB 方程可以做线性化近似，我们可以得到 Debye-Hückel (DH) 理论。

$$\nabla^2\phi \simeq -\frac{8\pi e^2 n_0}{\varepsilon k_\mathrm{B}T}\phi(r) = \lambda_\mathrm{D}^{-2}\phi(r) \tag{6-8-5}$$

其中新参数 λ_D 具有长度量纲，称为 Debye-Hückel 屏蔽长度：

$$r_\mathrm{D} = \sqrt{\frac{\varepsilon k_\mathrm{B}T}{8\pi e^2 n_0}} = \left(8\pi l_\mathrm{B} n_0\right)^{-1/2} \tag{6-8-6}$$

这个屏蔽长度变化范围很大。比如，在高离子强度的 1M 浓度的 NaCl 溶液中，DH 屏蔽长度约为 3Å，而在纯水中，可达 1μm 左右。粗略地说，在 DH 理论中，静电势按 $r^{-1}\mathrm{e}^{-r/r_\mathrm{D}}$ 变化，当 $r \leqslant r_\mathrm{D}$ 时，库仑作用只是稍加改变，而当 $r > r_\mathrm{D}$ 时，长程的库仑作用被强烈屏蔽了。

溶解自由能和 Born 模型[21,22]：溶解自由能定义为将一个溶质分子从真空移入溶剂中当前位置所需的能量，包括极化部分和非极化部分的贡献。对于带电分子，极化部分有主要贡献，包括自能和该分子与溶液中的电荷偶极子的相互作用能。非极化部分包括对抗溶剂压力将分子嵌入溶剂内所需能量，如溶剂的范德瓦耳斯作用能和重组分子周围溶剂分子的熵代价。在考虑静电相互作用时，一般只要考虑极化部分。

Born 模型给出了将一个电荷从真空移到一个球形溶剂腔中的溶解自由能的近似公式：

$$\Delta G=-\frac{N_{A}q^{2}}{8\pi\varepsilon_{0}a}\left(1-\frac{1}{\varepsilon}\right) \tag{6-8-7}$$

其中，N_A 是阿伏伽德罗常量常数；q 是电荷；ε 是相对介电常数；a 是腔的半径。Born 模型在几乎所有计算蛋白质相关的静电相互作用模型中都会用到。Born 公式结合库仑定律不仅可以用于从真空移入溶剂的自由能计算，也可用于两个均匀介质直接的电荷移动。比如，把药物从溶液中移到目标蛋白质的指定位置的自由能变化的计算。

强库仑耦合[5,9,23]：对于稀离子溶液和弱库仑耦合的情形，DH 理论是一个很好的近似，因为此时与无关联极限的偏离很小。但是，在强库仑耦合（$\frac{l_B}{b}\gg 1$）的情况下，平均场理论不再适用，必须考虑离子之间的关联。以带电平面为例，两种独立的方法得到了类似的结果：

（1）Noreita 和 Netz 将场论方法用于带电流体，得到在强库仑耦合（$\frac{l_B}{b}\gg 1$）的极限下平衡离子分布按指数衰减：

$$\rho(z)=2\pi l_{B}\sigma_{s}^{2}e^{-z/b} \tag{6-8-8}$$

其中，σ_s 是电荷的数密度，以基本电荷为单位。

（2）Shklovskii 则利用不同的但基于直觉的方法，引入了 Wigner 晶格的概念来研究在有效低温下的带电软物质。在这个方法中，吸附在带电表面的平衡离子会形成一个 Wigner 晶格，格点上的平衡离子互相关联，如果晶格留下空穴，会吸引溶液中的平衡离子。基于低温物理中的 Wigner 晶格图像，也可以得到正确的按指数衰减的平衡离子分布 $\exp(-z/b)$。

电荷逆转或电荷过度补偿问题：只有离子之间的库仑耦合不太强（比如水溶液中的一价离子）PB 理论甚至 DH 理论都能很好地描述电子分布和热力学系统特性。其结果与计算机模拟和非平均场理论的预言以及实验数据比较，符合精度都相当高。但是一旦离子-离子关联变得重要（比如水溶液中两价以上的离子），就会发生电荷过度补偿现象，也叫电荷逆转。实验上观察到，电解质溶液中带电体系［包括生物大分子，如 DNA、胶体粒子、膜等，通称宏观离子（macroion）］会吸引平衡离子以中和其所带电荷，但当平衡离子是多价离子时，其所吸引的平衡离子的电荷会大于宏观离子本身的电荷，导致电荷逆转现象[24,25]。这个现象与我们平常的物理直觉不一致，无法用经典的 PB 理论描述。事实上，问题的关键是平衡离子是离散地而不是均匀地涂到带电表面。如果是均匀带电，可以证明在稳定构型下电荷被完全中和。对于离散带电的平衡离子，它们之间的间距会最大化以使离子-离子的排斥最小。基于 Wigner 晶格的概念，可以发展出一个解析模型，计算由于平衡离子过电荷补偿获得的能量：

$$\Delta E_n = -\frac{\alpha Z_c^2}{\sqrt{4\pi a^2}}\left[(N+n)^{\frac{3}{2}} - N^{\frac{3}{2}}\right] + \frac{Z_c^2 n^2}{2a} \tag{6-8-9}$$

其中，α 是常数；n 是过补偿电荷数（对于完全中性 n=0）；a 是平衡离子在表面占据 Wigner 晶格的半径；Z_c 是平衡离子价数；Z_m 是表面带电数（以基本电荷 e 为单位），$N=Z_m/Z_c$ 是平衡离子数。上式中的第一项吸引能来自于平衡离子与 Wigner-Seitz 晶胞的相互作用，第二项则是平衡离子之间的排斥能。

离散带电表面：在上面的讨论中，平衡离子是离散的，但假定表面是均匀带电的。但在自然界中，表面电荷也是离散的。表面电荷的离散化如何影响平衡离子的分布以及对电荷逆转的影响也是一个现实的课题。初步的研究表明，电荷密度越大，离散表面电荷的关联长度越短，越接近于连续分布的情形，和我们的物理直觉一致。具体来说：

（1）在零温条件下，表面约化电荷密度越大和 / 或平衡离子价数越大，表面平衡离子具有更高的序。

（2）当电荷逆转发生时，对于大的表面电荷密度，结果与均匀带电的情形相近。而对于小的表面电荷密度和单价平衡离子，因为离子配对困难，电荷逆转要比连续情形更弱。对于同样小的表面电荷密度但是高价平衡离子时，由于离子配对，离散表面电荷可以产生更强的电荷逆转。

（3）在有限温度下（对应于水溶液中），溶液中的平衡离子分布只在低表面电荷密度和高价离子有变化。

二、与电泳相关的静电相互作用

生物大分子在溶液中由于带电基团的离解通常都带电，如果施加电场，它们就会在电场的驱动下发生定向运动[26]。按照双电层理论，带电表面由扩散离子层屏蔽，而离子层所带的电荷数量与大分子的表面电荷相同，但符号相反。电场不仅施加静电力在大分子上，而且也作用于扩散层的离子，使其

做反向运动。这种反向运动会通过黏滞作用转移到大分子上，形成电泳迟滞力。在稳态条件下，作用在生物大分子上的合力为零，即

$$F_{tot}=0=F_{el}+F_{f}+F_{ret} \tag{6-8-10}$$

其中，F_{el} 是电场力；F_{f} 是摩擦力；F_{ret} 是电泳迟滞力。在低雷诺数和中等大小电场条件下，生物大分子的漂移速度与所加电场成正比，其比例系数定义为电泳迁移率：

$$\mu_{e}=\frac{v}{E} \tag{6-8-11}$$

其中，v 是电泳速率；E 是外加电场。电泳迁移率与介质的性质有关，Smoluchowski 给出

$$\mu_{e}=\frac{\varepsilon_{r}\varepsilon_{0}\zeta}{\eta} \tag{6-8-12}$$

其中，ε_{r} 是介质的相对介电常数；ε_{0} 是真空介电常数；η 是介质的动态黏滞系数，而 ζ 是大分子或粒子的 zeta 电位。

电转动和介电泳：在交变电场作用下，悬浮粒子或生物大分子也会做出相应的反应，其中一个重要的性质是电场极化效应。在电场的作用下，会在可极化介质中诱导出偶极矩。对于电转动，一个交变转动电场会使一个可极化的悬浮粒子感应出偶极矩，使其按外场的角频率转动。随着外场的转动频率的增加，转动场的周期随之变小，直到其与感应偶极矩的响应时间相当。为使能量极小，偶极矩试图沿电场方向排列，但对于高频外场，这种排列无法跟上外场的频率。这样，响应的滞后导致一个施加在粒子的扭矩。这个扭矩和粒子的转动速度正比于偶极因子的虚部。另外，由于粒子悬浮在黏性介质中经受了一个耗散拖拽力，粒子的转动速度反比于介质的动态黏滞系数。胶体粒子在交变电场作用下的运动称为介电泳。作用于粒子的介电泳力既可以是吸引力也可以是排斥力，与粒子相对于介质的可极化性有关。在非均匀交变电场的情况下，介电泳力的大小和方向依赖于电场频率、粒子表面电荷密度和粒子附近自由电荷的变化。对应介电泳力改变符号时的频率称为跨越频率，表征了粒子的介电特性。因此，介电泳可以用来操纵微粒、纳米粒子以及细胞等。

三、与蛋白质相关的静电相互作用

静电相互作用在蛋白质中在三个方面扮演重要角色[27-31]：蛋白质结构稳定，酶催化和生物分子识别。

（一）蛋白质结构稳定

蛋白质结构一般来说是由二级结构单元和蛋白质基序由氢键连接在一起的。一种重要的结构是蛋白螺旋盖帽，在这种结构中，螺旋顶端的非成键的酰胺质子（N端）和羟基氧（C端）提供了高电荷密度以形成氢键，同时还有侧面的疏水基团也发挥作用。螺旋顶端还倾向于和带电基团相互作用。另一个稳定蛋白结构的重要因素是盐桥。虽然由于溶解能和熵贡献的平衡，蛋白质表面的盐桥对蛋白质稳定性有直接贡献，但是埋藏在表面下的盐桥对稳定蛋白也有贡献，比如这种盐桥在嗜热菌的蛋白中大量存在。同时，在蛋白质中，盐桥和氢键都组成网络结构。

除了稳定蛋白质结构外，静电相互作用在决定蛋白质折叠路线上也有重要作用。在二级结构形成的早期，静电作用可以使多肽链无须探索整个能量面。由于静电作用的长程性，可以使随机运动链正确定位折叠结构。对某些蛋白质，最终折叠态中的静电作用在其过渡态中就已经出现了。另外，静电力还影响蛋白质折叠的运动学，比如亮氨酸拉链结构中螺旋之间由于无静电斥力导致其极快的折叠速度。

（二）酶催化

在酶催化反应中，降低反应能垒的物理机制一直是一个备受争议的课题，但静电作用一定扮演着重要角色。理论模型计算也说明了这一点，理由主要有：①静电作用的能量范围正好与反应能垒的数量级匹配，而其他作用有数量级的差异；②为了补偿过渡态中电荷的去溶能，酶必须提供一个适当的极性环境。蛋白质的极性环境还可以稳定反应过渡态。

（三）生物分子识别

生物分子识别原理是生物物理的主要课题之一。已有大量的研究试图从蛋白质-蛋白质界面，蛋白质-DNA 界面的结构数据库的分析中寻找答案。在这些界面中，平均每个包含 10 个氢键，其中 1/3 涉及带电基团。每个界面平均有一个盐桥。当然存在着大的涨落，比如在芽孢杆菌 RNA 酶复合体中有 4 个盐桥。同时研究表明，与蛋白质折叠比较，在蛋白质结合中静电相互作用扮演着不同的角色。在界面存在着氢键和盐桥预示着可能有静电互补性，但仍需进一步的研究。

对于蛋白质-DNA 复合体，DNA 的聚电解质性质使得静电作用非常强。

DNA 骨架上的磷酸根与蛋白质的精氨酸和赖氨酸频繁接触，在骨架和侧链之间形成大量的氢键。对大部分 DNA 结合蛋白质的特性研究表明，盐桥为蛋白质与 DNA 的结合提供了非特异性的亲和力。在蛋白质-DNA 识别中，静电相互作用的另一个重要角色是将蛋白质保持在 DNA 附近的一个小体积范围内，以限制蛋白质的扩散。同样，在蛋白质-RNA 复合体中，也有大量的涉及精氨酸和赖氨酸的盐桥。另外，静电势也在分子识别中起着重要作用，而这在简单结构分析中是难以觉察的。

四、用于蛋白质计算各种层级的静电学模型

对蛋白质静电学的研究涉及很多的建模方法[32-34]，这些方法可以大致分为微观全原子模型、简化极性模型（或者叫简化微观模型），以及连续或者宏观模型。

（一）微观全原子模型

微观模型可以进一步分为经典模型和量子化模型。一个全原子模型讨论的是一个粒子集合的系统，包括蛋白质和溶剂，这些粒子通过量子力学的势能表面相互关联，势能面可以通过力场实验进行估算。尽管这些方法是非常严谨的，但是不管是基于统计力学的方法还是基于分子力学（MD）的方法，在纳秒级的模拟中收敛极其缓慢，主要是静电相互作用是长程力，边界条件的选择也非常关键。但是，有些计算代价较小的全原子方法，比如线性响应近似（LRA）方法和线性相互作用能（LIE）方法，对于基于结构的药物设计非常有用。

（二）简化微观模型

在微观全原子模型的静电场计算中，处理大量的水分子是一项既耗时又复杂的工作。因此，发展了许多简化的模型。一个自然的简化就是将分子或原子基团看成偶极子。蛋白质偶极子-朗之万偶极子（PDLD）模型就是一个简化的微观模型。在这个模型中，靠近催化或结合位点的水分子用偶极子表示，叫作朗之万偶极子，而远离位点的水分子被看成是介电常数为 80 的连续介质。朗之万偶极子是点状的，其时间平均极化以一个朗之万型的函数表示。

在 PDLD 模型中，蛋白质中的每个原子的诱导偶极子是显示表示的，由迭代方法计算。溶质点电荷的值由力场获得，而溶液中移动离子的分布用玻

尔兹曼分布得到。由于电荷分布的变化引起的溶质重组通过分子动力学模拟计算。

由于在 PDLD 模型中电荷不是用一个介电常数隐含处理的，同时用分子动力学模拟来近似蛋白质结构的弛豫，因此所有相关静电作用能都是用库仑定律直接计算且取介电常数为 1 。如果溶质周围是朗之万偶极子，其溶解能也是按库仑定律计算的。如果溶质周围的水是以连续介质表示的，则溶解能利用推广的 Born 模型（GB）估算。溶质和离子之间的静电能也按库仑定律计算，但此时介电常数取 40～80 的值。

（三）连续模型

计算蛋白质静电相互作用的连续模型通过简化蛋白质周围的溶剂的细节以大幅度提高计算效率。在所有连续模型中，溶质相当于镶嵌了原子电荷的一个低介电常数的介质，浸没在高介电常数的离子溶液中。对于给定的离子强度和原子电荷分布，计算蛋白质内和周围每一点的静电势。对于很多生物大分子复合体，用连续模型计算的静电势的图形揭示了静电作用的重要性[35,36]。特别是，连续模型广泛用于处理配体特异性结合和扩散控制配体结合等领域。连续模型主要是基于 PB 方程和 GB 模型。

PB 方程求解器[37]：PB 方程求解器利用空间变化的介电常数、离子强度和电荷分布来计算复杂蛋白质分子的静电势。PB 方程可以用迭代的方法快速求解。溶质的永久偶极矩显示表示为原子中心点电荷，而溶液中的移动离子则利用玻尔兹曼分布表示。介电常数包含了没有在方程中显示表示的所有其他贡献，蛋白质的介电常数值和溶剂不同。溶剂的介电常数（ε_w=80）既包括了溶剂中的永久偶极矩，也包括了溶剂在蛋白质周围弛豫的贡献。蛋白质的介电常数 ε_p 取值范围在 2～20，取决于计算类型和蛋白质结构类型。当 ε_p 只是反映电子极化效应时，取值 2 左右。在 PB 方程模型中，电子极化和电荷重组同时隐含处理时，ε_p 的优化值有时会超过 20。

研究者们已经开发了多种 PB 方程求解器软件包，既有多用途的也有单独用途的。这些软件包被广泛用于预测各类过程中溶解能和静电能。

GB 模型：用 PB 模型来计算一个配体-蛋白质结合自由能通常需要几分钟到几十分钟的机时。在药物筛选计算中，潜在的药物数以千计，因此需要比 PB 模型更高效率的计算方法。一种基于 Born 方程的非常快速的分析模型，即 GB 模型广泛用于配体结合的自由能计算，计算机辅助药物设计（CADD）和蛋白质构型分析。

用于配体结合的 GB 模型假设两个电荷之间静电作用的屏蔽可以由电荷和溶剂之间的相互作用来估算。溶质由一组原子球体来描述，每个原子球体有一个固定半径，称为有效 Born 溶解半径。每个原子的点电荷放置在球的中心。与 PB 模型一样，大分子是一个低介电常数区，周围是高介电常数的水性环境。在 GB 模型中，电荷之间的作用能是由真空中的库仑定律计算的。把一个分子由蛋白质转移到水性环境中的静电溶解能（也叫极化能）由 Born 近似计算得到。每个原子的有效 Born 溶解半径可以看成是电荷与蛋白质-溶剂边界的距离。这个参数是可调的，其值可以由 PB 方程计算的溶解自由能来推算。GB 模型无法计算蛋白质的静电势图，同时模型参数通常要利用 PB 方程解算器来优化。

五、静电学原理和计算方法在生物大分子中的应用

（一）聚电解质周围的平衡离子分布

生物大分子大部分都是聚电解质，如双链 DNA、肌动蛋白、微管骨架等。每个聚电解质周围都分布着与之相关的平衡离子。以 DNA 分子为例，双链 DNA 是一种高带电的聚电解质，每个碱基对（bp）带电-2e，每个单元电量属于一个骨架上的磷酸基团。DNA 也是自然界中刚性最高的聚合物。由于高带电量的特性，当周围环境存在平衡离子时，棒状的 DNA 会被周围的平衡离子所凝聚。在 20 世纪 70 年代，Manning 基于 PB 方程引入了一种理论方法去计算引起凝聚的平衡离子的浓度。在 Manning 平衡离子凝聚模型中，聚电解质是一根无限长的带电细棒。运用简单的静电学和自由能参数预测了电荷被不同化合价离子补偿的比例，发现凝聚与 Manning 常数 $\xi=z\lambda l_B$ 有关，这里的 z 是平衡离子的价态，λ 是细棒的有效线电荷密度，l_B=7Å 是常温下的 Bjerrum 长度。如果 $\xi<1$，DNA 的电荷无法克服熵对平衡离子结合的影响，导致平衡离子不会结合到聚电解质上。另外，如果 $\xi>1$，一部分平衡离子就会凝聚到 DNA 上，这时电荷线密度可以重整为 $\lambda_{eff}=(ql_B)^{-1}$。在单价平衡离子的情况中，Manning 理论预言，由于 DNA 的高带电量，每个磷酸基团只有 0.76 个单价离子可以用于凝聚，与所加入的盐的浓度无关。DNA 周围平衡离子的浓度因此是三个正电荷对应于每两个碱基对。温度的影响，体现在 Bjerrum 长度和 Manning 常数 ξ 上。

可以看出，Manning 常数 ξ 与平衡离子的价态成正比，所以多价离子可以更容易使 DNA 凝聚。Manning 理论与平衡离子的化学性质无关，但实际

上 DNA 凝聚还是跟周围离子的化学性质有关的，即使是一价离子。

这些现象很可能归因于离子与溶剂的相互作用。有较高电荷密度的小的单价离子（如 Li^+ 或 Na^+）会与水分子结合较强，而低电荷密度的大的单价离子与水分子（如 K^+，Cs^+，以及精氨酸，组氨酸和赖氨酸的侧链）的结合力较弱，结果导致离子的可溶性差异，因此吸附在大分子的表面的能力也会有所差异。

Manning 的简单理论也已经被更多复杂的方法所证实。平均场理论，其中包括 PB 方程，可以定量地给出关于电荷棒的离子分布。但是，平均场理论忽视了离子的离散性质，包括离子的有限大小以及它们之间的相关性。聚电解质的一个特别的性质是在盐溶液中同类电荷会发生相互吸引。在平均场理论中，不管在何种盐溶液里，同类电荷物质（如聚电解质）均会互相排斥。改变离子的化合价会改变屏蔽作用和它们之间的排斥力，但不会导致吸引。非线性 PB 方程能够给出 1∶1 离子溶液中高带电量的聚电解质的离子分布，与实验值相符，但对于高价离子，偏差较大。

在多价离子或高表面电荷密度的情况下，聚合电解质之间的复杂相互作用是由聚电解质周围的凝聚离子的组织和它们的动力学决定的。蒙特卡罗模拟结果显示，类似电荷的 DNA 之间确实存在相互吸引。相互吸引有着不同的物理起源，比如范德瓦耳斯相互作用，或者是强带电体系的静电特性。定性上，多种方法，比如密度泛函理论、积分方程和场论计算，均得出相同趋势。一个重要的共识是凝聚在聚电解质上的平衡离子间的关联产生吸引。如果这些离子形成一个有序的晶格，就会吸引有类似电荷排布的相邻宏观离子。Rouzina、Bloomfieldand 和 Shklovskii 的理论中认为这种排布就是简单的 Wigner 晶格。近期发展的场论方法主要用于强耦合极限下离子分布预测，其中平衡离子电荷或表面电荷浓度均超出 PB 理论的适用范围，平衡离子间有非常强的关联。

（二）与实际相关的聚电解质的凝聚

由多价离子和 / 或宏观离子引起聚电解质凝聚对许多基本的生物学和生物医学过程是非常重要的。在这些过程中，随着带相反电荷的多价离子或者大分子的浓度增加，溶液中聚电解质链压缩或折叠成一个紧凑的结构，如病毒、细菌、染色体中 DNA 的凝聚和包装。在基因治疗中，人工基因传递系统也需要阳离子聚合物，树枝状大分子或者阳离子双亲性的细胞膜来包装 DNA。一般来说，带电两亲性膜可以利用聚电解质自组装为各种形状的结构。

生物聚电解质的静电聚集也与很多疾病的发生有关。例如，组蛋白促进人突触核蛋白的聚集和纤维化，在帕金森病的发病机制中起了重要作用。在囊性纤维化相关的疾病中，阴离子性炎症性聚合物（如 DNA、肌动蛋白）和细胞外的细菌纤维会结合并水解带正电的内源性抗菌蛋白（如溶菌酶、乳铁蛋白），从而使它们失活。这种结合会导致永久性的甚至致命的呼吸道感染。这方面的深刻理解，对开发预防和治疗气道炎症的抗菌剂有启发和指导意义。

六、应用实例

（一）DNA

1. 离子介入的 DNA-DNA 相互作用

实验和数值计算均表明，带相同电荷的生物大分子（如双链 DNA）之间会发生互相吸引。这种离子介入的类似电荷吸引无法用 PB 理论解释，因此发展了超越平均场的各种方法试图解释这一现象。实验表明，吸引力的大小以及可能的物理机制都与平衡离子的种类有关。目前主要有三种解释：

（1）在单价离子溶液中也可以观察到纳米级别的 DNA 团簇，同时维里系数的测量也与离子介入的排斥相互作用理论不一致。为了解释这些观察，Manning 推广了其凝聚理论，其中涉及两种平衡离子：凝聚的平衡离子和溶液中的平衡离子。这个理论解释了两个双链 DNA 分子在有限的距离上的相互吸引，其物理图像类似于共价键机制，两个临近的双链 DNA 共享凝聚离子导致了平动熵的增高。

（2）化合价大于 +3 的离子，或某些特殊的二价金属离子（如 Co^{2+} 和 Cd^{2+}）可以导致 DNA 的聚集。对于 DNA 聚集（或压缩），现有两类理论来解释这些现象。一种方案叫作 Kornyshev-Leikin（KL）理论。这个理论保留了 PB 理论的框架，引入平衡离子修饰 DNA 的概念。DNA 经平衡离子修饰后，其上的电荷分布发生了变化，导致 DNA 之间的相互吸引。其中假设 DNA 不仅包含负电荷，也有一部分正价离子不可逆转地吸附在 DNA 的沟槽中。根据这个假设，我们可以计算双链 DNA 之间的相互作用，相互作用力以指数形式衰减，衰减的距离几乎与盐的种类无关。这种静电吸引力来源于分布在 DNA 片段之间的偶极矩。同源双链 DNA 之间有最大的静电吸引力，当其中一个 DNA 相对于另一个 DNA 的轴移动一半螺距时，就形成一个理想的静电拉链。如果双链 DNA 分子不是同源的，仍然会产生吸引力，但相对较弱。在第二种理论中，裸 DNA 分子在溶液中只带有负电荷，但是系统中有很强的离子-离子关联和离子-DNA 关联，就是所谓的强耦合机制，其中离子-DNA 强关联导致大部分平衡离子凝聚在 DNA 表面，形成二维强关联液体结构。这种强关联二维液体形成一个二维 Wigner 晶格，这时 DNA 表面的电荷分布显示出一种偶极矩结构，这会导致与 KL 模型中“经典拉链”相似的指数衰减吸引。Wigner 晶格给出的结果与最近的蒙特卡罗模拟非常一致。

（3）大部分带电荷的凝聚剂是多个带正电的棒状分子（如多胺）。这种特殊的结构提供了另一种类似电荷吸引机制，不同于离子-离子关联机制。实际上，Bohinc 组的研究发现，利用平均场理论，棒状二价离子会引起两个同种带电板间相互吸引。当两个板块之间的距离接近杆的尺寸时，静电力会促使平衡离子形成垂直于大分子方向的连接桥。假设一个杆状离子处于两个板块的正中，与板平行放置，杆是不稳定的，而垂直放置是稳定的。最后，当二价棒的长度大于某个阈值时，这个“桥接”的能量收益可能超过同类电荷排斥的能量代价，就会发生相变，导致同类电荷吸引。

跟棒状聚合物对比，DNA 分子是一种具有三个独立自由度的半柔性聚合物：弯曲，扭转和收缩。DNA 的弯曲和扭转与其功能相关。在 DNA 凝聚中，弯曲的能量成本是一个重要的考虑因素。由于双链 DNA 具有长的弹性持久长度，大约是 50nm，因此长的 DNA 链可以抽象为蠕虫链。基因组中的 DNA 很长，所以需要有效地包装。例如，DNA 在 T4 噬菌体基因组中包含 160 个 kbp，伸展长度接近 54μm，然而它可以包装进一个直径为 100nm 的病毒衣壳。DNA 包装中需要考虑各种能量平衡。高线性电荷密度会导致强烈的相邻链之间的库仑排斥，还有刚性 DNA 变形导致的 DNA 构型熵损失和能量代价。另外，平衡离子对系统的总自由能中的焓和熵都有贡献。在噬菌体体内 DNA 包装需要有 ATP 驱动的分子马达将 DNA 泵入病毒衣壳，以对抗内部压力。在真核基因组中，有多个蛋白质参与 DNA 的包装，特别是 DNA 环绕的组蛋白。

DNA 在体外的凝聚或聚集，可以由多种凝聚剂引起。除了多价离子，还有基本蛋白质、拥挤高分子、阳离子脂质体和多醇类等。在多价离子环境中，凝聚后的 DNA 形成圆环或棒状结构。在这些高度密集结构里，一部分 DNA 会形成高度规整的六角形晶格。渗透压可用于推动 DNA 有序排列并且可以影响静电效应。

除了静电作用和渗透压，离子和 DNA 的具体结合位点之间的化学反应也十分重要。同价的不同离子可能有不同的结合模式（Mn^{2+}，Ca^{2+} 或 Mg^{2+}）和不同的 DNA 结合位点。离子大小和几何形状的差异也有明确的影响。离子水化和特定的结合也有一定的影响。另外的因素包括结合表面的结构，如 DNA 沟槽的几何形状和螺距等。最后，虽然和离子的分布没有明确的联系，水化力也必须要考虑。

2. DNA 与蛋白质相互作用

DNA 结合蛋白也是一种宏观离子，表面有带电区块，以利于与 DNA 结合。带电区块的形状和电荷分布决定其与双链 DNA 的相互作用模式。当然，区块一般带正电，但是某些酶也有负电的带电区块，比如 DNase I。蛋白质与 DNA 的结合也受多价离子控制，类似于离子介入的 DNA-DNA 吸引。有意思的是，识别和结合特定位点的蛋白质，比如转录因子和限制性内切酶，平均来说比非特异结合蛋白（比如组蛋白或类核结合蛋白）带电量要低。特异性结合蛋白带电量低的原因可能与其结合方式有关。带电量低有利于其在 DNA 上滑动，以便找到正确的结合位点。的确，这类蛋白都呈现典型的凹陷形，形成一个面积较大的与 DNA 接触的适配界面（面积在几十平方纳米量级）。对于 DNA 结合蛋白，由于电中性的要求，补偿离子集中在 DNA 与蛋白质的带电区块之间。当蛋白质接近 DNA 时，离子密度的增加引起局部升高的渗透压。这个渗透压导致的排斥作用使得蛋白质无法接近 DNA 表面，以阻止化学键的形成。这样，蛋白质的移动性就增加了，沿着 DNA 的滑动成为可能。简单的模型计算表明，带电区块越大，则局部束缚离子越少，蛋白与 DNA 的平衡距离越短。因此，高带电的非特异性结合蛋白就直接粘在 DNA 上了；相反，特异性结合蛋白由于排斥较强，与 DNA 的距离还无法形成化学键，因此有足够的灵活性可以在 DNA 上滑动。

（二）RNA

与 DNA 相同的是，RNA 也是带负电的并且吸引带正电荷的平衡离子。与 DNA 不同的是，在生理条件下 RNA 分子是单链的。RNA 结构包含较硬的、短的螺旋部分，连接着更加柔软的单链部分、环状部分、链状部分以及节点。RNA 通常折叠成紧凑的三级结构。RNA 的折叠通常需要镁离子，只要有毫摩尔浓度的镁离子存在就会发生折叠。RNA 的平衡离子可以分成两类，一类是少数与 RNA 特定位点结合的平衡离子，另一类是大量的流动平衡离子。

镁离子在 RNA 折叠中扮演着中心角色。首先，RNA 的结构是离子-RNA

作用的重要决定因素，因此对 RNA-离子静电学的严格处理必须要考虑 RNA 的构型特点。其次，离子的电荷和尺寸也是 RNA 折叠的重要决定因素。离子的尺寸不仅决定了离子与 RNA 的最小距离，也会影响到离子-离子的距离，进而影响到库仑作用的强度以及排空体积关联。实验发现，二价离子电荷密度（即电荷 / 离子体积）与 RNA 折叠的稳定性成正比。对于大的 RNA 分子，数值模拟还表明：①凝聚在 RNA 附近的离子构成类似液体的离子关联；②非特异结合的流动离子分布就可以决定 RNA 的折叠态，以及结构稳定性对离子电荷密度的依赖关系。对于小的 RNA 分子，特异结合离子对静电作用有较大影响。

镁离子导致的 RNA 高效折叠的驱动力一直是 RNA 静电学中理论和实验的主要课题。对于 RNA 折叠，镁离子只需要毫摩级浓度，而钠离子需要摩级浓度。这种差异无法用简单的离子强度解释。计算机模拟结果显示，用平均场的 PB 方程计算的 RNA 折叠所需的镁离子浓度要比实验值大 10 倍，因此离子关联和离子尺寸效应必须考虑在内。紧束缚离子模型提供了一种处理离子强关联的方法，预言了毫摩级的 RNA 折叠相变。

离子水化在离子与 RNA 的作用中扮演了重要角色。完全水化的离子通过屏蔽骨架电荷，使得 RNA 螺旋互相接近，这样使得已经折叠的 RNA 分子更加稳定。脱水的离子与 RNA 骨架有特异的接触，同时在链的各个部分之间产生牢固的连接。

第三节　生物大分子静电学的前沿问题

生物大分子的静电学是一个高带电体系的基础化学物理和多种生物过程以及生物医学过程的交叉学科。由于理论和计算技术的发展，以及新的实验手段的应用，现在可以开始对各类原型体系（如 DNA、RNA 和蛋白质等）利用统一的概念和理论来加以梳理和解释。归纳起来，生物大分子的静电学有以下五个方面的基础科学前沿问题。

（1）平衡离子相关的基本问题：平衡离子在产生生物大分子之间的相互作用力中扮演着中心角色。探测平衡离子在生物带电体周围的分布，平衡离子的关联以及动力学就显得非常关键。在实验方面，已经有小角度 X 射线散射、高分辨非弹性 X 射线散射以及同步辐射 X 射线衍射等方法用于探测平衡离子的分布。在理论和计算方面，也发展了非线性 PB 方程的求解方法、场论方法以及强关联理论等。

（2）同类电荷相互吸引和电荷逆转问题：这是软物质和生物大分子静电学的中心问题。由于平衡离子的存在，溶液中静电作用被一定程度地屏蔽，同时由于熵效应，带同类电荷的生物大分子会互相吸引，进而形成有序结构。虽然对这个问题已经作了广泛的研究，对其有了更深入的理解，但是仍然没有一个大家普遍接受的物理机制和理论能够给出与实验数据一致的结果。对于电荷过补偿或者电荷逆转问题，情况也十分类似。一般认为，平衡离子的尺寸效应和关联效应是电荷逆转问题的物理根源，但是也缺乏一个统一的理论。

（3）蛋白质相关的静电学问题：由于计算能力的限制，对于与蛋白质相关的静电学问题还不能全部用全原子的方法进行分析和计算。半微观的连续模型方法是蛋白质静电势和其他电学性质计算的可行手段。但是，由于蛋白质内部和周围水分子不能显示地处理，连续模型的结果依赖于蛋白质介电常数这个参数的获得。所以，水分子的显示处理和蛋白质介电常数的实验校准是发展新的计算方法的一个重要方向。

（4）DNA 相关的静电学：DNA 相关的静电学包括离子介入的 DNA-DNA 相互作用，DNA 与蛋白质的相互作用以及 DNA 高级结构形成的动力学。其中涉及多种物理机制：一是涉及多价离子导致的 DNA 凝聚和聚集，包括渗透压的贡献；二是在组装蛋白介入下 DNA 等级结构的形成；三是 DNA 结构的组装。这些过程涉及的基本物理机制都需要深入的研究。

（5）RNA 折叠中的静电学：RNA 折叠的显著特点是其平衡离子分成固定和流动两类。固定平衡离子是与化学键相关的，而流动离子分布是与物理相关的。为了搞清 RNA 折叠的机制，需要先搞清几个关键问题：静电力（包括关联）和非静电力（如熵力）的大小，离子的特定结合和非特定结合的贡献，镁离子的关联效应的大小以及溶剂效应等。

第四节 未来 5～10 年重点发展方向

一、生物大分子附近平衡离子分布和动力学

平衡离子的分布、关联和动力学在生物大分子的静电学中起着至关重要的作用。利用各种实验手段获得这些信息对于理解其中的物理机制至关

重要。

一个问题是区分结合的平衡离子和游离的平衡离子。可能的检测方法有:离子的局域化导致其核磁共振弛豫率发生变化;测量发光离子的能量转移，这种能量转移依赖于碰撞频率，可用于区分束缚离子和游离离子；应用离子敏感染料可以精确测量游离离子的浓度。另一个方法是离子计数实验，可以测量带电生物大分子附近的平衡离子数和伴随离子数。

为了获取离子在空间中分布的信息，此信息与平衡离子引起的类似电荷吸引机制密切相关，需要在空间测量中精确到几个埃的技术。小角度 X 衍射可以提供系统中各个组分分布的空间信息。对于 DNA-平衡离子系统来说，平衡离子的分布对 X 射线散射图谱有调制作用，而新的反常小角度 X 射线散射检测平衡离子效果更加明显。实验得到的平衡离子分布数据可以用来检验 PB 理论的精度。现有的数据表明，对于一价离子，用离子大小修正过的非线性 PB 方程与实验数据一致。但是，初步结果表明，对于高价离子，平均场理论的结果与实验数据有较大的误差。反常小角度 X 射线散射实验表明带电量越高的离子与 DNA 越接近。因此，需要更多的关于高价平衡离子分布的数据，包括各种单独平衡离子分布的数据，或者它们之间相互竞争的分布数据。带电大分子可以是 DNA、RNA、蛋白质以及其他带电高分子。这些数据一方面用于检验现有的超越平均场方法的各种理论，另一方面也为深化和发展更精确的理论提供线索和启发。最后，即使 DNA 与 RAN 分子序列完全相同，其周围离子的分布也是不同的，说明螺旋的几何结构对离子凝聚也有重要影响。

另一个重要问题是平衡离子关联的实际测量。关联的信息对于理解同类电荷吸引和电荷逆转的物理机制是非常重要的。已经有实验对肌动蛋白的平衡离子的关联用同步辐射 X 射线衍射进行了测量。有趣的是，平衡离子并不随肌动蛋白的对称性产生相应的晶格，而是平行于肌动蛋白组成一维平衡离子密度波。而且，这些平行的带电密度波与带相反电荷的聚电解质的扭转畸变相互耦合，以优化聚电解质之间的吸引力。另外，离子分布如何在不同的条件下进行调整也是一个非常有意义的课题。比如现已知道，在棒状的聚电解质周围，随着棒状的聚电解质之间的空间减小，平衡离子晶格的结构会经历一系列的剪切变换。

在理论和计算方面，主要课题有：①精确求解 PB 方程，包括解析方法和数值方法。事实上，进一步深化 PB 方程也可以包含电荷关联效应，比如重正化 PB 方法可以用于处理高浓度电解液的情形，并与大规模数值模拟结果一致，并预言了电解液中的电荷振荡现象，正确地刻画了高浓度电解液中的电荷关联性[38]；②发展和细化了现有的强关联理论。比如，对于 DNA 之间相互吸引的机制，到底是表面关联离子液体还是电荷交叉的拉链机制，还需要进一步地深入研究。再如，关联离子晶格模型是普遍适用还是有其他的机制。对于一些体系的实验结果表明离子关联有多种形式。

二、蛋白质相关的静电效应

静电相互作用是理解蛋白质功能的关键因素之一。连续模型在处理静电相关过程，如酶催化、蛋白质稳定性、配体结合和质子化平衡等，获得了与实验值定量符合的结果。连续模型的主要问题是对于不同的应用需要调整蛋白质的介电常数，而介电常数又不能先验获知。连续模型的一个改进方向是内部水分子的显示处理。另一个方向是与实验结合，利用核磁共振滴定实验和振动 Stark 谱探测蛋白质内部的电场来校准计算参数。

连续模型中水是按介电常数为 80 的连续介质来处理的，这样就可以高效地处理水的溶解性质而不用显示地计算数以千计的水分子之间的静电相互作用，然后再平均到整个介质。但是，水分子有时并不是完全平均的，导致连续模型给出不准确的结果。比如在配体-蛋白质复合体的计算中，在它们之间的水分子紧束缚在作用点和空腔中，对这些水分子就需要显示地处理，而不能用连续介质代替。但是，选择哪些水分子做显示处理同时确定它们在蛋白质周围的位置仍然是一个有挑战的课题。

连续模型的精确计算依赖于蛋白质介电常数的校准。pH 相关的核磁共振滴定曲线可以测定蛋白质中很多原子核的电场效应，提供了蛋白质内部的高分辨率的电场强度数据，甚至可以作出原子分辨率的 3D 介电常数图谱。更进一步，可利用局域介电常数的关联以及其他局域性质，如电荷分布、局域弹性、与蛋白溶剂边界距离、原子堆积密度和氨基酸类型等，进一步改进和发展连续模型。

另一个探测蛋白质内部电场的方法是 Stark 效应谱。选取适当的探针不仅可以探知蛋白质内部电场大小的变化，还可以检测电场方向的变化。事实上，当蛋白质残基发生突变，蛋白质构象发生变化以及与配体结合时都可以检测到局部电场的变化。实验数据与经典的 PB 方程计算的结果比较发现，理论计算有较大的偏差。因此，利用核磁共振和 Stark 谱的实验数据来优化现有的静电学模型是深入理解蛋白质静电学的一个可行且有成果的研究方向。

三、RNA 折叠相关的静电学问题

对于 RNA 折叠及其中的离子效应，关于大分子拥挤环境影响的机制尚不明确。实验表明，溶液中 PEG 分子的加入可以极大地提升 Mg^{2+} 在 RNA 折叠中的效率，20% 体积比的 PEG 分子可以将 Mg^{2+} 的效率提高 4 倍以上。拥挤分子对 RNA 折叠中离子效应的影响与拥挤分子的性质和尺寸相关，一个理论问题就是预测大分子拥挤环境下 RNA-离子相互作用。

另外，由于 RNA 结构以及 DNA/RNA 人工结构研究的发展，短链螺旋间相互作用引起了人们极大的关注。与长链螺旋相比，短链螺旋间相互作用较弱，转向自由度变大，且由于分子小，螺旋的结构细节和螺旋尾端效应得以凸显，使短链核酸间的相互作用与长链有很大区别。例如，近来小角 X 射线衍射实验显示，高浓度 Mg^{2+} 可以导致短链 DNA 间相互吸引，这与长链 DNA 情形有着明显不同。截至目前，对 Mg^{2+} 能否导致（短链）螺旋间相互吸引依然存在争议，对短链核酸螺旋-螺旋相互作用以及 RNA 静电轰塌态中离子效应，特别是 2+ 离子效应也还缺乏全面深入的理解和预测，而短链核酸螺旋-螺旋相互作用中离子效应也是一个焦点问题。

四、电泳相关的静电学问题

电泳是一种广泛用于生物大分子分离和检测的重要手段。电转动和介电泳是传统电泳在近年来的进一步发展，在癌症研究、生物组织分离和鉴别以及生物芯片设计等方面有广泛的应用，但其物理基础涉及多极电荷效应和交变电场效应等，需要进一步的模型构建和数值模拟。

第五节　结　　语

本章主要讨论了生物大分子静电学的基本物理原理、相关的物理模型和计算方法，分析了生物过程中重要的典型应用，包括离子介入的 DNA-DNA 相互作用、DNA-蛋白质的相互作用、蛋白质相关的静电学问题、RNA 折叠中的物理机制以及电泳相关的静电学问题等，指出了今后一段时间内可能的发展方向。

杨光参（温州大学物理与电子信息工程学院）

参考文献

[1] 陆坤权，刘寄星．软物质物理学导论．北京：北京大学出版社，2006.

[2] Daune M, Duffin W J. Molecular Biophysics: Structures in Motion. Oxford: Oxford University Press Oxford, 1999.

[3] 菲利普·纳尔逊．生物物理学：能量、信息、生命．黎明，戴陆如译．上海：上海科学技术出版社，2006.

[4] 菲利普斯 R，康德夫 J，塞里奥特 J．细胞的物理生物学．涂展春，王伯林，等译．北京：科学出版社，2012.

[5] Wong G C, Pollack L. Electrostatics of strongly charged biological polymers: ion-mediated interactions and self-organization in nucleic acids and proteins. Annual Review of Physical Chemistry, 2010, 61: 171-189.

[6] Wong G C. Electrostatics of rigid polyelectrolytes. Current Opinion in Colloid & Interface Science, 2006, 11: 310-315.

[7] Wang Y, Ran S, Man B, et al. Ethanol induces condensation of single DNA molecules. Soft Matter, 2011, 7: 4425.

[8] Wang Y, Ran S, Man B, et al. DNA condensations on mica surfaces induced collaboratively by alcohol and hexammine cobalt. Colloids and Surfaces B: Biointerfaces, 2011, 83: 61.

[9] Carrivain P, et al. Electrostatics of DNA compaction in viruses, bacteria and eukaryotes: functional insights and evolutionary perspective. Soft Matter, 2012, 8: 9285-9301.

[10] Wang Y, Ran S, Yang G. Single molecular investigation of DNA looping and aggregation by restriction endonuclease BspMI. Scientific Reports, 2014, 4: 5897 (1-8).

[11] Vizcarra C L, Mayo S L. Electrostatics in computational protein design. Current Opinion in Chemical Biology, 2005, 9: 622-626.

[12] Zhang Z, Witham S, Alexov E. On the role of electrostatics in protein-protein interactions. Physical Biology, 2011, 8: 035001(1-8).

[13] Chen S J. RNA folding: conformational statistics, folding kinetics, and ion electrostatics. Annual Review of Biophysics, 2008, 37: 197.

[14] Dong L, Huang J, Yu K. Theory of dielectrophoresis in colloidal suspensions. Journal of Applied Physics, 2004, 95: 8321-8326.

[15] Huang J, Karttunen M, Yu K, et al. Electrokinetic behavior of two touching inhomogeneous biological cells and colloidal particles: Effects of multipolar interactions. Physical Review E,

2004, 69: 051402(1-6).

[16] Levin Y. Electrostatic correlations: from plasma to biology. Reports on Progress in Physics, 2002, 65: 1577.

[17] Grochowski P, Trylska J. Continuum molecular electrostatics, salt effects, and counterion binding-a review of the Poisson-Boltzmann theory and its modifications. Biopolymers, 2008, 89: 93-113.

[18] Chu V B, Bai Y, Lipfert J, et al. A repulsive field: advances in the electrostatics of the ion atmosphere. Current Opinion in Chemical Biology, 2008, 12: 619-625.

[19] Botello-Smith W M, Cai Q, Luo R. Biological applications of classical electrostatics methods. Journal of Theoretical and Computational Chemistry, 2014, 13: 1440008.

[20] Andelman D. Introduction to electrostatics in soft and biological matter. Proceedings of the Nato ASI & SUSSP on Soft Condensed Matter Physics in Molecular and Cell Biology, 2006: 97-122.

[21] Jönsson B, Lund M, Barroso da Silva F L. in Conference on Food Colloids, 2006: 129-154 (Royal Society of Chemistry).

[22] Warshel A, Sharma P K, Kato M, et al. Modeling electrostatic effects in proteins. Biochimica et Biophysica Acta (BBA)-Proteins and Proteomics, 2006, 1764: 1647-1676.

[23] Grosberg A Y, Nguyen T, Shklovskii B. Colloquium: the physics of charge inversion in chemical and biological systems. Reviews of Modern Physics, 2002, 74: 329.

[24] Besteman K, Van Eijk K, Lemay S. Charge inversion accompanies DNA condensation by multivalent ions. Nature Physics, 2007, 3: 641-644.

[25] Qiu S, Wang Y, Cao B, et al. The suppression and promotion of DNA charge inversion by mixing counterions. Soft Matter, 2015, 11: 4099-4105.

[26] Garfin D E. Gel Electrophoresis of Proteins. Oxford1, 2003, 197-268.

[27] Ren P, Chun J, Thomas D, et al. Biomolecular electrostatics and solvation: a computational perspective. Quarterly Reviews of Biophysics, 2012, 45: 427-491.

[28] Cisneros G A, Karttunen M, Ren P, et al. Classical electrostatics for biomolecular simulations. Chemical Reviews, 2014, 114: 779-814.

[29] Sheinerman F B, Norel R, Honig B. Electrostatic aspects of protein-protein interactions. Current Opinion in Structural Biology, 2000, 10: 153-159.

[30] Koehl P. Electrostatics calculations: latest methodological advances. Current Opinion in Structural Biology, 2006, 16: 142-151.

[31] Tobias D J. Electrostatics calculations: recent methodological advances and applications to membranes. Current Opinion in Structural Biology. 2001, 11: 253-261.

[32] Kukic P, Nielsen J E. Electrostatics in proteins and protein-ligand complexes. Future

Medicinal Chemistry, 2010, 2: 647-666.

[33] Sheinerman F B, Honig B. On the role of electrostatic interactions in the design of protein-protein interfaces. Journal of Molecular Biology, 2002, 318: 161-177.

[34] Warshel A, Dryga A. Simulating electrostatic energies in proteins: perspectives and some recent studies of pKas, redox, and other crucial functional properties. Proteins: Structure, Function, and Bioinformatics, 2011, 79: 3469-3484.

[35] Simonson T. Macromolecular electrostatics: continuum models and their growing pains. Current Opinion in Structural Biology, 2001, 11: 243-252.

[36] Collins K D. Why continuum electrostatics theories cannot explain biological structure, polyelectrolytes or ionic strength effects in ion-protein interactions. Biophysical Chemistry, 2012, 167: 43-59.

[37] Li C, Li L, Petukh M, et al. Progress in developing Poisson-Boltzmann equation solvers. Molecular Based Mathematical Biology, 2013, 1: 42-62.

[38] Lu B S, Xing X. Correlation potential of a test ion near a strongly charged plate. Physical Review E, 2014, 89: 032305(1-23).

第九章
癌细胞信号处理的生物物理建模

第一节　引　　言

癌症是严重威胁人类健康的重大疾病，全球每年约 800 万人死于癌症。我国已成为世界第二大癌症高发国。2017 年度的《中国肿瘤登记年报》公布的关于 2013 年癌症统计数据显示[1]，在男性人群中，发病率排在前三的癌症分别是肺癌、胃癌和肝癌；死亡率排在前三的分别是肺癌、肝癌和胃癌。在女性人群中，发病率排在前三的癌症分别是乳腺癌、肺癌和结肠癌；死亡率排在前三的分别是肺癌、胃癌和肝癌（图 6-9-1 和图 6-9-2）。据预测，2025 年我国肺癌病人数将居世界之首，每年新增死亡人数将超过 100 万。治疗癌症成为人们迫切的期盼。

肿瘤分为良性和恶性两种，恶性肿瘤就是癌症。癌症是对 200 多种疾病的总称，所有癌症都起始于正常细胞的失控性生长。正常细胞转化为肿瘤细胞是一个复杂的多步骤过程，化学因素、物理因素、生物因素和遗传因素等是其诱因。肿瘤研究关注的重大问题包括肿瘤的发生发展机制，早期诊断的标志物，靶向治疗与综合预防等。

1971 年，美国总统尼克松颁布《国家癌症法》，宣称要在十年内消除癌症带来的死亡，吹响了美国向癌症宣战的号角。几十年来，基于生物技术的各种新方法和突破，为阐明肿瘤细胞产生、发展、侵袭和转移行为等的分子机制，提供了许多新视角、新思路，使我们在肿瘤生物学、肿瘤诊断和医治预后等方面取得了许多重大的进展。但是，这些研究成果并未有效地为临床

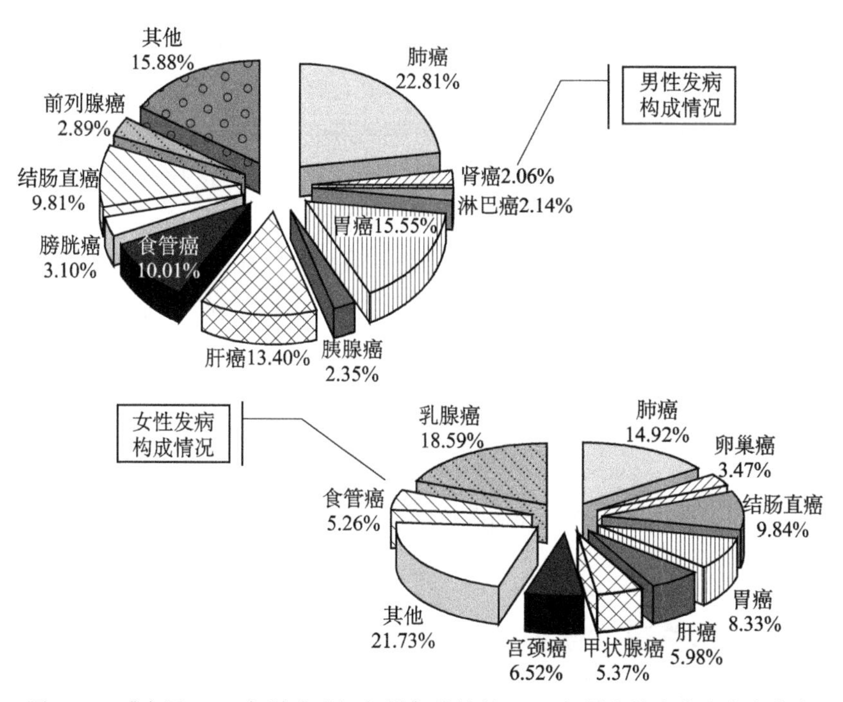

图 6-9-1 《中国 2017 年肿瘤登记年报》总结的 2013 年男女前十位发病率分布

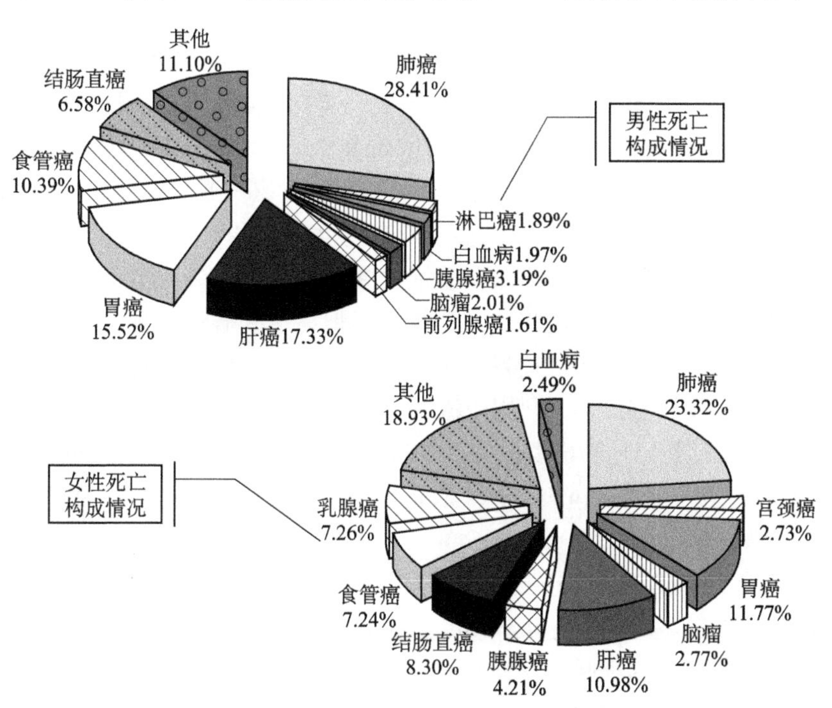

图 6-9-2 《中国 2017 年肿瘤登记年报》总结的 2013 年男女前十位死亡率分布

治疗提供实质性突破。当前，癌症的主要治疗手段是手术切除、放疗和化疗，目的是根除癌细胞。但上述治疗方法没有解决病因问题，癌症复发是必然的。化疗药物还诱发癌细胞产生抗药性，最终让癌症患者陷入无药可救的境地。尽管靶向精准治疗和免疫治疗现在逐渐成为肿瘤治疗的重要方法，且是从肿瘤发生病因上解决问题，但效果依然不那么显著。

自 2010 年起，美国国立卫生研究院（NIH）美国 NIH 已花费 1000 多亿美元用于肿瘤研究，但癌症的死亡率没有显著下降。这说明经过几十年的生物医学研究，癌症治疗仍然面临严峻的挑战，当前的癌症研究急需寻找新的突破[2]。

第二节　癌细胞物理学研究发展历程

一、肿瘤物理学研究的兴起和必要性

在 20 世纪 50 年代以前，癌症研究还是一个生命科学家和物理学家联合攻关的领域，但随着分子生物学的兴起和蓬勃发展，癌症的研究逐渐聚焦到其细胞、分子和基因层面，先前的生物学和物理学间的协作渐渐被遗弃。为鼓励交叉研究，2009 年美国国家癌症研究所专门成立了一个物质科学-肿瘤学中心（US National Cancer Institute Physical Sciences-Oncology Centers），联合并推动物理、化学、数学及工程科学家与生物学家一起研究癌症。

事实上，癌症的物理学研究这些年已成为癌症研究领域的新兴力量，在许多研究领域，物理学研究方法和技术发挥着独特且重要的作用。例如，用纳米技术制造的药物载体能将化疗药物准确地运送到病灶；基于物理学原理的单细胞成像技术应用于肿瘤发病早期诊断鉴定中；利用激光成像技术可预测实体瘤浸润及转移倾向；利用微流芯片技术能够更精确地诊断癌症突变类型和侵袭能力；利用计算机模拟可以定量理解癌细胞的发生和发展机制等。

2011 年 *Nature Reviews Cancer* 期刊发表综述文章《物理学能为癌症做什么？》（*What does physics have to do with cancer*）[3]。2012 年 11 月 22 日《自然》杂志发表含 13 篇综述和研究文章的专辑，聚焦物理学家关于癌症研究的物理方法和成果[4]。其中，美国 Moffitt 癌症研究所的 Robert Gatenby 教授指出，将研究焦点集中在癌细胞的突变基因上，可能会只见树木，不见森林。这是因为越来越多的研究表明，癌症不是一种单纯的基因突变疾病。单细胞水平上，癌症可以看成是信号转导通路疾病，而其群体增殖和侵袭转移，更与癌细胞间及其微环境的相互作用有密切关系。在物理学家看来，分子生物学关于癌症的研究过分注重了基因“细节”，反而忽略了“主体”，导致无法知道“生命机器”的崩溃具体哪儿出了故障。海量的基因组、转录组、蛋白质组、代谢物组等数据，并未帮助生物学家彻底厘清肿瘤发生的头绪。

二、癌细胞信号处理的物理建模研究的特点和历程

癌症生物学家主要专注于癌细胞内部的结构认知和功能调控，如 DNA 的结构变异和蛋白分子的功能调控等；病理学家关心的是癌症的临床特征和治疗。而癌细胞信号处理的物理建模重点是通过理论建模和计算机模拟，综合运用系统生物学、非平衡统计力学、非线性动力学、复杂系统自组织理论和生物信息学等，从动力学角度定量且系统地研究肿瘤细胞内信号转导机制、癌细胞群体迁移动力学以及癌细胞的药物治疗作用原理等，揭示相关的动力学规律，更准确地预测肿瘤发展趋势，探索最佳用药方式和剂量，为癌症的治疗和抗癌药物的设计提供新的线索和思路。

癌细胞具有很强的个体差异性，每个癌细胞也都具有无限维的复杂性，而生物实验结果一般都是对一类癌细胞的定性描述和抽样研究，对不同类细胞经常得出不同的实验结果，让人无所适从。以细胞信号网络为例，细胞内存在众多的蛋白分子，具有高度复杂且动态变化的相互作用，在时空上传递、处理和执行各种信息[5]。从系统生物学角度看，癌症不仅是基因突变疾病，更是涉及诸如增殖、分化、死亡和转移等多条细胞命运抉择的信号转导通路疾病。不同的细胞受体经常激活同样的重要通路和下游关键效应蛋白，关键的细胞命运抉择都是由关键效应蛋白的动力学所决定的[6]。这使我们能够合理地忽略多种信号，而专注于关键蛋白的信号处理，定量地研究关键信号通路的动力学与功能。如图 6-9-3 所示，考虑癌细胞中的主要信号通路（如细胞凋亡等），并对其建模，让我们的研究能透过纷繁复杂的表象得到统一的本质规律。

国内外已经有一些科学家相继从理论建模的角度来研究癌细胞的各种性质，加深了我们对细胞瘤化机制的认识。2003 年 Lee 等对与发育途径相关的 Wnt 通路进行建模，对 Wnt 信号网络进行鲁棒性分析，对癌症发生机制做出了一个理论解释[7]。2007 年 Stites 等在 *Science* 上发表了一篇有关激发细胞生长、分裂和分化的 Ras 信号通路与癌症突变的工作，研究网络参数变化对 Ras 激活的影响，讨论了 Ras 基因的热点突变[8]。哈佛大学的 Sorger

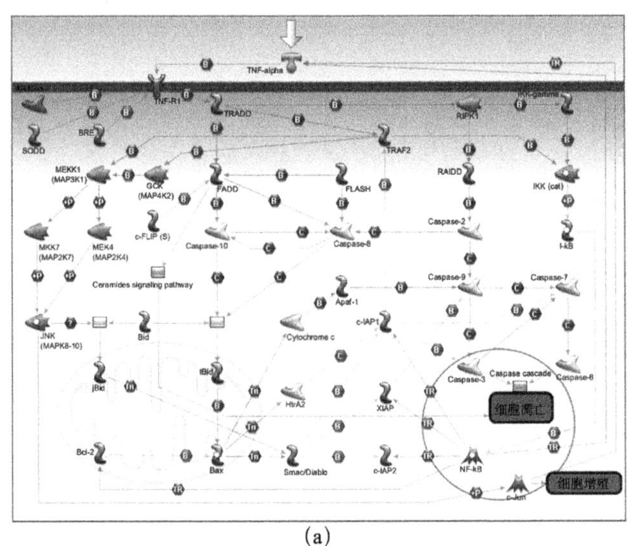

(a)

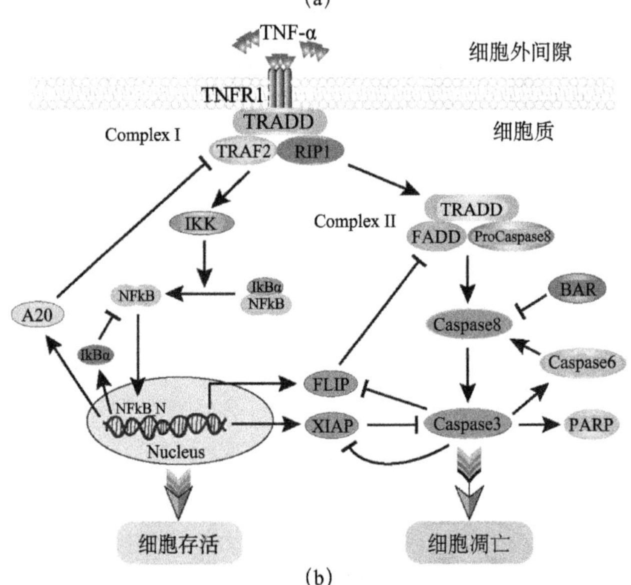

(b)

图 6-9-3　癌细胞凋亡信号网络。复杂网络图（a）可简化为关键分子相互作用网络（b），用于动力学理论研究

研究组 2009 年在 *Nature* 杂志发表文章，结合生物实验与数学模型，成功解释了不同细胞对肿瘤坏死因子诱导细胞凋亡的应答差异 [9]。2010 年 Cerami 等，利用细胞增殖的分子相互作用网络，讨论与脑瘤发生有关的特定功能模块 [10]。麻省理工学院的 Lauffenburger 教授长期研究细胞信号通路建模，讨论细胞的增殖、迁移和死亡等动力学行为 [11]。2014 年，普林斯顿大学

Austin 院士等则运用博弈理论，研究肿瘤细胞和间质成纤维细胞动态协同进化和能量代谢共生行为 [12]。这些研究极大地丰富了我们对癌症成因和发展的认识。

2000 年 Carlson 等在 *Phys. Rev. Lett.* 上发表论文，分析了控制细胞信号网络不同动力学状态的能力，从系统的鲁棒性出发，对肿瘤治疗的药物设计开发进行了理论研究 [13]。结合生物医学，人们通过对表皮生长因子受体信号 EGF 转导通路系统的理论和实验研究，提出癌症治疗的新一代靶标 [14]。Huang 等提出癌细胞动力学的概念，认为基因突变使得癌细胞动力学域增大，或是降低了进入癌细胞吸引子的阈值，进而解释细胞无限增殖的癌变机制，并从吸引子动力学角度讨论了多靶向药物治疗的一些局限 [15]。

在我国，北京大学的欧阳颀院士等利用非线性分岔分析，揭示了癌基因突变与细胞凋亡网络敏感参数的相关性等 [16,17]。上海交通大学的敖平教授等提出，癌细胞是细胞信号网络动力学的一种鲁棒态，正常细胞则是处于信号网络动力学的全局稳定态，这两种状态可以相互转化 [18]。南京大学的王炜教授等长期关注 p53 调控细胞命运抉择的动力学机制，阐明了 p53 肿瘤抑制功能的内在机制 [19,20]。

第三节　最近十年研究进展

下面我们结合具体的理论建模工作，包括单细胞层次上的癌细胞信号网络动力学、细胞群体层次上的癌细胞增殖和迁移行为和临床医学层次上的癌症治疗等模拟研究，介绍理论生物物理如何综合运用系统生物学、非线性动力学和生物信息学等，研究肿瘤细胞发生发展的各种机制，为癌症的治疗和抗癌药物的设计提供新的线索和思路。

一、癌细胞信号网络动力学建模

2008 年 *Science* 发表了几篇大规模癌基因组的测序工作，表明癌症不仅是基因突变疾病，更是细胞信号通路发生改变而导致的复杂疾病 [21,22]。对于同一种癌症的不同患者样本，其癌突变谱虽有很大不同，但这些繁杂的突变却总可以被归结到十几条细胞核心功能通路上。尽管不同癌症的基因突变各不相同，但这些突变基因主要影响的仍然是那些核心功能通路，它们的改变则导致细胞瘤化的产生。这些结果给癌症研究指出了一个新方向，即不单纯从基因，而是着重从细胞主要信号通路的角度去研究癌症及其治疗。

根据癌细胞的基本特征，癌细胞信号网络至少应该包括如下五个重要的功能模块：生存通路、增殖通路、凋亡通路、侵袭通路和能量代谢过程。图 6-9-4 列出了重要信号通路的部分节点，如生存通路包括 PI3K-AKt-Akk 和 PI3K-PKC 信号通路，增殖通路包括 Ras-Raf-MEK 和 Ras-Ral-MAC 信号通路，凋亡通路包括 XIAP 和 p53-线粒体凋亡通路，侵袭通路包括上皮细胞间质化（EMT）过程中的 Notch 和 Wnt 信号通路，能量代谢主要包括 Glut1-PFK-LDHA 过程等。

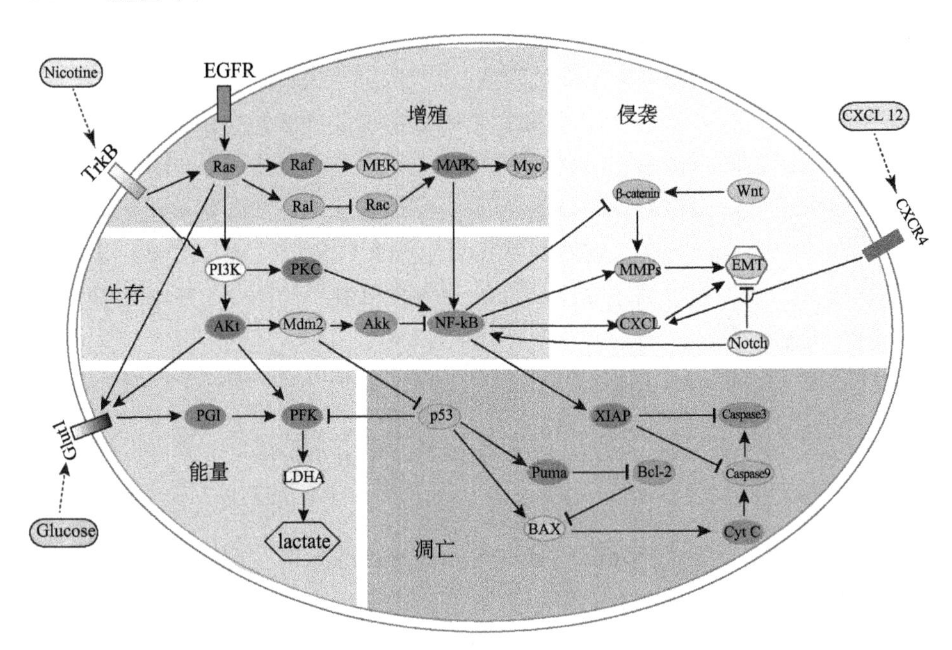

图 6-9-4　癌细胞信号通路图，包括生存、增殖、凋亡和侵袭通路，以及能量代谢过程

生物信号网络的建模，主要涉及贝叶斯网络、布尔网络、常微分方程、偏微分扩散反应方程以及马尔可夫随机动力学模型等。在此基础上，人们可以对模型进行动力学分析，包括网络拓扑结构分类、吸引子稳定性讨论、非线性分岔分析、不同振荡模式、时空斑图、微扰控制等讨论，而对大规模生物网络则可以用流分析方法。通过这些研究，我们能定量且系统地研究癌细胞信号网络的拓扑结构、动力学、生物功能及其联系等。

（一）p53 信号转导网络的研究

p53 是最重要的肿瘤抑制因子之一，被誉为“基因组卫士”。人类 50%

以上的肿瘤与 p53 基因突变有关。作为转录因子，p53 可选择性调控下游数百个靶基因的表达，进而调控细胞应答 DNA 损伤、组织缺氧、端粒缩短和癌基因活化等应激信号，参与调控 DNA 损伤修复、细胞周期阻断、细胞衰老和凋亡等过程，决定细胞的生死，抑制肿瘤的发生发展。因此，在细胞内存在一个以 p53 为中心的信号转导网络。人们曾经认为，p53 蛋白以分级响应的模式应答 DNA 损伤：轻度 DNA 损伤仅在细胞内激发少量 p53 蛋白，介导促细胞周期阻断基因（如 p21）的表达，在短暂的细胞周期阻断后，细胞恢复到正常增殖状态；而当损伤较严重时，大量的 p53 蛋白在细胞内积聚，进而表达促凋亡基因（如 Bax 等），细胞最终走向死亡。因此，细胞命运是由依赖于损伤强度的 p53 蛋白水平决定的。但近年来，人们逐渐认识到 p53 是以脉冲动力学调控着细胞的命运抉择。

2007 年，美国弗吉尼亚理工大学的 Tyson 研究组构建了 p53 对 DNA 损伤响应的理论模型，探讨 p53 脉冲的形成机制及其生理作用[23]。在该模型中，p53 脉冲是由耦合的正负反馈结构产生的。根据不同的磷酸化状态，激活的 p53 被分为三种形式。基于数值模拟结果，他们猜测，细胞命运可能与 p53 脉冲的数目有关。对于轻度损伤，少量 p53 脉冲引起细胞周期阻断，细胞可在修复完成后回归正常状态。对于严重损伤，持续激发的 p53 脉冲激活下游凋亡程序杀死不可修复细胞。该研究在 p53 脉冲生物功能方面做了非常有意义的探索，有助于人们理解 p53 调控细胞命运抉择的动力学机制。

2009 年，南京大学王炜课题组构建了更为全面的 DNA 损伤响应的 p53 网络模型[19]。基于乳腺癌细胞 MCF7 的信号通路组成，该模型整合了 DNA 损伤产生和修复、ATM 损伤感知、p53-Mdm2 振荡发生器和细胞命运抉择四个模块，可全面刻画 p53 网络对 DNA 损伤做出响应的动力学过程。特别地，模型中假设 p53 对损伤修复过程存在双重调控效应，即 p53 可促进轻度损伤的修复，而抑制严重损伤的修复。该假设可调和以往 p53 对 DNA 损伤修复调控方面看似矛盾的实验结果。在该模型中，p53 脉冲是由耦合的正负反馈回路在损伤驱动下产生的。研究结果表明，对于轻度损伤，仅有 p53-arrester 形式的脉冲被激发，细胞经历暂态生长停滞后回到正常状态；而当损伤较严重时，p53-killer 的脉冲在一定数目的 p53-arrester 脉冲之后被激活，并引起细胞凋亡。该模型提出的 p53 对损伤修复的双重调控机制，可最大程度地减少细胞命运抉择过程中的个体差异性，使得低剂量辐射下大部分细胞得以存活，而高剂量辐射下大部分细胞则被杀死，这对于癌症的放疗具有重要意义。相比于 p53 水平，p53 脉冲数目在决定细胞命运时显得更加鲁棒，可避免随机因素引起的细胞非正常死亡。

随后，他们又针对未发生严重基因突变的 MCF-10A 细胞系，构建了一个 p53 网络对 DNA 损伤响应的整合模型[20]（图 6-9-5）。该模型可实现一种更优化的 p53 动力学响应模式：p53 脉冲和持续高水平激活分别引起细胞周期阻断和细胞凋亡。在该模型中，主要考虑了三个 p53 相关的反馈回路：p53-Mdm2 负反馈、p53-Wip1-ATM 负反馈和 p53-PTEN-Mdm2 正反馈，对于塑造 p53 动力学起着关键作用。研究发现，轻度损伤的细胞中 p53 呈现少数几个脉冲；在严重损伤的细胞中 p53 动力学可呈现出两阶段行为。在第一阶段，两个负反馈占主导，p53 蛋白呈现一系列 p53-arrester 的脉冲。在第二阶段，p53 转变为具有促凋亡活性的 killer 形式，PTEN 被转录激活，p53-PTEN-Mdm2 正反馈开始占主导，p53 水平进一步升高到一个较高的稳态。高水平的 p53 导致细胞快速凋亡。因此，p53 水平和活性在响应过程中通过渐进的方式被调控，初步激活的低幅度脉冲引起细胞周期阻断，而完全激活的高水

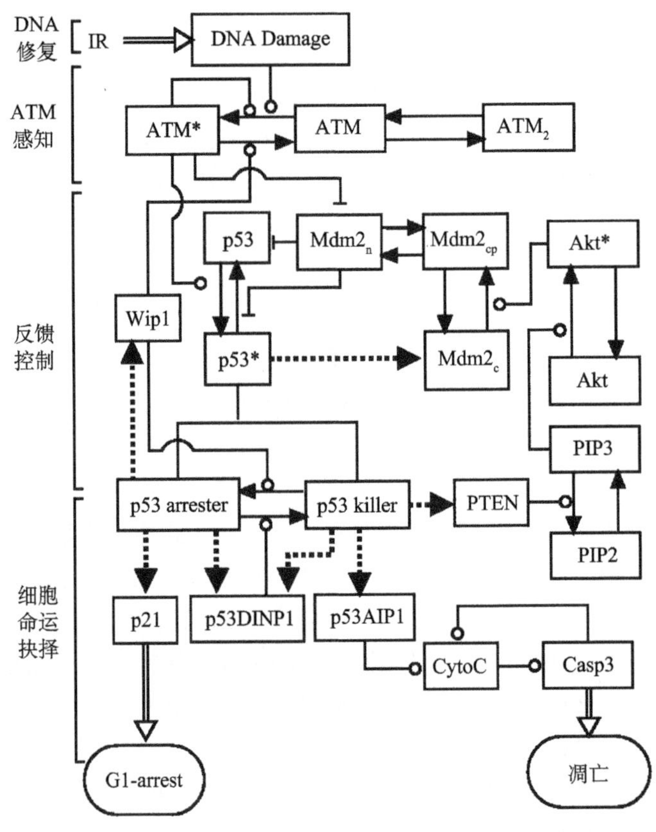

图 6-9-5　关于 p53 信号网络决定细胞命运的工作机制示意图[20]

平 p53 则引起细胞凋亡。p53 网络结构的时序性调控导致了 p53 动力学的两阶段性转变，负反馈主导下呈现 p53 脉冲，正反馈占主导则呈现出开关动力学，p53-arrester 到 p53-killer 的转变则是以上变化的内在原因。该研究提出了一种全新的 p53 动力学调控模式，该机制使得 p53 响应兼具灵活性与鲁棒性，可以看成是某种优化的应激响应模式。

（二）细胞线粒体凋亡调控网络

人们对癌症的一个普遍认知是，“基因突变改变并扰乱了正常细胞的各项生理调控功能，进而驱动了癌症的发生和发展”。但在系统层面上，基因突变诱导癌症的机制尚不清晰。细胞凋亡功能的缺失是癌细胞最主要的特征之一，其过程的实现与线粒体密不可分[24]。

北京大学的欧阳颀院士等通过对细胞信号网络进行动力学分析，并结合蛋白质相互作用动力学，从理论上对基因突变导致癌症病发的机制进行了详细阐述[17]。他们构建了外界信号诱导细胞线粒体凋亡调控信号网络，得到细胞凋亡的分岔模型，并将目标蛋白质 Casp8 的鞍节点分岔行为与细胞死亡的理论相对应（图 6-9-6）。

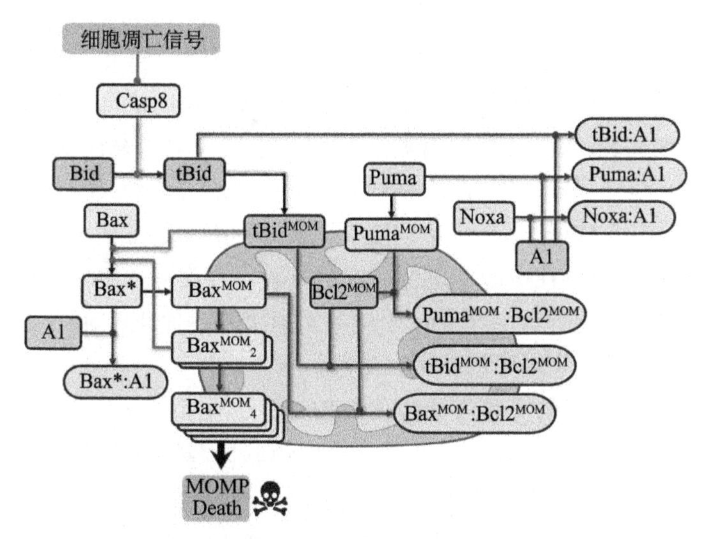

图 6-9-6 细胞凋亡信号通路[17]

他们通过分岔点的参数敏感性分析，得出模型中不同参数对分岔点位置影响程度的敏感性；进一步，他们统计了网络中所涉及的致癌突变在蛋白质结构上的分布情况，发现致癌的相关突变倾向于分布在敏感参数所对应的蛋白质相互作用结构域上。随后，他们利用分子动力学模拟分析了突变对网络中蛋白质相互作用动力学参数的影响。计算 29 对野生型与突变型蛋白复合物的结合自由能，发现其变化与网络中参数的扰动方向具有确定性的联系。

基于上述理论与计算分析，他们提出了癌症病发的一种可能机制，即基因突变改变了蛋白质相互作用的动力学参数，进而对细胞信号网络中的蛋白信号参数产生了扰动，影响了细胞信号网络动力学行为，诱导了癌症的发生和发展。这一研究为从动力学角度理解癌症的病发以及有效选择和发现抗癌药物靶点提供了新的启示。

（三）癌细胞 Ras 信号转导通路模拟

Ras 基因是人类癌细胞中最容易突变的基因之一。在人类肿瘤细胞中，Ras 基因的变异占 20%～30%，Ras 变异发生率最高的是胰腺癌（90%），其次为结肠癌（50%）和肺癌（30%）。Ras 蛋白通过结合鸟核苷酸（GTP 和 GDP）控制细胞信号转导，调节细胞分化、增殖和凋亡过程。目前发现 Ras 信号传导通路与人类绝大多数肿瘤的发生发展过程密切相关，该通路中的任何组分发生突变都会影响细胞的命运，因此发展以 Ras 信号转导通路为靶点的抗肿瘤抑制剂，具有很好的药学前景。

Ras 的活性主要受两种蛋白控制[25]：鸟苷交换因子（guanine nucleotide exchange factor, GEF）和 GTP 酶激活蛋白（GTPase activating protein, GAP）。GEF 促使 GDP 从 Ras 蛋白上释放出来，取而代之的是 GTP，从而激活 Ras；GEF 的活性受生长因子及其受体的影响。GAP 激活 Ras 蛋白的 GTP 酶，将结合在 Ras 蛋白上的 GTP 水解成 GDP，使 Ras 蛋白失活。正常情况下，Ras 蛋白基本上都与 GDP 结合在一起，定位在细胞质膜内表面上。Ras 基因的突变可扰乱正常状态下活化与非活化 Ras 蛋白的平衡机制，使正常非活化的 Ras 蛋白转变成活化形式，从而导致癌症的发生。

Stites 等结合相关的实验数据构建出 Ras 信号调控网络（图 6-9-7），基于对网络动力学的分析，考察信号通路中蛋白质的稳态浓度随参数的变化，解释癌细胞中常见的点突变[26]。他们首先考虑了两种常见的 Ras 点突变类型：Ras^{G12V} 和 Ras^{G12D}，并与野生型进行了对比，发现 Ras 蛋白的激活主要被网络中的四个过程影响，这与目前的病理和生理发现相一致。此外，基于对参数的讨论及实验证实，他们引入了另一种点突变 Ras^{F28L}，阐述了其在临床癌症中较 Ras^{G12D} 很少发现的原因：在正常生理浓度范围，Ras^{F28L} 对 Ras 信号通路的激活作用较 Ras^{G12D} 小很多。

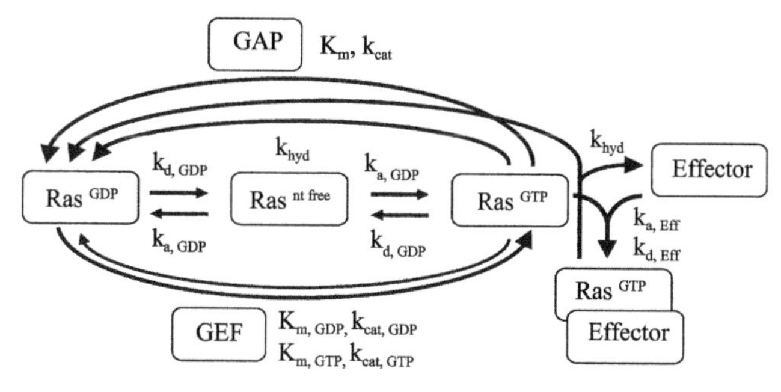

图 6-9-7 Ras 激活信号通路 [26]

最后，模型预测若存在一个与 Ras^{G12V} 及 Ras 有相同亲和力的药物，其不仅可以减小对正常细胞中的 Ras 信号通路抑制，而且更重要的是不影响对癌细胞中该信号通路的抑制，为癌症的治疗提供了一个新的设计思路。

（四）胃癌细胞内源性信号网络

胃癌是全球第二高发致死癌症，5 年生存率约 20%。胃癌细胞有两种异质性表型：胃上皮细胞和肠上皮细胞类型。无论是在正常还是异常的组织中，转录因子 Cdx2 是肠道分化过程中的关键蛋白。而 Sox2 和 Shh 是在胃分化过程中所必需的调控蛋白。临床上，Cdx2 的过表达和 Sox2 欠表达都在肠上皮化生（肠型胃癌的一种）中被观察到。但保持这两种异质性胃癌表型和调控表型转化的机制很不清楚。

上海交通大学的敖平课题组，基于内源性分子网络研究了胃癌的异质性问题 [27]。该网络由胃癌细胞中的内源性分子构成，包括转录因子、生长因子、细胞因子及其相互作用等。随后，对该网络进行量化，数值模拟结果很好地重现了正常和胃癌上皮细胞的主要特性，如正常胃上皮细胞主要表现出细胞周期阻滞和分化表型，而胃癌上皮细胞则显现了增殖、凋亡、炎症及异常分化等表型。

最后，他们提出了产生胃癌细胞异质性的两种机制：一种是在胃癌细胞中存在着负责维持肠道和胃的表型的特定正反馈回路；另一种是在胃癌细胞发展过程中，存在 16 条从正常吸引子到胃癌细胞吸引子的转化途径。关键分子的动力学行为能够表征不同转化途径的特征，表明胃癌细胞的发展过程可能是异质性的。

二、癌细胞群体相互作用、增殖和迁移理论

肿瘤是包含不同表型的细胞种群的生态系统，只研究单个肿瘤细胞是不

够的，还要研究癌细胞的群体行为，包括癌细胞之间的相互作用、癌细胞的分裂增殖、不同种类癌细胞的相互转化和细胞的侵袭迁移行为等。

（一）癌细胞博弈理论

不同细胞类型的肿瘤彼此处在类似合作与竞争的状态中，并在不同状态间转换。近年来生物物理学家也从生态协同演化角度，运用博弈论，研究癌细胞间的相互作用。肿瘤微环境包含两个主要的区域：外侧区域接近血管系统，供氧充分；内侧区域几乎没有可以利用的氧（图 6-9-8）。研究表明，缺氧细胞用糖酵解产生乳酸盐作为新陈代谢的产物，而有氧细胞则通过氧化磷酸化产生 ATP。肿瘤细胞可以通过在含氧低的细胞种群和含氧高的细胞种群之间的乳酸运送，使糖酵解的乳酸盐作为其能量来源。2014 年，约翰霍普金斯大学的 Pienta 教授等提出，对于富氧和缺氧癌细胞的相互作用，可用演化博弈论来描述（图 6-9-8）。他们考虑肿瘤细胞成本付出和收益回报的博弈规则，讨论了癌细胞增殖动力学行为[28]。

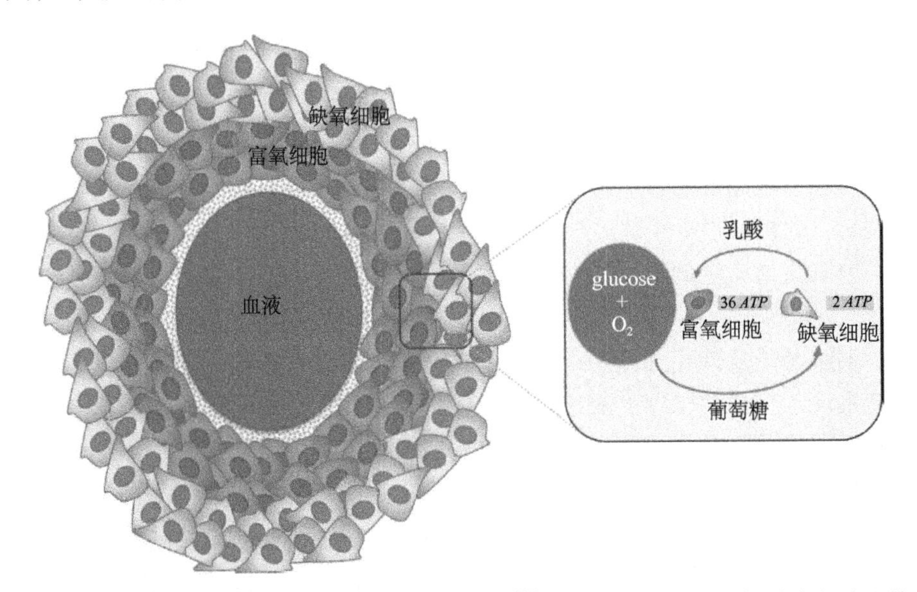

图 6-9-8　富氧和缺氧癌细胞之间的新陈代谢共生[28]。葡萄糖和乳酸分别是富氧细胞和缺氧细胞中使用的能量来源，这是新陈代谢共生博弈的纳什平衡。在纳什平衡中缺氧细胞用每摩尔葡萄糖产生 2mol ATP 和 2mol 乳酸，然而富氧细胞每 2mol 乳酸产生 36mol ATP

细胞通常用脂肪酸、谷氨酸盐、单羟酸（如乳酸）和葡萄糖等作为其能量来源，但为简单起见，他们仅仅考虑两类癌细胞的能量来源：葡萄糖和乳酸。因此在他们的模型中，只有含氧低的和含氧高的细胞种群两类参与者。每一个细胞可以选择不同的能量代谢路径，采用葡萄糖或乳酸。博弈的收益矩阵是基于 ATP 的产量，反映了细胞的新陈代谢率（表 6-9-1）。富氧细胞可以使用乳酸用于产生能量，则其能量产值是 L，富氧细胞和缺氧细胞也可以使用葡萄糖来产生能量，则其能量产值分别用 G_0 和 G_h 表示。

表 6-9-1　肿瘤代谢共生中作为两参与者的博弈收益矩阵 [28]

缺氧细胞	*glucose* 葡萄糖	*lactate* 乳酸	富氧细胞	*glucose* 葡萄糖	*lactate* 乳酸
glucose	$G_h/2$	G_h	*glucose*	$G_0/2$	G_0
lactate	0	0	*lactate*	L	0

注：左边矩阵表示缺氧细胞的能量产物值，右边矩阵表示富氧细胞的能量产物值

这个模型揭示了多个中间稳定态的存在，以及肿瘤内混合的能量策略。它预测了在两个亚种群之间癌细胞的非线性博弈作用，导致了肿瘤的关键转变，并且肿瘤可以获得在糖酵解和呼吸作用之间的不同中间态。这个模型表明，一个简单的细胞间信号，例如在肿瘤微环境中的乳酸分泌，可以诱导癌细胞在高葡萄糖消耗和低葡萄糖消耗之间产生临界转变。

（二）肿瘤干细胞分裂转化模型

实验发现，肿瘤组织中存在少量的癌细胞，具有自我更新、增殖和分化的潜能，因其许多特性与干细胞相似，被称为肿瘤干细胞。肿瘤干细胞对肿瘤的存活、增殖、转移及复发有着重要作用，通过自我更新和无限增殖维持着肿瘤细胞群的生命力。肿瘤干细胞可以长时间处于休眠状态，并对多种药物产生耐药性，对杀伤肿瘤细胞的外界因素不敏感，导致在常规肿瘤治疗方案中肿瘤复发。肿瘤干细胞的发现为我们重新认识肿瘤的起源和本质，以及设计治疗方案提供了新的方向和角度。

肿瘤干细胞能产生两种异质细胞，一种是与之性质相同的肿瘤干细胞，另一种是组成肿瘤大部分的肿瘤非干细胞。传统的理论认为，肿瘤干细胞分裂是一种单向的层级分裂过程。而最近提出的肿瘤细胞表型的可塑性机制（phenotype plasticity）则指出，肿瘤干细胞与肿瘤非干细胞之间存在双向转化的关系 [29]。可塑性机制是近年来肿瘤干细胞的研究热点之一，仍存在很多争议。通过理论建模，比较可塑性机制和经典机制的并同，具有重要的研究

价值。

厦门大学的周达等通过建立带有细胞可塑性机制的多表型分枝过程模型，针对前人实验数据中报道的表型均衡现象（phenotypic equilibrium）与暂态激增现象（transient increase），阐明了可塑性机制的优势[30]。一方面，经典肿瘤干细胞模型虽能在一定参数条件下重复表型均衡现象，但是其参数值可能没有生理意义；反观可塑性模型则能很好地再现表型均衡现象，参数取值也更为合理。另一方面，经典肿瘤干细胞模型无法重复暂态激增现象，而可塑性模型则能很好地模拟肿瘤干细胞比例迅速恢复的现象（图 6-9-9）。这些结果表明，表型可塑性机制可能通过对肿瘤干细胞的保护促进肿瘤细胞群体的异质性。

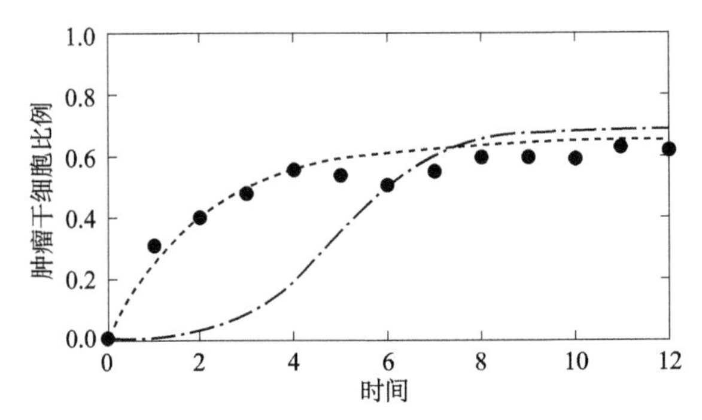

图 6-9-9　癌症干细胞的可塑性机制模型与经典模型比较[30]。黑点表示 SW620 结肠癌细胞系数据，短虚线表示带有可塑性机制模型的拟合曲线，点划线表示没有可塑性机制模型的拟合曲线。可见，只有短虚线可以拟合前期肿瘤干细胞（CSCs）比例的迅速恢复

（三）癌细胞迁移模型

癌细胞由其原发部位脱落，侵入血管或淋巴管或体腔，通过血液或淋巴系统流到另一部位，少部分存活的癌细胞滞留、黏附在管壁，然后穿出管壁，侵入其他器官或组织，并繁殖生长，形成与原发瘤同样类型的肿瘤，这一过程即为癌细胞的侵袭转移。转移是恶性肿瘤最重要的特征之一，也是影响患者预后的主要因素。关于细胞侵袭、细胞穿越基底膜或穿越内皮血管内壁的物理学概念还很少研究，除了对癌细胞力学性质的各种理论建模研究，对癌细胞侵袭迁移的建模也促进了我们对癌细胞转移的理解。

中国科学院物理研究所的刘雳宇等讨论了一个二维元胞自动机（CA）随机迁移模型[31]。模拟区域的尺寸为 4000μm×2000μm，元胞自动机的元胞尺

寸为 10μm，而每个元胞可以被细胞外基质大分子、癌细胞或空隙空间占用。类似于癌细胞的微流管迁移实验，他们考虑了两个具有不同密度的细胞外基质区域。在模型中考虑了 3 种可诱导癌细胞侵袭行为的特征机制，包括微环境的异质性、通过细胞间通信远程同型的化学吸引力、浓度梯度驱动的定向细胞迁移。得到随机迁移的癌细胞在 3 个不同时间点的分布结构如图 6-9-10 所示。研究结果表明，癌细胞的集体迁移行为不仅与细胞内在性质紧密相关，而且也深受周围微环境的影响。因此，用元胞自动机模型高度自由地构建和调控微环境的力学与化学条件，从理论角度描述细胞复杂多变的单细胞或者集体细胞行为，对于定量研究微环境控制细胞行为具有较为重要的指导意义。

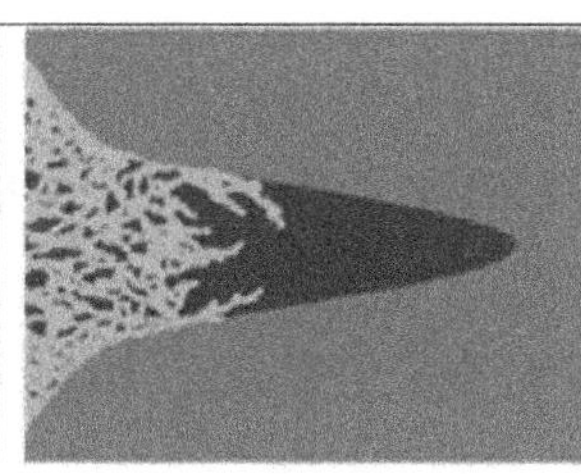
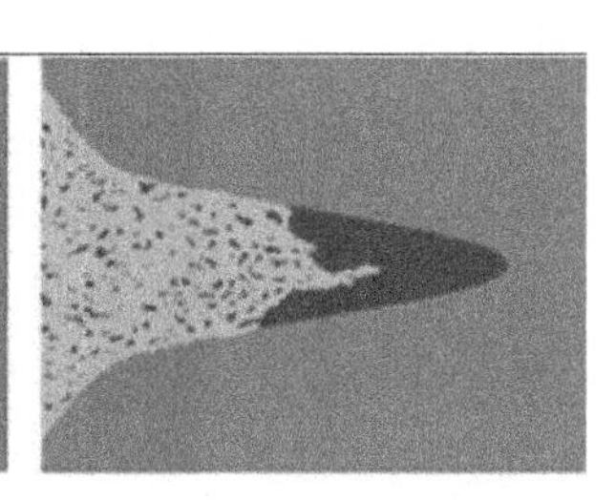

图 6-9-10 利用元胞自动机模拟癌细胞集体随机迁移行为。在 3 个不同时刻，癌细胞呈现不同迁移分布结构 [31]

从原发瘤脱离的癌细胞会寻找血管和淋巴结，以便转移到机体的其他器官。生物学家们曾长期认为，这些细胞通过随机游走到达血管，癌细胞在机体中的扩散是一个漫无目的的缓慢过程，被称为随机行走（random walk）行为。但是，约翰·霍普金斯大学的 Wirtz 课题组研究发现，这种随机游走模型只是癌细胞在二维培养皿上的运动方式，并不适用于人体内的三维空间。他们提出了一个新的计算模型，能够更好地展现细胞在三维环境中的迁移行为 [32]。他们观察到，癌细胞的运动更有方向性，其移动轨迹几乎呈直线能让癌细胞更有效地到达血管，是癌细胞扩散的有效途径。这个发现指出，癌细胞离开原发组织的时间比先前人们的预期更短。

三、肿瘤临床治疗方案及其疗效模拟

肿瘤研究的最终目的是揭示肿瘤细胞的成因，设计有效的肿瘤治疗手段和抗癌药物，实现对癌细胞的抑制或消除。理论生物学家也通过构建数学模型，研究药物和治疗方法对癌细胞的影响，寻求可能的最佳疗法。

（一）肿瘤化学疗法模拟

由于癌细胞抗药性的演化，针对转移癌的化学疗法通常会失败。因此，科学家把癌症治疗看成一场博弈；科学家选择一种疗法，而肿瘤细胞选择一种适应策略。Orlando 等因此提出关于癌细胞面对化学治疗的博弈理论[33]。

他们考虑两种化疗药物，其化疗效果或者具有对抗性，或者具有协同性，而癌细胞可遗传的策略、群体的抗药性策略、细胞的种群大小和两种化疗药物的浓度对瘤细胞的药物适应性产生影响。作者建立癌细胞演化的动力学模型，一方面，癌细胞面临着两种药物的抗药性配置，通过自然选择来决定策略；另一方面，肿瘤学家通过解决一个控制问题来选择治疗策略。

研究表明，当肿瘤细胞仅对一种药物具有强抗药性时，最佳的疗法是同时递送这两种药物。但如果癌细胞采用一种全方面的策略，对两种药物同时具有强抗药性，最佳的治疗方案则是随时间交替递送这两种药物。所以，在多种药物的联合治疗中，我们必须要考虑癌细胞对不同药物的不同强度反应，制定不同的用药方式。

（二）胶质瘤放射疗法模拟

胶质细胞瘤（glioblastomas）是最常见的恶性原发性中枢神经系统肿瘤，占所有颅内原发肿瘤的一半，生存率很低。目前对其的治疗方案以手术为主，但由于肿瘤浸润性生长，难以做到全部切除，后期还需配以放射治疗和化学治疗。尽管这种治疗方案在过去 50 年一直被广泛使用，但后期放疗方案并不相同，且仍伴有较高的复发率。随着对胶质瘤的深入研究，人们发现其形成与血小板衍生生长因子（PDGF）信号通路的异常密切相关。另外，利用基因工程对小鼠进行改造，可以更加清晰地认识胶质瘤在组织上的形成及迁移等分子机制，促进对胶质瘤的有效治疗。

尽管前人已构建了数学模型来研究放疗对疗效胶质瘤的影响，相关医疗方案并没有达到预期疗效。Kevin 等[34]最近构建 PDGF 导致胶质瘤的小鼠模型，分析了小鼠对动态辐射的响应过程，提出了一种更有效的治疗方案。他们考虑了两类细胞群体，即具有抗放射性的干细胞样胶质瘤细胞和对辐射敏感的已分化胶质瘤细胞。在辐射过程中，部分对辐射敏感的已分化胶质瘤细胞会迅速恢复到具有抗辐射性的干细胞样胶质瘤细胞，这可能与胶质瘤的高复发率有关。基于该模型，他们预测了一种可有效延长小鼠生存期的放疗方案。

更重要的是，当实验上对患有胶质瘤的小鼠进行测试时，该方案比传统治疗方案明显提高了小鼠的生存期。通过增加时间约束来更准确地得出细胞基于上次辐射所获得的抗辐射能力，为下次治疗提供依据。具体而言，如果已分化的胶质瘤细胞无法迅速获得抗辐射能力，这种治疗方案将不会起作用。该结果表明已分化的胶质瘤细胞获得抗辐射能力在治疗中发挥了重要作用，至少对 PDGF 导致的胶质瘤，传统的治疗方案并不是最有效的。

第四节　癌细胞理论研究的前沿学科问题和发展方向

肿瘤研究相关的物理学手段不仅为肿瘤学家研究肿瘤细胞功能特性提供了新技术，还开拓了人们对肿瘤进展的认识，例如区分恶性和良性肿瘤、确定肿瘤疾病发生发展相关的功能性信号通路。2009 年，一个由生物学家和物理学家团队组建的物质科学肿瘤中心，联手研究肿瘤的物理学特征。该计划取得了许多成就，包括对癌症进化的理解，预测细胞癌变时间和发现癌症生物标志物等。目前，新型的“肿瘤物理学”（physics of cancer）以生物物理学和软物质物理学为主导，将物理学、分子生物学和生物化学有机结合起来。

在我国，生物物理领域的科学研究日益受到重视，国家自然科学基金委员会数理学部物理一处近年来在项目申请指南中明确提到“要鼓励与生命科学相关的物理和实验方法”，也正是这一趋势的体现。但国内癌症的生物物理研究还相对薄弱。

2011 年 3 月，著名的生物学家 Weinberg 和 Hanahan 在 *Cell* 杂志发表综述文章，总结出肿瘤细胞的十大特征：持续的增殖信号，对生长抑制不敏感，逃避免疫摧毁，无限的复制能力，促瘤炎症反应，抵抗细胞死亡，持续的血管生成，组织浸润和转移，异常的能量调控，基因组不稳定和突变[35]。这为深入阐述癌症发生发展和侵袭转移的分子机制提供了一个好的框架。下面，我们列出一些可望在较短时间内取得突破，围绕癌细胞理论建模的可能研究方向。

一、关于细胞信号转导机制的研究

长期以来，肿瘤细胞基因突变引起 DNA 损伤和基因组不稳定性是肿瘤研究的焦点，大量的癌基因和抑癌基因被发现，它们的作用被鉴别。但它们不是孤立地起作用，而是通过广泛的信号转导来实现功能。有充分证据指出，从正常细胞到肿瘤细胞的转变是由细胞的信号调控机制发生紊乱造成的。研究肿瘤细胞恶性演化过程中信号转导机制的变化，既可以揭示肿瘤细胞演化的机制，也为抗肿瘤药物的开发提供靶点。所以，研究细胞的信号转导是肿瘤研究一个非常重要的方向。建立可靠的生物网络模型，通过讨论网络的动力学来阐述细胞信号处理过程的机制，已被证明是一条行之有效的研究途径。

目前，生物学实验积聚了海量的数据，但这些数据往往是离散、碎片化的，无法对某一个生物过程或现象提供相对完整的图像，急需将不同的信息整合起来，理论建模在这方面可以发挥很大的作用。细胞内的信号传导涉及多个层面，从基因的表达调控，蛋白的合成及其翻译后修饰，细胞内的定位，复合物的形成，到信号的级联传递，各种细胞命运的竞争抉择等，每一个层面都涉及网络调控。因此，我们需要根据所研究的问题选择恰当的近似方法。由于生物网络结构的复杂性，会含有很多的参数，仅作参数鲁棒性分析还不够；也不能只提供网络成分随时间变化的动力学，还要发展新的刻画手段，拓宽分析能力，提供更有效的定量刻画方法。

二、关于基因表达调控机制的研究

基因表达是指从 DNA 转录 mRNA 再翻译合成蛋白质的过程，是高度受控的过程。转录起始是调控的第一步，对基因表达起着至关重要的作用。真核基因转录涉及由 RNA 聚合酶、通用转录因子、媒介子、转录激活子等组成的转录机器的运转。尽管转录机器的基本架构已大致知晓，但其运转机制依然很不清楚。转录机器是如何感知时变的信号，以合适的速率起始 mRNA 的合成？分子的无规运动与基因表达的精确性是如何协调的？要回答这些问题，必须深入研究转录机器的动力学。

有不少的转录因子是重要的肿瘤抑制因子，如 p53。在肿瘤细胞中，p53 基因发生突变，不仅丧失了抑癌基因的功能，还获得了一系列类似癌基因特性的功能（即功能获得）。突变的 p53 与正常的 p53 调控的靶基因有很大的不同，直接影响了肿瘤的进程。因此，阐明突变的转录因子对基因转录的调控机制将具有重要的意义。

基因沉默与基因激活一样，也是重要的基因表达调控模式，涉及 RNA 干扰和非编码 RNA。在众多调节性 RNA 分子中，微 RNA（miRNA）最引人注目。miRNA 比信使 RNA(mRNA) 更稳定，相对容易检出。miRNA 促进靶

mRNA的降解或抑制其翻译。与正常细胞相比，肿瘤细胞中许多miRNA都有表达差异，广泛影响肿瘤生长过程。因此，探究miRNA自身的动力学及其对基因转录的调控机制，将有助于揭示miRNA在肿瘤发生发展过程中的作用。

三、关于生物网络结构、动力学和功能的研究

尽管细胞的信号转导通路往往非常复杂，但常常可以大致分解成一些功能性模块。这些模块又由一些经常出现的基序（motif）组成。因此，系统地研究这些基序和功能模块能帮助人们了解更复杂的生物系统。用非线性科学的方法研究具有不同生物功能、不同拓扑结构的生物网络的稳定性、多态性、动态平衡点的动力学性质及演化路径的动力学性质等。借助定量的刻画方法，总结生物网络动力学的普适性规律，揭示生物网络动力学性质与拓扑性质的一般关系以及它们对生物功能的影响。选择一些具有代表性的生物网络，比较细胞癌变前后网络拓扑结构、动力学和功能的变化，进而为疾病干预、药物设计提供理论基础和研究手段。

癌症的理论生物研究并不能独立于生物和医学领域之外，真正富有意义的研究应该是基于物理学家、生物学家和病理学家的良好沟通，整合各方面的优势，从基础到临床都实现多学科交叉。结合生物实验、生物理论建模以及临床医学应用，综合运用多种手段和方法，定量刻画细胞内的稳态行为和应激响应，比较其在正常细胞与肿瘤细胞发生、发展等不同阶段的异同，为揭示肿瘤细胞的成因、设计有效的肿瘤治疗手段和抗癌药物提供重要的线索和思路，并终将攻克癌症难题。

帅建伟（厦门大学），刘锋（南京大学）

参考文献

[1] 国家癌症中心. 2017年中国肿瘤登记年报. 2017.

[2] Siegel R L, Miller K D, Jemal A. Cancer Statistics, 2017. CA: A Cancer Journal for Clinicians, 2017, 67: 7-30.

[3] Michor F, Liphardt J, Ferrari M, et al. What does physics have to do with cancer? Nature Reviews Cancer, 2011, 11: 657-670.

[4] Gravitz L. Physical scientists take on cancer. Nature, 2012, 491: S49.

[5] Kholodenko B N, Hancock J F, Kolch W. Signalling ballet in space and time. Nature Reviews Molecular Cell Biology, 2010, 11: 414-426.

[6] Purvis J E, Lahav G. Encoding and decoding cellular information through signaling dynamics. Cell, 2013, 152: 945-956.

[7] Lee E, Salic A, Kruger R, et al. The roles of APC and Axin derived from experimental and theoretical analysis of the Wnt pathway. PLoS Biology, 2003, 1: E10.

[8] Stites E C, Trampont P C, Ma Z, et al. Network analysis of oncogenic Ras activation in cancer. Science, 2007, 318: 463-467.

[9] Spencer S L, Gaudet S, et al. Non-genetic origins of cell-to-cell variability in TRAIL-induced apoptosis. Nature, 2009, 459: 428-432.

[10] Cerami E, Demir E, Schultz N, et al. Automated network analysis identifies core pathways in glioblastoma. PLoS One, 2010, 5: e8918.

[11] Dang M, Armbruster N, Miller M A, et al. Regulated ADAM17-dependent EGF family ligand release by substrate-selecting signaling pathways. Proc. Natl. Acad. Sci. USA, 2013, 110: 9776-9781.

[12] Wu A, Liao D, Tlsty T D, et al. Game theory in the death galaxy: interaction of cancer and stromal cells in tumour microenvironment. Interface Focus, 2014, 4: 20140028.

[13] Carlson J M, Doyle J. Highly optimized tolerance: robustness and design in complex systems. Physical Review Letters, 2000, 84, 2529-2532.

[14] Gschwind A, Fischer O M, Ullrich A. The discovery of receptor tyrosine kinases: Targets for cancer therapy. Nature Reviews Cancer, 2004, 4: 361-370.

[15] Huang S, Kauffman S. How to escape the cancer attractor: rationale and limitations of multi-target drugs. Seminars in Cancer Biology, 2013, 23, 270-278.

[16] Chen J, Yue H, Ouyang Q. Correlation between oncogenic mutations and parameter sensitivity of the apoptosis pathway model. PLoS Computational Biology, 2014, 10: e1003451.

[17] Zhao L, Sun T, Pei J, et al. Mutation-induced protein interaction kinetics changes affect apoptotic network dynamic properties and facilitate oncogenesis. Proceedings of the National Academy of Sciences of the United States of America, 2015, 112: E4046-4054.

[18] Ao P, Galas D, Hood L, et al. Cancer as robust intrinsic state of endogenous molecular-cellular network shaped by evolution. Medical Hypotheses, 2008, 70: 678-684.

[19] Zhang X P, Liu F, Cheng Z, et al. Cell fate decision mediated by p53 pulses. Proceedings of the National Academy of Sciences of the United States of America, 2009, 106: 12245-12250.

[20] Zhang X P, Liu F, Wang W. Two-phase dynamics of p53 in the DNA damage response.

Proceedings of the National Academy of Sciences of the United States of America, 2011, 108: 8990-8995.

[21] Jones S, Zhang X, Parsons D W, et al. Core signaling pathways in human pancreatic cancers revealed by global genomic analyses. Science, 2008, 321: 1801-1806.

[22] Parsons D W, Jones S, Zhang X, et al. An integrated genomic analysis of human glioblastoma multiforme. Science, 2008, 321: 1807-1812.

[23] Zhang T L, Brazhnik P, Tyson J J. Exploring mechanisms of the DNA: damage response-p53 pulses and their possible relevance to apoptosis. Cell Cycle, 2007, 6: 85-94.

[24] Tait S W, Green D R. Mitochondria and cell death: Outer membrane permeabilization and beyond. Nat. Rev. Mol. Cell. Biol., 2010, 11: 621-632.

[25] Bos J L, Rehmann H, Wittinghofer A. GEFs and GAPs: critical elements in the control of small G proteins. Cell, 2007, 129: 865-877.

[26] Stites E C, Trampont P C, Ma Z, et al. Network analysis of oncogenic ras activation in cancer. Science, 2007, 318: 463-467.

[27] Li S, Zhu X, Liu B, et al. Endogenous molecular network reveals two mechanisms of heterogeneity within gastric cancer. Oncotarget, 2015, 6: 13607-13627.

[28] Kianercy A, Veltri R, Pienta K J. Critical transitions in a game theoretic model of tumour metabolism. Interface Focus, 2014, 4: 20140014.

[29] Gupta P, Fillmore C, Jiang G, et al. Stochastic state transitions give rise to phenotypic equilibrium in populations of cancer cells. Cell, 2011, 146: 633-644.

[30] Zhou D, Wang Y, Wu B. A multi-phenotypic cancer model with cell plasticity. Journal of Theoretical Biology, 2014, 357: 35-45.

[31] Zhu J, Liang L, Jiao Y, et al. Enhanced Invasion of Metastatic Cancer Cells via Extracellular Matrix Interface. PLoS One, 2015, 10: e0118058.

[32] Wu P H, Giri A, Sun S X, et al. Three-dimensional cell migration does not follow a random walk. Proceedings of the National Academy of Sciences of the United States of America, 2014, 111: 3949-3954.

[33] Orlando P A, Gatenby R A, Brown J S. Cancer treatment as a game: integrating evolutionary game theory into the optimal control of chemotherapy. Phys. Biol., 2012, 9: 065007.

[34] Leder K, Pitter K, Laplant Q, et al. Mathematical modeling of PDGF-driven glioblastoma reveals optimized radiation dosing schedules. Cell, 2014, 156: 603-616.

[35] Hanahan D, Weinberg R A. Hallmarks of cancer: The next generation. Cell, 2011, 144: 646-674.

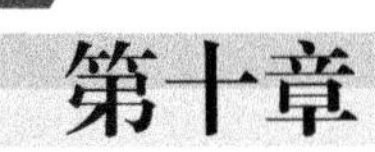

第十章 细菌运动的物理生物学

细菌几乎遍布地球的每一个角落，是地球上生存至今最古老的生命物种。21 世纪初微观显微镜技术的迅猛发展为物理学家研究细菌生命系统中的物理现象提供了难得的机遇。本章统一介绍细菌运动的生物物理学和物理生物学的概况和基本研究方向。本章侧重研究细菌动力学，主要讨论课题包括分子尺度上的细菌鞭毛马达的研究动向；细菌在界面附近的游动，积聚和黏附等复杂行为研究；细菌群体的集群涌动扩散行为和自组织斑图的形成；以及细菌的趋化性和细菌体内传感网络。最后讨论了最新的研究动向和今后 5～10 年可能取得的重大进展。

第一节 引 言

细菌对地球环境及人类健康有广泛的影响。地球上的大型生物种类正在逐年减少，很多生物物种已经濒临灭绝。而细菌的种类从 400 万年前演化至今却在地球上分布得越变越多。这些长度只有微米大小的生命体适应环境生存的能力极强。研究人员发现，细菌甚至可以在超高压、超冷、超热、极端缺氧的地球环境中生存[1]。细菌是地球生态系统中分布最广泛的物种，它们既可以依附在大气的尘埃中，也可以潜伏在冰冷黑暗的极深海底。人类身体中细菌细胞的数量是普通细胞数量的十倍之多[2]。人体中大多数的细菌由于受到人体免疫系统的管制是无害的，而且已经成为人体生态系统正常运转不可或缺的一部分。这些微生物帮助人体消化食物，抵御病菌，甚至能影响大脑的活动。有害细菌也会引起人类疾病。例如，肠道内的大肠杆菌导致腹泻，进入人类胃里的幽门螺杆菌能导致胃炎、胃溃疡。人类持续地开发各种新型药物来对付有害细菌，而有害细菌则能在一两年内迅速演化出新的种类，使得原来的药物失去了作用。

细菌的游动能力是其成功生存和繁衍的关键功能。作为微小的原核生物，细菌体内的 DNA 分散在细胞质中，外面由细胞膜和细胞壁包裹。细菌的外表大多有鞭毛、菌毛和一些脂多糖类长链分子形成的表面皮毛。细菌鞭毛最大的功能是驱动，使得细菌在溶液中游动[3]。鞭毛是由鞭毛蛋白形成的螺旋结构的细杆[4-6]，并通过其根部的鞭毛马达与细菌体连接起来。细菌体通过鞭毛马达的转动实现游动，也通过对鞭毛马达转动的控制实现趋化、趋氧、趋光，或者其他各种定向的运动。细菌的游动也使得它们积聚（accumulation）于界面附进，有助于形成微生物膜（biofilms）[7-11]。微生物膜是微生物和环境科学中的一个较新的名词。它专指在各种界面上彼此黏附在表面上的微生物群体。微生物膜可以形成于活的或无生命的表面，并且可以是在天然的工业和医院环境中普遍存在。微生物的游动性对微生物膜的生长有决定性的作用，但微生物膜里的细菌通常与来自相同物种的自由浮动的细菌有显著不同的性质。在生物膜中一群密排的外层的细胞和分泌物可以有效地保护社区的内部。这种环境允许微生物膜里的细菌合作，并以各种集体反应模式应对周围的生存环境。比如，有些细菌在这种环境中提高了耐洗涤剂和耐抗生素特性。细菌分泌物也是微生物膜的重大组成成分。微生物膜中含有大量的多糖和细胞外的 DNA，其中有些 DNA 能有一种叫横向基因转移的机制（lateral gene transfer）[12]，有些微生物膜会通过这些机制导致更稳定的结构，并大大促进对抗生素的抵抗性。

细菌生物物理学是现代微生物学与物理学的前沿交叉学科。随着近几年纳米科学的迅速发展，通过物理手段战胜细菌的方法也有了快速进展[13]，比如，在人体内定点释放药物，微纳米机器人在人体内精确清除病变细胞和疏通血管，用电磁波和超声等物理工具操纵微型医疗马达等。一个不断运动生长的细菌通常是在黏滞系数很大的血液和体液中游动，其驱动机制和运动方式都与宏观机械截然不同。比如，大肠杆菌的运动实际上涉及跨越细胞膜的多个蛋白质的组装及构型变化、信号的传递、能量的来源、速度和方向的控制，以及对环境变化的响应和演化等。人类如果想制造出能够在微观世界长期稳定运转的微纳米机器，需要对细菌的除生物学意义之外的其他科学方面有更深入精细的了解，这其实涉及纳米化学、微纳米尺度电磁相互作用、控制论、微观工程技术、多体自组装，以及低雷诺数液体中的流体力学行为，非传统热机的大分子形变引擎等，为基础物理学的研究提出了新的要求。

本章报告的几项主要战略研究课题有重要基础研究价值和应用前景。研究细菌的分子马达运行机制，为设计微观引擎，研制超灵敏的生物传感器，设计新型微纳米自推仿生器等奠定研究基础，也在微观医学、微纳米工程技术等方面有潜在的应用前景。研究分子与复杂界面相互作用对于工业制造各种有特殊要求的薄膜材料也有启迪作用，如透气、防毒、过滤、防水、医学防护等。通过对细菌运动行为和趋化性能，及其调控机制过程深入研究，有助于开发潜在的药物靶点，阻断细菌的交流渠道，从而杀死细菌。细菌喜欢集群生长形成微生物膜，共同应对环境的威胁。了解细菌的群体运动策略有助于人类找到应对大自然危机的有效方案。微生物膜本身就是一个微小的生态系统，不仅其中的细菌个体之间相互影响[14]，而且它们与周围的微观环境也有密切的相互作用。微生物膜和人类健康以及工业应用也有很大关系。相关的微生物膜研究能够帮助我们理解微生物和人体环境的相互作用。利用微生物膜来处理环境有机物和重金属污染，以及用来制备生物能源电池，是环境工程的新领域。

第二节　细菌动力学的背景和现状

本节将分别介绍：细菌的自驱动运动，细菌与界面的相互作用，细菌的集体运动及自组织行为，细菌趋化性能、信息交流及群体响应。

一、细菌的自驱动运动

细菌依靠尾部鞭毛的驱动在溶液中游动。细菌在溶液中游动并不像无生命的粒子那样无规则地随机行走。多鞭毛细菌的表面生长着长短不一的鞭毛，并指挥这些鞭毛旋转或摆动来控制自己游向更适合其生存的区域。在营养充足的条件下，细菌快速生长到一定尺寸之后，就会分裂成两个同样的生命体。这种无性倍增的繁殖方式可以让细菌在短短几小时内形成一个细菌群落。

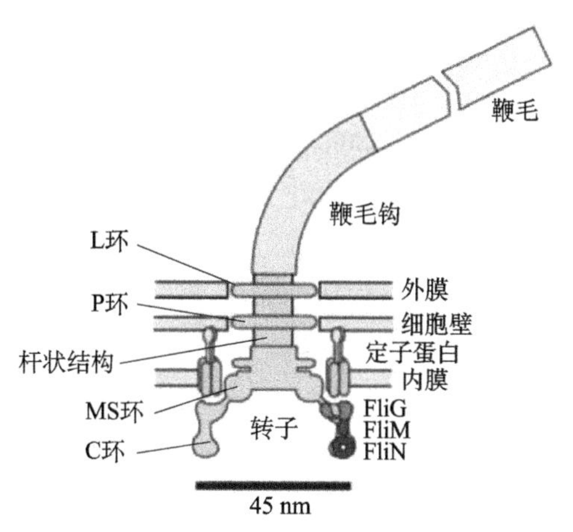

图 6-10-1 鞭毛马达微观结构示意图

图片来源：中国科学技术大学学报，第 44 卷第 5 期（2014）

细菌的鞭毛依靠镶嵌在细胞膜内的分子马达的转动来驱动。鞭毛马达（motor）大约 45 nm 宽。鞭毛（filament）根部是近十种蛋白形成的转子（rotor），包含 4 个环状结构（C 环，MS 环，P 环和 L 环）及 1 个杆状结构（rod）[3,15,16]。包围着这个转子的是一圈定子蛋白（stator units），它们一端结合在坚硬的细胞壁（cell wall）上，另一端横跨细胞内膜（inner membrane）与转子 C 环相互作用。每个定子蛋白包含两个跨膜的质子通道，当质子进入通道吸附在定子蛋白上时，定子蛋白会发生构象变化。变形后的定子蛋白与转子蛋白产生很强的局域静电相互作用，从而推动转子转动。当质子通过质子通道并脱离定子蛋白后，定子会恢复原来构型。转子 C 环的不同构象决定了转子带动的鞭毛是顺时针旋转还是逆时针旋转。当马达逆时针旋转时，众多鞭毛同步运动凝聚形成螺旋状，推动细菌直线游动[17]；当马达顺时针旋转时，众多鞭毛披散开，细菌原地转动，等待重新定向游动的时机。

细菌的微观马达工作原理完全不同于宏观热蒸汽马达和机电马达。这一点对于未来微观引擎的设计很重要。通常意义上的宏观马达是一个各种部件组合成的稳固不变的机械结构，因此微观的马达也常常被认为具有类似的静态结构。然而，最近的实验表明鞭毛马达的构成是动态的。首先，定子在不停地随机地与马达结合或脱离。以往的马达复活实验间接地证明了这一点[18]。最近的单分子荧光标记定子实验直接证明了这一点。其次，最近的实验表明转子 C 环的构成也是动态的：转子 C 环蛋白 FliM 也在不停地与马达结合或脱离，而且这种动态构成使马达自身具有适应性：当细菌内 CheY-P 浓度降低时，C 环 FliM 蛋白数增加，从而使马达对 CheY-P 的感应更灵敏，这使得细菌趋化过程中信号转导系统输入（受体蛋白组）和输出（鞭毛马达）之间的耦合具有很强的鲁棒性（robustness）[19]。鞭毛马达作为分子机器的这种适应性重组机制可能具有普遍性，在其他分子机器上也有类似现象。鞭毛马达的定子构成的动态性也是一种适应性重组：最近的实验表明，马达定子的稳定性跟外界负载相关，当负载增加时，定子跟马达的结合更稳定，马达的定子数目增加，从而能产生更大的力矩[20,21]。

多鞭毛细菌和单鞭毛细菌运动方式遵循不同的物理机制。自然界中很多细菌只有单个鞭毛，如溶藻弧菌[22-25]、绿脓杆菌[26-29]和新月柄杆菌[30-32]。单鞭毛的细菌和其他多鞭毛菌一样，如大肠杆菌[6,33-36]，都有趋化性（chemotaxis）[37-39]。最近的研究表明单鞭毛的细菌采用一种叫“正反甩尾”的方式在水中游动[40,41]。这种运动模式与改变鞭毛马达转向及化学响应诱导能力结合起来之后，导致其采用与偏随机游走（run-and-tumble）[3,42]不同的过程来实现其趋化机制。

细菌体表面其他附件对细菌运动有辅助和调节作用。细菌的外层膜的外表面仍然种植着一层毛茸茸的脂多糖类长链分子形成的表面皮毛。在这些表面皮毛中也会长出 2～3μm 长的菌毛（pilus）[28,43]。雄性细菌可以通过这些菌毛向雌性细菌转移 DNA。菌毛的数量远远多于长达 10μm 左右的鞭毛。因此，在近界面处，菌毛其实是最先与界面接触的。细菌可以利用菌毛在界面上爬行，对于细菌在界面上的运动和黏附（adhesion) 都起着决定性作用[28,43-45]。

二、细菌与界面的相互作用

界面是细菌最常见的聚集生存环境。寄生于生命体中的细菌常常在由各种界面分隔开的有限空间中运动，其生存特性受到固体界面、液体界面和生物膜的强烈影响。如何在界面附近的高度黏滞液体中迅速游动，并有效转变方向来保证追寻适合自己生存的环境是细菌面临的首要任务。科学家发现在界面上爬动的细胞可以感知衬底界面的弹性强度和刚性，而且会根据末端反馈回的界面的信息对细胞内部的很多生物功能进行生化调制和重新适应[46]。界面附近有很特异的物理性质，有利于细菌屯集。在界面附近，局域的电磁势能、化学势能和表面张力与远离界面的区域相比较经常显示出特殊分布。这些局域的势场会束缚诸多大分子和各种离子，而这些大分子是很多微生物的养分食物来源。空气中的氧在空气和水的界面浓度最高。有机分子也喜欢吸附在空气-水[34,47]或水-固界面。细菌的细胞膜表面也和空气-水或者固体界面之间存在很强的范德瓦耳斯相互作用[48-50]。上述各种物理化学因素的综合作用增强细菌在表面附近聚积的倾向[48,51]。

当细菌游动在固体表面附近时，它们往往形成一些圆形轨道[25,52-54]。近界面的流体力学相互作用一直被认为是游动细菌在界面附近积累的一个重要原因[51,55]。如果距离缩小至鞭毛的长度范围之内，流体动力相互作用则强烈依赖于细菌鞭毛丝的长度，细菌鞭毛通过顺时针或逆时针的转动“推进”或“拉进”等具体过程[55,56]。最近研究显示，游动细菌与固体表面的碰撞和旋转布朗运动随机变化都会对细菌在表面附近积累有很大影响[51,57,58]。

细菌体附件对细菌在界面附近的行为也起着至关重要的作用。细菌体表面除了分布有较长的鞭毛之外，也生长有不少其他的细菌体附属生长物，如菌毛，这些附属生长物对细菌运动也起了很大的辅助及调节作用。有些细菌的菌毛，如绿脓杆菌，可以黏附在界面上，进行伸张缩回运动，诱导出细菌的爬行运动[28]。新月柄杆菌在跟固体表面相互作用时，也是先通过菌毛黏附表面和缩回，继而导致细菌体永久黏附到固体表面上[59]。

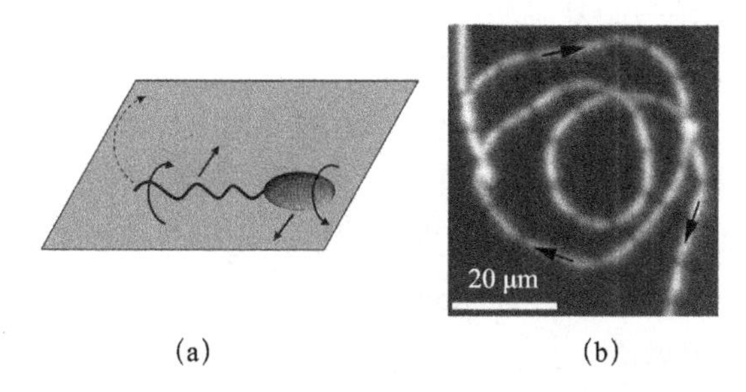

图 6-10-2 (a)Caulobacter swarmer 细胞的鞭毛转动方向及其细胞头部转动受力图分析示意；(b) 在表面附近沿着环形轨迹游动。图像取自（Li G, Tam L K, Tang J X, 2008，PNAS，105: 18355-9）

三、细菌的群体运动及自组织行为

细菌在地球上常常集体行动生长。地球上的大多数细菌以群落的形式生活在各种表面上，形成覆盖在有机体的或者无机体的表面[60]的微生物膜（biofilm）(图 6-10-3)。微生物学和临床医学研究表明，微生物膜对生态系统和人体健康起着重要的作用[7-9,60,61]。微生物膜也展现了很多有趣的物理现象，在非平衡态斑图（pattern）动力学、自驱动系统和复杂流体的物理研究中吸引了越来越多物理学研究者[11,62-64]。细菌的构造、功能和行为相对简单，其运动行为具有很高的可控性，而且又具有明显不同于无生命粒子的特殊表现，因此为传统物理学中多体集体运动和自组织现象的研究提供了更新颖的物理机制。

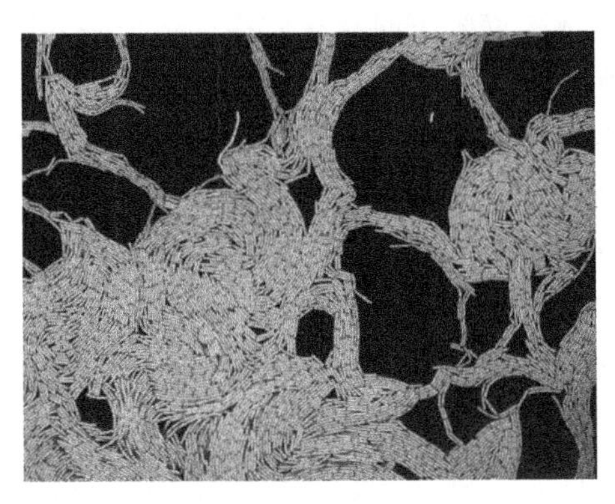

图 6-10-3 共聚焦荧光显微镜下观察到的形成初期的微生物膜。图片中的细菌是枯草杆菌（bacillus subtilis），在土壤中十分常见，也是一种益生菌。图片来源：英国剑桥大学 Haseloff 实验室

微生物膜是细菌与胞外基质的集合体。微生物膜中可以包含一种或多种细菌。细菌分泌的多糖和蛋白质形成胶状胞外基质[17,18]，把细菌包裹在一起。在实验室条件下，微生物膜通常生长在液-固界面上（如流体腔室 flow cell 的内表面）或者气体-半液半固体界面上（如琼脂表面）。细菌通过细菌触觉感知这两种不同界面的表面环境，其进行细胞分化和分裂生长表现出两种不同的行为：①浮游-定植分化（planktonic-to-sessile differentiation）。当浮游状态的细菌接触到液-固界面时，它们的基因表达会发生显著变化，分化成定植状态，进而形成微生物膜[65]。在这个分化过程中，它们会抑制运动能力[66]，表达菌毛（pili）[37,65]和纤维（curli）[37]，并产生胞外基质[65,67]。②集群运动分化（swarm-cell-differentiation）。当带鞭毛的细菌，如大肠杆菌（escherichia coli）被接种到合适硬度的琼脂表面时，它们会首先分化成能够在表面环境上活跃运动的状态，然后转变为静止的定植状态并产生胞外基质，进而形成微生物膜[68]。尤其是在合适的湿度环境下，这些细菌会表达更多的鞭毛[69,70]，同时在琼脂表面制造一层几微米厚的液体，然后它们在这层液体薄膜中做剧烈的集体运动；这个过程被称为细菌集群（bacterial swarming），形成的菌落被称为集群菌落（bacterial swarm），相关的文献综述参见[62], [71-76]。靠近集群菌落中心的细菌会逐渐失去运动活性，转变为定植状态并产生胞外基质，进而形成微生物膜[77]。

细菌集群是大量细菌组成的自驱动多体系统。细菌集群运动与微生物膜的形成[10,76]、细菌子实体的发育[78]、微生物的扩散和共生[79]以及病菌的侵入[80]等过程密切相关，因而具有生态学和医学意义。各种细菌的集群运动行为相似。集群菌落中细菌紧密排布，以团簇、涡旋和束流等形式呈现高度有序流畅的集体运动[81-83]。尽管集群中细菌之间碰撞频繁，但单个细菌的运动速度和细菌在不受限的三维液体环境中运动速度相近[82]。这是细菌集群的一个重要特征。有趣的是，细菌在集群中倾向于做往复运动[82,84]。例如，大肠杆菌大约每隔 1.5s 反转一次运动方向[82]。这种往复运动可能是棍状物体在高密度环境下的最自然的运动方式[85]。细菌集群运动不需要趋化性，只要鞭毛马达能够反转即可[86]。细菌集群动力学和自组织的主导因素之一是空间排斥作用[36,87]，另一个重要因素是流体动力学相互作用[58,82]。细菌通过渗透压梯度从基质中提取水分，形成一层几微米厚的液体层包裹整个集群菌落[83,88,89]。细菌运动剧烈搅动这层液体薄膜，产生每秒几十微米的流体运动；流体运动进而影响细菌之间的相互作用[85,88]。除了带鞭毛的细菌，使用其他运动机制的细菌也可以做集群运动，比如一大类使用隐性运动器官的滑动菌[78,90]，这些滑动菌的集群行为与带鞭毛细菌类似，特别是它们也倾向于做往复运动[78,90]。比如，黏菌（myxobacteria) 大约每 8min 周期性往复运动一次[90,91]。基于单细菌的生物力学模型结果表明，这种周期性往复运动对黏菌集群菌落的扩张是必要的，而且有助于集群在拥挤的环境下建立取向序并维持有序流畅的运动[91]。

细菌集群生长扩张会形成多样的自组织图案（图 6-10-4）。通常来说，如果环境养分越少越不适合细菌的生存，细菌集群生长形成的图案就越复杂多样。这些复杂图案的形成与单体细菌之间如何相互作用紧密关联。细菌集群中的个体之间会通过长链分子感知邻居细胞的行为。细菌细胞也会分泌一些

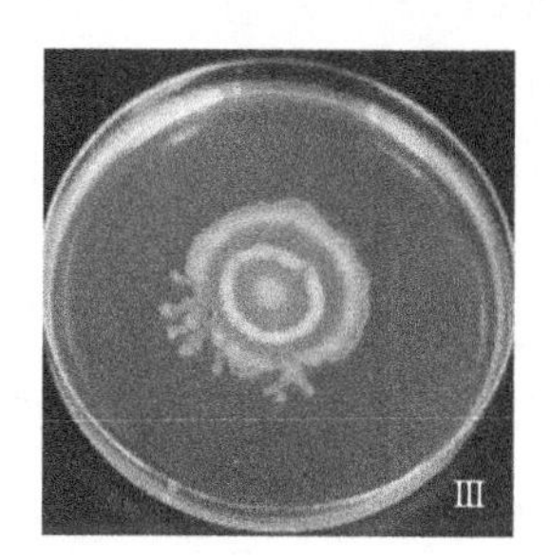

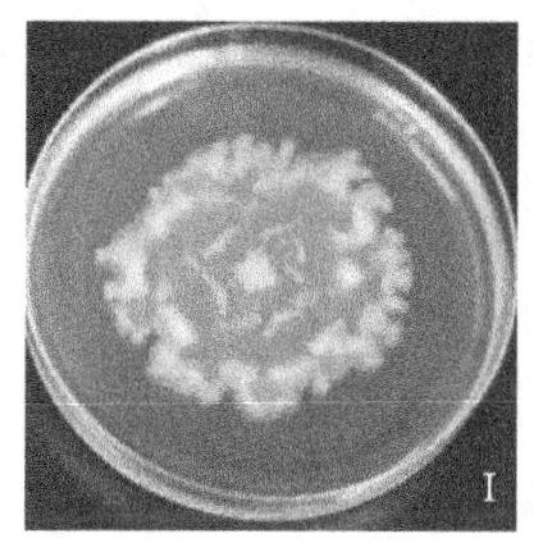

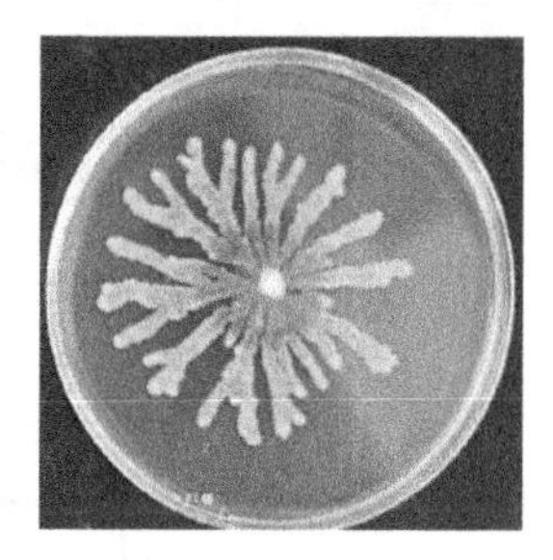

图 6-10-4　绿脓杆菌群体生长扩散过程中形成的自组织斑图。
图片来源：Yang A，Thesis for Bachelor of Science, Brown University，2015

表面活性剂或者从琼脂衬底上吸取水分使得自己更容易移动。细菌的自组织扩张曾经被生物学家认为是发生了基因突变。而实际上不少没有生命的自驱动粒子也表现出自组织的行为。如何从物理原理的角度出发来解释这些复杂的斑图也是值得深入研究的问题。目前已经有人开始从反应扩散方程、非平衡统计物理学和元胞自动机理论模型出发来研究水分在细菌集群运动形成斑图的过程中起到什么作用[92-94]。

四、细菌趋化性能、信息交流及群体响应

细菌趋化性（chemotaxis）是研究生物信号转导的一个经典范例。生命科学（尤其是系统生物学）的核心问题之一是理解细胞如何探测外界物理及化学环境的改变并做出响应。这通常是由细胞内各种信号转导系统完成的。控制细菌运动的趋化信号转导系统由于其系统架构的简洁性，以及它探测外界环境时展现的高灵敏度、宽动态范围、鲁棒的完美适应性，被引为信号系统的一个经典范例[95,96]。对它的研究和理解有助于理解其他更复杂的信号系统。

大肠杆菌（E. coli）及相关的沙门氏菌（salmonella enterica）被普遍用来研究细菌趋化信号传导系统，其架构如图 6-10-5 所示。当受体蛋白（receptor）探测到化学物质时，将调节自己的活性，与受体结合的蛋白酶 CheA 的活性也随即改变，从而改变细菌内小信号蛋白 CheY 的磷酸化水平（CheA 将 CheY 磷酸化）。磷酸化的 CheY(也即 CheY-P) 和鞭毛马达底部控制转向改变的蛋白 FliM 结合，提高马达顺时针转向的概率。另一种蛋白酶 CheZ 将 CheY-P 去磷酸化，大大降低 CheY-P 的寿命，从而加快这个系统的响应速度。在稳定状态下，CheY 被 CheA 磷酸化的速度与 CheY-P 被 CheZ 去磷酸化的速度相同，CheY-P 的浓度从而保持在一个稳定水平。受体蛋白的活性也受其甲基化水平控制，两种蛋白酶 CheR 和 CheB 能分别增加和降低受体蛋白的甲基化水平，同时 CheR 和 CheB 的活性又受受体蛋白和 CheA 的活性影响，形成一个反馈网络。这个反馈网络使细菌趋化系统具有鲁棒的完全自适应性[39,96,97]。

近年来对细菌趋化信号系统的研究积累了丰富的定量数据，基于这些数据的定量模型也相继被提出。进而这一系统的许多有代表性的系统特性相继被发现，其中包括高灵敏度[98-100]、宽动态范围[38,100]、鲁棒的完全适应性[39,96,97]以及输出端自身的适应性[19]。

关于细菌趋化系统的定量模型，可参见最近的一篇比较全面的综述文

章 [101]。最近对细菌趋化系统鲁棒性的研究进一步发现此系统对温度的鲁棒性[102]。在对趋化信号转导系统定量理解的基础上，基于信号通路的对细菌群体趋化运动的多尺度模型也被提出[103]。

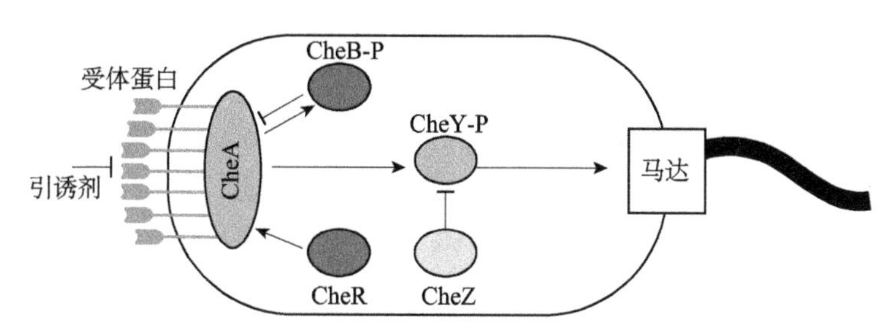

图 6-10-5　细菌趋化网络示意图。图片来源：中国科学技术大学学报，第 44 卷第 5 期（2014）

细菌趋化信号系统通过控制其输出端（鞭毛马达）的转向来实现趋化运动。马达转动方向的改变受趋化信号转导系统调控。马达的行为在不同负载下很不一样。由于实验手段的限制，过去对细菌鞭毛马达行为的研究集中在中高负载范围，这限制了对马达动力学性质的进一步探索。近年来新技术的发展使得对马达在低负载下行为的系统研究成为可能[19,104,105]。近年来的实验发现，鞭毛马达的转向改变的性质不仅取决于细胞内趋化信号系统中信号蛋白（CheY-P）的浓度，也与负载有关[88,105]。这个发现揭示了细菌对力感应的一个可能机制。

多鞭毛与单鞭毛驱动细菌运动中受到化学浓度吸引而重新定向的过程也有很大不同。多鞭毛细菌可以通过鞭毛合作运动成鞭毛束或者分散开转动来实现转向。而单鞭毛细菌则会通过鞭毛马达反转实现逆向游动，然后停留到某一地点后马达又恢复到原来的转向继续向前游动[41]。鞭毛根部的力学响应对鞭毛马达转向起关键性的作用，是通过一个叫欧拉屈曲（Eular buckling）的力学机制实现新的定向[41]。单鞭毛细菌这一独特的转向机制也造成其与多鞭毛细菌实现驱化性的方式迥异。最近美国匹兹堡大学（University of Pittsburgh) 的吴小伦教授研究组在这个专题上发表了一系列的理论模型[106-108]。

第三节　目前细菌动力学的前沿问题

一、细菌马达的生物及物理机制

通过三十多年对细菌鞭毛马达的研究，人们已经比较了解马达的基本结构。电子显微镜已经能够重构出鞭毛根部转子的微观结构，并对看到的每一蛋白部件进行命名。但是C环蛋白以及定子蛋白都是膜蛋白分子，很难通过结晶来探测其完整结构。即便是完全清楚其分子组成之后，仍然面临着如何理解这些蛋白分子在马达运转过程中具体动作和作用。鞭毛马达并不像宏观的齿轮那样，每一时刻与周围咬合的凹槽数目是固定的，马达C环是以一个动态随机变化的方式与定子蛋白结合或脱离的。目前人类还没有超精细显微照相机来拍摄到纳米尺度的高达每秒300转的转动过程，所以生物物理学家们只能通过目前可以观测到的介观现象，结合已知部件的微观结构来推测鞭毛马达的运作过程和原理。实验观测发现，鞭毛马达顺时针和逆时针转向的概率几乎是随机的。目前转子上的FliG蛋白结构已经解析清楚。在此基础上，一个完整环的转换机制模型被提出来用来解释马达力矩的转向[109]。该模型认为FliG构象的变化导致力矩产生过程中的静电电荷产生逆转。Ising类型的经典二态统计物理模型，酶变构协调模型或者酶变构时序模型也用来解释围成一圈的定子蛋白分子是如何在酶的催化作用下相互合作，最终实现集体逆转马达转向[110]。马达顺时针转动的概率如何依赖于细菌体内的蛋白酶CheY-P的浓度、温度、溶液pH，负载强度等仍然需要深入研究。同一细菌上的多个鞭毛马达之间会相互协调，同步成一束鞭毛，它们之间是通过何种方式耦合在一起的，比如除了与流体力学耦合作用相关[111]，是否还有酶开关控制在两者之间通信等问题，需要大量的实验和理论研究。

二、细菌集群运动的生物及物理机制

细菌集群中的单个细菌的行为和集体行为的关系通常难以用直觉理解。比如，黏菌通过控制单个细菌往复运动的周期可以呈现多种截然不同的集体运动模式[112,113]。细菌群落在生长过程中有时候会长出各种粗细不同的分支，类似章鱼的爪向四面扩张，有时候只是一圈一圈地扩散，但是密度分布沿着半径呈现周期性分布。即便在同样的半径圆环上，细菌有时候也聚合成一团一团，并均匀间隔开分布。目前反应扩散方程模型能够理论计算出大多数的斑图形状，然而离物理实际有一定距离。反应扩散方程模型假设了细菌是受到食物的吸引向四周扩散。有实验证据表明这个假设可能不成立，因为即使在食物匮乏的情况下，细菌仍然会活跃爬行。斑图的形成似乎只与细菌本身的活跃状态相关。目前大概有三种理论模型试图解释细菌群聚斑图动力学：基因突变理论[114]，反应扩散方程模型[93]，Maragoni流模型[115]。这些理论模型都有各自的局限性。例如，Maragoni流模型虽然考虑了细菌群落的厚度和边缘几何，却无法得出细菌生长的分支形状。最近，又有人从元胞自动机出发计算模拟研究水在细菌群聚斑图形成中的作用[116]。而关于细菌群聚斑图背后的统计物理学和几何力学行为的研究尚处于刚刚开展的状态。这是一个值得深入探讨的问题。它涉及的很多物理因素的重要性程度需要通过系列的实验来确认。由于涉及活体的细菌，细菌的很多状态都在随着时间变化，其中涉及细菌如何消耗溶液中的养分以及如何通过分泌化学物质应对环境的威胁和变化。复杂生物化学调控网络使得细菌的集体行为变得复杂而难以预测。目前已经有实验证据表明，两个邻近的细菌群落之间存在竞争关系。比如，处在两个细菌群落之间的空白地带之间有化学分泌物，可以杀死活的细菌[117]。关于细菌集群的研究中一个关键问题是单个细菌的行为如何导致集体行为。除了细菌运动本身，细菌集群菌落和周围微观物理化学环境的相互作用也是一个重要问题。细菌的生长和运动会使周围的物理化学环境变得非均一，而非均一性的环境又会反过来影响细菌的行为。在微生物膜形成过程中，细菌的集体生长和集群运动如何改变周围的微观环境？细菌的能量代谢模式如何适应这些变化？当周围环境充满各种物质的浓度梯度时，细菌如何调节运动模式以寻求最佳的生存条件？细菌的集群运动如何影响微生物膜的扩张和三维结构？因此，有必要开展系列的实验和理论分析，对细菌的个体行为及其所处的微观物理化学环境进行操控。

三、细菌动力学的研究与活性软物质领域的交叉关系

细菌群落的集体迁徙扩张，鸟群的运动和海洋中鱼群的游动都非常类似于液晶材料和铁电材料等软物质体系。值得注意的是，生物活性软物质体系是一个不停消耗能量的体系，已经超出了以往非平衡态统计物理学理论体系能够描述的范畴。细菌集群是最简单的呈现集体运动的生物体系[118]。关于细菌集群的研究为发展用来描述广义自驱动体系和活性物质的理论提供了丰富的实验基础[56,63,118,119]，尤其是细菌集群的简单和高度可控性有助于验证理论模型的预测结果。活性软物质物理已经制备了很多人工制造的自驱动粒子[56]。比如，螺旋形的微米粒子有磁性并可通过交流电磁场促成其运动[120]。也有材料科学家让纳米管一半镀金，另一半镀铂。把这样的纳米管放入掺加过氧化氢的溶液中，它们就会像细菌一样游动[121-123]。无论是人工自驱动粒子还是自然界存在的棒状细菌，其运动速度都很低，属于典型的低雷诺数流体中运动。流体力学研究表明，在低雷诺数的流体运动中是不存在湍流或者涡旋等形态的。然而，在研究动量守恒系统中长波稳定性的数值模拟结果

中，低雷诺数的涡旋结构可以在某些时刻出现。而一些实验图像表明，细菌在水中集体运动时的确看到了类似湍流的图案[124]。这些现象违背了一直以来很成功的流体力学。而活体细胞中成千上万的蛋白纤维丝构成的网络也表现出了和长链分子构成网络系统完全不同的性质，包括伤口愈合、细胞分裂等活体系统特有的现象。总之，活体软物质系统对于科学家反思无生命粒子的流体，无生命粒子的多体系统的局限性提供了新的启示。譬如湍流、长程序、稳定性理论、斑图动力学等，在活体软物质系统中需要重新考虑，生命系统的复杂性也启发了无生命粒子的多体系统研究新的出发点。

四、细菌动力学研究引发的仿生学及生物材料方面的应用

细菌的集体形态运动对于工程、医学、生物传感仿生材料方面也有诸多的启示作用。和大海中的鱼群一样，微生物喜欢聚集在一起行动。似乎群体的感知行为与单个个体有很大不同。微生物膜形成机制的一个悬而未决的核心问题是在微生物膜形成初期细菌如何感知并适应表面环境。许多观察发现，在与表面初始接触时，细菌的基因表达和行为会发生显著变化，这些变化直接导致微生物膜的形成。这种表面感应现象可以称为“细菌触觉”[74]。细菌触觉是许多细菌种属共有的一种保守行为。不同于多细胞生物依赖神经系统的触觉，细菌触觉发生在单细胞层次上，因此和细菌所处的微观物理化学环境有关。尽管近年来报道了一系列的基因筛选结果[125-129]，细菌触觉的分子生物学通路和激发这种触觉的物理化学信号都尚不明确。理解细菌触觉的机制将有助于深入理解微生物膜的形成机制，并能为防治病菌侵染提供新的思路。大多数的人造水下航行器都需要严格防水，即使是没有驾驶员的无人航行器也要严格防水。然而像细菌这样的水中微生物却只能在水里生存运动。如何制造出微纳米尺度的水中航行器，是仿生工业未来需要突破的难题[130]。在基础理论研究方面，需要把复杂的生物机器分解成基本部件，研究这些部件之间如何跨越不同的尺度相互合作，如何用更可靠的材料代替改进这些生物机器。从相对来说比较简单的细菌入手，不失为一个好的选择。

第四节　展望今后十年可能实现的重大研究进展

一、弄清细菌马达的力学及统计物理机制

鞭毛马达在自然界已经存在十亿多年，其工作效率远远超越了宏观引

擎。细菌马达究竟如何高效地把化学能转成机械转动呢？这一根本问题今后5～10年随着对马达蛋白结构的了解及模拟有望会有实质进展。由于每个细菌马达上的同种蛋白有30多个，它们必须协调一致地旋转。有些统计物理机制对理解细菌马达如何控制顺逆时针旋转至关重要。这是今后5～10年可能解决的一个重要研究课题。新的成像技术和超分辨技术，马达标记技术，细菌荧光标记技术将会进一步发展起来，有力推动分子马达领域的进展。

二、建立细菌集群运动的清楚的物理机制

关于细菌集群运动的物理机制有两个重要问题：单个细菌的行为如何导致集体行为，以及细菌集群菌落和周围微观物理化学环境如何相互作用。要回答这些问题，需要对细菌的个体行为及它们所处的微观物理化学环境进行操控。近年来，基因工程和微流体控制技术的进步使得精确操控细菌的个体行为及它们所处的微观环境成为可能。随着这些技术的应用，我们预期将能更深刻地理解细菌集群运动的物理机制。

三、细菌动力学的研究成果会给活性软物质领域展示多种机制

最近活性软物质领域涌现了众多理论模型来描述广义自驱动体系和活性物质的非平衡性质[56,63,118,119]。这些模型结果表明个体的行为参数，如噪声[131]、运动能力[132]和记忆性[133]，对自驱动体系和活性物质的整体行为至关重要。然而，这些模型的不少结论都缺乏实验验证。作为用来研究自驱动和活性物质的最简单的模型体系之一，细菌集群运动可以验证理论模型的预测结果。我们预期通过控制细菌的个体行为及它们所处的微观环境，可以在更广的参数空间中检验描述自驱动体系和活性物质的理论模型。

四、细菌动力学的研究引发一些仿生学及生物材料方面的应用

科学家受到植物叶子的启发设计出了一种人造叶子。这种人造叶子利用真氧产碱杆菌将二氧化碳加氢气转化成液体燃料异丙醇，从而将太阳能转化成液体燃料[134]。长有鞭毛的细菌也被设计用来推动直径约100μm的人造齿轮[135]。而在比细菌更小的尺度上，一些生物分子马达也被成功用来推动大分子公交车，使之行进在刻好的约100nm宽的沟槽内[136]。今后十多年里可能会涌现出更多的仿生微纳米航行器或者利用细菌治疗疑难疾病的例子。

第五节　结　　语

20世纪下半叶生命科学发生了根本性的变革：DNA的双螺旋结构引发的一系列科学研究大爆发，引导人类对病毒、细菌、动植物和人类的基因有了深入了解，信息生物学直接引导了21世纪的精准医疗技术革命。这些变革也带来21世纪的多学科交叉研究的兴起。很多重要的生物现象都是在微纳米尺度的液体、固体和气体的界面以及其他微纳米尺度的大分子等组成的复杂环境中进行的，有很多通常采用生物基因遗传突变等生物进化理论来解释的现象其实可以从物理学的角度给出简单合理的解释。最近十多年里，从事于复杂系统和软物质凝聚态物理研究的物理学家们已经发现大量自组织的复杂系统，而这些系统表现的自组织斑图是没有生命的粒子形成的。而类似于斑图，生物力学之类的物理现象却不是生物研究关心的课题。事实上物理理论预测的多粒子反应扩散行为有助于人们理解一些生命体的斑图现象，因此生物系统中的物理现象和复杂行为为物理学的研究提供了新奇的研究课题。最近十年，法国国家科学研究院和德国马克斯-普朗克科学促进学会等建立了多个软物质或复杂体系研究中心。美国国家健康基金会也于近几年资助建立了十多个用物理手段治疗癌症的研究中心（参阅physics.cancer.gov）。细菌生物物理学研究的尺度介于宏观和微观之间，在大尺度上与宏观流体力学有关联，在微观尺度上与生物化学、微观力学、生物传感系统等有关联，我们相信细菌领域值得多学科的研究人员合作深入挖掘研究。细菌可以依附空气、洋流、动物等传播繁衍。细菌领域的跨学科交叉研究对环境保护及防止并消灭瘟疫，促进人类健康发展都会起到积极的推动作用。

唐建新（美国布朗大学物理系），
袁军华（中国科学技术大学物理系），
吴艺林（香港中文大学物理系及香港中文大学深圳研究院），
司铁岩（哈尔滨工业大学）

参考文献

[1] Glud R N. Wenzhöfer F, Middelboe M, et al. High rates of microbial carbon turnover in

sediments in the deepest oceanic trench on Earth. Nature Geoscience, 2013, 6: 5.

[2] Sears C L. A dynamic partnership: Celebratingour gut flora. Anaerobe., 2005, 11: 5.

[3] Berg H C. The rotary motor of bacterial flagella. Annu. Rev. Biochem., 2003, 72: 19-54.

[4] Bardy S L, Ng S Y, Jarrell K F. Recent advances in the structure and assembly of the archaeal flagellum. J. Mol. MicroBiol., Biotechnol., 2004, 7: 41-51.

[5] Driks A, Bryan R, Shapiro L, et al. The organization of the Caulobacter crescentus flagellar filament. J. Mol. Biol., 1989, 206: 627-636.

[6] Turner L, Ryu W S, Berg H C. Real-time imaging of fluorescent flagellar filaments. J. Bacteriol., 2000, 182(10): 2793-2801.

[7] Costerton J W, Stewart P S, Greenberg E P. Bacterial biofilms: a common cause of persistent infections. Science, 1999, 284: 1318-1322.

[8] Donlan R M. Biofilms: microbial life on surfaces. Emerg. Infect. Dis., 2002, 8: 881-890.

[9] Hall-Stoodley L, Costerton J W, Stoodley P. Bacterial biofilms: from the natural environment to infectious diseases. Nat. Rev. Microbiol., 2004, 2: 95-108.

[10] López D, Vlamakis H, Kolter R. Biofilms. Cold Spring Harb. Perspect. Biol., 2010,2: a000398.

[11] Wilking J N, Angelini T E, Seminaraa A, et al. Biofilms as complex fluids. MRS Bulletin, 2011, 36: 385-391.

[12] Molin S, Tolker-Nielsen T. Gene transfer occurs with enhanced efficiency in biofilms and induces enhanced stabilisation of the biofilm structure. Curr. Opin. Biotechnol., 2003, 14(3): 255-261.

[13] Kaewkamnerdpong B M K. Modelling nanorobot control using swarm intelligence: A pilot study. Innovations in Swarm Intelligence, Volume 248 of the series Studies in Computational Intelligence, Springer Berlin Heidelberg., 2009, 248: 175-214.

[14] Yang A. Effects of Physical Factors on the Swarming Motility of Pseudomonas Aeruginosa. Thesis for Bachelor of Science, Brown University, 2015.

[15] Kojima S, Blair D F. Conformational change in the stator of the bacterial flagellar motor. Biochemistry, 2001, 40: 13041-13050.

[16] Thomas D R, Morgan D G, DeRosier D J. Rotational symmetry of the C ring and a mechanism for the flagellar rotary motor. Proc. Nat. Acad. Sci. USA, 1999, 96(18): 10134-10139.

[17] Turner L, Zhang R, Darnton N C, et al. Visualization of flagella during bacterial swarming. J. Bacteriol., 2010, 192: 3259-3267.

[18] Leake M C, et al. Stoichiometry and turnover in single, functioning membrane protein complexes. Nature, 2006, 443: 355-358.

[19] Yuan J, Branch R W, Hosu B G, et al. Adaptation at the output of the chemotaxis signaling pathway. Nature, 2012, 484: 233-236.

[20] P P L, Hosu B G, Berg H C. Dynamics of mechanosensing in the bacterial flagellar motor. Proc. Natl. Acad. Sci. USA, 2013, 110: 11839-11844.

[21] Tipping M J, Dalelez N J, Berry R M, et al. Load-dependent assembly of the bacterial flagellar motor. mBio, 2013, 4: 00551-00513.

[22] Atsumi T, et al. Effect of viscosity on swimming by the lateral and polar flagella of vibrio alginolyticus. J. Bacteriol, 1996, 178(16): 5024-5026.

[23] Sowa Y, Hotta H, Homma M, et al. Torque-speed relationship of the Na+-driven flagellar motor of vibrio alginolyticus. J. Mol. Biol., 2003, 327(5): 1043-1051.

[24] Kudo S, Imai N, Nishitoba M, et al. Asymmetric swimming pattern of Vibrio alginolyticus cells with single polar flagella. FEMS Microbiol. Lett., 2005, 242(2): 221-225.

[25] Goto T, Nakata K, Baba K, et al. A fluid-dynamic interpretation of the asymmetric motion of singly flagellated bacteria swimming close to a boundary. Biophys. J., 2005, 89(6): 3771-3779.

[26] Carnazza S, Satriano C, Guglielmino S, et al. Fast exopolysaccharide secretion of Pseudomonas aeruginosa on polar polymer surfaces. J. Colloid Interface Sci., 2005, 289(2): 386-393.

[27] Sauer K, Camper A K, Ehrlich G D, et al. Pseudomonas aeruginosa displays multiple phenotypes during development as a biofilm. J. Bacteriol., 2002, 184(4): 1140-1154.

[28] Skerker J M, Berg H C. Direct observation of extension and retraction of type IV pili. Proc. Nat. Acad. Sci. USA, 2001, 98(12): 6901-6904.

[29] O'Toole G A, Kolter R. Flagellar and twitching motility are necessary for Pseudomonas aeruginosa biofilm development. Mol. Microbiol., 1998, 30(2): 295-304.

[30] Brun Y V, Janakiraman R. Dimorphic life cycles of Caulobacter and stalked bacteria. ASM Press, 2000: 297-317.

[31] Li G L, Smith C S, Brun Y V, et al. The elastic properties of the Caulobacter crescentus adhesive holdfast are dependent on oligomers of N-acetylglucosamine. J. Bacteriol., 2005, 187(1): 257-265.

[32] Poindexter J S. Biological properties and classification of the Caulobacter crescentus group. Bacteriol. Rev., 1964, 28: 231-295.

[33] Wu M, Roberts J W, Kim S, et al. Collective bacterial dynamics revealed using a three-dimensional population-scale defocused particle tracking technique. Appl. Environ. Microbiol., 2006, 72(7): 4987-4994.

[34] Lemelle L, Palierne J F, Chatre E, et al. Counterclockwise Circular motion of bacteria swimming at the air-liquid interface. J. Bacteriol., 2010, 192(23): 6307-6308.

[35] Hodges K, Hecht G. Bacterial infections of the small intestine. Current Opinion in

gastroenterology, 2013, 29(2): 159-163.

[36] Swiecicki J M, Sliusarenko O, Weibel D B. From swimming to swarming: escherichia coli cell motility in two-dimensions. Integrative Biology, 2013, 5(12): 1490-1494.

[37] Pratt L A, Kolter R. Genetic analysis of Escherichia coli biofilm formation: roles of flagella, motility, chemotaxis and type I pili. Molecular microbiology, 1998, 30: 285-293.

[38] Mello B A, Tu Y. Effects of adaptation in maintaining high sensitivity over a wide range of backgrounds for escherichia coli chemotaxis. Biophys. J., 2007, 92: 2329-2337.

[39] Hansen C H, Endres R G, Wingreen N S. Chemotaxis in escherichia coli: a molecular model for robust precise adaptation. PLoS Comput. Biol., 2008, 4: e1.

[40] Xie L, Altindal T, Chattopadhyay S, et al. Bacterial flagellum as a propeller and as a rudder for efficient chemotaxis. Proceedings of the National Academy of Sciences, 2011, 108(6): 2246-2251.

[41] Son K, Guasto J S, Stocker R. Bacteria can exploit a flagellar buckling instability to change direction. Nature Physics, 2013, 9: 5.

[42] Locsei J T. Persistence of direction increases the drift velocity of run and tumble chemotaxis. J. Math. Biol., 2007, 55(1): 41-60.

[43] Bodenmiller D, Toh E, Brun Y V. Development of surface adhesion in caulobacter crescentus. J. Bacteriol., 2004, 186(5): 1438-1447.

[44] An Y H, Friedman R J. Concise review of mechanisms of bacterial adhesion to biomaterial surfaces. J. Biomed. Mater. Res., 1998, 43(3): 338-348.

[45] Busscher H J, van Der Mei H C. Use of flow chamber devices and image analysis methods to study microbial adhesion. Meth. Enzym., 1995, 253: 455-477.

[46] Discher D E, Janmey P, Wang Y L. Tissue cells feel and respond to the stiffness of their substrate. Science, 2005, 310: 1139-1143.

[47] Morse M, Huang A, Li G, et al. Molecular adsorption steers bacterial swimming at the air/water interface. Biophysical J., 2013, 105(1): 21-28.

[48] Vigeant M A, Ford R M. Interactions between motile Escherichia coli and glass in media with various ionic strengths, as observed with a three-dimensional-tracking microscope. Appl. Environ. Microbiol., 1997, 63(9): 3474-3479.

[49] Vigeant M A S, Ford R M, Wagner M, et al. Reversible and irreversible adhesion of motile Escherichia coli cells analyzed by total internal reflection aqueous fluorescence microscopy. Appl. Environ. Microbiol., 2002, 68(6): 2794-2801.

[50] Vigeant M A S, Wagner M, Tamm L K, et al. Nanometer distances between swimming bacteria and surfaces measured by total internal reflection aqueous fluorescence microscopy. Langmuir, 2001, 17(7): 2235-2242.

[51] Li G, Tang J X. Accumulation of microswimmers near a surface mediated by collision and rotational Brownian motion. Phys. Rev. Lett., 2009, 103(7): 078101.
[52] Frymier P D, Ford R M, Berg H C, et al. 3-dimensional tracking of motile bacteria near a solid planar surface. P. Natl. Acad. Sci. USA, 1995, 92(13): 6195-6199.
[53] Magariyama Y, et al. Difference in bacterial motion between forward and backward swimming caused by the wall effect. Biophys. J., 2005, 88(5): 3648-3658.
[54] Li G L, Tam L K, Tang J X. Amplified effect of Brownian motion in bacterial near-surface swimming. P. Natl. Acad. Sci. USA, 2008, 105(47): 18355-18359.
[55] Berke A P, Turner L, Berg H C, et al. Hydrodynamic attraction of swimming microorganisms by surfaces. Phys. Rev. Lett., 2008, 101(3).
[56] Marchetti M C, Ramaswamy S, Liverpool T B, et al. Aditi Simha. Hydrodynamics of soft active matter. Rev. Mod. Phys., 2013, 85(3): 47.
[57] Li G, et al. Accumulation of swimming bacteria near a solid surface. Physical Review E, 2011, 84(4): 041932.
[58] Drescher K, Dunkel J, Cisneros L H, et al. Fluid dynamics and noise in bacterial cell-cell and cell-surface scattering. Proceedings of the National Academy of Sciences, 2011, 108(27): 10940-10945.
[59] Li G, et al. Surface contact stimulates the just-in-time deployment of bacterial adhesins. Molecular microbiology, 2012, 83(1): 41-51.
[60] Costerton J W, et al. Bacterial biofilms in nature and disease. Annu. Rev. Microbiol., 1987, 41: 435-464.
[61] Parsek M, Singh P. Bacterial biofilms: an emerging link to disease pathogenesis. Annu. Rev. Microbiol., 2003, 57: 677-701.
[62] Copeland M F, Weibel D B. Bacterial swarming: a model system for studying dynamic self-assembly. Soft Matter, 2009, 5: 1174-1187.
[63] Ramaswamy S. The mechanics and statistics of active matter. Annu. Rev. Condens. Matter Phys., 2010, 1: 323-345.
[64] Nadal J P, Papaioannou E, Wahjudi M, et al. Role of PvdQ in Pseudomonas aeruginosa virulence under iron-limiting conditions. Microbiology, 2010, 156(1): 11.
[65] Prigent-Combaret C, et al. Developmental pathway for biofilm formation in curli-producing Escherichia coli strains: role of flagella, curli and colanic acid. Environ. Microbiol., 2000, 2: 450-464.
[66] Wolfe A J, Visick K L. Get the message out: cyclic-Di-GMP regulates multiple levels of flagellum-based motility. J. Bacteriol., 2008, 190: 463-475.
[67] Danese P N, Pratt L A, Kolter R. Exopolysaccharide production is required for development

of Escherichia coli K-12 biofilm architecture. J. Bacteriol., 2000, 182: 3593-3596.

[68] Vlamakis H, Aguilar C, Losick R, et al. Control of cell fate by the formation of an architecturally complex bacterial community. Genes. Dev., 2008, 22: 945-953.

[69] Allison C, Lai H C, Gygi D, et al. Cell differentiation of Proteus mirabilis is initiated by glutamine, a specific chemoattractant for swarming cells. Molecular microbiology, 1993, 8(1): 53-60.

[70] Harshey R M, Matsuyama T. Dimorphic transition in Escherichia coli and Salmonella typhimurium: surface-induced differentiation into hyperflagellate swarmer cells. Proc. Natl. Acad. Sci. USA, 1994, 91: 8631-8635.

[71] Allison C, Hughes C. Bacterial swarming: an example of prokaryotic differentiation and multicellular behavior. Sci. Prog., 1991, 75: 403-422.

[72] Harshey R M. Bacterial motility on a surface: many ways to a common goal. Annu. Rev. Microbiol., 2003, 57: 249-273.

[73] Henrichsen J. Bacterial surface translocation: a survey and a classification. Bacteriol. Rev., 1972, 36: 478-503.

[74] Kearns D B. A field guide to bacterial swarming motility. Nat. Rev. Microbiol., 2010, 8: 634-644.

[75] McCarter L L. Dual flagellar systems enable motility under different circumstances. J. Mol. Microbiol. Biotechnol., 2004, 7: 18-29.

[76] Verstraeten N, et al. Living on a surface: swarming and biofilm formation. Trends Microbiol., 2008, 16: 496-506.

[77] Hamze K, et al. Single-cell analysis in situ in a Bacillus subtilis swarming community identifies distinct spatially separated subpopulations differentially expressing hag (flagellin), including specialized swarmers. Microbiology, 2011, 157(9): 2456-2469.

[78] Kaiser D. Bacterial swarming: a re-examination of cell-movement patterns. Curr. Biol., 2007, 17: 561-570.

[79] Kerr B, Riley M A, Feldman M W, et al. Local dispersal promotes biodiversity in a real-life game of rock-paper-scissors. Nature, 2002, 418(6894): 171-174.

[80] Struthers J K, Westran R P. Clinical Bacteriology. CRC Press, 2003: 192.

[81] Zhang H P, Be'er A, Smith R S, et al. Swarming dynamics in bacterial colonies. EPL, 2009, 87: 48011.

[82] Darnton N C, Turner L, Rojevsky S, et al. Dynamics of bacterial swarming. Biophys. J. 2010, 98: 2082-2090.

[83] Zhang H P, Be'er A, Florin E L, et al. Collective motion and density fluctuations in bacterial colonies. Proc. Natl. Acad. Sci. USA, 2010, 107: 13626-13630.

[84] Be'er A, Strain S K, Hernández R A, et al. Periodic reversals in paenibacillus dendritiformis swarming. J. Bacteriol., 2013, 195(12): 2709-2717.

[85] Wu Y, Hosu B G, Berg H C. Microbubbles reveal chiral fluid flows in bacterial swarms. Proc. Natl. Acad. Sci. USA, 2011, 108(10): 4147-4151.

[86] Mariconda S, Wang Q, Harshey R M. A mechanical role for the chemotaxis system in swarming motility. Mol. Microbiol., 2006, 60: 1590-1602.

[87] Peruani F, Deutsch A, Bär M. Nonequilibrium clustering of self-propelled rods. Physical Review E, 2006, 74(3): 030904.

[88] F B Minamino T, Wu Z, Namba K, et al. Coupling between switching regulation and torque generation in bacterial flagellar motor. Phys. Rev. Lett., 2012, 108: 178105.

[89] Ping L, Wu Y, Hosu Basarab G, et al. Osmotic pressure in a bacterial swarm. Biophysical Journal, 2014, 107(4): 871-878.

[90] McBride M J. Cytophaga-Flavobacterium gliding motility. J. Mol. MicroBiol., Biotechnol., 2004, 7: 63-71.

[91] Wu Y, Kaiser D, Jiang Y, et al. Periodic reversal of direction allows myxobacteria to swarm. Proc. Natl. Acad. Sci. USA, 2009, 106: 1222-1227.

[92] Mimuraa M, S H, Matsushitac M. Reaction diffusion modelling of bacterial colony patterns. Physica A, 2000, 282.

[93] Matsushita M W J, Itoha H, Watanabea K, et al. Formation of colony patterns by a bacterial cell population. Physica A, 1999, 274: 10.

[94] Mezanges C R, Gerin C, Deroulers C, et al. Modeling the role of water in Bacillus subtilis colonies. Phys. Rev. E., 2012, 85(4): 9.

[95] Hazelbauer G L, Falke J J, Parkinson J S. Baterial chemoreceptors: high-performance signaling in networked arrays. Trends Biochem. Sci., 2008, 33: 9-19.

[96] kollmann M, Lovdok L, Bartholome K, et al. Design principles of a bacterial signalling network. Nature, 2005, 438: 504-507.

[97] Alon U, Surette M G, Barkai N, et al. Robustness in bacterial chemotaxis. Nature, 1999, 397: 168-171.

[98] Segall J E, Block S M, Berg H C. Temporal comparisons in bacterial chemotaxis. Proc. Natl. Acad. Sci. USA, 1986, 83: 8987-8991.

[99] Duke T A J, Bray D. Heightened sensitivity of a lattice of membrane receptors. Proc. Natl. Acad. Sci. USA, 1999, 96: 10104-10108.

[100] Sourjik V, Berg H C. Receptor sensitivity in bacterial chemotaxis. Proc. Natl. Acad. Sci. USA, 2002, 99: 123-127.

[101] Tu Y. Quantitative modeling of bacterial chemotaxis: Signal amplification and accurate

adaptation. Annu. Rev. Biophys., 2013, 42: 337-359.

[102] Oleksiuk O, Jakovljevic V, Vladimirov N, et al. Thermal robustness of signaling in bacterial chemotaxis. Cell, 2011, 145: 312-321.

[103] Si G TW, Ouyang Q, Tu Y. Pathway-based mean field model for Escherichia coli chemotaxis. Phys. Rev. Lett., 2012, 109.

[104] Yuan J, Berg H C. Resurrection of the flagellar motor near zero load. Proc. Natl. Acad. Sci. USA, 2008, 105: 1182-1185.

[105] Yuan J, Fahrner K A, Berg H C. Switching of the bacterial flagellar motor near zero load. J. Mol. Biol., 2009, 390: 394-400.

[106] Xie L, Wu X L. Bacterial motility patterns reveal importance of exploitation over exploration in marine microhabitats. Part I: theory. Biophysical Journal, 2014, 107(7): 1712-1720.

[107] Xie L, Lu C, Wu X L. Marine bacterial chemoresponse to a stepwise chemoattractant stimulus. Biophysical Journal, 2015, 108(3): 766-774.

[108] Yang Y, He J, Altindal T, et al. A non-poissonian flagellar motor switch increases bacterial chemotactic potential. Biophysical Journal, 2015, 109(5): 1058-1069.

[109] Lee L K, Ginsburg M A, Crovace C, Donohoe M, et al. Structure of the torque ring of the flagellar motor and the molecular basis for rotational switching. Nature, 2010, 466: 6.

[110] Bai F, Branch R W, Jr N D, et al. Conformational spread as a mechanism for cooperativity in the bacterial flagellar switch. Science, 2010, 327: 5.

[111] Kim M, Powers T R. Hydrodynamic interactions between rotating helices. Phys. Rev. E, 2004, 69(6): 061910.

[112] Berleman J E, Scott J, Chumley T, et al. Predataxis behavior in Myxococcus xanthus. Proceedings of the National Academy of Sciences, 2008, 105(44): 17127-17132.

[113] Thutupalli S, Sun M, Bunyak F, et al. Directional reversals enable Myxococcus xanthus cells to produce collective one-dimensional streams during fruiting-body formation. Journal of The Royal Society Interface, 2015, 12(109).

[114] Davies D G, et al. The involvement of cell-to-cell signals in the development of a bacterial biofilm. Science, 1998, 280(5361): 295-298.

[115] Fauvart M, et al. Surface tension gradient control of bacterial swarming in colonies of Pseudomonas aeruginosa. Soft Matter, 2012, 8(1): 70-76.

[116] Mezanges X, et al. Modeling the role of water in Bacillus subtilis colonies. Physical Review E, 2012, 85(4).

[117] Be'er A, et al. Lethal protein produced in response to competition between sibling bacterial colonies. Proc. Natl. Acad. Sci. USA, 2010, 107(14): 6258-6263.

[118] Vicsek T, Zafeiris A. Collective motion. Physics Reports, 2012, 517(3-4): 71-140.
[119] Koch D L, Subramanian G. Collective hydrodynamics of swimming microorganisms: living fluids. Annual Review of Fluid Mechanics, 2010, 43(1): 637-659.
[120] Ghosh P K, Li Y, Marchesoni F, et al. Pseudochemotactic drifts of artificial microswimmers. Phys. Rev. E Stat. Nonlin. Soft Matter Phys., 2015, 92(1): 012114.
[121] Wang Y, H R M, Bartlett D J, et al. Bipolar Electrochemical Mechanism for the Propulsion of catalytic nanomotors in hydrogen peroxide solutions. Langmuir, 2006, 22.
[122] Wu Z L X, Wu Y, Si T, et al. Near-infrared light-triggered "On/Off" motion of polymer multilayer rockets. ACS Nano, 2014, 8(6): 9.
[123] Shao J X, Dai L, Si T, et al. Near-infrared-activated nanocalorifiers in microcapsules: vapor bubble generation for in vivo enhanced cancer therapy. Angewandte Chemie, 2015.
[124] Dombrowski C, Cisneros L, Chatkaew S, et al. Self-concentration and large-scale coherence in bacterial dynamics. Phys. Rev. Lett., 2004, 93(9): 098103.
[125] Prigent-Combaret C, Vidal O, Dorel C, et al. Abiotic surface sensing and biofilm-dependent regulation of gene expression in Escherichia coli. J. Bacteriol., 1999, 181: 5993-6002.
[126] Wang Q, Frye J G, McClelland M, et al. Gene expression patterns during swarming in Salmonella typhimurium: genes specific to surface growth and putative new motility and pathogenicity genes. Molec. Microbiol., 2004, 52: 169-187.
[127] Kim M, Powers T R. Hydrodynamic interactions between rotating helices. Phys. Rev. E., 2004, 69(1):061910.
[128] Inoue T, et al. Genome-wide screening of genes required for swarming motility in Escherichia coli K-12. J. Bacteriol., 2007, 189: 950-957.
[129] Girgis H S, Liu Y, Ryu W S, et al. A comprehensive genetic characterization of bacterial motility. PLoS Genet., 2007, 3: 1644-1660.
[130] Vogel V. Soft robotics: Bionic jellyfish. Nat. Mater., 2012, 11(10): 841-842.
[131] Chaté H, Ginelli F, Montagne R. Simple model for active nematics: quasi-long-range order and giant fluctuations. Phys. Rev. Lett., 2006, 96(18): 180602.
[132] McCandlish S R, Baskaran A, Hagan M F. Spontaneous segregation of self-propelled particles with different motilities. Soft Matter, 2012, 8(8): 2527-2534.
[133] Nagai K H, Sumino Y, Montagne R, et al. Collective motion of self-propelled particles with memory. Phys. Rev. Lett., 2015, 114(16): 168001.
[134] Torella J P, Gagtiardi, Chen J S, et al. Efficient solar-to-fuels production from a hybrid microbial water splitting catalyst system. PNAS, 2014, 112(8): 2337.
[135] Darnton N T L, Breuer K, Berg H C. Biophys. J., 2004, 86.
[136] Vogel V, Hess H. In Controlled Nanoscale Motion. Nobel Symposium, 2007, 131: 711.

第十一章 纳米颗粒与蛋白以及细胞膜等的相互作用

第一节 引　　言

近几十年来，随着纳米科学与纳米技术的飞速发展，著名物理学家费曼于 1959 年提出的微观世界的可视图像正逐渐变为现实。1985 年科学家们首次合成了富勒烯 C_{60}，自此以后，人们便逐步具备了合成几纳米到几百纳米的各种纳米材料的能力。这些纳米材料的特性与其体相材料相差甚远，由此催生了广阔的纳米科学与纳米技术领域[1,2]。过去的二十年间[3]，人们在纳米材料（NMs）和纳米粒子（NPs）的表征及制备技术上所取得的巨大进步使其超越了基础科学逐渐渗透到人们的日常生活中，并同时极大地改变了生物医药研究领域的方方面面。理论与实验科学前沿的巨大进步推动纳米科学与技术成为新世纪三大最为突出的研究领域之一（另外两项分别为生命科学和空间技术）。

纳米材料在热学、力学和物理化学等方面拥有诸多的优良特性，如小尺寸效应、量子效应和巨大的比表面积等，使得它们被成功应用到消费品以及加工制造过程当中，如电子元器件、防晒霜、食品色素添加剂、表面涂料等。纳米粒子已经渗透到越来越多的日常商业产品中。此外，由于纳米材料在生物条件下的结构稳定性和其较大的比表面积，它们被广泛应用于各类疾病的诊断和治疗，比如作为生物传感器、细胞成像造影剂、药物和基因的载体等[4,5]。在本章中，纳米材料和纳米粒子特指天然的或合成的尺度范围在1～1000nm的结构或分子。

纳米粒子的广泛应用增加了其由于偶发或医疗暴露而进入人体的可能性，并由此引发了关于其生物相容性和生物安全性的讨论[6-9]。具体而言，虽然纳米粒子的某些特性对消费者和医疗产品极具吸引力，但同时也有可能会对人体健康或者环境造成危害[10-16]。比如，使用纳米粒子防晒霜时会使人眼产生刺痛感。纳米粒子可以通过与皮肤接触，渗透、吸入、食入或药物注射/植入进入机体，随后被输运到各个器官，或驻留，或进行代谢。因此，理解纳米粒子和纳米材料与机体各层次的生命体，如细胞膜、蛋白质以及其他相关生物系统的相互作用，对于我们洞悉其生物功能及机制尤为重要。

在体内，纳米粒子会表现出各种行为，有些是独特的、肉眼不可见的。细胞是生物体维持正常生命活动的基本结构单元和功能单元——“任何一个生命科学问题的关键都必须在细胞中寻找”（著名生物学家Wilson）。有实验证据表明纳米粒子能够破坏细胞膜的完整性，导致细胞裂解[17-20]。另外，与金属离子相似，纳米粒子和蛋白质发生相互作用，可改变其构象、调节其活性或者竞争性地与其结合、螯合等[21-26]。然而，纳米粒子表面共价与非共价钝化的技术运用，可以调节其生物毒性和相关生物应激反应，并最终改变纳米粒子的可适性和功能[27-29]。其中非常有意义的是这些导致纳米材料吸附蛋白质的相互作用对纳米粒子生物相容性的影响。具体来说，通过吸附，纳米粒子-蛋白质会形成动态的分子复合物，即“蛋白冠”。蛋白冠的形成给纳米粒子提供了“生物相似性”，当纳米粒子在被“蛋白冠”屏蔽的状态下再与其他生物分子发生相互作用时，其输运性质、相互作用和细胞毒性等特性会发生显著改变。

纳米材料的细胞生物学效应，与其穿越生物屏障过程、和细胞膜发生的作用、进而进入细胞质甚至细胞核的过程密切相关。但是它们通过什么途径跨越细胞膜，如何与执行各种细胞生理功能的生物分子和细胞器发生相互作用，这些相互作用如何影响细胞的结构和功能？同时纳米材料在细胞这一特殊微环境中其物理化学性质、聚集状态、组装结构等会发生什么变化，纳米材料如何在细胞中分布、停留、清除等，都是相关学科领域内的核心问题。实验研究虽然已经在纳米材料和生物系统相互作用的细节上提供了非常丰富的重要信息，但是限于仪器分辨率等固有的局限性，目前的实验能力仍然无法从分子水平上描述这些机制。为此，计算机模拟便成了相关实验的最佳补充，成为联系实验结果与原子分子机制（细节）的完美桥梁，并由此创造了一个全新的研究领域即“纳米毒理学分子模拟”——该领域在过去的短短十年内就成了新的研究前沿。通过实验研究与计算机仿真模拟的有机融合，取长补短，人们能够精确展现纳米粒子在与靶向生物分子相互作用过程中的动、静态行为，并揭示其深层的分子机制。

这些问题的研究和解决，不仅有助于理解纳米材料的生物学效应，深入洞悉细胞的功能，揭示其在纳米尺度上发生的生命过程的本质，同时也为实现在纳米尺度上对细胞或其他生物体系的人工调控，促进纳米材料在细胞生物学、分子生物学、生物化学等基础研究相关领域的应用和疾病检测、医学成像、药物递送、组织工程、人造器官等医学诊断治疗中的应用提供重要依据，孕育着在纳米科学和生命科学若干前沿科学问题上取得重大突破的机遇。

第二节 最近二三十年的研究进展

随着纳米材料制备技术的迅速发展，纳米材料在生物医药卫生领域的广泛应用和产业化研究已经成为纳米科学领域的热点，并出现了许多新的前沿科学问题：一方面，一些功能化的纳米材料所拥有的高效治疗和高效诊断的生物医学功能是由纳米材料与细胞和生物分子的相互作用所产生的，但是，纳米材料与细胞和生物分子的相互作用过程却是一个研究相对滞后的领域，这无疑将会限制纳米医学的发展；另一方面，如何避免或降低由纳米材料的某些特殊性质导致的纳米毒理学效应，更有效更安全地实现纳米材料在生物医学领域中的应用；这些问题已引起人们的关注和重视。目前，国内外已在积极地开展纳米材料的生物学效应研究，尤其是对于具有生物医学应用前景的纳米材料。从生物整体水平、细胞水平、分子水平等各方面开展研究，研究内容涉及纳米材料的整体生物效应，纳米材料的毒理研究，纳米材料在体内的吸收、分布、代谢和清除，以及各种纳米材料与生物靶器官、靶细胞和靶分子等的相互作用。从研究进展和发展趋势来看，有以下特点：

（1）研究重点已经由生物整体效应逐步深入纳米材料与细胞和功能性生物大分子的相互作用。2001～2005 年世界主流研究方向主要是关于各种纳米材料的生物学效应（医学功能与毒理学性质）的研究；2006～2010 年主要集中在各种纳米材料在生物体内的行为，以及纳米材料的特性与生物学效应（医学功能与毒理学性质）的相关性；2010 年以来，纳米材料生物学效应（医学功能与毒理学性质）的细胞生物学本质与分子生物学本质成为研究的热点。其中，由于纳米粒子与细胞膜的相互作用是决定其如何启动下游一系列细胞内生物化学反应过程的关键，纳米粒子的跨膜过程及其分子机制的研究已经成为一个重大的基础科学问题。随着科学家对纳米材料生物效应研究的日益重视，虽然在关于纳米材料与细胞相互作用机制研究方面已取得了一系列进展，但距离完整地理解纳米-生物过程的本质还相差尚远。例如，Bianco等研究碳纳米管进入细胞的机制及途径时发现碳纳米管是以直接插入的方式通过细胞膜进入细胞的[30,31]，但是Dai等的研究结果则表明碳纳米管是以内吞方式进入细胞的[32,33]。还有，纳米材料所带电荷对其细胞摄入量以及摄入途径的影响，研究表明不同纳米材料具有显著的差异性。例如，Dausend和Harush-Frenkel等的研究都发现带正电的D,L-polylactide (PLA) 纳米颗粒比带负电的更容易进入细胞，并且主要是通过网格蛋白介导的内吞途径来实现的[34,35]；对于量子点的研究则相反，结果表明经COOH修饰的带负电的量子点比中性或带正电的量子点更容易进入细胞，并且通过脂筏介导的内吞途径进入细胞[36]。加拿大Maysinger研究小组通过对未经修饰的CdTe量子点的细胞效应进行研究，认为量子点在光照下会在细胞中产生活性氧导致细胞毒性[37]；而另一些研究小组则认为是量子点表面降解产生的镉离子引起细胞损伤，因而可以通过表面修饰有效地阻止镉离子的扩散，消减毒性[38-40]。此外，赵等的研究还表明同一种纳米材料在不同类型细胞内的定位和分布存在差异，并最终导致不同的纳米生物学效应[41]。因此，要全面正确地认识纳米材料的生物效应，还需要从亚细胞水平和分子水平进行更深入的研究。

（2）研究发现纳米材料的毒性效应与其自身的物理化学性质紧密相关。纳米材料的毒性效应与其自身的物理化学性质具有紧密的联系，这些性质包括尺寸、形状、表面电荷、化学组成、表面修饰、金属杂质、团聚与分散性以及降解性能等。细胞是一个开放的结构体系，在进行各种生命活动过程中不断与环境发生物质交换关系。与生物分子尺寸大小相当的纳米材料，有可能通过这种生物体系自身的已发现的交换途径进出细胞，也有可能利用新的过程或机制。Nel等将氧化应激和炎症反应归结于自由基过量产生的结果，认为在低剂量ROS时，主要通过nrf-2转录因子激活细胞抗氧化元件；较高浓度ROS则激活NF-κB信号通路产生炎症反应；而更高的ROS水平则启动细胞凋亡信号。最近的研究发现，纳米材料对细胞自噬的抑制或激活也是纳米材料毒性效应的一个重要方面。自噬是细胞内成分的一种降解途径，通过双层膜将需要降解的生物大分子和细胞器包裹，形成自噬体（autophagosome）；自噬体进一步与溶酶体融合，形成自噬溶酶体（autolysosome），将包裹的内容物降解。据报道，多种纳米材料都可以诱导细胞自噬，包括各种金属氧化物、贵金属 Au 纳米材料及碳纳米材料（如富勒烯 C60, SWCNTs 等）。自噬与多种细胞功能密切相关，包括免疫、炎症和细胞凋亡等。因此，纳米材料可能通过对细胞自噬功能的抑制或激活，对细胞或机体产生毒副作用[42]。

纳米材料可以通过呼吸道、消化道、皮肤渗透以及注射的方式进入血液，进入血液的纳米颗粒迅速与血清蛋白结合，形成“蛋白冠”。蛋白冠的形成将会改变纳米颗粒的尺寸和表面组成，进而影响纳米颗粒的吸收、转运以及毒性。通过血液蛋白的调理作用，纳米材料迅速被血液以及组织中的单核 / 巨噬细胞摄取，分布于网状内皮系统，这导致纳米材料被快速从血液中清除，并在肝、脾中富集[43]。我们通过理论与实验合作研究发现，SWCNT 与血液蛋白质形成的“蛋白冠”大大降低了 SWCNT 的急性毒性，具有解毒效果。血液中的四种主要蛋白（如纤维蛋白原、免疫球蛋白、白蛋白、转铁蛋白等）会在碳纳米管的表面进行竞争性吸附。SWCNT 与血浆蛋白的结合主要取决于碳纳米管与蛋白质芳香氨基酸（Trp, Phe, Tyr）之间的 π-π 堆积作用[44]，因此，通过合成过程设计和调控纳米材料表面的物理化学特性，能够有效调控其蛋白质吸附特性，进而降低其毒性，提高生物利用度。

（3）理论计算与实验研究相结合，有利于解决在研究复杂生物体系过程中面临的科学问题。纳米-生物的表面 / 界面发生的化学或生物学过程非常复杂。因为在纳米尺度下有多个重要的特征参数可以直接影响纳米粒子的生物学效应，而细胞生物学过程本身也极其复杂。面对这样的复杂体系，单靠实验研究很难明确阐述纳米-生物界面发生的真实过程。这也是导致在迄今关于纳米材料的生物医学功能和毒理学效应的研究结果中大量结论相互矛盾的根本原因。如何发挥实验研究与理论建模两者有机结合的优势解决这些问题，是本领域极具挑战性的课题。随着软、硬件和能量函数的发展进步，大规模的分子动力学（MD）模拟已经成为一种解决纳米粒子靶向生物分子的动、静态行为难题的工具。

第三节　代表性研究案例简介——理论与实验的结合

最近，Zhou 和合作者结合体内、体外实验和计算机模拟研究了单壁碳纳米管和四种代表性血液蛋白［牛血清白蛋白（BSA）、纤维蛋白原（BFg）、γ免疫球蛋白（γ-Ig）和转铁蛋白（Tf）］的相互作用（图 6-11-1）。实验发现，四种血液蛋白都会与碳纳米管形成“蛋白冠复合物”，而蛋白冠复合物的形成能够显著降低碳纳米管的细胞毒性。原子力显微镜（AFM）获得的数据显示 Tf 和 BSA 能够在数分钟内达到饱和吸附，而 BFg 和 Ig 则需要较长的时间。通过运用分子动力学仿真模拟进一步发现，四种血液蛋白的吸附能力差异导致了不同的吸附模式，但是所有的吸附模式都与芳香族氨基酸有很大的关联。每种蛋白质结构的变化都由单壁碳纳米管和芳香族氨基酸之间的 π-π 堆积相互作用而引发。

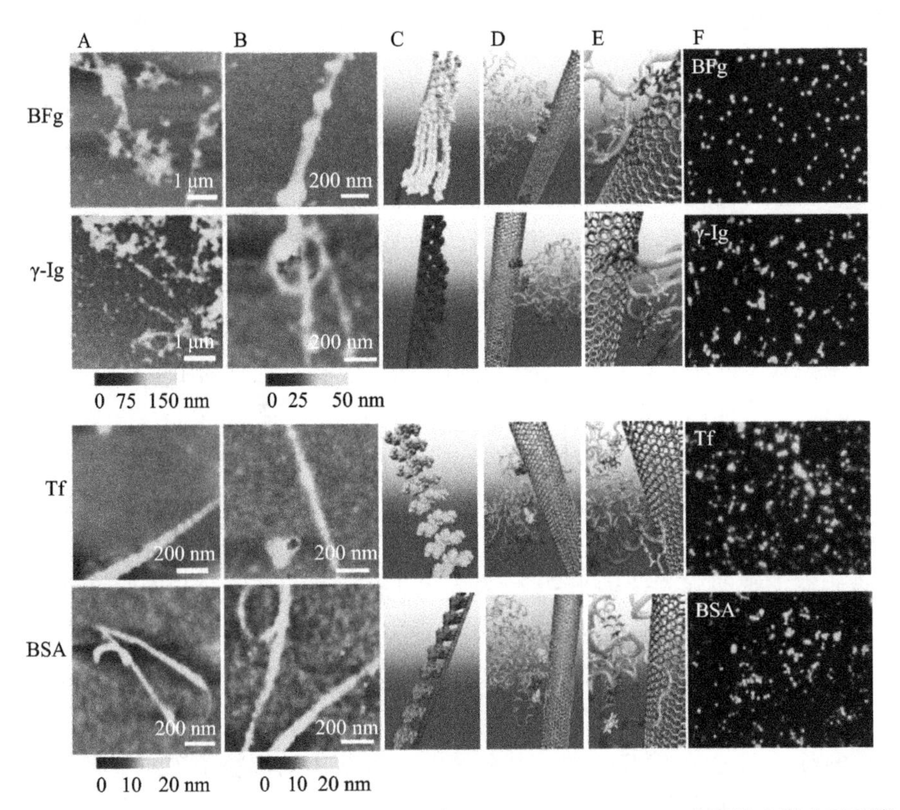

图 6-11-1 人血液中丰度最高的四种血液蛋白 BFg、γ-Ig、Tf、BSA 在碳纳米管表面吸附模式的 AFM 图样（A 和 B），分子动力学模拟图样（C～E）和细胞毒性的变化（F）

氨基酸亚结构域 villin headpiece（HP 35）蛋白是一种不含二硫键或非天然氨基酸的广泛用于蛋白质折叠和去折叠研究的模型蛋白，其结构为三个 α-螺旋束。通过对其与一系列碳基纳米材料，包括富勒烯（C60）、（5, 5）-单壁碳纳米管和石墨烯片层等的相互作用进行仿真模拟研究发现，在任一情况下，蛋白质都会在几十纳秒内被快速地吸附到这些碳基纳米材料上，同时伴随着二级结构的破坏、去折叠。特别是随着 α-螺旋含量的大幅减少，三级结构也开始遭到破坏。但是，对于石墨烯片层、单壁碳纳米管、富勒烯这三种纳米材料，蛋白质与之的结合模式不尽相同。对于单壁碳纳米管和富勒烯这两种纳米材料，疏水相互作用被认为是驱动 HP35 吸附到它们表面的主导因素。同时，在吸附过程中也发现了 π-π 相互作用。然而，对于石墨烯片层，人们却广泛接受 π-π 相互作用是导致 HP35 吸附到它表面的主要因素。这个有趣的发现可以归因于碳同素异形体的刚性和表面曲率的差异性，平面性更好，更柔软的石墨烯能够更好地调整与芳香族氨基酸残基之间的相对位置，如此以形成更强的 π-π 堆积相互作用。另外，Balamurugan 等通过模拟仿真研究发现，在与平面型的碳的同素异形体的相互作用中，螺旋的破坏程度最为明显，这表明石墨烯片层破坏螺旋结构更有效 [45]，甚至会引起蛋白质的构象从 α-螺旋到 β-折叠片层的转变。Wei 等对两条肽链和石墨烯片层进行了模拟，虽然没有观察到 β-折叠片层结构的形成，但是发现蛋白质会发生去折叠并重新组装成一个无定形结构的二聚体 [46]。鉴于蛋白质二级结构对蛋白质功能至关重要，因此，纳米吸附或者其他任何干扰或破坏蛋白二级结构的行为都可能会影响蛋白质的结构稳定性和功能。

从前面所展示的例子我们可以看到，分子间的非键相互作用力，即静电相互作用和色散相互作用（包括与水分子互动产生的疏水相互作用）被认为是碳纳米颗粒与蛋白质相互作用的驱动力。为了确定结合能的主要来源，仔细研究这种长程静电力和色散力的分别贡献非常重要。Tomásio 等对两种富含色氨酸的蛋白质与单壁碳纳米管和石墨烯的相互作用进行了 MD 模拟。值得注意的是，他们使用了经过修改的极化力场，该力场能够充分地考虑纳米材料的多偶极到四偶极以及极化效应 [47]。为了比较结合的亲和力，他们把双位点的色氨酸突变成单位点的芳香族氨基酸苯丙氨酸和酪氨酸，并对蛋白质和碳纳米颗粒再次进行模拟。Tomasio 等的实验结果强调：由于曲率和纳米材料结构的缺陷，在与石墨片层与氨基酸残基选择性地相互作用过程中，相对于苯丙氨酸和酪氨酸，色氨酸与石墨残基的结合能力更强 [48]。我们都知道，在固定电荷力场中使用的平滑的非键相互作用势的精度要低于 QM 计算中的相互作用势。Tomasio 等指出，如果芳香族基团与石墨材料表面的 π-π 相互作用是吸附的驱动力，那么静电相互作用的可极化处理就应当优先考虑，尽管这会增加计算的开销。细致的可极化处理更具优点，但是有充分的证据显示[49]，现今的固定电荷力场能够恰如其分地对芳香族氨基酸残基与石墨材料之间的 π-π 相互作用进行描述。尽管相比极化力场，固定电荷力场在预测细致的结合模式上还并不太令人满意。

贵金属（如黄金和白银）纳米材料在纳米技术和纳米药物[5,50,51]中也得到了广泛的应用。黄金纳米粒子（AuNPs）和纳米棒（AuNPs）已经成为流感和其他疫苗的组分，长期扮演着医疗角色。与石墨纳米颗粒类似，这些贵金属纳米材料的表面化学性质和整体结构与其蛋白质吸附、细胞摄取和细胞毒性等行为密切相关[9,52]。受到之前关于纳米材料的不同包被对细胞活性影响的启发，Wang 等[50]最近报道了包被 AuNPs 硬的 BSA 蛋白冠的特征。Wang 等判定，BSA 中的 12 个硫原子能够促进蛋白质吸附到 AuNPs 表面。Au-S 配位键是一种非常强的非共价相互作用，他们在模拟中将其近似成一个弹簧势。如此，他们可以研究结合的稳定性，结合的关键氨基酸残基以及蛋白质吸附在硬蛋白冠上的构象变化。最终，他们发现蛋白质包被的纳米棒（BSA-AuNPs）能够有效地降低裸露的金纳米粒子的急性毒性，并有效地抑制了对细胞膜的潜在破坏性。

纳米银离子和粒子（AgNP）已被确认为有效的抗菌剂[17,53]和抗真菌剂[54]，因而，刻画相关的生物应激反应对于探索纳米毒性就显得特别重要且非常有趣。Cho 研究组[55]使用一种 GPU 优化的 GO-模型的 MD 模拟方法研究了 AgNP 生物分子蛋白冠的形成过程。在他们的方法中，将蛋白质-纳米粒子的相互作用近似简化为一种依赖于粒子浓度和分子电荷的 Debye-Hückel 静电势。在一个特例中，他们将一个被柠檬酸分子钝化的带负电荷的 AgNP 暴露于粗粒化的载脂蛋白中。他们观察到在与 AgNP 发生相互作用后，蛋白质中的 α-螺旋含量急剧下降。类似于 Cho 等的研究，Ding 等结合全原子分子动力学模拟与粗粒化模型（CG）方法补充描述了被柠檬酸包被的 AgNP 与多达 50 个泛素蛋白分子的相互作用[56]。该 AgNP 被近似看成是疏水性球体聚积物，同时还分布有一些带正电的粒子，类似于残留的银离子。全原子的分子动力学模拟表明，泛素和柠檬酸分子能够竞争性地结合到 AgNP 上。结合过程的驱动力来自于 AgNP 和泛素中 11 个带负电荷的基团之间的特异性的静电相互作用。Ding 等将一种粗粒化的、把每个氨基酸残基粗粒化成两颗珠子的相互作用势来描述蛋白质，并以此扩展了他们先前的全原子模拟。利用该模型，他们能够模拟多聚体泛素的蛋白冠的形成。他们观察到泛素分子在结合到 AgNP 后仍然保持折叠状态，蛋白质的螺旋结构朝向纳米粒子。较高浓度的泛素分子能够形成多层的蛋白冠，其中，第一层（硬质的）蛋白冠由上述的静电相互作用维持，而第二层（柔软的）蛋白冠由蛋白质与蛋白质之间的相互作用维持。

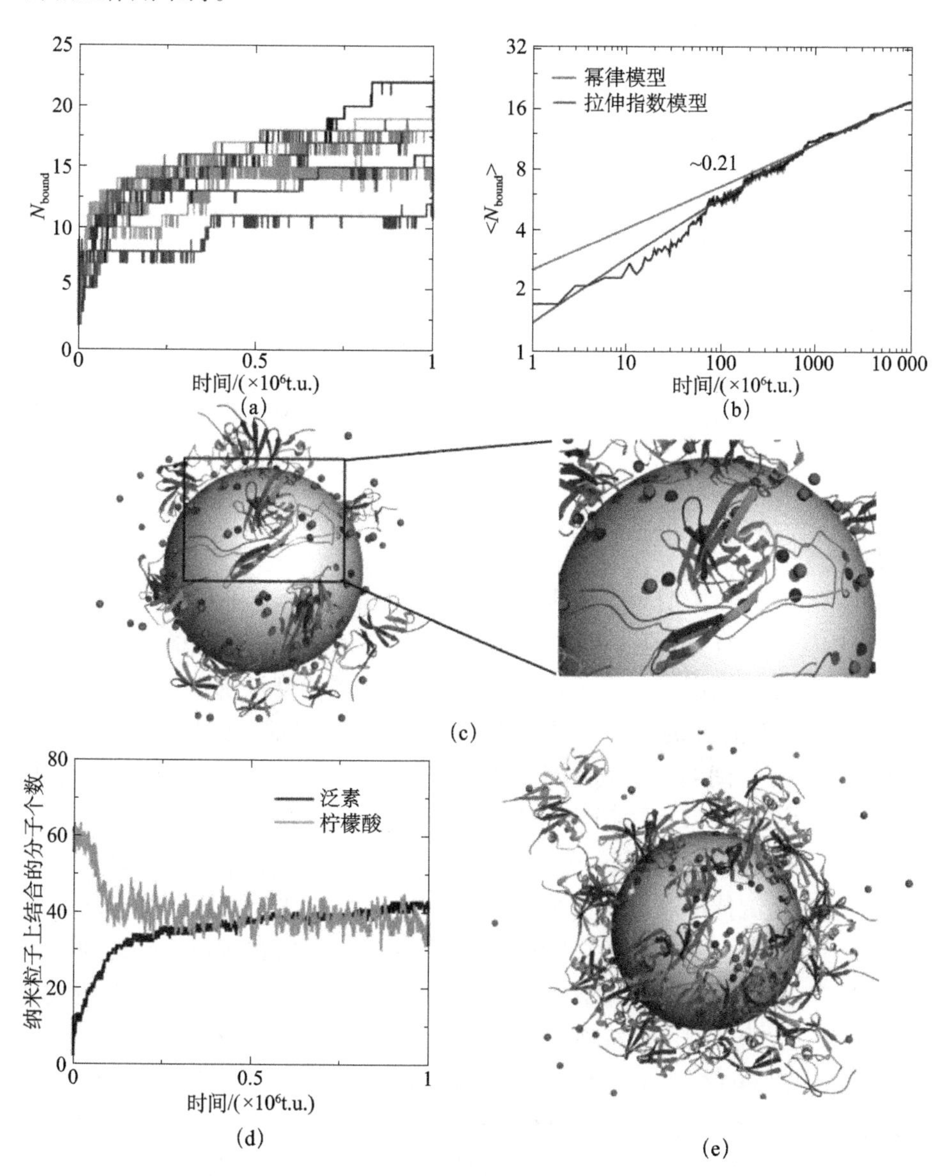

图 6-11-2 泛素 -AgNP 蛋白冠的形成

(a) 与 AgNP 结合的泛素分子的数目。(b) 泛素结合到 AgNP 的平均数目。(c) 从所述模拟中的一个（对应于具有最高结合数目）的最终结构。(d) 粗粒度的 DMD 模拟与结合在 AgNP 表面的柠檬酸盐数目。(e) 计算表明泛素可以多层沉积在 AgNP(e) 的表面上（转载自文献 [54]。版权所有：英国皇家化学学会，2013 年）

另一大类纳米粒子是量子点（QDs）。它们是具有独特的电学、光学性质的半导体纳米晶体（2～100 nm）[57,58]，被广泛应用于生物医学成像和电子工业。其独特的荧光光谱使它们成为优良的生物医学成像荧光探针[59,60]。量子点固有的物理化学性质和其在外部环境下的反应活性决定了QD的毒性更为复杂[61]。QD的尺寸、电荷、浓度、外层包被材料、氧化性、光解性和机械稳定性中的每个单独的性质，或者多个性质协同都能催生QD的毒性。例如Lovrić等[62]发现，10μg/mL的，包被有巯基丙酸（MPA）和半胱胺的CdTe量子点对大鼠嗜铬瘤细胞（PC12）会产生细胞毒性。没有任何包被的CdTe量子点，则在1μg/mL时就会产生细胞毒性。在相同浓度的条件下，带正电荷的尺寸较小的量子点（2.2±0.1）nm要比带相等电荷但尺寸更大的量子点（5.2±0.1）nm的细胞毒性更加明显[62]。QD的大小还会影响亚细胞水平的分布，较小的阳离子量子点分布于细胞核，而较大的阳离子量子点则分布于细胞质。虽然目前QD导致细胞凋亡的机制尚不清楚，但有一种猜测认为是自由镉（QD核降解）、自由基的形成或者量子点与胞内组分相互作用而导致其功能的丧失。

目前的模拟主要针对有机物表面包被对QD的毒性和电子稳定性的影响。最近的理论研究表明，钝化配体可以在QD表面形成稳定的配位，维持QD的化学结构。Azpiroz等[63]运用QM计算中的密度泛函理论（DFT）确认CdSe量子点的有机钝化会引起分子间形成一种弱耦合的静电相互作用，而不是一种分子间弱的范德瓦耳斯相互作用。而就在最近，Zhou和合作者们[64]也运用DFT理论研究了被不同长度和分支数目的$OPMe_2(CH2)_nMe$配体（Me = 甲基基团，n=0,1～3）进行过钝化处理的QD $(CdSe)_{13}$的电子结构。他们发现通过与配体的结合，在紫外-可见（UV-VIS）光谱的吸收峰出现了尺度约为100 nm的明显蓝移。但是一旦结合到（CdSe）$_{13}$的配体总数达到饱和状态时（9或10），吸收峰就不会再发生蓝移，而配体的脂肪链的长度对QD核的光学特性的影响可以忽略。通过对结合特性的进一步分析证实了光转换主要是由中央QD核支配，而不是有机钝化。有趣的是，尽管配体与中心的QD之间没有一致的振动模式，但是态密度（DOS）与振动光谱具有类似的特征。这些发现为在生物医学应用中设计更加安全的量子点有机钝化方案提供了新的见解。总的来说，这些最近的研究成果表明，广泛的理化特性、功能性包被和QD核稳定性可能是评估QD在现实暴露环境下毒性风险的最显著因素。

细胞膜容易受到石墨纳米材料的直接物理切入和内吞作用的影响[19,65]。针对革兰氏阴性和革兰氏阳性细菌细胞，石墨纳米片显示出强烈的细胞毒性。如上述讨论，如果石墨烯纳米颗粒被蛋白包被，那么纳米颗粒对哺乳动物的细胞膜的毒性就会降低。在精准给药情形下，这种对细菌的毒性行为可能被运用到新型抗生素的生产。Zhou 和合作者[66]报道了在石墨烯和石墨烯氧化物纳米片层对于大肠杆菌细胞膜的影响的实验和理论研究，目的是深入探讨细胞膜-纳米颗粒的相互作用行为，了解潜在的纳米颗粒裂解细胞膜的性质。在 TEM 图像中，随着纳米材料的浓度变化，在孵育期间细胞形态大致分为三个阶段。第一阶段：初始形态；第二阶段：在没有任何明显切口的情况下，观察到磷脂密度有所降低；第三阶段：细胞的完整性基本丧失，细胞质可能产生泄漏。随后，他们运用全原子分子动力学模拟，对石墨烯和石墨烯氧化物对大肠杆菌的外膜和内膜的相互作用进行了系统的仿真模拟。他们在模拟中观察到了细胞膜被严重插入和断裂的现象，并发现石墨烯通过两种不同的机制破坏细胞膜完整性。机制 A：石墨烯直接刺入细胞膜，将细胞膜切出一个伤口，该现象与之前讨论的实验结果一致。机制 B：石墨烯直接将磷脂分子从膜中提取出来，这种直接提取行为在之前从未被报道到，这也解释了在 TEM 图像中发现的磷脂密度降低这一现象。这些膜-石墨烯系统的细致动态过程表明，石墨烯纳米片能够强有力地提取脂质，达到杀死细胞的作用。而引发该过程的是石墨烯与磷脂分子之间强大的色散相互作用，该相互作用远强于细胞膜中磷脂分子之间的相互作用。

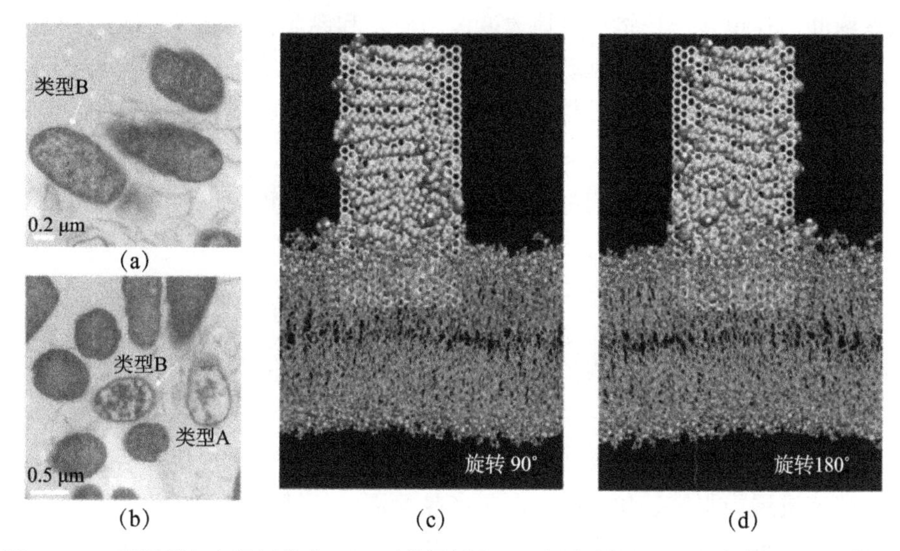

图 6-11-3 石墨烯与大肠杆菌在 37℃下共同孵育 2.5 小时后的 TEM 照片［(a) 和 (b)］，石墨烯明显能使大肠杆菌细胞变薄。分子动力学模拟轨迹截图［(c) 和 (d)］，石墨烯能够将大肠杆菌细胞膜上的磷脂分子从膜中抽取出来

第四节　学科前沿问题

纳米粒子进入细胞的驱动力是什么，小于 100nm 或更小的纳米粒子是否存在新的跨膜过程和机制；纳米粒子与蛋白、DNA、RNA 等生物大分子的相互作用分子机制是什么，有什么特异性；这些相互作用的特异性是否能解释不同纳米材料的特异生物兼容性与安全性。这些都是亟待解决的学科前沿问题。同时，纳米材料的细胞生物学效应与其物理化学特性密切相关。通过改变具有生物医学应用前景的纳米材料的不同物理特性（如不同的尺寸、电荷、形状、结构）和不同化学修饰等，结合实验研究与理论建模的系统研究，实验研究它们与细胞的作用方式，进入细胞的过程及其对细胞功能（如生长、分化、迁移、增殖等）的影响；以及在进入细胞以后，在生物微环境下，研究它们与一些重要功能蛋白、核酸的相互作用动态过程及它们对这些重要生命物质结构和功能的影响，揭示不同物理特性和化学修饰等对纳米材料细胞生物效应的影响规律，探索如何利用这些规律对纳米材料生物效应进行人工调控，为纳米材料在生物医学领域的实际应用奠定理论基础。

另一个学科前沿问题是如何利用纳米-生物分子的相互作用机制来设计制造纳米药物。未来 10 年或更长一些时间内，纳米医药不仅会变革传统的临床诊断和治疗模式，还将极大地改变医药研发和商业化的路径。因此，揭示促使纳米药物进入细胞的驱动力，探索纳米药物与传统小分子药物的不同药代动力学，以及探究其是否与有机小分子存在不同的跨膜过程和机制，具有重大的科学意义，也是目前面临的重大学科前沿之一。纳米药物的广泛研究也可以进一步协助揭示它们与重要功能蛋白、核酸等重要生命物质的相互作用机制。

第五节　未来 5～10 年学科发展趋势

总之，纳米材料的合成、开发和表征已经彻底改变了现代生活。纳米技术的研究进展为服装、化妆品、电子产品等消费领域带来了新的可能性，更重要的是，纳米材料和纳米粒子在医学预防、治疗和诊断中的应用潜力有利于更加有效地、创伤小且个性化地治疗。随着纳米材料和纳米粒子的广泛使用，其生物效应以及对人类健康、环境的潜在危害也需要进行仔细的评估。

为此，必须通过理论和实验的共同努力，提高我们对纳米材料与生物机体相互作用的进一步理解。随着人们对生物分子/纳米粒子相互作用的分子模拟的兴趣的提高，使得对复杂的体外和体内实验中得到的往往看似矛盾的发现有了一个详尽的原子层面的理解，同时还为纳米药物的从头设计提供了指导方针和启发。

理论与实验的有机融合，联合攻关是学科内的重要趋势。尽管实验研究已经能够在有关纳米材料和一些重要生命物质之间相互作用的细节上提供非常丰富的重要信息，但是限于仪器分辨率固有的局限性，目前的实验能力仍然不能从分子水平上描述这些机制。近年来，随着软、硬件和能量函数的发展进步，大规模的分子动力学模拟已经成为一种研究纳米粒子与靶向生物分子的动、静态相互作用过程的有效工具，并成为链接实验结果与纳米粒子-生物分子复合物的原子细节之间的完美桥梁。通过实验研究与计算机仿真模拟的有机融合，人们能够精确展现纳米材料与靶向生物分子的特异性相互作用并揭示其分子机制。这些分子机制的突破将为纳米材料在生物医学领域的更广泛应用奠定重要的科学基础。

周如鸿（国际商业机器公司托马斯·约翰·沃森研究中心/哥伦比亚大学）

参 考 文 献

[1] Kroto H W, Heath J R, Obrien S C, et al. C-60-Buckminsterfullerene. Nature, 1985, 318(6042): 162-163.

[2] Iijima S. Helical Microtubules of graphitic carbon. Nature, 1991, 354(6348): 56-58.

[3] Rao C N R, Sood A K, Subrahmanyam K S, et al. Graphene: the new two-dimensional nanomaterial. Angew Chem Int Edit, 2009, 48(42): 7752-7777.

[4] Noon W H, Kong Y F, Ma J P. Molecular dynamics analysis of a buckyball-antibody complex. P. Natl. Acad. Sci. USA, 2002, 99: 6466-6470.

[5] Jain P K, El-Sayed I H, El-Sayed M A. Au nanoparticles target cancer. Nano Today, 2007, 2(1): 18-29.

[6] Smart S K, Cassady A I, Lu G Q, et al. The biocompatibility of carbon nanotubes. Carbon, 2006, 44(6): 1034-1047.

[7] Dietz K J, Herth S. Plant nanotoxicology. Trends. Plant. Sci., 2011, 16(11): 582-589.

[8] Mahmoudi M, Azadmanesh K, Shokrgozar M A, et al. Effect of nanoparticles on the cell life cycle. Chem. Rev., 2011, 111(5): 3407-3432.
[9] Aillon K L, Xie Y M, El-Gendy N, et al. Effects of nanomaterial physicochemical properties on in vivo toxicity. Adv. Drug. Deliver. Rev., 2009, 61(6): 457-466.
[10] Brayner R. The toxicological impact of nanoparticles. Nano Today, 2008, 3(1-2): 48-55.
[11] Fischer H C, Chan W C W. Nanotoxicity: the growing need for in vivo study. Curr. Opin. Biotech., 2007, 18(6): 565-571.
[12] Kagan V E, Bayir H, Shvedova A A. Nanomedicine and nanotoxicology: two sides of the same coin. Nanomedicine-UK, 2005, 1(4): 313-316.
[13] Nel A E, Madler L, Velegol D, et al. Understanding biophysicochemical interactions at the nano-bio interface. Nat. Mater., 2009, 8(7): 543-557.
[14] Shvedova A A, Kisin E R, Porter D, et al. Mechanisms of pulmonary toxicity and medical applications of carbon nanotubes: Two faces of Janus? Pharmacol. Therapeut., 2009, 121(2): 192-204.
[15] Suh W H, Suslick K S, Stucky G D, et al. Nanotechnology, nanotoxicology, and neuroscience. Prog. Neurobiol., 2009, 87(3): 133-170.
[16] Sharifi S, Behzadi S, Laurent S, et al. Toxicity of nanomaterials. Chemical Society Reviews, 2012, 41(6): 2323-2343.
[17] Sotiriou G A, Pratsinis S E. Antibacterial activity of nanosilver ions and particles. Environ. Sci. Technol., 2010, 44(14): 5649-5654.
[18] Hu W B, Peng C, Luo W J, et al. Graphene-based antibacterial paper. Acs Nano, 2010, 4(7): 4317-4323.
[19] Wong-Ekkabut J, Baoukina S, Triampo W, et al. Computer simulation study of fullerene translocation through lipid membranes. Nature Nanotechnology, 2008, 3(6): 363-368.
[20] Yang K, Ma Y Q. Computer simulation of the translocation of nanoparticles with different shapes across a lipid bilayer. Nature Nanotechnology, 2010, 5(8): 579-583.
[21] Vaitheeswaran S, Garcia A E. Protein stability at a carbon nanotube interface. J. Chem. Phys., 2011, 134(12).
[22] Lynch I, Dawson K A. Protein-nanoparticle interactions. Nano Today, 2008, 3(1-2): 40-47.
[23] Kang Y, Wang Q, Liu Y C, et al. Dynamic mechanism of collagen-like peptide encapsulated into carbon nanotubes. J. Phys. Chem. B, 2008, 112(15): 4801-4807.
[24] Shi B Y, Zuo G H, Xiu P, et al. Binding preference of carbon nanotube over pro line-rich motif ligand on SH3-domain: a comparison with different force fields. J. Phys. Chem. B, 2013, 117(13): 3541-3547.
[25] Shen J W, Wu T, Wang Q, et al. Induced stepwise conformational change of human serum

albumin on carbon nanotube surfaces. Biomaterials, 2008, 29(28): 3847-3855.
[26] Gao H J, Kong Y. Simulation of DNA-nanotube interactions. Annu. Rev. Mater. Res., 2004, 34: 123-150.
[27] Thanh N T K, Green L A W. Functionalisation of nanoparticles for biomedical applications. Nano Today, 2010, 5(3): 213-230.
[28] Partha R, Conyers J L. Biomedical applications of functionalized fullerene-based nanomaterials. International Journal of Nanomedicine, 2009, 4: 261-275.
[29] Bitar A, Ahmad N M, Fessi H, et al. Silica-based nanoparticles for biomedical applications. Drug. Discov. Today, 2012, 17(19-20): 1147-1154.
[30] Pantarotto D, Briand J P, Prato M, et al. Translocation of bioactive peptides across cell membranes by carbon nanotubes. Chemical Communications, 2004, 1: 16-17.
[31] Shi X H, Von Dem Bussche A, Hurt R H, et al. Cell entry of one-dimensional nanomaterials occurs by tip recognition and rotation. Nature nanotechnology, 2011, 6(11): 714-719.
[32] Kam N W S, Jessop T C, Wender P A, et al. Nanotube molecular transporters: Internalization of carbon nanotube-protein conjugates into mammalian cells. J. Am. Chem. Soc., 2004, 126(22): 6850-6851.
[33] Porter A E, Gass M, Muller K, et al. Direct imaging of single-walled carbon nanotubes in cells. Nature Nanotechnology, 2007, 2(11): 713-717.
[34] Harush-Frenkel O, Debotton N, Benita S, et al. Targeting of nanoparticles to the clathrin-mediated endocytic pathway. Biochem. Bioph. Res. Co., 2007, 353(1): 26-32.
[35] Dausend J, Musyanovych A, Dass M, et al. Uptake mechanism of oppositely charged fluorescent nanoparticles in HeLa cells. Macromol. Biosci., 2008, 8(12): 1135-1143.
[36] Zhang L W, Monteiro-Riviere N A. Mechanisms of quantum dot nanoparticle cellular uptake. Toxicol. Sci., 2009, 110(1): 138-155.
[37] Lovric J, Cho S J, Winnik F M, et al. Unmodified cadmium telluride quantum dots induce reactive oxygen species formation leading to multiple organelle damage and cell death. Chem. Biol., 2005, 12(11): 1227-1234.
[38] Derfus A M, Chan W C W, Bhatia S N. Probing the cytotoxicity of semiconductor quantum dots. Nano Lett., 2004, 4(1): 11-18.
[39] Hoshino A, Fujioka K, Oku T, et al. Physicochemical properties and cellular toxicity of nanocrystal quantum dots depend on their surface modification. Nano Lett., 2004, 4(11): 2163-2169.
[40] Kirchner C, Liedl T, Kudera S, et al. Cytotoxicity of colloidal CdSe and CdSe/ZnS nanoparticles. Nano Lett., 2005, 5(2): 331-338.
[41] Wang L M, Liu Y, Li W, et al. Selective targeting of gold nanorods at the mitochondria of

cancer cells: implications for cancer therapy. Nano Lett., 2011, 11(2): 772-780.
[42] Izvekov S, Voth G A. A multiscale coarse-graining method for biomolecular systems. J. Phys. Chem. B, 2005, 109(7): 2469-2473.
[43] Monopoli M P, Aberg C, Salvati A, et al. Biomolecular coronas provide the biological identity of nanosized materials. Nature Nanotechnology, 2012, 7(12): 779-786.
[44] Ge C C, Du J F, Zhao L N, et al. Binding of blood proteins to carbon nanotubes reduces cytotoxicity. P. Natl. Acad. Sci. USA, 2011, 108(41): 16968-16973.
[45] Balamurugan K, Singam E R A, Subramanian V. Effect of curvature on the alpha-Helix breaking tendency of carbon based nanomaterials. Journal of Physical Chemistry C, 2011, 115(18): 8886-8892.
[46] Ou L C, Luo Y, Wei G H. Atomic-level study of adsorption, conformational change, and dimerization of an alpha-helical peptide at graphene surface. J. Phys. Chem. B, 2011, 115(32): 9813-9822.
[47] Tomasio S M, Walsh T R. Modeling the binding affinity of peptides for graphitic surfaces. Influences of aromatic content and interfacial shape. Journal of Physical Chemistry C, 2009, 113(20): 8778-8785.
[48] Walsh T R, Tomasio S M. Investigation of the influence of surface defects on peptide adsorption onto carbon nanotubes. Mol. Biosyst., 2010, 6(9): 1707-1718.
[49] Yang Z X, Wang Z G, Tian X L, et al. Amino acid analogues bind to carbon nanotube via pi-pi interactions: Comparison of molecular mechanical and quantum mechanical calculations. J. Chem. Phys., 2012, 136(2).
[50] Wang L M, Jiang X M, Ji Y L, et al. Surface chemistry of gold nanorods: origin of cell membrane damage and cytotoxicity. Nanoscale, 2013, 5(18): 8384-8391.
[51] Tiwari P M, Vig K, Dennis V A, et al. Functionalized gold nanoparticles and their biomedical applications. Nanomaterials-Basel, 2011, 1(1): 31-63.
[52] Sarikaya M, Tamerler C, Jen A K Y, et al. Molecular biomimetics: nanotechnology through biology. Nat. Mater., 2003, 2(9): 577-585.
[53] Ahamed M, Alsalhi M S, Siddiqui M K J. Silver nanoparticle applications and human health. Clin. Chim. Acta., 2010, 411(23-24): 1841-1848.
[54] Kim K J, Sung W S, Suh B K, et al. Antifungal activity and mode of action of silver nano-particles on Candida albicans. Biometals, 2009, 22(2): 235-242.
[55] Li R Z, Chen R, Chen P Y, et al. Computational and experimental characterizations of silver nanoparticle-apolipoprotein biocorona. J. Phys. Chem. B, 2013, 117(43): 13451-13456.
[56] Ding F, Radic S, Chen R, et al. Direct observation of a single nanoparticle-ubiquitin corona formation. Nanoscale, 2013, 5(19): 9162-9169.

[57] Bruchez M, Moronne M, Gin P, et al. Semiconductor nanocrystals as fluorescent biological labels. Science, 1998, 281(5385): 2013-2016.

[58] Dabbousi B O, Rodriguezviejo J, Mikulec F V, et al. (CdSe)ZnS core-shell quantum dots: synthesis and characterization of a size series of highly luminescent nanocrystallites. J. Phys. Chem. B, 1997, 101(46): 9463-9475.

[59] Alivisatos P. The use of nanocrystals in biological detection. Nat. Biotechnol., 2004, 22(1): 47-52.

[60] Chan W C, Maxwell D J, Gao X, et al. Luminescent quantum dots for multiplexed biological detection and imaging. Curr. Opin. Biotech., 2002, 13(1): 40-46.

[61] Hardman R. A toxicologic review of quantum dots: toxicity depends on physicochemical and environmental factors. Environ. Health. Persp., 2006, 114(2): 165-172.

[62] Lovrić J, Bazzi H S, Cuie Y, et al. Differences in subcellular distribution and toxicity of green and red emitting CdTe quantum dots. J. Mol. Med-Jmm., 2005, 83(5): 377-385.

[63] Azpiroz J M, Matxain J M, Infante I, et al. A DFT/TDDFT study on the optoelectronic properties of the amine-capped magic (CdSe)(13) nanocluster. Physical Chemistry Chemical Physics, 2013, 15(26): 10996-11005.

[64] Gao Y, Zhou B, Kang S G, et al. Effect of ligands on the characteristics of (CdSe)(13) quantum dots. Rsc Advances, 2014, 4(52): 27146-27151.

[65] Pogodin S, Baulin V A. Can a carbon nanotube pierce through a phospholipid bilayer? Acs Nano., 2010, 4(9): 5293-5300.

[66] Tu Y S, Lv M, Xiu P, et al. Destructive extraction of phospholipids from Escherichia coli membranes by graphene nanosheets. Nature Nanotechnology, 2013, 8(8): 594-601.

第十二章 生物信息大数据挖掘

第一节 引　　言

人类对生命科学的研究经历了从宏观到微观的过程。现在，人们已经可以从分子层面来观察生命的活动过程。随着现代高通量数据获取技术的不断突破，生物数据海量呈现并持续快速增长[1-7]，多源性、异质性已经成为现代生物数据所固有的特性。所谓多源性是指同一个生命活动现象（如蛋白质功能）可以从不同角度来体现，由不同来源、不同层次的数据（如蛋白质相互作用网络、基因表达等）来表达；所谓异质性是指不同来源的数据在信息表达上存在着差异性、多样性以及互补性。

然而，数据的累积并不等于知识的增加。伴随着生物数据在广度、深度和时间三个维度方向上的海量涌现，生物数据逐渐展现了独特的数据特征，传统的数据分析与挖掘算法面临着极为严峻的考验，迫切需要建立与发展针对性强的生物数据挖掘理论新框架，对加深人类生命过程的认知，帮助人们改善生存环境和提高生活质量有着重要的意义。

第二节 学科发展背景和现状

一、发展背景

通过传统的生物实验来分析理解生物数据存在着诸如耗时、昂贵等问题。随着高通量下一代测序技术（next-generation sequencing）、自动化分子荧光显微成像、冷冻电镜（CryoEM）等技术的飞速发展，生物数据产生的能力与数据理解的速度之间的鸿沟已经越来越大：以当前已测序的蛋白质数据为例，只有约 0.6% 的蛋白质具有生物实验功能注释，0.1% 的蛋白质被求解出三维结构，0.3% 的真实蛋白质相互作用被实验验证。这个差距更会随着技术的不断进步和成熟而不断扩大。因此，研发先进的、能从海量生物数据中挖掘出有用知识的数据挖掘方法与技术，已经成为生物信息领域的热点问题 [8]。

二、研究现状

通过调研文献，可以发现生物数据挖掘领域已取得了丰硕的成果，发表了不少具有很高理论意义及实用价值的论文。但是，通过进一步地梳理分析，当前的研究工作还主要是把生物数据当成一个验证应用背景，即简单移植现有成熟算法框架，而忽略了生物数据所特有的特征，从而导致虽然报道算法的结果很好，但在大规模应用中效果很差 [9]，发展生物数据特征驱动的新型数据挖掘和学习理论框架成为迫切要求。目前已有数据挖掘方法存在的若干不足如下。

（一）生物数据是海量且持续增长的，现有静态挖掘模型适用性差、更新困难

现代生物数据是海量且不断增长的。以蛋白质序列数据为例，目前在著名的蛋白质数据库 UniProtKB[10] 中，已经存有超过 8000 万条（89 089 768）蛋白质的一级序列信息（截至 2015 年 1 月 31 日），这比 1986 年初建时的数据（4782，1986-12-31）增长了 18000 余倍。回顾已有的生物数据挖掘方法可以发现，一个重要的不足之处在于这些方法通常是基于静态框架的，即从一个有限规模的数据集训练得到相应的计算模型，然后用于对未标定的数据进行预测，并不关注该模型在后续新认识增加背景下的更新问题。当代数据的海量性将不可避免地导致数据分布空间的复杂性和不均匀性，数据的不断增长也必将带来已获知识需要不断修正的问题，因此，基于静态框架的挖掘模型不可避免地会存在以下几个方面的问题。

1. 可扩展性低（low scalability）

静态模型方法通常在一个固定的数据集上训练得到预测和挖掘模型，当有新的标定数据可用时，需要利用原有的数据及新的数据来重新训练。由于

新的可用数据是持续出现的，因此，重新训练的过程亦需要不断重复，效率较低。换句话说，静态模型方法的可扩展性较低。

2. 过拟合 / 过优化（over-fitting/over-optimization）现象严重

通常，研究人员提出一种新的计算模型后需要和其他已有的方法进行比较，以验证所提方法的有效性。这种对比往往基于某个(些)固定的标准测试数据集，所提出的方法可能会被过拟合到测试数据集或是在测试数据集上进行过优化[11]。例如，一个普遍的导致过优化的方式是挑选有利于新方法的特征、遍历参数以得到最优的参数集，等等。因此，一个问题就随之产生了：当数据集变大时，上述被过拟合或是过优化的静态预测模型还有效吗？

已有的研究表明，中小规模数据集上训练得到的预测模型，在大规模数据集上的评测效果往往不尽如人意，甚至结果很差；此外，研究人员还发现，在大数据集上，很多“简单的”方法比“复杂的”方法反而更为有效（simple is better than complex）[9]。有两个方面的理由可以解释上述现象：一方面，“复杂的”静态方法被过拟合或是过优化的概率更高；另一方面，静态方法不能有效利用新可用数据集中的知识。正如 Rajaraman 所述，“更多的数据往往胜过更好的算法”[12]。

3. 低可用性（low applicability）

当数据集非常庞大的时候，训练一个全局的静态挖掘模型很可能是不现实的。这是由于海量数据空间中样本分布的不均匀性、非凸性本质所带来的模型低性能问题，或是出于对内存过多的要求，或是出于优化时间太长等其他的原因。因此传统的做法是在小规模的数据集上建立一个计算模型，但由于训练数据集的覆盖范围有限，且没有包含最新的数据样本信息，当把这些静态模型实际应用在大规模真实测试数据时，结果往往会很差，导致低可用问题。

（二）生物数据具有多源异质性，需要跨源知识共享与迁移挖掘

生物系统是一个庞大、复杂但却有序的系统。任何现有的生物数据获取手段都只能记录生物系统在某一时刻、某一方面的表现。在生物学家和信息科学家的努力下，借助于最新的数据获取技术，生物学研究还产生了类型繁多的数据，如不同物种数据、基因本体数据、转录因子结合位点数据、蛋白质序列数据、蛋白质结构数据、蛋白质相互作用数据以及各种文献数据等。

物种多样性、多源性、异质性正成为生物数据的固有特性。这些数据具有以下几个特征：①数据形式、格式复杂，数据可以以多种形式来表现，如关系数据库、文本、表格以及图像等；②数据在动态地变化并不断地增加；③数据包含很高的噪声，具有不完整性和不确定性。表 6-12-1 简要概括了部分异质生物数据的来源。

表 6-12-1　多源异质生物数据资源示例

数据类型	网络资源
蛋白质 序列数据	http://www.ebi.ac.uk http://www.ebi.ac.uk/uniprot http://www.ncbi.nlm.nih.gov/protein http://www.ncbi.nlm.nih.gov/gene http://www.sanger.ac.uk http://www.jcvi.org/cms/home
蛋白质相互作用数据库	http://www.hprd.org http://mips.helmholtz-muenchen.de/proj/ppi http://string-db.org http://thebiogrid.org
蛋白质图像数据	Human Protein Atlas (HPA) http://www.proteinatlas.org/
基因数据库	EMBL、2DDBJ、GenBank、dbEST、GSDB、SGD、UniGene、TDB
基因本体数据	http://www.geneontology.org
转录因子结合位点数据	http://www.gene-regulation.com/pub/databases.html http://jaspar.genereg.net/cgi-bin/jaspar_db.pl http://www.factorbook.org/mediawiki/index.php/Welcome_to_factorbook
文献数据	http://www.ncbi.nlm.nih.gov/pubmed

以基因表达数据为例：基因表达数据所记录的是基因在一定条件下的特定表达以及不同基因之间的相互关系。而在生物的生命活动过程中，基因的表达以及基因之间的关系是在动态变化的，存在着复杂的基因表达调控网络；更进一步，基因表达数据和其他数据（如蛋白质相互作用）之间存在着密切的调控关系。要揭示生物活动的本质，单纯使用某一类型的数据是远远不够的。因此，充分利用多种异质生物数据，从中挖掘所蕴含的信息就尤为必要。

传统的数据挖掘研究往往基于单类型的数据（single-data-type）展开。以基因型-表现型相互作用（genotype-phenotype interactions）的研究为例，可以通过对家族数据连锁分析（linkage analysis in family-based data）[13]、对家

族数据 [14] 或是种群数据 [15] 的关联分析来辨识 DNA 序列变异。虽然这些基于单类型数据的挖掘方法研究取得了重要的进展，但是，一个重要的不足是：这种基于单类型数据的挖掘方法忽略了不同类型数据之间所蕴含知识的关联性、互补性，导致挖掘得到的结果深度及广度都还远远不够 [15]。例如，Schadt 等 [16] 的研究表明，融合大规模多源异质数据可以有效地提高数据挖掘的性能：通过关联蛋白质组与基因组（DNA）、转录组（RNA）等大规模数据，可以揭示 DNA 与表观性状之间的关系，对复杂疾病的表现形式（如癌症）做出正确的预测 [16,17]。

因此，为了了解细胞活动的全貌，进而揭示生命活动的本质，有必要将这些多源生物数据所蕴含的信息进行融合分析，实现知识的跨源共享和信息互补。

（三）生物数据具有层次性，需要跨层级联和深度学习挖掘

正如图 6-12-1 所示，从基因组、表观遗传组、转录组、蛋白质组及代谢组等不同层次研究所得到的生物数据之间呈现出典型的层次特性。实际上，即便是同一类型的数据，如蛋白质，也具有"残基-序列-家族"这样的层次关系。因此，研究能从生物层次角度出发并能体现生物层次特性的学习算法就具有重要的意义。

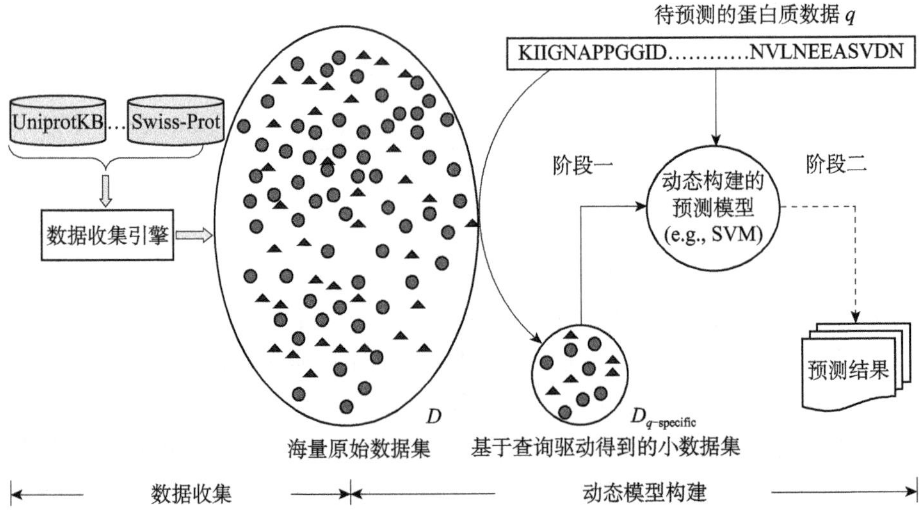

图 6-12-1　两阶段基于查询驱动的动态学习示意图

深度神经网络算法与生物数据存在的层次特性之间具有天然的契合度。近年来，深度神经网络算法在生物信息学中的应用受到了广泛关注 [18-22]。在

蛋白质结构预测[23-25]、蛋白质相互作用预测[26,27]、遗传变异致病性预测[28]、人类长链未编码 RNA 预测[29]等问题上都成功应用了深度神经网络相关算法来构建计算模型，且在预测精度上取得了长足的进步。

遗憾的是，这些已采用的深度学习算法和模型还主要是简单移植已有的机器学习领域的深度神经网络算法和模型，很难取得好的效果，其原因主要在于：①传统的非监督初始化方法并不能保证深度神经网络在复杂生物数据中学习到最优的模型；②深度神经网络在学习生物计算模型的过程中容易出现过拟合现象，从而导致计算模型的推广能力不足；③深度神经网络模型的学习机制很难给出合理解释，使得生物学家很难理解和使用。

第三节 前 沿 问 题

为了解决上述的生物数据挖掘的困难，领域专家已经开始致力于发展针对性强的生物数据挖掘算法来推进该领域的进展。处于前沿的问题如下。

一、如何在不断增长的海量生物数据下构建动态学习模型

为了应对生物数据持续动态增长的问题，增量式学习日益得到重视。增量式学习通常基于一个小的或是中等规模的数据集训练一个初始预测模型；当有新数据可用时，使用新数据来对预测模型进行增量更新[30,31]。在生物信息领域，增量式学习近年来已经得到应用：Andonie 等[32]利用 Fuzzy ARTMAP (FAM) 增量式学习方法，基于一个小分子数据集，在预测潜在的 HIV-1 蛋白酶抑制剂的生物活性实验中取得了较高的准确度；Miyanishi 和 Ohkawa[33]利用了增量式学习的方法，从文献中自动学习并提取出蛋白质的功能信息；Zheng 等[34]提出了一种在线增强学习的 SVM 来解决在一般 SVM 对大规模数据集处理中存在的存储和时间消耗巨大的问题。

上述已有的研究已经表明，增量式学习模型确实能在一定程度上缓解海量数据下静态学习模型所具有的不足，提高模型的可扩展性及可用性。但是，通过进一步分析可以发现，还存在如下的不足之处。

首先，增量式学习得到的模型从本质上来讲仍然是一个全局模型，也就是试图用单一的整体计算模型来统一体现蕴含在所有数据中的知识，不能解决海量复杂生物数据存在的分布不均匀、非凸等典型问题。

其次，在预测阶段未能充分考虑待预测（查询）数据本身的特性：对于

任何待预测的数据，和传统的静态模型一样，均统一使用当前的整体计算模型。

因此，基于现有的增量式学习方法，研究能够有效应对生物数据海量、非凸并且持续增长等诸多特性的动态挖掘模型与算法，以解决传统静态模型和增量式模型的不足，始终是前沿问题之一。

为了解决上述问题，建议可以从以下几个层面展开：①基于查询驱动的动态学习框架；②层次性动态学习框架；③动态学习框架的有效性评价；如图 6-12-1 所示。

（一）基于查询驱动的动态学习框架

为解决传统静态学习模型的不足，可尝试基于查询驱动的动态学习框架，它与传统静态学习方法使用全体数据构建全局预测模型有本质的不同，是一种“局部最相关特性”的模型动态生成方法，即依据查询数据输入的特性，从全体数据中选取部分样本来动态地构建一个查询驱动（query-driven）的预测模型。该新框架的突出优势在于把数据收集和模型构建独立分开：一方面，数据收集部分可以不断包含新增长的样本信息，但是否用到模型构建中则取决于待预测的数据；另一方面，模型构建由于是数据驱动动态生成的，是最适合该查询数据的特异性模型，因而能取得更好的预测结果。使用新的“局部最相关”动态模型构建框架有望能解决海量数据带来的信息分布不均匀和非凸性建模困难，避免海量数据中的不相关样本或噪声带来的模型预测偏差问题，提供一个解决在数据持续增长情况下更新静态模型困难的可行性方案。

（二）层次性动态学习框架

生物数据往往有着明显的层次性特征，不同层次的数据蕴含着不同的生物信息抽象。因此，可以在基于查询驱动的动态学习框架的基础上进一步引入层次结构，得到层次性动态学习框架，使得用户可以灵活地在不同层次上展开动态模型的构建。例如，可以从残基级、序列级、家族级以及更高层次来分析蛋白质的结构与功能。

（三）动态学习框架有效性评价

动态学习框架的核心思想是查询驱动、动态生成、局部最相关，依据查询输入选择性地从海量数据集合中使用部分训练样本来构建有针对性的预测模型。其中有几个关键问题需要解决：如何度量生物数据相似性、如何评价所选择样本的有效性、如何评价动态生成模型的有效性、如何平衡模型生成的时间和实时性的要求等。

其中的关键之一是，需要研究出适合于生物数据特性的新型相似性度量指标以及海量数据的组织存储新方案，以便快速有效地从海量生物数据中高效地提取查询驱动的局部训练集，这将显著影响到动态学习模型的性能和效率。

二、多源异质生物数据迁移学习模型一直备受关注

近年来，迁移学习及相关的多任务学习在生物序列分析、系统生物学以及生物医学等挖掘领域中已得到了一定的研究与应用。例如，Herndon 等[35]利用域自适应方法探讨了蛋白质定位；Kshirsagar 等[36]利用多任务学习研究了病原体宿主蛋白的相互作用；Zhang 等[37]使用任务正规化和提升的多任务学习研究了多目标蛋白质的化学相互作用预测；Mei 等[38]还基于 AdaBoost 多实例迁移学习研究了沙门氏菌与人体蛋白质之间的全蛋白质组（Proteome-Wide）相互作用预测；在生物医学方面，Wiens 等[39]利用多个医院的数据来增强特定医院的预测；Jain 等[40]探讨了信号多任务学习和调控网络应用于人类对流感的响应；Liu 等[41]基于异构数据域迁转移学习研究了硅片目标特定的 siRNA 设计；Zhang[42]等研究了用于老年痴呆症的基于多模式多任务的多元回归和分类变量联合预测。

根据上述迁移学习及相关的多任务学习在生物数据挖掘中的系统分析，我们发现虽然迁移学习及相关技术在生物数据挖掘中已得到研究利用，但目前还处于比较初级的阶段，远远达不到实际的生物数据挖掘需求。主要原因在于生物数据的多源异质性没有得到充分考虑。生物数据的多源异质性主要体现在如下方面。

（1）不同类型和不同介质存储的生物数据（cross-data），如蛋白质序列数据、蛋白质相互作用数据、基因本体数据、转录因子结合位点数据、文献数据、生物图像等。

（2）不同层次存储的生物数据（cross-level）。例如，以残基-序列-家族递进层次存储的生物数据。

（3）不同物种对应的生物数据（cross-species）。例如，不同物种衍生的生物数据既具有相似性，又具有差异性。

针对如上三种典型的多源异质生物数据，需要探讨如下几个方面的生物数据的迁移学习机制和相应的智能模型构建方法，如图 6-12-2 所示。

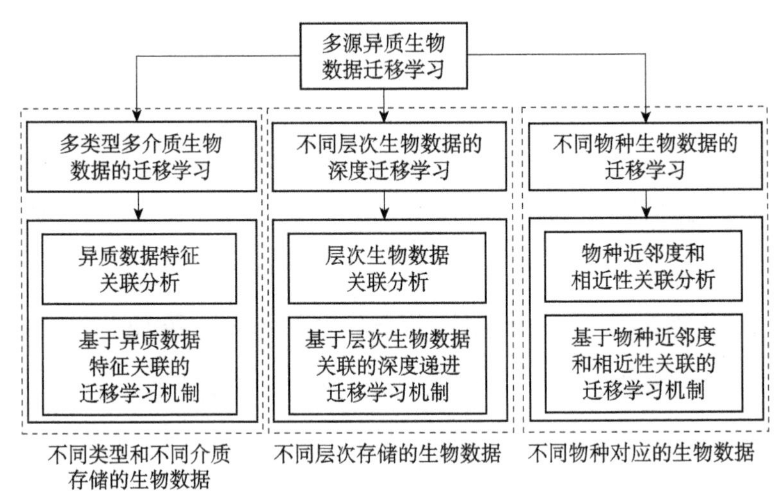

图 6-12-2　多源异质生物数据迁移学习的研究内容

（一）基于多类型多介质生物数据的迁移学习（cross-data transfer learning）

目前的经典迁移学习技术主要是基于同质数据和相同特征空间进行学习的，针对不同类型和不同介质存储的生物数据，通常具有不同的特征空间，这需要发展出新型的能处理不同特征空间的迁移学习算法。基于多类型多介质生物数据的迁移学习的关键挑战是如何实现在不同类型和介质生物数据学习时的知识迁移机制的构造。需要探讨不同介质存储的相异特征空间的数据间潜在的共性隐空间之学习构建技术，其突出的优势是通过潜在的隐空间找到不同介质数据的共性信息，从而通过共性信息来增强多类型多介质生物数据的学习能力和提高所构建的预测模型的性能。

（二）基于不同层次生物数据的深度迁移学习（cross-level transfer learning）

生物数据相关性的一个重要表现是存在层次性。不同层次上的生物数据之间表现形式等虽然具有明显的不同，但又存在重要的联系。因而，基于不同层次生物数据的迁移学习也是非常重要的。当某个层次标注的数据稀少时，通过其他相关层次丰富可用的标注数据来实现该层次有效的学习，从而实现不同层次生物数据的知识迁移学习。这种迁移框架在领域内还未发现相关的报道，是一种在新问题驱动下将要产生的新迁移框架。需要深入探讨不同层次生物数据的关联关系，提出基于不同层次生物数据关联信息的迁移机制，进一步把此类机制应用于智能模型的构建学习，从而提出基于不同层次生物数据深度迁移学习的有效智能预测模型构建技术和方法。

（三）基于不同物种生物数据的迁移学习（cross-species transfer learning）

由于各种条件的限制，在不同生物物种上的研究进展是不平行的，某些物种的可用标注数据是非常稀少的，但有些物种的研究已经比较深入，已经形成了很多的可用标注数据。众所周知，从稀少的标注数据中很难学习到高性能的智能预测模型，但由于不同物种，特别是相似物种的生物数据具有一定的相似性，因此借鉴相似物种已有的充足的标注数据来指导具有稀少标记的物种数据的学习是非常有用的。需要对不同物种生物数据，研究物种之间的关联信息，基于物种之间的关联性有针对性地控制物种间的迁移强度，最终实现面向不同物种生物数据的自适应迁移学习技术和方法。

三、面向生物数据层次隐含信息挖掘的深度学习理论与方法

深度学习的重要创新是建立层次模型来逐步学习和推理，它的出现为解决深层复杂结构相关的优化难题带来了新的解决途径。而生物数据本身具有典型的层次性、多样性、复杂性，因此，在生物数据中运用深度学习策略具有天然的契合度和合理性。在当前研究中，针对生物数据与深度学习的层次特征，如何研发一套面向生物数据层次特性的深度学习的理论与方法，用以有效地挖掘生物数据中的深层次隐含信息是主要的前沿问题之一。具体来说，可以展开的研究内容包括以下几方面。

（一）优化深度神经网络的初始化

传统的深度学习策略，在初始化步骤中通常使用非监督方式来学习数据的分布信息，并用以初始化深度网络的权值。但是，由于非监督初始化方法并不与学习目标直接相关，使用它们来初始化不一定是有效的，特别是当面对复杂生物数据应用时。因而，如果能采取某种与学习目标直接相关的策略（如先验生物知识）来帮助深层网络结构的确定与初始化，那么我们有理由相信，这样的深度网络能以更高的概率与更快的速度学习逼近最优解。

（二）提升深度神经网络在生物计算上的泛化能力

泛化推广能力是一个机器学习系统的重要评价指标。理论上，一个有足

够数目隐节点的三层神经网络就可以对学习的函数做到任意精度的逼近，但很容易出现过拟合现象，并不能保证学习得到的神经网络有良好的推广能力。这种过拟合现象，在现今生物数据爆炸性增长的情况下会显现得更加明显。正则化策略已经被验证是提升传统机器学习模型推广能力的有效途径，然而在面对不断增长的生物数据情况下，传统的正则化策略并不一定能满足生物数据复杂多变的条件。因此，根据生物数据增长的条件与规律，研究适合深度神经网络进行生物计算的正则化理论与方法，从而保证在生物大数据中深度神经网络计算方法拥有良好的推广泛化能力。

（三）提升面向生物数据的深度神经网络模型的可解释性

模型可解释性一直是机器学习与生物信息学领域中有待解决的关键问题之一，特别在面对生物学和信息科学的交叉生物信息学研究中，可解释性更是两个学科相互沟通的基础。在现有的生物计算方法中，模型大都呈现“黑箱”特性，虽然在计算精度与推广能力上取得了长足的进展，但是算法的通用性、可解释性等都有待进一步的提高，而且深度神经网络本身也是一种可解释性不强的神经网络模型，所以如何提升深度神经网络学习模型的可解释性是一个亟待解决的问题。未来的研究中，需要结合生物数据的动态性、层次性、多样性等特征，着重考虑生物数据中隐含的语义信息，从而准确反映生物数据的本质特性，加强深度神经网络算法决策推理过程的可解释性，研究针对生物数据的基于规则的推理及衍化模型，从大量学习样本中学习总结相关决策规则，提升深度神经网络学习模型在生物计算中的可解释性、通用性，并形成可以推广的可解释性透明学习策略。

四、多源异质生物数据挖掘平台的建设

如何设计实现多源异质生物数据挖掘平台是生物信息领域中的重要前沿问题之一。它将为广大生物学家及相关学者提供高效、易用的在线挖掘服务，从而促进生物实验科学的发展，加快对生命活动现象的认识。在实现多源异质生物数据挖掘平台时，以下几个方面值得重点关注。

（一）挖掘平台整体架构的设计

架构设计对于一个系统的成败有着至关重要的作用。需要从系统的稳定性、可靠性、可扩展性以及跨平台等多个角度全面考虑，选用已被业界证明为成熟高效的信息技术，来完成多源异质生物数据挖掘平台的架构设计。

（二）多源异质生物数据的整理及预处理

在海量的生物数据中存在着大量的噪声与冗余，这些无效数据的存在给生物数据的存储、分析、传播都带来了严峻挑战。以基因表达数据为例，由于数据获取技术手段的原因（如芯片的质量、探针的质量、样本质量以及实验条件等），得到的数据必然会包含大量的噪声。然而，直接从带有噪声与冗余的数据中学习到的生物学知识是不可信的、不足以指导生物学实验与生物制药的开展（rubbish in, rubbish out）。因此，需要针对生物数据中含有噪声与冗余这一特点，研发一套生物数据质量的评估方法，并形成适合于生物数据质量控制的数学理论基础。

考虑到生物数据是富含噪声、不完整以及误标定的，因此，一个重要的方面是对原始的生物数据做必要的数据质量评估和提纯，以提高数据挖掘的可靠性。可以从数据的拓扑结构分析、完整性分析以及基于学习算法的分析三个方面来加以实施。

首先，从生物数据的拓扑结构出发，结合生物数据分层的特性，研究数据与数据之间的分层拓扑关系，从而进行生物数据的质量评估。在生物体进化的过程中，不同时间与不同地点所产生的生物数据所表示的结构与功能信息会存在很大的差异性。通过引入完备的拓扑理论与方法来研究生物数据之间的拓扑关系，进行生物数据质量评估方法的构建，发现并移除离群生物数据，发现生物数据拓扑群，降低生物数据中的噪声。

其次，生物数据所包含信息一方面存在着冗余性，另一方面完整性又可能得不到保障。需要在仔细分析生物数据拓扑关系后，深入研究生物数据所表达生物信息的完整性与冗余性，去除生物数据中冗余数据，补充不完备数据的信息表达。

最后，经过上述数据的处理以后，为了进一步对生物数据进行质量评估，需要结合生物计算的学习算法进一步分析每一个生物数据对生物计算过程中的作用，将作用小的生物数据视为质量不高的生物数据去除。通过以上步骤构建一个优秀的生物数据评估框架与方法，使得随后的生物计算理论与方法的研究得到数据的保障。

（三）面向多源异质生物数据挖掘算法的实现和集成

实现动态学习、迁移学习以及深度学习算法，部署到挖掘平台。研发界面友好的用户接口，为用户提供快速、鲁棒、高效的专业化挖掘服务。强化所提机器学习算法的可解释性以及挖掘结果的可视化，解决传统挖掘算法的“黑盒”特性，提高挖掘平台的易用性和可解释性。

可以在软件和硬件两个层面来展开高通量多源异质生物数据挖掘平台的建设工作。一个可行的挖掘平台的总体架构如图 6-12-3 所示。

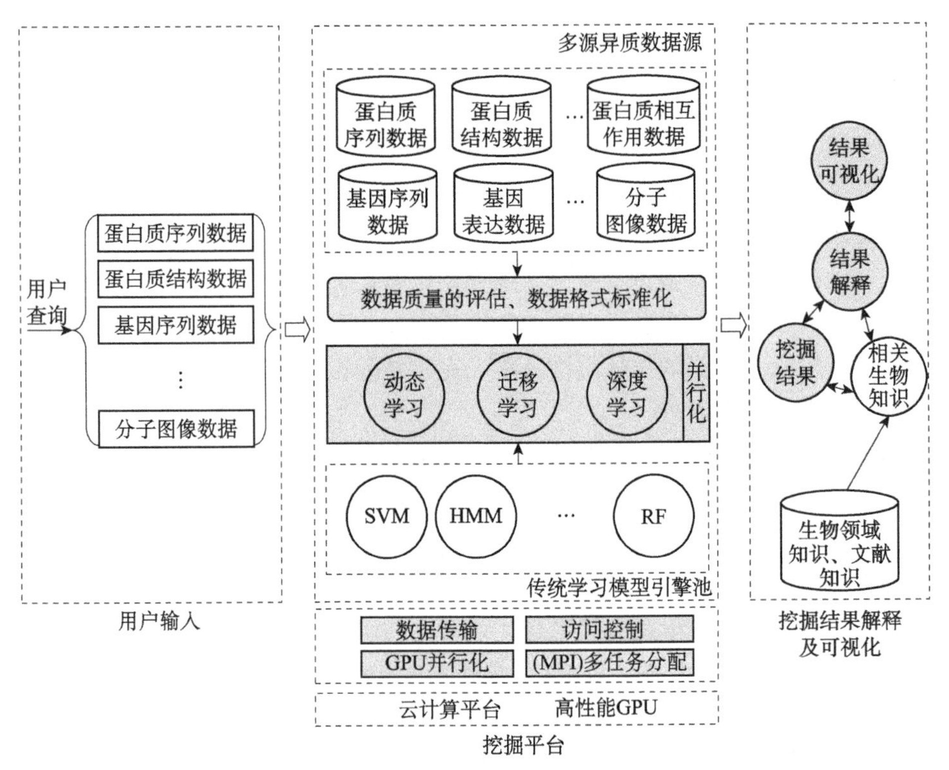

图 6-12-3　多源异质生物数据挖掘平台总体架构

在硬件层面，可以依托云计算平台来提供海量生物数据的存储与计算服务，以 GPU 为硬件支持来进行挖掘算法的并行化加速，以消息传递接口技术（message passing interface, MPI）来进行挖掘计算的多 GPU 计算节点任务分配。

在软件层面，深入探索和了解生物数据的特性，根据生物数据的大小和复杂性寻找有效的海量数据存储、传输及维护方法，提供快速安全的数据和服务访问控制；根据数据的复杂度和所提算法（动态学习、迁移学习和深度学习）的特性，研究有效的算法并行化方案。算法的并行化对于提升挖掘性能、缩短计算时间具有重要意义，在本挖掘平台的构建中占有举足轻重的地位。

在建设挖掘平台时，还需要特别考虑挖掘结果的可解释性和可视化，以提高挖掘平台的易用性，为相关的生物科学家提供具有高可解释性的挖掘结果并能可视化展现。

第四节　未来5～10年重点发展方向

针对上述几个生物信息数据挖掘的前沿问题，对未来5～10年重点发展方向建议如下。

一、面向海量生物数据的动态学习新框架

如前所述，传统基于机器学习的生物计算和挖掘模型通常是基于静态学习框架的：从一个有限规模的数据集，训练得到相应的计算模型，然后用于对未标定的数据进行预测。基于静态学习框架的方法不可避免地会遇到诸如可扩展性低、过拟合 / 过优化以及低可用性等问题。

因此，充分考察传统静态学习模型不足，在此基础上进一步研究与利用海量生物数据中固有的分层、同源、异质等性质来搭建面向海量生物数据的增量、动态学习新框架，将有助于解决拥有多源异质性的生物数据的海量并持续快速增长现状，应当是未来的一个重点发展方向。

二、多源异质生物数据迁移学习算法

随着高通量测序等技术的发展，生物数据呈爆炸式增长趋势，不断有新的数据被发现、新的领域被开拓，这给传统的生物计算方法带来了巨大的挑战。传统的生物计算方法需要对每一个领域中作为训练集合的数据都进行标定，这将会耗费大量的人力与物力，而大量没有标注的数据，往往会造成模型构建的困难；另外，传统的生物计算学习方法一般的假设是训练集合数据与待预测数据集会服从相同的数据分布，然而，在许多实际应用情况下，这种同分布假设并不满足。所以当面对缺乏标注数据的新兴领域、新数据与旧数据不是同一数据分布时，传统的计算方法就无法学习到有效的生物计算模型。迁移学习技术将是解决这一问题的有效方案。迁移学习可以从现有的数据中迁移知识，用来帮助将来的学习，其目标是有效利用一个环境中学到的知识来帮助学习另一个环境中的有效知识，并不需要像传统的生物计算方法一样做出同分布的假设。因此，需要重点研究面向生物数据的多源异质的特

征，提出一套能够面对生物数据复杂环境的新迁移学习生物计算方法。

虽然迁移学习及相关技术在生物数据挖掘中已得到研究利用，但目前还处于比较初级的阶段，远远达不到实际的生物数据挖掘需求。主要体现在如下几个方面。

（1）目前的研究还主要集中于多任务学习的相关应用研究。虽然多任务学习可视为一种特殊的迁移学习，但是它不具备迁移学习的一些独特的特点。例如，迁移学习注重于目标域性能的提高，对源域的性能则不太关注，而多任务学习则通常把不同域视为同样重要的平行任务，这就使得多任务学习在生物数据挖掘中对于目标域的学习任务之针对性较差。

（2）生物数据的异质性在目前的迁移学习和多任务学习中还没有得到重视，已有的技术通常假设不同的域具有相同的特征空间，这在实际的应用中是不现实的，更为实际的情况是生物数据通常具有多源异质性，即具有多样的数据存储载体，从而存在不同的特征描述空间。

（3）目前的研究主要集中于相似生物组织间的迁移学习，缺少深度迁移方面的探讨。例如，以残基-序列-家族递进层次存储的生物数据为例，不同层次间的迁移学习还未得到重视。

因此，结合生物数据自身的特点，发展针对性强的多源异质生物数据间的迁移学习技术，构建更可靠和更有效的生物迁移挖掘模型是非常必要的，将是未来重点发展方向之一。

三、优化生物信息领域的深度学习方法，提升深度网络模型的可解释性

最近几年来，深度学习在图像处理、语音识别等方面取得了巨大成功。重要原因之一在于深度学习可以从零知识开始做自主的特征学习，逐层构建强大的分类网络。但是，已有的研究也表明，在有些领域中，如自然语言处理、生物信息处理，直接应用深度学习与已有的方法相比性能提升并不是十分显著。原因在于深度学习虽然可以从零知识进行自主的特征学习，但是没有有效利用研究对象本身存在的层次关系（如单词-词汇-语句-段落，残基-序列-家族-物种）及相应先验知识。

因此，在生物信息大数据挖掘领域中，深度学习的一个潜在的重要发展方向是，如何针对研究背景，在深度学习框架中高效地融合人们已经汲取出来的丰富先验知识，并提升深度神经网络学习模型在生物数据挖掘中的可解释性、通用性，形成可以推广、可解释性透明学习策略。

第五节　结　　语

面向生物大数据理解分析的智能生物信息学理论与方法研究是当前的国际学术前沿，这是因为近年来高通量数据采集技术的飞速发展产生了海量的生物数据，但由于各种实验手段的昂贵和耗时等限制，生物数据产生的能力与数据理解的速度之间的鸿沟正在迅速拉大。生物分子的多源、异质、分层等特有复杂特征也使得简单移植已有的人工智能和知识发现算法很难能取得好的效果，促使研发生物数据特征驱动的先进算法和方法成了一个重要课题，这在促进社会进步和生命健康研究，且对信息理论的完善及与实际应用更紧密的结合都具有重要的意义。同时，如何建设和维护易用性强的高性能生物计算平台对高效地发现和提取知识，构架生命科学和信息科学的交叉学科桥梁也具有重要的价值。

沈红斌（上海交通大学图像处理与模式识别研究所），
於东军（南京理工大学计算机科学与工程学院）

参 考 文 献

[1] Ritchie M D, Holzinger E R, Li R, et al. Methods of integrating data to uncover genotype-phenotype interactions.Nature Reviews Genetics, 2015, 16: 85-97.

[2] Metzker M L. Sequencing technologies—the next generation. Nature Reviews Genetics, 2010, 11: 31-46.

[3] Laird P W. Principles and challenges of genome-wide DNA methylation analysis. Nature Reviews Genetics, 2010, 11: 191-203.

[4] Shulaev V. Metabolomics technology and bioinformatics. Briefings in Bioinformatics, 2006, 7: 128-139.

[5] Shapiro E, Biezuner T, Linnarsson S. Single-cell sequencing-based technologies will revolutionize whole-organism science. Nature Reviews Genetics, 2013, 14: 618-630.

[6] 李衍达．信息与生命．化学通报，2001, 10: 601-607.

[7] Chou K C, Shen H B. Cell-PLoc: a package of Web servers for predicting subcellular localization of proteins in various organisms, Nat. Protoc., 2008, 3: 153-162.

[8] Yang J, Jang R, Shen H B. High-accuracy prediction of transmembrane inter-helix contacts and application to GPCR 3D structure modeling. Bioinformatics, 2013. 29(20): 2579-2587.
[9] Salichos L, Rokas A. Evaluating ortholog prediction algorithms in a yeast model clade. PLoS One, 2011, 6: e18755.
[10] O'Donovan C, Apweiler R. A guide to UniProt for protein scientists. Methods. Mol. Biol., 2011, 694: 25-35.
[11] Sterlinga T D. Publication decisions and their possible effects on inferences drawn from tests of significance-or vice versa. Journal of the American Statistical Association, 1959, 54: 30-34.
[12] Rajaraman A. More data usually beats better algorithms. Datawocky, 2008.
[13] Almasy L, Blangero J. Multipoint quantitative-trait linkage analysis in general pedigrees. The American Journal of Human Genetics, 1998, 62: 1198-1211.
[14] Horvath S, Xu X, Laird N M. The family based association test method: strategies for studying general genotype-phenotype associations. European Journal of Human Genetics: EJHG, 2001, 9: 301-306.
[15] Devlin B, Roeder K, Bacanu S A. Unbiased methods for population-based association studies. Genetic Epidemiology, 2001, 21: 273-284.
[16] Schadt E E, Linderman M D, Sorenson J. et al. Computational solutions to large-scale data management and analysis. Nature Reviews Genetics, 2010, 11: 647-657.
[17] 王丙强，邱爽，郭贵鑫，等. 应对生命科学的大数据挑战 . 中国计算机学会通讯，2012, 8: 29-33.
[18] Greene C S, Tan J, Ling M. et al. Big data bioinformatics. Journal of Cellular Physiology, 2014, 229: 1896-1900.
[19] de Ridder D, de Ridder J, Reinders M J. Pattern recognition in bioinformatics. Briefings in Bioinformatics, 2013, 14: 633-647.
[20] Huang D S, Han K, Gromiha M M. Intelligent Computing in Bioinformatics: 10th International Conference, ICIC 2014, Taiyuan, China, August 3-6, 2014, Proceedings. Springer Publishing Company Incorporated, 2014, 6216(4): 9-43.
[21] Chicco D, Sadowski P, Baldi P. Deep autoencoder neural networks for gene ontology annotation predictions //Proceedings of the 5th ACM Conference on Bioinformatics, Computational Biology, and Health Informatics: 2014. 533-540.
[22] Nguyen S P, Shang Y, Xu D. DL-PRO: A novel deep learning method for protein model quality assessment // Neural Networks (IJCNN), 2014 International Joint Conference on, 2014: 2071-2078.
[23] Jesse E, Cheng J. DNdisorder: predicting protein disorder using boosting and deep networks.

Bmc Bioinformatics, 2013, 14(1): 88.

[24] Lena P D, Nagato K, Baldi P F. Deep spatio-temporal architectures and learning for protein structure prediction // Advances in Neural Information Processing Systems, 2012: 512-520.

[25] Zhou J, Troyanskaya O G. Deep supervised and convolutional generative stochastic network for protein secondary structure prediction. Computer Science, 2014: 745-753.

[26] Di Lena P, Nagata K, Baldi P. Deep architectures for protein contact map prediction. Bioinformatics, 2012, 28: 2449-2457.

[27] Eickholt J, Cheng J. Predicting protein residue–residue contacts using deep networks and boosting. Bioinformatics, 2012, 28: 3066-3072.

[28] Quang D, Chen Y, Xie X. DANN: a deep learning approach for annotating the pathogenicity of genetic variants. Bioinformatics, 2015, 31(5): 761-763.

[29] Fan X N, Zhang S W. lncRNA-MFDL: identification of human long non-coding RNAs by fusing multiple features and using deep learning. Molecular BioSystems, 2015, 11(3):892-897.

[30] He H B, Chen S, Li K, et al. Incremental learning from stream data. IEEE Transactions on Neural Networks, 2011, 22: 1901-1914.

[31] Wang Z L, Jiang M, Hu Y H, et al. An incremental learning method based on probabilistic neural networks and adjustable fuzzy clustering for human activity recognition by using wearable sensors. IEEE Transactions on Information Technology in Biomedicine, 2012, 16: 691-699.

[32] Andonie R, Fabry-Asztalos L, Abdul-wahid, C B et al. Fuzzy ARTMAP prediction of biological activities for potential HIV-1 protease inhibitors using a small molecular data set. IEEE/ACM Transactions on Computational Biology and Bioinformatics (TCBB), 2011, 8: 80-93.

[33] Miyanishi K, Ohkawa T. A Method of Extracting Sentences Containing Protein Function Information from Articles by Iterative Learning with Feature Update. Springer Berlin Heidelberg, 2012, 7845: 81-94.

[34] Zheng J, et al. An online incremental learning support vector machine for large-scale data. Neural Computing and Applications, 2013, 22: 1023-1035.

[35] Herndon N, Caragea D. Predicting protein localization using a domain adaptation approach. Communications in Computer & Information Science, 2013, 452: 191-206.

[36] Kshirsagar M, Carbonell J, Klein-Seetharaman J. Multitask learning for host–pathogen protein interactions. Bioinformatics, 2013, 29: i217-i226.

[37] Zhang J, Lushingtor G H, Hunan J. Multi-target protein-chemical interaction prediction using task-regularized and boosted multi-task learning // Proceedings of the ACM

Conference on Bioinformatics. Computational Biology and Biomedicine, 2012: 60-67.
[38] Mei S, Zhu H. AdaBoost based multi-instance transfer learning for predicting proteome-wide interactions between salmonella and human proteins. PLoS One, 2014, 9: e110488.
[39] Wiens J, Guttag J, Horvitz E. A study in transfer learning: Leveraging data from multiple hospitals to enhance hospital-specific predictions. Journal of the American Medical Informatics Association, 2014, 21: 699-706.
[40] Jain S, Gitter A, Bar-Joseph Z. Multitask learning of signaling and regulatory networks with application to studying human response to flu. PLoS Computational Biology, 2014, 10: e1003943.
[41] Liu Q, Zhou H, Zhang K, et al. In Silico target-specific siRNA design based on domain transfer in heterogeneous data. PLoS One, 2012, 7: e50697.
[42] Zhang D, Shen D, Initiative As D N. Multi-modal multi-task learning for joint prediction of multiple regression and classification variables in alzheimer's disease. Neuroimage, 2012, 59: 895-907.

第七篇
软物质交叉领域

第一章 抗污染、智能、便携式可穿戴微流控器件

本章首先叙述了微流控的定义和与软物质的关系，简要回顾了微流控的发展和应用历程，介绍了目前微流控设计和应用发展涉及的三个方面的挑战和最新研究进展：在微尺度下如何解决流体输运的污染问题，开发具有抗污染性质的微流控系统，将在生物医学、节能系统、高效物质分离等领域带来巨大的经济效益；在微尺度下如何实现通道的功能化，将为智能流体系统在生物检测、药物筛选、微颗粒制备等领域带来全新的研究平台；如何实现柔性可拉伸的微流控系统，将在开发液态天线、柔性机器人、便携式流体装置等领域带来巨大的技术推动作用。最后，针对目前国家对软物质凝聚态物理学发展战略，建议未来在微流控器件的抗污染、功能化和柔性可拉伸等方面加强研究，为中国微尺度流体系统的发展奠定技术和应用基础。

第一节 引　言

软物质（soft matter）是指处于固体和理想流体之间的物质，又称软凝聚态物质，如液晶、聚合物、胶体、膜、泡沫、颗粒物质、生命体系等。软物质，在美国曾被称为“复杂流体”（complex fluid），1991 年，诺贝尔奖获得者皮埃尔·吉勒·德热纳（Pierre-Gilles de Gennes，1932—2007）在获奖典礼上使用“软物质”一词代替之前的专业术语——“复杂流体”，并在其后得到了学术界的广泛认同。对于不同尺度流体的控制存在着各种不同的控制

技术和应用领域（图 7-1-1）。微流控（microfluidics）是一种精确控制和操控微尺度流体，一般来说控制的流体应该在一个维度上至少小于 1 mm。由于在生物、化学、医学等领域的巨大潜力，微流控芯片已经发展成为一个涉及材料、生物、化学、医学、流体、电子、机械等学科交叉的前沿研究领域，其中流体主要包括液体（有机的 organics，水成的 aqueous，碳氟化合物 fluorocarbons 等）和气体（如氧气、氮气、氦气等）。

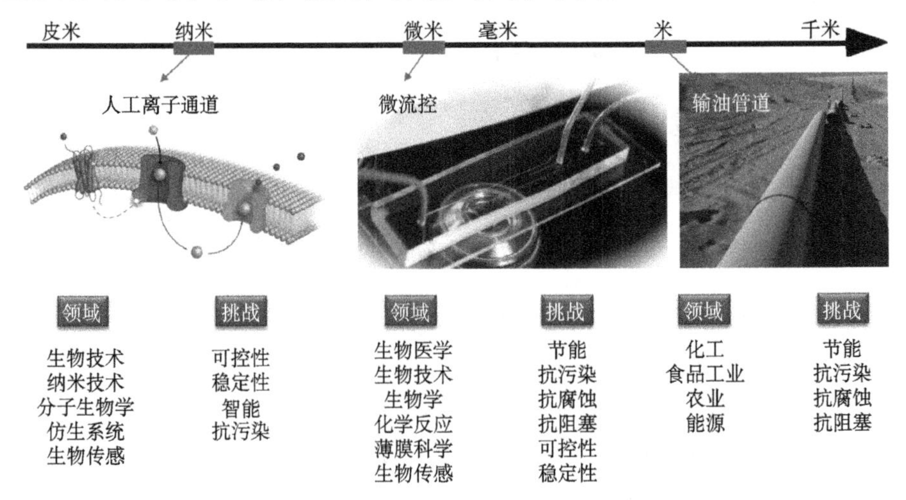

图 7-1-1　多尺度流体控制的应用领域与挑战（插图均来自：http://jonlieffmd.com/blog/brain-receptors-just-got-even-more-complex; https://darwin-microfluidics.com/blogs/reviews/microfluidic-chips-a-quick-overview; https://baike.baidu.com/item/ 输油管道）

微流控芯片具有样品用量少（纳升、皮升级别）、能量消耗低、规模集成等特点，微流控装置本身占用体积很小。为了实现复杂流体在微尺度下控制，微流控是一个包括材料学、物理化学、工程学、微纳技术和生物医学工程等的多交叉学科。目前最热门的研究之一是基于微流控芯片技术，把物理、化学、生物、医学分析过程的样品制备、反应、分离、检测等基本操作单元集成到微纳米尺度的芯片上，自动完成分析检测分离等全过程，它的空间特征尺度范围在 1 μm 至 1mm。

目前，微流控的主要研究内容包括：①微通道的设计制造（当通道的特征尺寸缩小到微纳米尺度，通道表面积与通道内部空间的体积之比变得很大，通道的形状和结构与通道壁物理化学性质都将对输运的流体流动状态带来了很大的影响，所以设计并制造出尺寸大小精确、形状结构合理、通道流体界面性质可控的微通道是微流控发展的前提条件）[1]，以及制备材料选择，包括：无机材料：主要有玻璃 [硅酸盐类，主要成分二氧化硅（silicon dioxide）]，有机材料：主要有聚二甲基硅氧烷 [polydimethylsiloxane]、聚甲基丙烯酸甲酯 [polymethyl methacrylate]、聚乙烯 [polyethylene]、聚碳酸酯 [polycarbonate]、聚苯乙烯 [polystyrene]、聚氯乙烯 [polyvinylchloride]、聚四氟乙烯 [polytetrafluoroethylene]，金属材料：有不锈钢 [stainless steel] 等；②微流控系统的集成封装（随着微纳米技术的不断发展升级，微流控系统也日趋微型化和多功能化）与移动（基于微流控在环境分析和生物医学的应用中，越来越多的实际要求，需要微流控系统可以适应不同环境而实现系统和器件的随时移动）；③微尺度下对流体的输运控制（与宏观流体相比，在微尺度下的流体的流动状态和输运性质存在很大的区别，具有明显的尺寸效应（随着通道尺寸的减小，流体的体积减小，重力基本可以忽略），此时流体的比表面积增加，通道与流体界面的表面张力（宏观尺度下基本忽略其对输运的影响）成为主要控制因素，从而影响通道中流体流动状态（包括液-固、液-液及液-气界面的形态、大小和所处空间位置）。

微流控技术存在以下优点：

（1）动态的减少样品和试剂的消耗；

（2）缩短实验的时间（同时也缩减了相关应用的成本）；

（3）有望制备出小型便携式检测分离高敏感系统；

（4）增加单位时间的实验数量，为大量的生物平行实验提供基础。

微流控将极大地简化工作流程，与半导体工业一样，由于微纳制备技术的高速发展，将大大节约使用空间、节省劳动力和缩短工作时间，并且有望更加自动化地实现生物、化学工程等的工作流程。例如，包裹细胞技术，利用传统的溶液搅拌法无法得到均匀的球状液滴，而利用微流控技术能轻松地实现形状均匀包裹细胞的液滴制备（图 7-1-2）。[2]

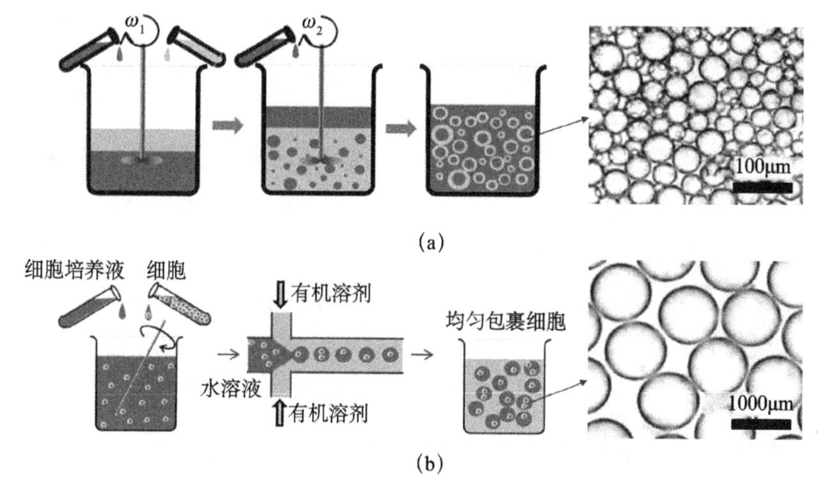

图 7-1-2 传统的溶液搅拌法与基于微流控技术的包裹细胞技术的对比 [2]
（a）利用传统的溶液搅拌法；（b）利用微流控技术

目前微流控装置主要的发展方向如下（图 7-1-3）。

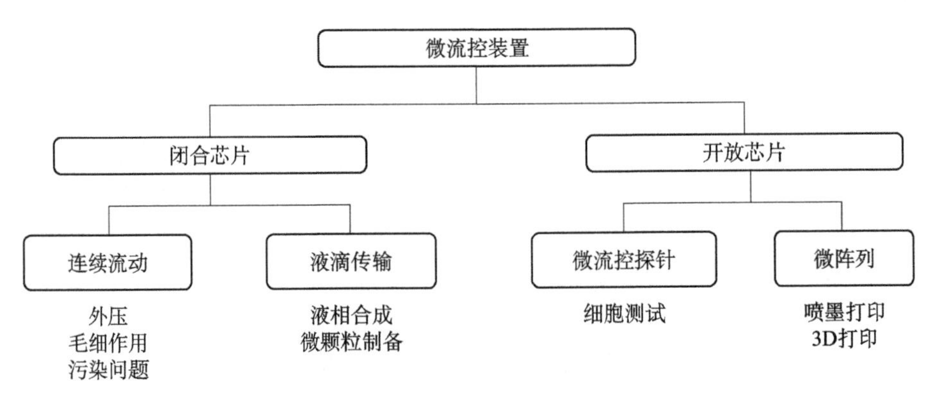

图 7-1-3　微流控装置的发展方向

第二节　发 展 历 程

目前微流控是一个具有数十亿美元价值的技术产业，它所涉及的交叉学科十分广泛，健康医疗是其中一个主要的应用领域，包括分子生物学、药物开发、诊断和鉴定、药物传输、分析等。与此同时，微流控技术也可以应用于环境分析，高价值流体制备（如化妆品、食物、香料等），微纳米颗粒的制备，石化产品开发等。

关于微流控的发展历程，我们可以从哈佛大学 Whitesides 教授 2006 年在 *Nature* 杂志上的一篇微流控综述，即微流控的起源和未来（*The origins and the future of microfluidics*）讲起。[3]

Whitesides 教授提出微流控的四个重要研究和应用，包括分子分析、生物防卫、分子生物学、微电子学（图 7-1-4）。他首先谈到了微流控的起源是微全分析的发展［气相色谱法（gas-phase chromatography，GPC）、高压液相色谱法（high-pressure liquid chromatography, HPLC）和毛细管电泳法（capillary electrophoresis, CE）等基于毛细现象的方法］，同时它结合了激光光学探测技术实现了对微量样品的高敏感、高分辨的实时检测。从微量分析方法的成功应用，研究人员显而易见地看到了开发新型更加复杂多功能的系统将为化学和生物化学领域带来更多的微尺度分析方法的应用前景。

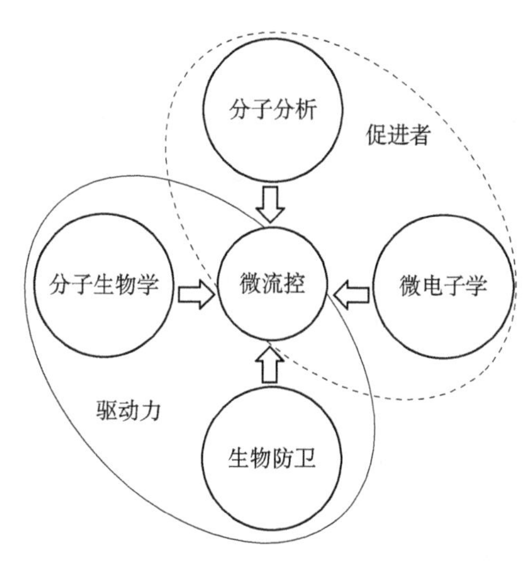

图 7-1-4　微流控发展的促进者和驱动力

随后，他又谈到发展微流控的一个非常重要的推动力是冷战结束以后，生化武器成了来自当今军事和恐怖分子最主要的威胁之一，如何对付这些威胁，美国国防部（the Defense Advanced Research Projects Agency）在 20 世纪 90 年代资助了一系列科研项目来开发基于微流控系统的反恐探测器。这些资助带来了微流控技术学术研究的迅速发展。

接着，他又谈到分子生物学对微流控发展起到了又一个重要的推动作用。早在 20 世纪 80 年代，基因组学在出现了高通量 DNA 测序技术（high-throughput DNA sequencing）后迎来了爆炸式发展，因此对分子生物学在微尺度上测试数据量有了更大的需求，同时对更高灵敏度和更高分辨率以及大量分析研究有了更多的需求，而微流控技术为其带来了可能性。

最后，他谈到微电子学的发展也为微流控的发展带来了很大的促进作用。在硅微电子学的光刻胶及相关技术的发展为制备更多具有微尺度结构微流控系统带来了更大的材料选择范围，也带来了许多新的性能，为微流控器件的更多的潜在应用带来了材料基础。比如，对于传统的玻璃和硅材料，气体不能通过，而采用有机的聚二甲基硅氧烷制备的微流控系统具有很好的透气性，这为在微流控中实现活细胞的培养所需要的气体环境带来了条件。

根据以上 Whitesides 教授的论文观点，我们可以通过图 7-1-4 来全面了解它们之间的逻辑关系。简而言之，分子分析和微电子学领域的技术发展对微流控的发展起到了促进作用，而生物防卫和分子生物学的迫切需求成为微流控发展的驱动力。

第三节 最近二三十年的研究进展

通过文献调研，早在 1964 年，Eringen 就在国际杂志 *Int. J. Engng. Sci.*[4] 中提出了微流体的概念，他提到“一种简单微流体，概略地讲，是一种流动的介质，它的性质和行为是受到在它的各个体积元内附近运动材料颗粒的性质影响”。

1984 年，Jaromir Ruzicka 等发表了第一篇描述在一个基底上构筑通道用于整合的微流体装置在注入流体的分析的应用[5]。

1998 年，哈佛大学夏幼南博士和 Whitesides 教授首次把软蚀刻术（soft lithography）应用于微流控领域，使该领域得到了飞速的发展[6]。可以说，对于微流控从 1990 年到 2010 年的二十年期间的发展使它成了一个具有很大希望的新兴的热门技术。

根据麦姆斯咨询的 2015 年微流控应用报告（图 7-1-5），可以看出微流控产业呈现连续多年增长的良好势头。展望未来，微流控市场还将保持快速增长的趋势，从 2015 年的 25.6 亿美元增长到 2020 年的 59.5 亿美元，18% 的

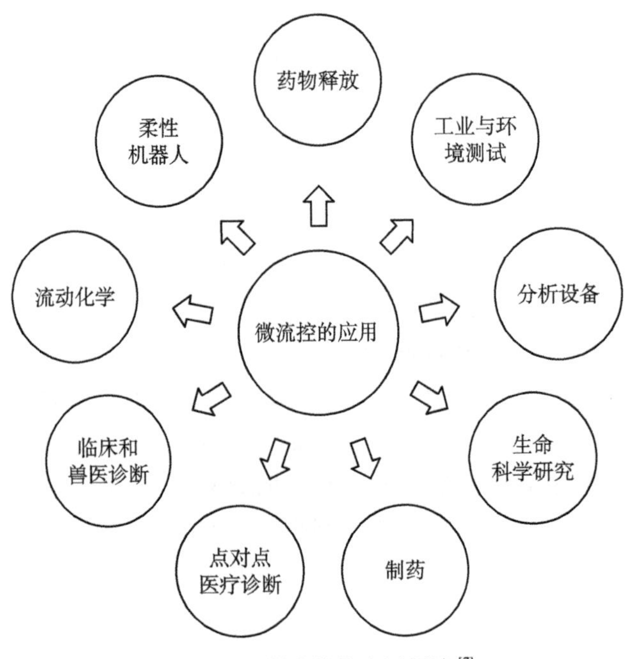

图 7-1-5 微流控的应用领域[7]

年复合增长率主要由生命科学和体外诊断驱动。报告同时也提到，随着医疗保健行业朝着个性化医疗方向发展，需要微流控技术不断发展以适应新的市场需求。报告还指出，由于基因组学的突破，制药技术有了显著提升，最近流行的埃博拉病毒也迫切需要廉价的检测方法，以满足现场快速诊断需求。为了避免未来流行病的爆发，创新的微流控技术将越来越多地应用到即时诊断领域。与此同时，近期的卫生丑闻使得农产品和水行业不得不加强法律法规，从而大幅增加了专业细菌检测实验的需求。所以，微流控技术在工业和环境检测领域发展得如火如荼。鉴于创新的解决方案和成本优势，微流控技术吸引了越来越多的厂商加入进来，许多大企业都在探索采用微流控技术来提升产品的竞争力[7]。

2015 年微流控应用报告还指出，新的微流控技术也逐渐“渗透”于成熟应用和新兴应用。他们同时也提供微流控技术 / 应用路线图（图 7-1-6）来介绍微流控技术的未来发展趋势[7]。

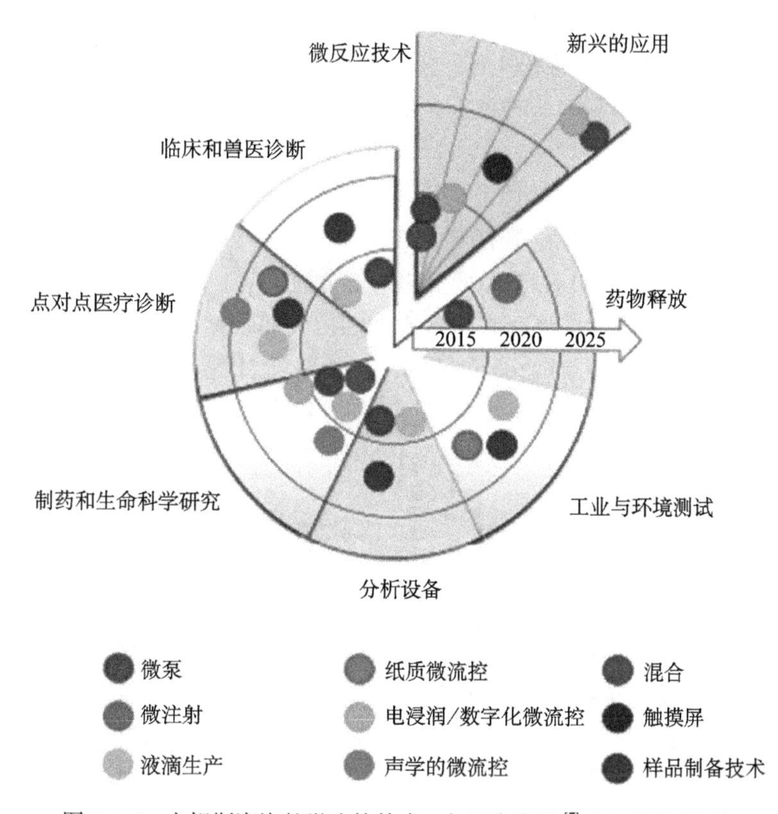

图 7-1-6　麦姆斯咨询的微流控技术 / 应用路线图[7]（文后附彩图）

第四节　学科前沿问题

人们越来越清醒地认识到，虽然微流控和芯片实验室的研究得到高速发展，带来了许多学术论文的发表和令人激动的时刻，但是对于微通道的稳定性、可靠性，以及在微尺度下如何解决流体输运的污染问题，在微尺度下如何实现微通道的功能化，以及如何实现具有环境适应性的柔性可拉伸的微流控系统等，仍然存在着很多理想的情况和技术瓶颈。因此，如何真正地解决这些问题对于微流控技术本身及其应用领域的发展具有重大的意义和作用。

人们曾经预期到了 2000 年，微流控技术将在许多应用领域逐步成为关键性技术。芯片实验室（lab on a chip）也被看成是几乎所有科学研究和诊断的发展必然趋势。可是直到近期，早前对微流控预期商业应用产品仍然还停留在理论和实验室阶段，这给人们带来了不小的失望[8]。

在过去许多年里，几乎所有有关微流控的非学术报告都会提及目前微流控所面临的问题主要是缺少实实在在在工业界的关键性应用（killer application）[9,10]。

除了在打印机和糖尿病检测的微流控应用，为什么到目前为止没有一个完全基于微流控的应用产品？以上所面临的问题，一部分是受到了应用瓶颈的限制，比如在微尺度下如何解决流体输运的污染问题；另一部分是如何在微尺度下实现微流控通道的智能化，告别仅有单一功能的微流控系统应用局限性；以及如何实现对于动态环境变化的可适应性等。

即使微流控应用目前还存在着各方面的挑战和困难，但在 2013 年，全球微流控的市场价值也达到了 15.9 亿美元，预期到 2016 年这个数值会增加到 30 亿美元（图 7-1-7）[7]。

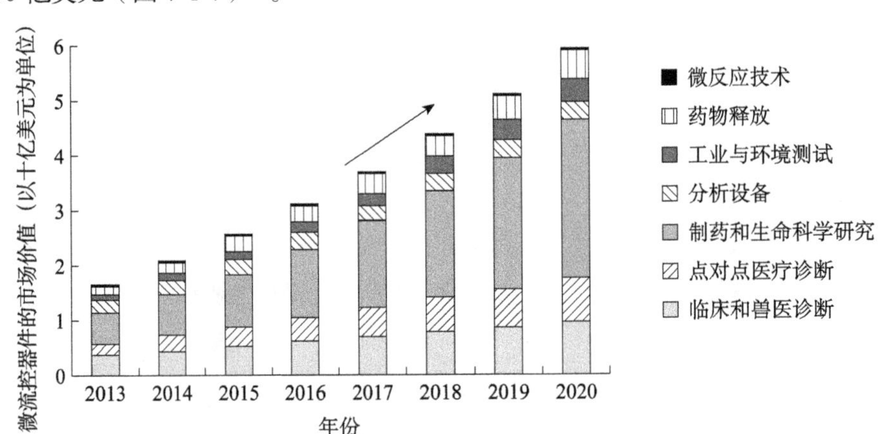

图 7-1-7　YOLE 法国市场调研公司的全球微流控的市场价值预测[7]

第五节 与实际需求结合的重大问题

对于微流控的应用，国际著名学术杂志分析化学（*Analytical Chemistry*）副主编 Rajendrani Mukhopadhyay 曾在 2005 年专题讨论了如何应对微流控的污染问题以及它的重要性[11]。其中，美国伊利诺伊大学香槟分校的 Deborah Leckband 指出微流控中污染问题是一个难以置信的复杂问题，几乎没有很好的办法来解决它（It's an incredibly complex problem, and there's no magic bullet）[11]。美国 Caliper Life Sciences（该公司主要为生命科学领域内致力于挽救生命、快速有效地提高药效和诊断测试水平的研究人员提供尖端技术）的 Josh Molho 指出，我们不得不正视微流控中的污染问题，因为如果我们的装置不能保证能可靠地工作，没人愿意去买它（Fouling is critical if you're talking about making a real device that you want to sell to someone who wants it to work reliably. That's where the strongest motivation is—to go to that extent and prove your device will survive long enough or over the variety of samples）[11]。因此，要全面实现微流控技术在生物医学、节能系统、高效物质分离等领域实际应用，如何解决微通道的污染问题是现实情况下急需解决的重要问题。

目前，解决微流控的污染问题主要有两种思路（图 7-1-8）：第一种思路是选择低表面能的材料来直接制备微通道，从而提升微通道的抗污性能［图 7-1-8（a）[12]］；另一种思路是通过对微通道材料内表面的化学修饰改性，来实现微通道的抗污性能的提升［图 7-1-8（b）[13]］。但这些传统的方法均存在一些缺陷，往往还带来制备工艺的复杂化，同时带来抗污染性能不稳定等情况。所以，如何在微尺度下解决流体输运的污染问题，与实际需求相结合开发具有抗污染的微流控系统，将为生物医学、节能系统、高效物质分离等领域带来巨大的经济效益。

材料智能化是现代高技术新材料发展的重要方向之一，它将支撑未来前沿技术的发展，实现结构功能化、功能多样化。科学家预言，智能材料的研制和大规模应用将引起材料科学发展的重大革命[14]。在微尺度下如何实现通道的功能化、智能化，将为现实情况下流体系统在生物检测、药物筛分、微颗粒制备等领域带来全新的智能平台。

NOVEMBER 1, 2005 / ANALYTICAL CHEMISTRY

WHEN MICROFLUIDIC DEVICES GO BAD

当微流控装置遇到污染问题后，我们能做些什么？

选择低表面能材料来制备微通
以提升抗污染性能

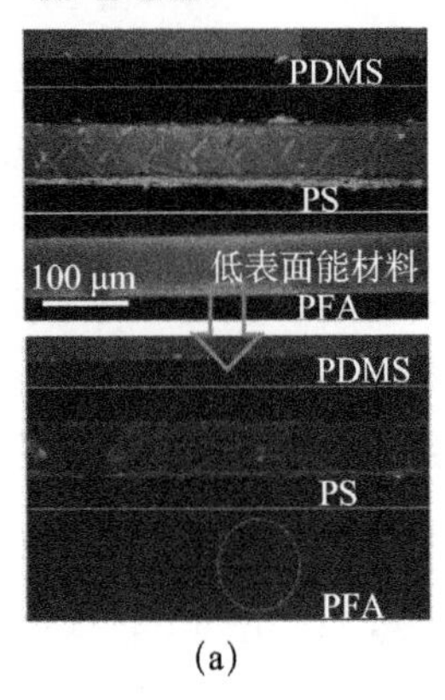

(a)

通过材料表面化学修饰改性
以提升抗污染性能

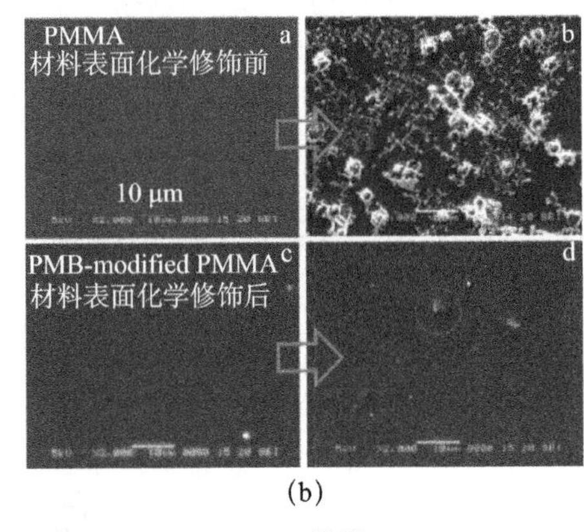

(b)

图 7-1-8　目前主要解决微流控污染问题的两种方法 [11-13]

目前对于微流控微通道的功能化、智能化方法，主要包括微通道材料的设计与选择 [1] 和材料表面的物理化学修饰（图 7-1-9[15]）。比如，设计和开发

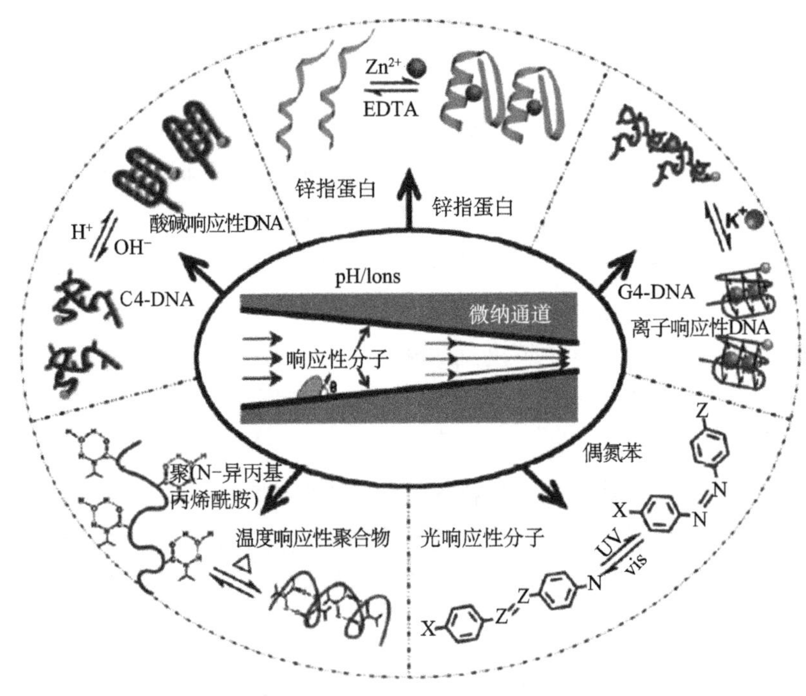

图 7-1-9　智能门控微纳米通道 [15]

修饰有智能响应性分子的微通道，不仅可以实现通道内表面智能响应外场的可控性，而且还可能通过对不同外场响应性分子的选择和设计，实现 pH、温度、特定离子等多外场协同作用的智能响应，这将为设计和开发智能响应性微流控检测器件提供一种新的思路。

“穿戴设备”被评为 2013 年的十大科技前沿之一，微流控芯片是否可以应用于可穿戴便携式医疗检测领域，目前还属于早期探索阶段，但是可以预见它将成为微流控技术与实际需求结合的重要研究方向（图 7-1-10）。如何实现柔性可拉伸的微流控系统，将在开发液态天线、柔性机器人、便携式流体装置在生物医学与环境检测等领域带来巨大的技术推动作用。国际 Infineon 公司曾发布了一种可穿戴生物芯片，它能直接织入衣服等纺织品中，可以应用于监测病患的健康情况的变化。

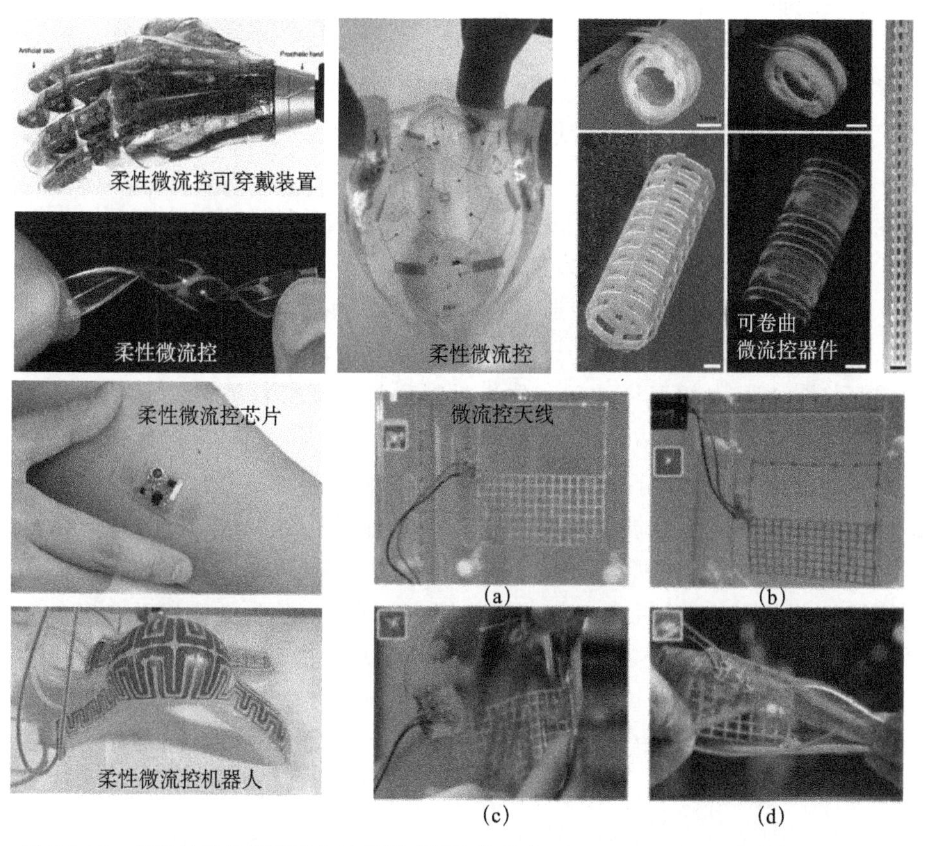

图 7-1-10 柔性材料在微流控中的应用（图片均来自：https://newatlas.com/sweat-skin-patch/46605/; https://www.technologyreview.com/s/428921/flexible-robot-comes-with camouflage/; https://www.researchgate.net/publication/227342949_Microfluidic_elect ronics/figures?lo=1）

第六节　未来5～10年学科发展趋势

综合考虑微流控技术的发展现状和与实际需求结合的重大问题分析，未来5～10年微流控学科发展趋势应该针对目前国家对软物质凝聚态物理学发展战略，建议未来在微流控的抗污染、功能化和柔性可拉伸等方面加强研究，为中国微尺度流体系统的发展奠定技术和应用基础。微流控即将迎来新一轮的高速发展时期。目前，许多新的应用正在实验室阶段开展研究，比如智能仿生材料的设计和开发、分子生物学、多孔介质的模拟、连续流动化学有机合成和微纳米颗粒的制备等。在这些新兴研究中，微流控技术将在很大程度上控制研究成本并节约研究时间。

微流控经过了初期不现实的期望阶段得到了飞速的发展，新一轮的高速发展阶段将更加关注于以它作为核心技术实现商品化应用。那么如何突破微流控应用的瓶颈问题，如微流控的抗污染、功能化以及柔性可拉伸等方面，使其具备和发挥独有的性能，不再是仅仅考虑装置的大小以及复杂程度，它应该是节能、智能、便携通用的微尺度智能高效的系统，可以说新一轮将很可能带来更大的应用市场高速发展。

如果未来能很好地解决微通道的污染和功能化问题，那么在海水淡化、油水分离处理、药物及生物分子筛分检测和复杂流体中的高效除气技术等领域都将带来巨大的经济效益。仿生材料科学是新材料领域的前沿交叉领域，它将为微流控未来5～10年的发展带来新的动力。自然界的生命体经过亿万年的进化几乎完成了智能操纵的所有过程，向自然学习物理化学的性质是智能材料和新体系发展的永恒主题[16]。正如具有自清洁功能的荷叶，具有定向吸附能力的壁虎脚，具有特殊光学性能的蝴蝶翅膀，具有流动减阻功能的鲨鱼皮和具有超滑几乎无摩擦的猪笼草叶子等，仿生微流控系统将会为微流控的设计与应用带来取之不尽的新设计研发灵感。例如，基于仿生孔道的抗污设计的仿生微流控系统[17, 18]，实现了通过毛细管稳定的流体能够可逆地将微孔密封在闭合状态，并可在压力下迅速重新配置，以生成内壁上有流体的开启微孔。因为每种运输物质都有一个不同的可以在微流控中合理调节的门控阈限压力，所以该系统能够为气-液分选从而实现全新的压强控制排气系统（图7-1-11），以及为对一个“微流体流”中的一种气-水-油三相混合物进行分离而动态地被调制，并同时使微通道具有了优异的抗污染性能（图 7-1-12），实现了构建单独一个能够选择性处理和控制复杂多相运输的抗污系统，为未来高效的污水处理和油水分离等领域提供了新的技术和方法。

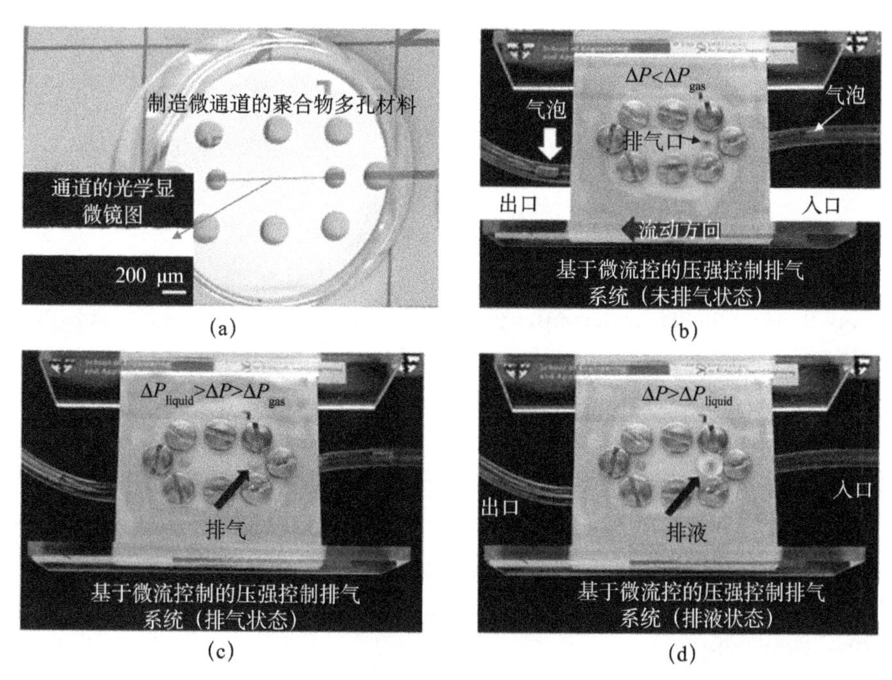

图 7-1-11　基于微流控的压强控制排气系统 [17]

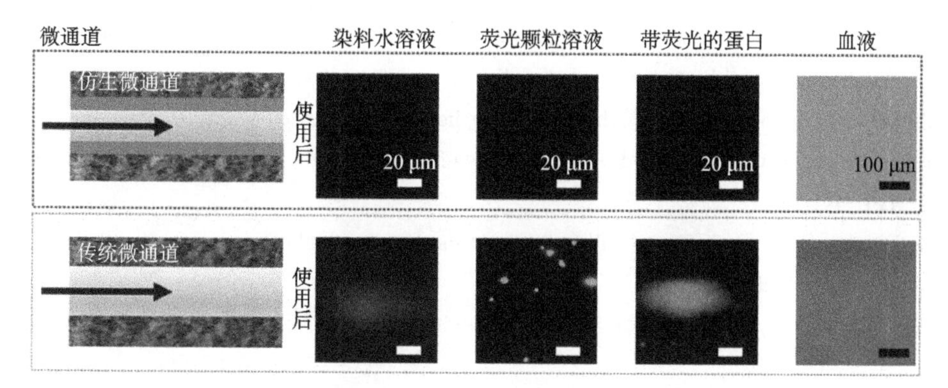

图 7-1-12　仿生微流控系统与传统微流控系统的抗污比较 [17]

未来 5～10 年微流控柔性可拉伸系统的研究和发展，将会为微流控的便携式实时检测器件的开发和应用带来巨大的机遇。例如，未来基于微流控的便携式诊断仪将可以直接分析败血症等血液疾病的炎症痕迹，从而解决发展中国家诊断设备缺少的困境。[19]

侯旭（厦门大学生物仿生与软物质研究院、物理科学与技术学院、
化学化工学院、能源材料化学协同创新中心）

参考文献

[1] 李宇杰，霍曜，李迪 . 等. 微流控技术及其应用与发展. 河北科技大学学报，2014, 35(1): 11-19.

[2] Gray R. Microfluidics review-history & key applications. the Dolomite Centre, 2015, Dolomite Workshop.

[3] Whitesides G M. The origins and the future of microfluidics. Nature, 2006, 442(7101): 368-373.

[4] Eringen A. Cemal. Simple microfluids. Int. J. Engng. Sci., 1964, 2: 189-203.

[5] Růžička J, Hansen E H. Integrated microconduits for flow injection analysis. Analytics Chimica Acta, 1984, 161: 1-25.

[6] Xia Y N, Whitesides G M. . Soft lithography. Angewandte Chemie International Edition, 1998, 37(5): 550-575.

[7] Développement Y. Microfluidic Applications in the Pharmaceutical, Life Sciences, In-Vitro Diagnostic, and Medical Device Markets. 2015. http://www. i-micronews. com/medtech-report/product/p2015-microfluidic-applications-in-the-pharmaceutical-life-sciences-in-vitro-diagnostic-and-medical-device-markets. html.

[8] Blow N. Microfluidics: in search of a killer application. Nature Methods, 2007, 4: 665-670.

[9] Mukhopadhyay R. Microfluidics: on the slope of enlightenment. Anal. Chem., 2009, 81(11): 4169-4173.

[10] Beckera H. Hype, hope and hubris: the quest for the killer application in microfluidics. Lab. Chip., 2009, 9: 2119-2122.

[11] Mukhopadhyay R. When microfluidic devices go bad. Anal. Chem., 2005, 77(21): 429A-432A.

[12] Ren K N, Dai W, Zhou J H, et al. Whole-Teflon microfluidic chips. Proceedings of the National Academy of Sciences of the United States of America, 2011, 108(20): 8162-8166.

[13] Bi H Y, et al. Construction of a biomimetic surface on microfluidic chips for biofouling resistance. Anal. Chem., 2006, 78(10): 3399-3405.

[14] 百度百科：智能材料 . http://baike. baidu. com/link?url=0GRDEj9rUqsXqyuYo5i9Syv11zluV38IGzCJXA-Kj2xTxqXWo4e9WY7zaq_xvsiWYdP3w37Io0z74SS7ERI3Yq.

[15] Wen L P, et al. Bioinspired smart gating of nanochannels toward photoelectric-conversion systems. Adv. Mater., 2010, 22(9): 1021-1024.

[16] 侯旭，江雷．仿生智能单纳米通道的研究进展．物理，2011, 40(5): 304-310.

[17] Hou X, Hu Y H, Grinthal A, et al. Liquid-based gating mechanism with tunable multiphase selectivity and antifouling behaviour. Nature, 2015, 519: 70-73.

[18] Hou X, Li J, Tesler A B, et al. Dynamic air/liquid pockets for guiding microscale flow. Nat Commun, 2018, 9(1): 733.

[19] Bose S, et al. Affinity flow fractionation of cells via transient interactions with asymmetric molecular patterns. Scientific Reports, 2013, 3: 2329.

第二章
活性物质动力学

第一节　引　　言

活性物质所研究的非平衡现象和体系在自然界中广泛存在，特别是在生命现象中扮演着重要的角色。随着物理学家对生命系统的研究日益关注，活性物质作为统计物理、软物质物理和生命科学的新兴交叉学科，近十多年取得了飞速的发展。

活性物质所包含的系统很广，大到宏观上的鸟群、鱼群、陆地上的动物群落，小至自由游动的细菌形成的细菌群落，生长迁移的上皮细胞皮层，细胞内部的细胞骨架等。这些系统的共同特点定义了活性物质：一类典型的非平衡态系统，由大量活性粒子组成，这些粒子通过将其他形式的能量转化为动能获得自我推进的能力[1]。由于粒子与粒子、粒子与媒介之间的相互作用，整个系统能够在宏观上表现出异常丰富的动力学现象（图 7-2-1）。从物理的

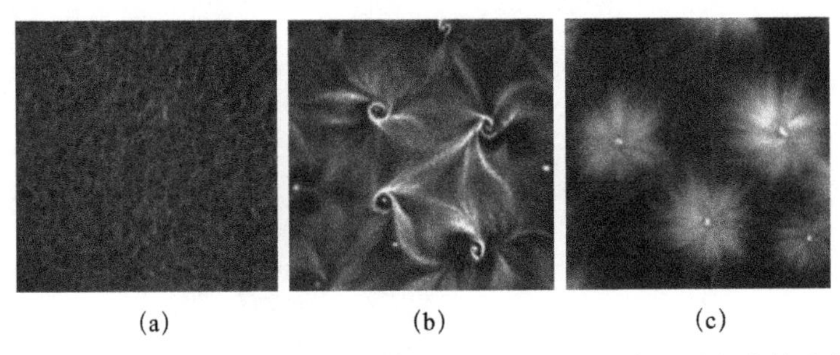

图 7-2-1　固定在基板上的肌球蛋白驱动微丝体系随着浓度增加形成的集体动力学行为
(a) 无序; (b) 螺旋形; (c) 星形

角度去解释这些生命系统中出现的现象是十分独到、新颖和有趣的，因此，吸引了越来越多的物理工作者投入这个领域的研究中来。

第二节 学科发展背景和现状

根据活性粒子所处环境的不同，活性物质可以分为干活性物质和湿活性物质两大类。在理想的干活性物质中，活性粒子所处的背景是固定表面或多孔的媒介，由于活性粒子与背景媒介之间的相互作用，活性粒子系统的动量不守恒。干活性物质的具体例子包括固定表面上运动的细菌群落[2]，陆地上的动物群[3]，还有一些人造系统，如平板上的振动颗粒[4-9]。在湿活性物质中，活性粒子处于背景流体中，活性粒子与背景流体之间有动量的交换，但两者总动量是守恒的。在湿活性物质的理论研究中，背景流体的动力学影响需要考虑在内，因此，湿活性物质的理论模型通常比干活性物质复杂。在本章，我们仅限于对干活性物质的动力学理论研究进展的介绍。事实上，在有些情况下，一些似乎并不太“干”的活性物质也可以近似当成干活性物质来处理，如高密度的在流体中的细菌群[10]和马达纤维[11]，在这类系统中，背景流体的存在通常被忽略，这是由于活性粒子受到的流体动力学作用被多体作用和随机作用所掩盖。当研究鸟群时，尽管鸟群的密度比细菌的密度要小得多，但鸟与鸟之间的长程相互作用也可以忽略[12]，取而代之是所谓的 metric-free 短程相互作用[13]，因此，仍把鸟群简化为干活性物质来处理。

对于活性物质，根据粒子本身和相互排序作用的对称性，又可以划分为极性和非极性两大类。极性活性物质由头尾可以区分的粒子（极性粒子）组成，如鸟、鱼等，每个粒子的取向由从尾指向头的箭头表示，箭头的方向同时也是粒子的运动方向，并且粒子与粒子之间的相互作用倾向于使箭头指向同一个方向，如 Vicsek 模型[14]，当系统处于有序相时，具备类似于铁磁体的指向有序性。非极性活性物质可以由头尾不可区分的非极性粒子组成（如黑色素细胞[15]），粒子之间的相互作用使得粒子的取向相互平行，或者，也可以由极性粒子组成，如自我推进的小硬杆[16-18]，但粒子之间的相互作用没有极性，只是倾向于使粒子的取向相互平行，当系统处于有序相时，具备向列型液晶的指向有序性。上述各种情形详见卡通图（图 7-2-2）。

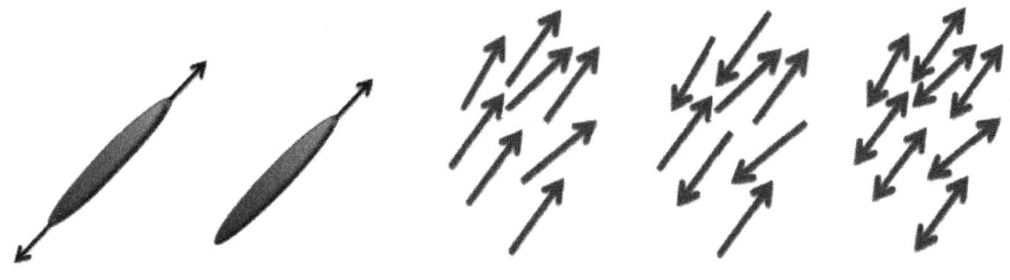

(a) 非极性粒子　(b) 极性粒子　(c) 极性粒子，极性指向有序　(d) 极性粒子，非极性指向有序　(e) 非极性粒子，非极性指向有序

图 7-2-2　活性粒子种类及粒子间相互作用情形

一、Vicsek 模型

Vicsek 模型[14]是最著名、最基本的研究活性物质集群运动行为的微观模型，虽然规则很简单，却揭示了一些非常有趣的非平衡态统计物理现象。这个模型由大量具有自我推进能力，局限于二维平面运动的点粒子组成，并采用周期性边界条件。模型中的每个粒子（体积）大小忽略、速率恒定，它在当前时刻的运动方向受上一时刻近邻的平均运动方向的影响，可以用以下的式子表示：

$$\theta_i(t+1)=\langle\theta_i(t)\rangle_{r_0}+\eta_i(t)$$

式中，θ_i 代表了第 i 个粒子的运动方向与某个参考方向之间的夹角；$\langle\ \rangle_{r_0}$ 是指对以粒子 i 为中心，r_0 为半径的圆内的其他粒子作的平均；η_i 则代表处于某个区间范围内的一个随机数。另外，每个粒子的位置随时间变化的规律为

$$\boldsymbol{r}_i(t+1)=\boldsymbol{r}_i(t)+v_0\left(\cos\theta_i(t),\sin\theta_i(t)\right)$$

由此可见，粒子与近邻的排序相互作用与随机噪声之间存在竞争的关系（图 7-2-3）。当前者处于主导地位时，粒子平均速度不为零，系统整体能够产生定向移动，处于有序相；当后者处于主导时，粒子运动方向随机，系统不能产生定向移动，处于无序相。指向有序相的存在是 Vicsek 模型所发现的一个非常重要的结果，在二维平衡态系统中，不可能形成自发对称破缺，即不能存在真长程的有序相，这就是 Mermin-Wagner 定理[19]。该定理在 Vicsek 模型中不再成立，根本原因是 Vicsek 模型模拟的系统属于非平衡态。

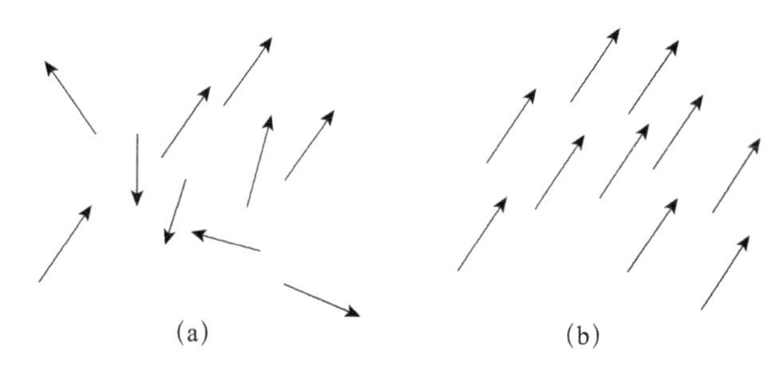

图 7-2-3　(a) 无序相，对应粒子数密度小、噪声强；(b) 有序相，对应粒子数密度大、噪声弱

Vicsek 模型中的有序-无序之间的转变在过去的几年 [20-22] 受到了很大的关注，起初被认为是连续相变，后来对大尺寸系统的计算机模拟则证实，该转变是不连续转变（一级相变）。当系统在有序相一侧接近临界点时，粒子的均匀密度分布变得不稳定，形成带状结构（图 7-2-4），这样一种新状态的出现标志了相变的不连续。类似的密度分布带状结构在实验中也有观测到 [23]。

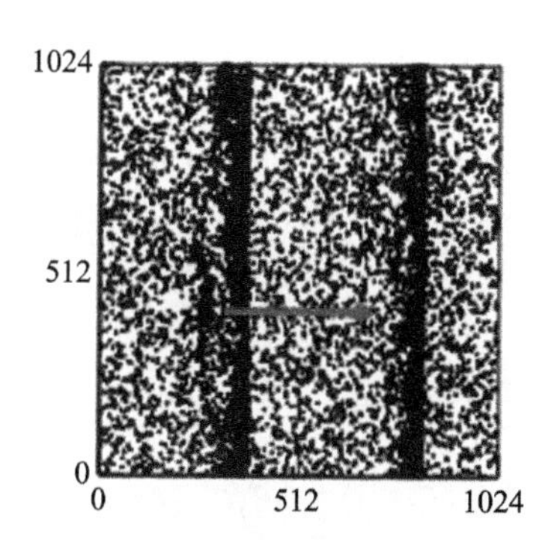

图 7-2-4　Viscek 模型在有序相一侧接近临界点处呈现出的带状结构

另外，一些基于 Vicsek 模型改进的其他微观模型也得到了相应的研究。例如，在无穷大的系统区域中，为了使粒子不会扩散开去，也为了使粒子不会过度汇聚而塌陷，引入了短程相互作用（吸引力和排斥力）[24]；又例如所谓的 metric-free 模型 [13]，在原始的 Vicsek 模型里，粒子之间的相互作用只有在指定的作用距离内才发生，因此，排序作用与局域的粒子数密度密切相关，而在 metric-free 模型中，无论粒子之间的距离远近，每个粒子总是与固定数目的最近邻粒子发生相互作用，对 metric-free 模型的计算机模拟发现，有序-无序相变类型具备连续相变的特征 [13]。

二、极性活性物质: Toner-Tu 连续体模型

1995 年，Toner 和 Tu 提出了适用于干极性活性物质的连续体理论模型[3,25,26]。他们从经典的动力学理论 Navier-Stokes 方程中获得启发，着眼于系统在大尺度上所表现出的普适的动力学规律，采用唯象的思想方法，完全基于对称性的考虑推导出系统的连续体理论模型。后来，Bertin 等直接对 Vicsek 模型进行粗粒化[20,21,27]，得到了在形式上和 Toner-Tu 理论完全相同的连续体理论模型。后者的推导方式通过玻尔兹曼方程，能把连续体理论中的参数与 Vicsek 模型的具体微观细节定量联系起来。接下来对 Toner-Tu 连续体理论模型作简单的介绍。

对于干极性活性物质，由于动量不守恒，粒子数是唯一的守恒量（假定粒子不发生分裂或死亡），另外，系统的有序性需要通过序参量（即粒子速度）来描述，因此，粗粒化的数密度场 $\rho(t,\boldsymbol{r})$ 和速度场 $v(t,\boldsymbol{r})$ 是该连续体模型的两个动力学变量，这里的 t 和 $\boldsymbol{r}$ 分别代表时间和空间的坐标。假定系统处于各向同性的环境中，基于对称性（主要是旋转不变性），该连续体模型可以用以下两个动力学方程来描述:

$$\begin{aligned}&\partial_t \boldsymbol{v}+\lambda_1(\boldsymbol{v}\cdot\nabla)\boldsymbol{v}+\lambda_2\boldsymbol{v}(\nabla\cdot\boldsymbol{v})+\lambda_3\nabla\left(\boldsymbol{v}^2\right)\\&=\alpha\boldsymbol{v}-\beta v^2\boldsymbol{v}-\nabla P+D_B\nabla(\nabla\cdot\boldsymbol{v})+D_T\nabla^2\boldsymbol{v}+D_2(\boldsymbol{v}\cdot\nabla)^2\boldsymbol{v}+\boldsymbol{f}\end{aligned} \tag{7-2-1}$$

$$\partial_t\rho+\nabla\cdot(\rho\boldsymbol{v})=0 \tag{7-2-2}$$

其中，P 代表压强，$\boldsymbol{f}$ 代表系统噪声，它的统计规律由下式表示。

$$\left\langle f_i(t,\boldsymbol{r})f_j(t',\boldsymbol{r}')\right\rangle=\Delta\delta_{ij}\delta^d(\boldsymbol{r}-\boldsymbol{r}')\delta(t-t')$$

方程（7-2-1）与 NV 方程非常类似，两者的区别主要在于: NV 方程满足伽利略不变性，而方程（7-2-1）则不满足，活性物质存在于某种媒质之中，方程（7-2-1）只在该媒质静止的参考系中才成立，因此，与 NV 方程相比，方程（7-2-1）的对称性要低，形式更复杂，能包含更多的项，如 λ_2、λ_3、α、β 项，其中后两项能够解释为什么活性物质能发生无序–有序（即整体静止–整体定向移动）的转变，而普通流体则不能。方程（7-2-2）是连续性方程，来自于粒子数守恒。从理论上讲，方程（7-2-1）和（7-2-2）中的参数的具体取值由系统的微观细节决定。

在平均场近似下，如果忽略系统的涨落，依据 α 取值的不同，方程（7-2-1）能够展示两种截然不同的稳态。为了保证方程（7-2-1）物理意义的合理性，

必须有 $\beta>0$。当 $\alpha<0$，系统达到稳态时，$v=0$，这对应系统的无序相；当 $\alpha>0$，系统达到稳态时，$|\boldsymbol{v}|=\sqrt{\alpha/\beta}$，这对应系统的有序相。显然，该平均场近似预测在临界点 $\alpha=0$ 处，两相的转变应当是连续的。然而，基于微观模型的理论计算发现[20,21]，在有序相一侧接近临界点处，系统出现了密度分布的不稳定性。基于动力学方程（7-2-1）和（7-2-2）的数值解则发现[21,28]，在不稳定区域，粒子密度分布均匀的有序相消失，被密度分布成带状的结构所代替，基于 Vicsek 模型的计算机模拟也在临界点附近位于有序相的一侧观察到了带状的结构[24,29]，如图 7-2-3 所示。这些研究结果表明，该相变应当是不连续相变。由于密度的不稳定性是造成该相变不连续的主要原因，最近提出在不可压缩的活性物质中实现该连续相变[30]。

Toner-Tu 理论的一个很重要的预测是，当活性物质处于有序相并远离临界点时，系统中存在着巨大的（与平衡态系统相比显得异乎寻常大）粒子数密度涨落。将方程（7-2-1）和（7-2-2）在稳态附近作线性展开，计算出静态结构因子，进而可以得到粒子数涨落满足的标度关系：

$$\Delta N\sim\langle N\rangle^{\alpha},\quad a=\frac{1}{2}+\frac{1}{d} \tag{7-2-3}$$

其中，$\langle N\rangle$ 为某区域的平均总粒子数；ΔN 为标准偏差；d 为系统所处空间的维数。当系统处于二维空间（即 $d=2$）时，式（7-2-3）预测 $\Delta N\sim\langle N\rangle$，这表明，当 $\langle N\rangle\to\infty$ 时，$\Delta N/\langle N\rangle$ 趋于有限值，这与平衡态系统的结果有着本质的区别；在平衡态系统中 $\Delta N\sim\sqrt{\langle N\rangle}$，当 $\langle N\rangle\to\infty$ 时，$\Delta N/\langle N\rangle$ 趋于 0。式（7-2-3）对指数 a 的预测基于对方程（7-2-1）和（7-2-2）的线性近似，运用重正化群计算把非线性效应考虑在内得到的 a 的值与线性近似的结果有明显差异[3,25,26]。活性物质中的粒子数的巨涨落现象已经在颗粒物质的振动实验中得到了证实，极性颗粒物质的振动实验测得 $2a=1.45\pm0.05$[10]，非极性颗粒物质的振动实验测得 $a\sim1.0$[7]，前者与重正化群的理论结果吻合得很好，后者与线性理论的结果非常接近。粒子数巨涨落现象也在对 Vicsek 模型的计算机模拟中观察到[29,31,32]。在细菌群落实验中（图 7-2-5）[33]，测得 a 的值明显比线性理论的预测值小，但与重正化群的理论结果很接近；不过，要彻底了解活性物质中的粒子数涨落现象还需要付出更多的努力，粒子数巨涨落原来被认为是活性物质有序相才具有的现象，但后来的一些研究工作还发现了能导致活性物质在无序相产生粒子数巨涨落的机制[34-36]。

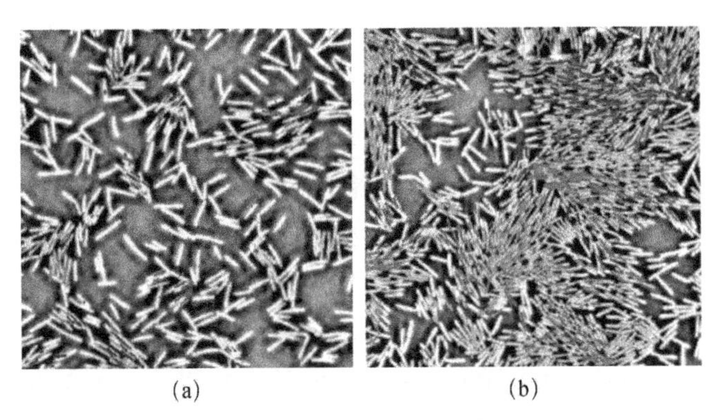

(a) (b)

图 7-2-5 游动的细菌表现出很强的极性指向有序；低密度时（a）的粒子数涨落比高密度时（b）（图中细菌个数为 718）的小；高密度时的粒子数涨落指数是 0.75，明显大于平衡态系统的涨落指数 0.5

Toner-Tu 理论的另一个很重要的预测是处于二维极性活性物质能具备真长程指向有序，这与 Vicsek 模型的计算机模拟结果一致。而与之相对应的平衡态系统——二维 XY 模型，依据 Mermin-Wagner 定理[19]，只能具备准长程指向有序。

在二维的情况下只能形成准长程的指向有序相。活性物质之所以能够突破 Mermin-Wagner 定理的限制，是因为活性物质属于非平衡态系统，这在系统的理论模型中也能够得到反映。与 XY 模型的含时金兹堡-朗道方程相比，方程（7-2-1）中多出了系数为 λ_i 的非线性项，运用重正化群对这些非线性项的影响进行定量处理，在二维的情况下可以计算出系统处于有序相时序参量关联函数的精确标度行为，最终结果表明序参量的涨落是有限的，即系统具备真长程指向有序；相比较而言，XY 模型中的序参量涨落随系统尺寸增大呈对数发散，这表明系统具备准长程指向有序。对三维活性物质的有序相，目前连续体理论还没能给出确定的结果，不过对粒子数不守恒的系统，序参量涨落的精确标度行为可以求解[37]。连续体理论对二维活性物质序参量关联函数的标度行为、声波传播模式等的预测在基于微观模型的模拟实验中得到了测试，在定量上也吻合得很好[38]。

总的来说，对粒子数密度比较高，粒子分布比较均匀的活性物质，连续体理论预测的结果在实验上和数值模拟中得到了比较好的验证。但对于粒子数密度比较小的系统，实验和数值模拟得到的结果，与连续体理论的预测可能存在比较大的差异[22]。

三、非极性活性物质：连续体理论

非极性活性物质能够形成的最简单的序是向列序。非极性活性物质由棒状的非极性粒子或极性粒子组成，粒子之间的近邻作用使得粒子长轴取向相同，形成与向列型液晶相同的序，但这近邻作用与极性活性物质中的不同，它是非极性的，它使近邻粒子的运动方向倾向于相互平行，但同向和反向的概率相同。因此，非极性活性物质的粒子平均速度为零，整体没有定向移动。从表面上看，非极性活性物质的有序相与平衡态系统的向列相似乎没有区别，但理论研究却发现这两个系统中的涨落行为有着很大的区别[4]，后来的实验研究[7]和数值模拟[31,39]也都证实了这一点。与极性活性系统相同，非极性活性物质的连续体模型可以用唯象的思想方法推导出来[40]，也可以从微观模型出发经过粗粒化得到[41-44]，接下来简单介绍该连续体模型。

处于有序相并远离临界点时，系统的动力学变量是粒子数密度 $c(\boldsymbol{r},t)$ 和序参量。动量不是动力学变量，因为对于干活性系统它不守恒。与极性系统不同，向列相的序参量是一个张量，对于单轴的 d 维系统，该张量用矩阵表示为

$$Q=\left[\hat{n}\hat{n}-\left(1/d\right)\right]S$$

其中，$\hat{\boldsymbol{n}}$ 是单位矢量，代表局部粒子运动的取向，S 代表系统有序的程度，它的值越大，有序性越高。首先，由粒子数守恒得到

$$\frac{\partial c}{\partial t}=-\nabla\cdot\boldsymbol{j}=-\nabla\cdot\left(c\boldsymbol{v}\right) \tag{7-2-4}$$

其中，粒子流密度 $\boldsymbol{j}=c\boldsymbol{v}$，$\boldsymbol{v}$ 代表局部粒子速度。对动量密度 $m\boldsymbol{j}$ 应用牛顿第二定律：

$$m\frac{\partial \boldsymbol{j}}{\partial t}=-\Gamma\boldsymbol{v}-\nabla\cdot\boldsymbol{\sigma}+\boldsymbol{f}_R=-\Gamma\boldsymbol{v}-w_0\nabla c-w_1\nabla\cdot\left(Qc\right)+\boldsymbol{f}_R \tag{7-2-5}$$

其中，Γ 项代表活性物质与固定表面或媒介之间的摩擦作用所带来的动量损耗；$\boldsymbol{\sigma}$ 是应力张量，包含了粒子之间的相互作用；w_0 项类似于压强的效果；w_1 项是活性物质所特有的，它表明序参量分布的不均匀性导致局部的极性，从而引发粒子的流动（图 7-2-6）；$\boldsymbol{f}_R$ 是随机高斯白噪声。

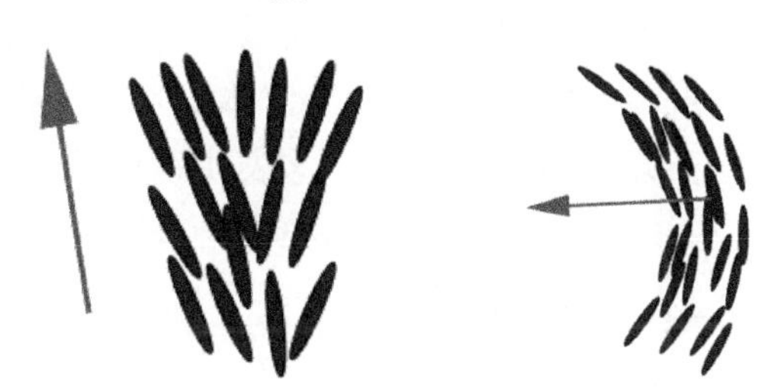

图 7-2-6　非极性系统中活性粒子取向的弯曲导致局部显极性，从而引发粒子流

因为系统处于远离平衡点的有序相，序参量的大小（即 S）可以近似为常量，因此序参量的涨落主要来自于粒子运动取向（即$\hat{\boldsymbol{n}}$）的变化。定义系统的平均取向为 $\hat{z}$，将$\hat{\boldsymbol{n}}$关于 $\hat{z}$ 展开，记作$\hat{\boldsymbol{n}}=\hat{z}+\delta\boldsymbol{n}$，由于$\hat{\boldsymbol{n}}$是单位矢量，$\delta n_z \sim \left|\delta\boldsymbol{n}_\perp\right|^2$，这里“⊥”代表垂直于 $\hat{z}$ 的方向，因此，$\hat{\boldsymbol{n}}$的涨落取决于 $\delta\boldsymbol{n}_\perp$。通过简单的推导可以证明，活性系统与平衡态向列型液晶的 $\delta\boldsymbol{n}_\perp$满足的动力学方程形式相同，均由下式给出：

$$\partial\delta\boldsymbol{n}_\perp=\lambda_+\partial_z\boldsymbol{v}_\perp+\lambda_-\nabla_\perp\boldsymbol{v}_z+K_1\nabla_\perp(\nabla_\perp\cdot\delta\boldsymbol{n}_\perp)+K_2\nabla^2{}_\perp\delta\boldsymbol{n}_\perp+K_3\partial^2_z\delta\boldsymbol{n}_\perp+\boldsymbol{f}_\perp \quad (7\text{-}2\text{-}6)$$

其中，$\lambda_\pm=(\lambda\pm1)/2$；$\lambda$ 是所谓的“流动排序参数”；随机噪声$\boldsymbol{f}_\perp$的统计规律可以表示为

$$\left\langle f_{\perp i}(\boldsymbol{r},t)f_{\perp j}(\boldsymbol{r}',t')\right\rangle=\delta_{ij}\Delta_n\delta^d(\boldsymbol{r}-\boldsymbol{r}')\delta(t-t')$$

在长时间尺度极限下，方程（7-2-5）中的 $\partial\boldsymbol{j}/\partial t$ 可以忽略，求解 $\boldsymbol{v}$，把它写成 $\boldsymbol{c}$ 和 $\delta\boldsymbol{n}_\perp$的表达式，代入方程（7-2-5）和（7-2-6）中，同时将粒子数密度在平均值附近展开为 $c(\boldsymbol{r},t)=c_0+\delta c(\boldsymbol{r},t)$，如果仅保留线性项，将得到下面两个动力学方程：

$$\frac{\delta c}{\partial t}=\left(D_z\partial_z^2+D_\perp\nabla_\perp^2\right)\delta c+2c_0\alpha\partial_z\left(\nabla_\perp\cdot\delta\boldsymbol{n}_\perp\right)+\nabla\cdot\boldsymbol{f}_c \quad (7\text{-}2\text{-}7)$$

$$\frac{\partial\delta\boldsymbol{n}_\perp}{\partial t}=\left(K_z\partial_z^2+K_\perp\nabla_\perp^2+K_L'\nabla_\perp\nabla_\perp\cdot\right)\delta\boldsymbol{n}_\perp+D_{cn}\partial_z\nabla_\perp\delta c+\boldsymbol{f}_\perp \quad (7\text{-}2\text{-}8)$$

上面两式中的$\boldsymbol{f}_c$和各系数由方程（7-2-5）和（7-2-6）中的系数决定。要得到完整的理论模型，方程（7-2-7）和（7-2-8）中还应包含非线性项。对于平衡态的向列型液晶，方程（7-2-8）中的非线性项不会定性改变线性理论的结果，这意味着二维的向列型液晶具备准长程的指向有序性，遵循 Mermin-Wagnar 定理[19]。对于非极性活性系统，方程（7-2-8）中包含更多的非线性项，它们能否与极性活性物质的理论模型中的非线性项一样，定性改变线性理论的结果，使系统获得长程指向有序？这个问题目前还没有定论。

基于方程（7-2-7）和（7-2-8）可以预测系统的粒子数涨落情况。计算结果发现，对于二维的非极性活性物质，给定体积内粒子数的标准偏差正比于该体积内的平均粒子数，即 $\Delta N \backsim \langle N\rangle$，这表明系统呈现粒子数巨涨落现象。这么巨大的涨落应该可以在生命系统的实验中观测到，例如，二维的非极性活性物质可以通过将高浓度的黑素细胞悬浮液散播到某个平面衬底上形成[45]，不过到目前为止，在这种系统里测量粒子数涨落的尝试还没有进行过。在单层头尾对称的微米尺寸的小铜线振动实验中，观测到不同时间的数

密度关联函数随时间间隔呈对数衰减[7]，强力地支持了粒子数巨涨落的预测，不过这个实验结果也有争议[46,47]。

第三节 前沿问题

一、活性物质中的物相类型

截至目前，对活性物质有序相的研究主要集中在铁磁相和向列相，已经发现了很多不同于平衡态系统的奇特性质。从对称性的角度出发，所有平衡态系统中存在的物相类型，如各种液晶和晶体，都可能在活性物质中出现，因此，过去研究各种平衡态物相所关注的科学问题同样适用于活性物质，由此带来的研究课题将是非常丰富的。相比实验而言，这方面的解析理论研究和数值模拟更加容易开展，最近，活性层列型液晶和活性晶体的解析理论研究和数值模拟已经取得了一些进展[48-50]。

二、活性物质中的相变

活性物质中有可能存在类似于平衡态系统中的物相类型，这些相之间的转变属于非平衡态相变，它们的性质值得研究和探索。截至目前，对极性系统中的铁磁相变研究得最多，这应该是活性系统中最简单的相变。计算机模拟研究指出该相变属于不连续相变，这个结论已被广泛接受（可能还存在小范围的争议），不过关于这个相变的解析理论，如标准的重正化群理论分析，现在还没有完成。

三、团簇，相分离

团簇和相分离，以及与此相关的粒子数巨涨落，是活性物质中普遍存在的现象。起初粒子数巨涨落被认为是有序相才具有的现象，是与系统发生自发对称破缺紧密相关的。后来发现团簇和相分离（这当然意味着巨大的粒子数涨落）在一些无序的活性物质中也能出现，它们的产生似乎和非平衡态系统的细致平衡破缺有关。要彻底搞清楚活性物质中的团簇和相分离现象还需要更多的努力。

四、边界效应

边界和障碍物对活性物质的集群运动行为有很强的影响。实验和理论都证实在不施加外力的情况下，只通过改变边界或衬底的对称性就能实现对活性粒子集群运动的调控和引领，因此，清楚了解活性粒子与障碍物或被动粒子（passive particles）之间的相互作用机制有助于制造出由微生物驱动的微型机器。另外，如何将连续体理论适用于边界效应强的系统也是一个未知的难题。

五、活性物质的动力学和流变行为

除了研究活性物质的大尺度的集体动力学行为，考察体系中单个活性粒子的动力学行为也是至关重要的。由于活性驱动力和活性噪声的影响，粒子的迁移和扩散行为、粒子所经历的耗散和涨落将显著地不同于对应的被动系统。研究活性粒子的这些动力学特征，以及探讨非平衡态物理恒等式（如涨落耗散关系、涨落定理）在活性系统中的适用性，将有助于建立描述活性物质的非平衡统计物理方法。此外，对于活性物质溶液，特别是活性胶体溶液，自驱动的特点将显著地改变系统的流变行为。研究活性物质溶液的流变行为对系统特征参量（如浓度、活性能力等）的依赖性，是一个重要的基础问题。

六、活性物质对外场的相应

研究活性物质对外场的响应行为对理解活性物质本身是至关重要的。这样的外场包括力场、浓度场、温度场、流场等。在近平衡的系统中，线性响应理论为我们提供了一个理想的理论框架；然而活性物质是一个远离平衡的体系，因此如何建立一个适用于活性物质的响应理论就成为一个重要的开放的非平衡物理问题。更为重要的是为了实现活性物质的实际应用，利用外场来调控其非平衡的结构和动力学行为是最关键的一步。

七、活性物质的实际应用

活性物质不仅为我们提供了一个新颖的物理体系来研究丰富的非平衡态物理问题，活性物质本身也有着巨大的实际应用价值。比如，微观的活性粒子能够被用来进行药物输送，作为微型马达来驱动微观器件，在微流芯片中执行指定的任务。此外，活性物质不断地把环境的能量转化为机械功，因此构成了一个新型的能量转化和回收媒介。可以期待，通过自组装的途径，大量的活性粒子可以形成宏观的新型智能材料，甚至组装成具有特定功能的微型机器人。

第四节 未来5～10年重点发展方向

一、进一步发展连续体理论

在过去二十年的活性物质的理论研究中连续体理论取得了很大的成功，在未来相当长的一段时间内，它仍将是非常有效的理论手段，应当进一步发挥它在研究活性物质集群运动、涨落、相变临界现象等方面的独特作用。

二、活性物质体系的模拟和理论分析的结合研究

探索一系列有效的非平衡理论和技术手段来描述和刻画自驱动系统的非平衡态的大尺度动力学。通过动态蒙特卡罗、布朗动力学、分子动力学、动力学平均场理论以及流体动力学理论相互结合的研究方法，来发展一套可靠的行之有效的分析大尺度非平衡动力学的手段。根据微观模型建立相应的运动学方程理论，从而使理论和模拟可以紧密结合。在此基础上，如何通过稳定性分析理论、流场分析技术等手段，系统地确立分析体系的动力学特性，预言体系可能呈现的新的大尺度动力学结构，为实验验证提供可靠的理论结果。

三、对理论预言的实验验证

活性物质的理论研究已经取得了长足的发展，在若干预言上出现了争议。检验和甄别这些理论预言已经成为学科进一步发展的必由之路。为解决这一问题，实验工作者需进一步开发新的实验系统和测量手段，并与理论工作者保持紧密的合作。

四、利用先进的介观尺度的流体模拟方法来研究活性胶体溶液

活性胶体溶液是一类最有代表性的活性物质，有着最为直接和广泛的应用前景。其组成单元具有微观的空间尺度，包括细菌溶液和合成的自泳活性胶体等。活性粒子之间有着十分丰富的相互作用，如体积排除作用，流体力

学作用，泳效应等。现有的理论和模拟方法往往过于简化，不能满意地处理这些复杂的粒子间的相互作用。介观模拟方法能够恰当地处理多粒子之间的流体力学作用和泳效应，得到的模拟结果将有助于澄清活性胶体复杂现象背后的微观机制，进而为发展更加合理的理论模型和执行进一步的实验研究提供依据和指导。

五、开发活性物质的实际应用

活性物质远离热力学平衡态，表现出十分奇特的相行为和材料特性，对它的研究开发有望发现具有优越特性的新材料。例如，近来的实验已经展示活性胶体可以在微通道内靶向地输运货物，固定的活性粒子能够作为一个微流泵来驱动流体，转动的活性粒子可以作为一个微观马达从非平衡的环境中获取能量。这些研究表明活性胶体在微流芯片中有着巨大的潜在应用价值。

第五节　结　　语

作为软凝聚态物理的一个新兴研究方向，活性物质的研究方兴未艾。它开辟了从统计物理学的角度去研究生命体系和相关的复杂多体非平衡体系的新方向，这不仅能给物理知识带来极大的创新，同时在开发智能材料、药物输运等方面也有巨大的应用价值。在过去十多年中，活性物质的研究虽然在理论上和实验上都取得了很大的成功，但正如本章前面所述，仍然遗留了很多重要的问题亟待解决，所以充满了机遇和挑战。目前我国的活性物质研究规模尚小，但从发展趋势看，很有发展潜力，希望有更多的物理工作者能加入这个充满活力的新兴研究方向。

由于作者知识有限，在“学科发展背景和现状”的介绍局限于干活性物质的动力学理论发展，而这仅仅是活性物质研究进展的“冰山一角”。在“前沿问题”和“未来5～10年重点发展方向”这两部分的书写过程中征求了同行们的意见，涵盖的范围相对要广一些，但肯定还有遗漏的部分。任何一位对活性物质感兴趣的研究者都可以列出一些自己认为重要的前沿问题，这些问题很可能比本章列举的更有趣、更有科学意义，毕竟活性物质包含的研究内容是如此丰富、新颖和有趣。

陈雷鸣（中国矿业大学）

参考文献

[1] Schweitzer F. Brownian Agents and Active Particles: Collective Dynamics in the Natural and Social Sciences. Berlin: Springer, 2003.

[2] Wolgemuth C, Hoiczyk E, Kaiser D, et al. How myxobacteria glide. Curr. Biol., 2002, 12: 369.

[3] Toner J, Tu Y. Flocks, herds, and schools: a quantitative theory of flocking. Phys. Rev. E, 1998, 58: 4828.

[4] Ramaswamy S, Simha R A, Toner J. Active nematics on a substrate: giant number fluctuations and long-time tails. Europhys. Lett., 2003, 62: 196.

[5] Yamada D, Hondou T, Sano M. Coherent dynamics of an asymmetric particle in a vertically vibrating bed. Phys. Rev. E, 2003, 67: 040301.

[6] Aranson I S, Tsimring L S. Theory of self-assembly of microtubules and motors. Phys. Rev. E, 2006, 74: 031915.

[7] Narayan V, Ramaswamy S, Menon N. Long-lived giant number fluctuations in a swarming granular nematic. Science, 2007, 317: 105.

[8] Kudrolli A, Lumay G, Volfson D, et al. Swarming and swirling in self-propelled polar granular rod. Phys. Rev. Lett., 2008, 100: 058001.

[9] Deseigne J, Dauchot O, Chate H. Collective motion of vibrated polar disks. Phys. Rev. Lett., 2010, 105: 098001.

[10] Drescher K, Dunkel J, Cisneros L H, et al. Fluid dynamics and noise in bacterial cell-cell and cell-surface scattering. Proc. Natl. Acad. Sci. USA, 2011, 108: 10940.

[11] Liverpool T B. Anomalous fluctuations of active polar filament. Phys. Rev. E, 2003, 67: 031909.

[12] Ballerini M, et al. Interaction ruling animal collective behavior depends on topological rather than metric distance: Evidence from a field study. Proc. Natl. Acad. Sci. USA, 2008, 105: 1232.

[13] Ginelli F, Chate H. Relevance of metric-free interactions in flocking phenomena. Phys. Rev. Lett., 2010, 105: 168103.

[14] Vicsek T, Czirok A, Ben-Jacob E, et al. Novel type of phase transition in a system of self-driven particles. Phys. Rev. Lett., 1995, 75: 1226.

[15] Gruler H, Dewald U, Eberhardt M. Nematic liquid crystals formed by living amoeboid cells.

Eur. Phys. J. B, 1999, 11: 187.

[16] Peruani F, Deutsch A, Bar M. Nonequilibrium clustering of self-propelled rod. Phys. Rev. E, 2006, 74: 030904.

[17] Baskaran A, Marchetti M C. Hydrodynamics of self-propelled hard rods. Phys. Rev. E, 2008, 77: 011920.

[18] Yang Y, Marceau V, Gompper G. Swarm behavior of self-propelled rods and swimming flagella. Phys. Rev. E, 2010, 82: 031904.

[19] Mermin N D, Wagner H. Absence of ferromagnetism or antiferromagnetism in one- or two-dimensional isotropic heisenberg models. Phys. Rev. Lett., 1966, 17: 1133.

[20] Bertin E, Droz M, Gregoire G. Boltzmann and hydrodynamic description for self-propelled particles. Phys. Rev. E, 2006, 74: 022101.

[21] Bertin E, Droz M, Gregoire G. Hydrodynamic equations for self-propelled particles: microscopic derivation and stability analysis. J. Phys. A, 2009, 42: 445001.

[22] Gregoire G, Chate H. Onset of collective and cohesive motion. Phys. Rev. Lett., 2004, 92: 025702.

[23] Schaller V, Weber C, Semmrich C, et al. Polar patterns of driven filaments. Nature, 2010, 476(7311): 73-77.

[24] Grégoire G, Chaté H. Onset of collective and cohesive motion. Phys. Rev. Lett., 2004, 92: 025702.

[25] Toner J, Tu Y. Long-range order in a two-dimensional dynamical XY model: How birds fly together. Phys. Rev. Lett., 1995, 75: 4326.

[26] Toner J, Tu Y, Ramaswamy S. Hydrodynamics and phases of flocks. Ann. Phys. (Amsterdam), 2005, 318: 170.

[27] Ihle T. Kinetic theory of flocking: Derivation of hydrodynamic equations. Phys. Rev. E, 2011, 83: 030901.

[28] Mishra S, Baskaran A, Marchetti M C. Fluctuations and pattern formation in self-propelled particles. Phys. Rev. E, 2010, 81: 061916.

[29] Chate H, Ginelli F, Gregoire G, et al. Collective motion of self-propelled particles interacting without cohesion. Phys. Rev. E, 2008, 77: 046113.

[30] Chen L, Toner J, Lee C F. Critical phenomenon of the order–disorder transition in incompressible active fluids. New Journal of Physics, 2015, 17: 042002.

[31] Chate H, Ginelli F, Montagne R. Simple model for active nematics: quasi-long-range order and giant fluctuations. Phys. Rev. Lett., 2006, 96: 180602.

[32] Chate H, Ginelli F, Gregoire G, et al. Modeling collective motion: variations on the Vicsek model. Eur. Phys. J. B, 2008, 64: 451.

[33] Zhang H P, Beer A, Florin E L, et al. Collective motion and density fluctuations in bacterial colonies. Proc. Natl. Acad. Sci. USA, 2010, 107: 13 626.

[34] Tailleur J, Cates M E. Statistical mechanics of interacting run-and-tumble bacteria. Phys. Rev. Lett., 2008, 100: 218103.

[35] Cates M E. Diffusive transport without detailed balance in motile bacteria: does microbiology need statistical physics? Rep. Prog. Phys., 2012, 75: 042601.

[36] Fily Y, Baskaran A , Marchetti M C. Cooperative self-propulsion of active and passive rotors. Soft Matter, 2012, 8: 3002.

[37] Toner J. Birth, death, and flight: a theory of malthusian flocks. Phys. Rev. Lett., 2012, 108: 088102.

[38] Tu Y, Ulm M, Toner J. Sound waves and the absence of galilean invariance in flocks. Phys. Rev. Lett., 1998, 80: 4819.

[39] Mishra S, Ramaswamy S. Active nematics are intrinsically phase separated. Phys. Rev. Lett., 2006, 97: 090602.

[40] Simha R A, Ramaswamy S. Hydrodynamic fluctuations and instabilities in ordered suspensions of self-propelled particles. Phys. Rev. Lett., 2002, 89: 058101.

[41] Ahmadi A, Marchetti M C, Liverpool T B. Hydrodynamics of isotropic and liquid crystalline active polymer solutions. Phys. Rev. E, 2006, 74: 061913.

[42] Mishra S. Bangalore ：Indian Institute of Science, 2009.

[43] Shi X, Chate H, Ma Y. Instabilities and chaos in a kinetic equation for active nematicsar. New J. Phys., 2014, 16: 035003.

[44] Bertin E, Chate H, Ginelli F, et al. Mesoscopic theory for fluctuating active nematics. New J. Phys., 2013, 15: 085032.

[45] Gruler H, Dewald U, Eberhardt M. Nematic liquid crystals formed by living amoeboid cells. Eur. Phys. J. B, 1999, 11: 187.

[46] Aranson I S, Snezhko A, Olafsen J S, et al. Comment on long-lived giant number fluctuations in a swarming granular nematic. Science, 2008, 320: 612.

[47] Narayan V, Ramaswamy S, Menon N. Response to comment on long-lived giant number fluctuations in a swarming granular nematic. Science, 2008, 320: 612.

[48] Chen L, Toner J. Universality for moving stripes: a Hydrodynamic theory of polar active smectics. Phys. Rev. Lett., 2013, 111: 088701.

[49] Adhyapak T C, Ramaswamy S, Toner J. Live soap: order, fluctuations and instabilities in active smectics. Phys. Rev. Lett., 2013, 110: 118102.

[50] Romanczuk P, Chaté, H, Chen L M,et al. Emergent smectic order in simple active particle models. New J. Phys., 2017, 18(6): 1-34.

第三章 智能软聚合物及其应用

本章所介绍的智能软聚合物主要包括形状记忆聚合物（shape memory polymer, SMP）和电活性聚合物（electroactive polymer, EAP）两大类。形状记忆聚合物能在外界环境（如温度、光、电、磁、溶液等）变化的刺激下实现形状的记忆和回复。介电弹性体（dielectric elastomer, DE）是电活性聚合物的一种，它能够在外加电场下改变形状。两类智能软聚合物材料以其独有的特性，在智能仿生、航空航天、生物医学等领域都有广泛的应用潜力，已成为近年来智能材料与结构领域的研究热点之一。

本章对形状记忆聚合物和介电弹性体的研究和发展进行了回顾，重点介绍了国内外学者对其理论研究的进展，同时阐述了两类智能软聚合物材料在航空航天、生物医学、智能模具等领域的应用，最后对其相关研究中所存在的问题与发展前景进行了展望。

第一节 引　　言

长期以来，研究人员多致力于在工程领域广泛应用的硬质材料研究，然而，自然界中动、植物生物体的机械结构通常是软质材料。与传统的金属、陶瓷等“硬质材料”相比，高分子聚合物是“软质材料”的典型代表[1]。软质材料针对外界的刺激（如机械力、温度、电磁场、热场和化学场等）能够产生不同程度的形变而体现出活性。智能软质材料是一种能感知外部刺激、进行判断并适当处理本身可执行的新型功能材料，相比传统硬质智能材料具有可承受大变形、良好生物亲和性、轻质廉价等特点。智能软材料及应用器件有巨大的潜在应用前景，大到石油开采密封，小到药物输送，在航空航天、智能仿生、机械、生物、军工等领域都有广泛的应用潜力。智能软材料由于其变形能力和驱动力接近于生物肌肉，在机器人领域也有非常好的发展前景。智能软质材料在模拟生物特征方面展现出了无与伦比的特性，因此也被称为活性软质材料。

《国家中长期科学和技术发展规划纲要（2006—2020 年）》在“五、前沿技术”中指出:“智能材料与智能结构是集传感、控制、驱动（执行）等功能于一体的机敏或智能结构系统。”智能材料是继天然材料、合成高分子材料、人工设计材料之后出现的新材料，其构想来源于仿生学，目标在于研制具有类似于生物各种功能的“活”材料，具有感知功能、驱动功能、反馈功能、控制功能和自修复等功能，可实现结构功能化、功能多样化[2-5]。当前，研究应用较为广泛的智能材料主要有形状记忆聚合物（shape memory polymer, SMP）、电活性聚合物（electroactive polymers）、压电材料（piezoelectricity materials）、形状记忆合金（shape memory alloys）、电磁流变流体（electro/magneto rheological fluids）和胶体（gels）等。每一种智能材料都各具其特点，有着广阔的应用前景和发展潜力[2-3,6-7]。本章主要介绍形状记忆聚合物和电活性聚合物两大类智能软聚合物。

形状记忆聚合物作为一种新型智能聚合物材料，能在外界环境条件（如温度、光、电、磁、溶液等）变化的刺激下，实现材料和结构的形状记忆和回复特性[7-11]。由发现至今，国内外学者发展了多种不同形状记忆聚合物材料以满足应用需求，包括环氧、苯乙烯、氰酸酯、聚酰亚胺、聚乙烯、聚苯乙烯、聚丙烯酰胺等[12-19]。形状记忆聚合物作为一种有机高分子材料，刚度远低于形状记忆合金及形状记忆陶瓷材料。因此，通过将各种纤维及颗粒作为增强体掺杂到聚合物材料中，制备出形状记忆聚合物复合材料（shape memory polymer composite, SMPC），可提高材料刚度和恢复力[7-8,20-21]。通过特定的热力学循环过程，形状记忆聚合物及其复合材料可以实现形状记忆和恢复功能。形状记忆聚合物及其复合材料已经被应用于航空航天、生物医学、仿生学、纺织和结构工程等领域[22-25]。

介电弹性体是典型的 EAP 材料，由美国斯坦福研究院 20 世纪 90 年代开始研究。研究发现，在外加电场下，可以产生大变形，具有高弹性能密度、超短反应时间、高效率等优点，被广泛应用于人工肌肉、面部表情、驱动器、能量收集装置、传感装置、机器人和盲文显示装置等各个方面，在航空航天、智能仿生等领域具有巨大的应用潜力[1-26]。虽然介电弹性体在近几年来迅猛发展，但是现在多数还只限于实验室应用，包括易失效破坏、驱动力较低和电致变形达不到实际应用的要求等问题制约了其发展，因此商业应用仍然面临着巨大挑战。因此，迫切需要深入研究介电弹性体及其结构的力学特性，进而对其材料和器件的设计进行指导和优化。另外，由于介电弹性体及其复合材料或结构在力学上具有几何大变形与材料非线性的特征，而且在响应机制和工作环境上具有多物理场耦合（机械力场、电场和热场等）的特点，因此对于这类材料与结构的力学问题的研究具有十分重要的价值。

第二节 发展历程

一、形状记忆聚合物及其复合材料

形状记忆聚合物是通过感知外界环境变化而产生主动变形的一种新型智能聚合物材料，其在外界环境条件刺激下产生变形、回复的行为称为形状记忆效应（shape memory effect，SME）[2-5]。依据其形状记忆机制的差异，可分为固态形状记忆高分子材料和凝胶体系形状记忆高分子两大类；依据其激励条件的不同，可分为热敏型、磁敏型、光敏型等多种类型，目前研究较多、较成熟的为热敏型形状记忆聚合物[26-32]。

通过对形状记忆聚合物的结构分析可知，形状记忆聚合物的形状记忆效应主要由于材料内部的两相结构：保持宏观原始形状的固定相和可逆软化硬化的可逆相[12,33,34]。其中，固定相能够保证材料的宏观形状及刚度，可逆相能够保证材料在变形记忆状态过程中的大变形及保持临时形状。当材料受外部激励和载荷作用时，其内部分子链的取向和交联点将发生平移；撤去外部激励并保持载荷，可使这些重新取向的分子链段产生二次交联，直到分子链段的微布朗运动冻结，材料硬化成型；当材料再次受激励时，其内部可逆相软化，分子链段的二次交联被解除，宏观表现出形状回复。

以热致型形状记忆聚合物为例，其固定相的结晶熔融温度 T_m 或玻璃化转变温度 T_g 较高，在材料的使用温度范围内不会发生软化和松弛，可保证材料原始形状的记忆与回复;可逆相的 T_m 或 T_g 较低，能随温度变化相应发生软化、硬化，保证材料具有较高的变形能力。图 7-3-1 为形状记忆聚合物在变形和回复阶段的网络结构的变化示意图。其中，可逆相为长的线条，代表聚合物中的分子链段；固定相为黑色点，代表将分子链段连接在一起的网络交联点。当材料温度低于形状记忆聚合物的玻璃化转变温度 T_g 时，材料可逆相和固定相均处于冻结状态；当温度高于玻璃化转变温度 T_g 以上时，可逆相的微观布朗运动加剧，材料发生软化，在外载作用下可逆相由卷曲缠绕状态变为较为伸展的有序状态，保持外载作用将材料冷却至玻璃化转变温度 T_g 下，此时可逆相处于玻璃状态，材料保持变形形状。当温度再次升高至玻璃化转变温度 T_g 以上时，可逆相再次由玻璃状态逐渐过渡为活动状态，并在固定相的作用下重新回复到初始的卷曲缠绕状态。材料在宏观上表现为软化并发生形状回复[33-36]。

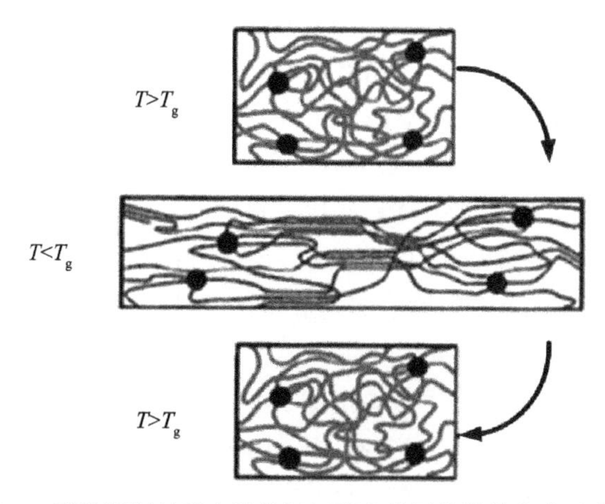

图 7-3-1 形状回复过程中形状记忆聚合物网络结构变化示意图[36]

（一）形状记忆聚合物

自 20 世纪 70 年代发现交联聚乙烯具有形状记忆效应以来，各种类型的形状记忆聚合物的开发和不同的驱动方式的探索一直是研究人员研究的重点，同时对形状记忆聚合物应用的研究也在不断深入。1984 年法国 CdF-Chimie 公司（现在的 ORKEM 公司）成功开发了聚降冰片烯型形状记忆聚合物，首先由乙烯和环戊二烯在 Dies-Aldeer 催化条件下合成降冰片烯，然后再开环聚合成双键和五元环交替键合的无定型聚合物；1988 年日本可乐丽公司开发了反式聚异戊二烯形状记忆聚合物；同年，日本旭化成公司开发了苯乙烯-丁二烯共聚形状记忆聚合物，其固定相为聚苯乙烯（T_g 为 120℃）可逆相为低熔点的聚丁二烯（T_m 为 60℃），而三菱重工业公司开发了第一例聚氨酯形状记忆聚合物。

现阶段形状记忆聚合物的研究主要集中于构建分子链段，发展新的形状记忆聚合物材料及材料的多功能开发，而对于微观结构在形状记忆过程中的动态分析及机制研究较少。形状记忆聚合物的研究主要包括热塑性形状记忆聚合物和热固性形状记忆聚合物两种，其中热塑性形状记忆聚合物的研究开展得较早，主要集中在热塑性聚氨酯材料的研究。例如，Jeong，Tobushi 及 Kim 等均以热塑性聚氨酯作为研究对象，系统研究了聚氨酯软硬段结构、组成比例、分子量、含量变化对性能的影响[37-44]。Jang 等研究了表面改性的二氧化硅微粒交联的形状记忆聚氨酯，复合材料显示出超过 99% 的形状固定率和超过四个循环的形状回复[45]。Xie 等研究了多形状记忆效应的形状记忆聚合物[11]。热塑性形状记忆聚合物材料具有加工方便的优势，因此对其研究仍有很大的发展空间。虽然如此，热塑性形状记忆聚合物材料依靠分子链缠结点作为形变的固定点，在实际应用中存在机械强度不高、记忆及回复精度不大的问题。

与热塑性形状记忆聚合物材料相比，热固性形状记忆聚合物能够大幅度提高材料的机械强度。Rule 等以结晶性 PCL 作为软段合成形状记忆聚氨酯，该材料具有较高的形状回复率、良好的记忆特性和生物相容性等特点[46]。Lendlein 等研究了多种具有生物降解性的交联型聚己内酯、聚丙交酯和聚氨酯等形状记忆聚合物材料[4,33,47]；Gall 等研究了交联型聚丙烯酸酯类形状记忆聚合物材料[48-51]。Leng 等研究了热性能和机械性能优良的苯乙烯、环氧和氰酸酯等一系列热固性形状记忆聚合物[52-55]。热固性形状记忆聚合物材料具有机械强度高、回复速率好以及回复驱动力大等较好的形状记忆性能。

（二）形状记忆聚合物复合材料

形状记忆聚合物具有可回复应变较大的特点，一般能达到 10%～100%[56]，但材料的模量、强度等力学和热-力学性能较差，导致材料变形回复输出力较小，运动稳定性和可靠性较差，蠕变和松弛现象较严重等缺点，影响了其应用，特别是在航天器上的应用[8,23]。通过形状记忆聚合物与其他增强材料（如颗粒、纤维）掺杂，经过复合工艺制备形成的形状记忆聚合物复合材料，具备了可回复应变大、变形回复输出力较大、比强度高、比刚度高和低成本等优点[22-27,56-58]。

颗粒填充形状记忆复合材料通过向聚合物中添加特定颗粒改善材料力学和热-力学特性，可作为功能材料。但颗粒填充相在形状记忆聚合物中多为随机均匀分布，可主动控制性较弱[59-60]。Gall 等研究了 SiC 颗粒增强的环氧形状记忆聚合物复合材料，发现 SiC 纳米颗粒增强了体系的弹性模量及回复力，同时 SiC 纳米颗粒含量对其形状恢复性能也有一定影响[60-61]。Huang 等研究了炭黑填充的形状记忆聚氨酯复合材料，该形状记忆复合材料具有良好的导电性和形状记忆性能[62,63]。Leng 等将极少量的镍粉添加于炭黑填充的形状记忆复合材料中，研究了 Ni 粉分布结构对该复合材料的热力学性质的影响，发现 Ni 粉分布成链结构对 SMP 的力学性能具有更强的增强作用[64-65]。此外，研究人员还研究了陶瓷颗粒、炭黑、碳纳米管填充的形状记忆聚合物材料。

纤维增强形状记忆复合材料的回复力较大，比强度和比刚度较高，综合力学性能好，可作为结构材料。但由于纤维增强相的有效应变较小（一般＜2%），其纤维增强方向一般不作为材料变形的主方向[5,66,67]。相对于常规的树脂基复合材料，纤维增强形状记忆复合材料是革命性的创新[66]。Wei 等分别用碳纤维、玻璃纤维和 Kevlar 纤维制备环氧形状记忆聚合物复合材料。力学性能测试结果表明，各纤维增强的复合材料的刚度、强度都较原基体显著提高[57]。Ohki 等以聚氨酯形状记忆聚合物为基体，制备了不同玻璃纤维质量含量的形状记忆聚合物复合材料，测试的这些样品的力学性能和形状记忆效果表明：玻璃纤维的加入使得聚氨酯形状记忆聚合物复合材料的拉伸强度得到了明显的提高，加入玻璃纤维后的聚氨酯形状记忆聚合物复合材料仍保持形状记忆特性[58]。美国的 Colorado 大学以及 CTD（composites technology development）、CRG（cornerstone research group）两个公司研究了大量长纤维和纤维布增强形状记忆聚合物复合材料。其制备工艺与普通复合材料制备工艺相同，纤维质量百分比含量一般在 10%～40%，但其强度、弹性模量等力学性能比纯形状记忆聚合物提高了 20% 以上[67,68]。目前，作为结构材料应用的形状记忆聚合物复合材料多为碳纤维增强型。

（三）形状记忆聚合物复合材料驱动方法

形状记忆聚合物需要吸收能量来克服其内部的应力，回复其初始形状。因此，如何高效、可控地驱动形状记忆聚合物形状回复成为研究的重点。目前形状记忆聚合物的驱动方式有电驱动、磁场驱动、光驱动、溶液驱动、水驱动等多种方式。

电致驱动形状记忆聚合物复合材料是将导电的炭黑、碳纳米管、碳纳米纤维、短切碳纤维或连续的碳纤维等导电增强相与形状记忆聚合物复合。

当一定的电压施加在这样的导电复合材料上时，通过导电增强相上的电流由于电阻热效应一部分电能转化为热能。这些阻热效应产生的热能使导电增强相的温度升高，而后这些热量由增强相向聚合物基体传递，使形状记忆聚合物到达其形状记忆转变温度以上，复合材料的形状记忆效应由此被触发[57,64,69]。

磁场和电磁场驱动形状记忆聚合物复合材料是通过在聚合物中加入磁性颗粒制备而成的。复合材料内部的磁性颗粒随着磁场场强的周期变化而发生往复运动，此时磁性颗粒与聚合物分子之间的相对运动产生摩擦和碰撞，进而产生热量[70]。

光敏性形状记忆聚合物，是指形状记忆聚合物中含有光响应官能团，在特定波长光照作用下产生可逆的交联和分解反应，而这种驱动方式仅局限于少数几种形状记忆聚合物[36]。然而对于热敏性形状记忆聚合物的光热驱动方式，是通过聚合物吸收光波的热量间接地达到其转变温度后驱动其形状记忆效应的。例如，将光纤埋入形状记忆聚合物中，通过光将特定波长的光能传递到聚合物中，而后光波在光纤末端被传到聚合物中，这些光能被吸收后加热聚合物到其形状转变温度以上，从而驱动形状记忆聚合物，实现形状记忆效应[71]。

溶液驱动形状记忆聚合物是将溶剂分子通过扩散作用渗透到聚合物材料中，被吸收的溶剂分子对聚合物网络结构产生增塑作用。增塑效应可以降低聚合物网络结构内部分子链段之间的相互作用力，提升聚合物分子链段的柔顺性和运动能力，进而与聚合物分子产生化学的或物理的相互作用，从而间接地降低聚合物材料的转变温度。当聚合物的转变温度降低至室温时，固定在形状记忆聚合物分子链段内部的弹性应变能得以释放，形状记忆效应因此触发[72]。

水驱动形状记忆聚合物是将水的氢键与形状记忆聚合物分子中的极性官能团结合，能够降低材料的玻璃化转变温度，且这种结合还可在高温下分离而不改变材料的性质，从而实现形状记忆效应的驱动[62,63,73,74]。

二、介电弹性体电活性聚合物

（一）电活性聚合物

电致活性聚合物是一种智能多功能材料，在受到外加电场作用时能够改变形状或体积，当外加电场撤掉后，又恢复成原来的形状或体积，可以用来设计和制造智能转换器件，如驱动器、传感器和能量收集器等[1,75-79]。EAP材料与压电、铁电材料相比，具有大变形和轻质量等优点，是一种具有重大发展潜力的智能多功能软质材料[80-99]。

按照其作用机制不同，EAP主要分为电子型和离子型[76-82]。电子型EAP材料主要包括介电弹性体（dielectric elastomer, DE）、铁电聚合物（ferroelectric polymers）、液晶弹性体（liquid crystal elastomers）和电致伸缩接枝弹性体材料（electrostrictive graft elastomers）等。离子型EAP材料主要包括离子聚合物金属复合材料（ionic polymer-metal composites, IPMC）、碳纳米管（carbon nano tube）、电致流变液体（electrorheological fluids）、离子聚合物凝胶体（ionic polymer gels）和导电型聚合物（conductive polymers）[76-79]。表7-3-1列举出生物肌肉和典型EAP材料的主要性能。

表 7-3-1　生物肌肉和典型 EAP 材料的性能比较[79]

驱动器材料	典型材料	优点	缺点
天然肌肉	哺乳动物肌肉	中等应力（350 kPa），较大应变（40%），高弹性能密度（20～40 MJ/kg），效率（40%），可循环再生，可变硬度	不是工程材料，工作温度范围小
介电弹性体	硅橡胶 丙烯酸	中等应力（几 MPa），大应变（120%～380%），高弹性能密度（10 kJ/m^3～3.4 MJ/m^3），高机电耦合效率（最大 90%），中等频率带宽（10 Hz～1 kHz），低消耗，低电流	高电场（150 MV/m） 低弹性模量（1 MPa）
铁电体聚合物	聚偏氟乙烯	高应力（45 MPa），中等应变（10%），高弹性能密度（1 MJ/m^3），高机电耦合效率，高弹性模量（400 MPa），低电流，可在空气、真空或水中应用	高电场（150 MV/m） 循环寿命短
液晶弹性体		大应变（45%），高机电耦合（75%）	高电场（1～25 MV/m）蠕变对其影响较大
碳纳米管	SWNT MWNT	高应力（>10 MPa），高弹性模量，低电压（2 V），温度范围大	小应变 低机电耦合效率 材料较昂贵
传导型聚合物	聚吡咯	高应力（最大 34 MPa），中等应变（2%），高弹性能密度（100 kJ/m^3），高弹性模量（1GPa），低电压（2V）	低机电耦合效率，响应慢，需要封装

续表

驱动器材料	典型材料	优点	缺点
分子型 EAP		高应力（>1 MPa），高应变（20%），高弹性能密度（100 kJ/m^3），低电压（2V）	响应慢，需要封装
离子聚合物金属复合材料（IPMC）	Nafion Flemion	低电压（<10 V），大位移	低机电耦合效率，需要封装
压电材料	压电陶瓷	高应力，响应速度快，高弹性能密度（100 kJ/m^3），高弹性模量	小应变

铁电聚合物可以自发极化，这是因为它的晶相中有一个无定形相，在施加高电场时，极化的方向将沿着电场的方向取向成为有规则的排列从而产生形变，具有高弹性模量、高应力、高弹性能密度、高机电耦合效率等优点[80,97]。电致伸缩接枝弹性体由柔性主链和支链构成，相邻的主链和支链相互交联形成晶体单元，施加外电场时，晶体单元产生旋转，带动柔性主链局部重新排列，宏观表现为产生明显的变形[81,98]。电致黏弹性聚合物硅橡胶、人造橡胶和电极共同合成。在交联耦合之前其形态是类似流体的，施加外电场时，呈现出剪切模量随着电场变化的类固体状态[82,99]。

碳纳米管可分为单壁型和多壁型两大类。它的力学性能非常突出，弹性模量高、抗拉强度和机械强度较高，而且还具有热稳定性高的特点，在高温下的性能超过其他 EAP 材料[83,97-99]。电致流变液体由介电颗粒悬浮于载液中组成。在外电场作用下，介电颗粒发生极化并沿电场方向结成链状结构，使流体的流变性能发生突变，从自由流动的液体变为类似凝胶状的固体[78]。它具有响应迅速、性能稳定等优点。离子聚合物凝胶体在电场作用下，氢离子进出时被聚合链上带负电离子吸引或排出，引起形状变化。凝胶可以产生与生物肌肉相当的力和能量密度，驱动性能优异[79]。

电子型 EAP 材料需要很高的驱动电场（>100V/μm），以保证其产生一定的电致变形，该电场接近材料的击穿电场[76, 77]。离子型 EAP 材料由电极和电解液组成。为保证其正常工作，需保持其表面湿润和周围环境湿润[78-79]。这类材料在较低电场下就可以产生稳定的伸长、缩短或弯曲等响应[80]。

总之，EAP 材料是一种具有重大发展潜力的智能多功能材料，具有大变形、高弹性能密度、高机电转化效率、快速响应、低噪声、轻质量、易于加工、易于成形、极佳柔性和回弹性等优点[81]。EAP 材料由于性能优异，被应用于驱动器、人工手臂、面部表情、仿生机器人、浮空器飞艇舵、固态飞行器、能量收集器、制冷器、传感器、盲文显示器和自适应光学系统等各个方面，在人工肌肉、航空航天、智能仿生和机械工程等领域显示出巨大的应用潜力[82-99]。

（二）介电弹性体

介电弹性体属于电子型EAP材料，在外加电场下可以产生大变形，具有高弹性能密度、高机电转化率、超短反应时间、轻质量和极佳柔性等特点[1,75-90]。硅橡胶和丙烯酸是最常见的介电弹性体[75-79,84-90]。表7-3-2列举出硅橡胶和丙烯酸材料性能的比较。

表7-3-2 硅橡胶和丙烯酸材料性能的比较[78]

材料性质	硅橡胶（TC-5005）	丙烯酸（VHB4910）
最大应变/%	150	380
应力/MPa	典型：0.3；最大3.2	典型：1.6；最大7.7
能量密度/（kJ/m^3）	典型：10；最大750	典型：150；最大3400
密度/（kg/m^3）	1000	960
最高功率/（W/kg）	5000	3600
可持续功率/（W/kg）	500	400
带宽/Hz	1400	10
最大适用范围	＞50kHz	＞50kHz
循环寿命（不同应变下）	$>10^7$：5%；10^6：10%	$>10^7$：5%；10^6：50%
机电耦合/%	最大：80；典型：15	最大：90；典型：25
效率/%	最大：80；典型：25	最大：90；典型：30
弹性模量/MPa	0.1～1	1～3
响应速度/（m/s）	＜30	＜55
热膨胀/（mm/℃）		1.8×10^{-4}
电压/V	＞1000	＞1000
最大电场/（MV/m）	110～350	125～440
相对介电常数	～3	～4.8
温度范围/℃	－100～250	－10～90

在介电弹性体两个相对表面均匀涂覆柔性电极（如石墨），当对其施加电压时，弹性体将发生厚度的减小和面积的扩大[1,75-99]。图7-3-2是电压驱动介电弹性体的原理图，由于在电极上施加电压，上下两层电极上的异性电荷

相互吸引，每层电极上的同性电荷相互排斥，当电场力足够大时，薄膜将产生明显的面积和厚度变化。

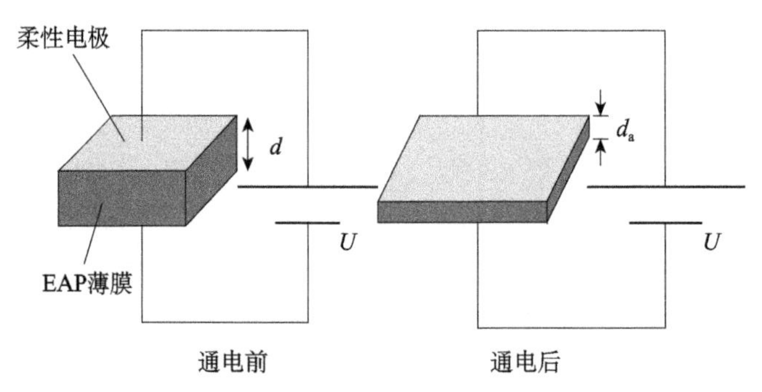

图 7-3-2 电压驱动介电弹性体的原理图 [75]

（三）介电弹性体复合材料

介电弹性体的驱动电场与其击穿电场是同一数量级的，这极大地影响了它的商业应用。研究人员通过制备性能优异的介电弹性体复合材料来解决这一问题 [76,100-102]。下面简要介绍介电弹性体复合材料。

大体上通过两种思路制备高性能的介电弹性体复合材料，一种是基于物理共混方法来制备颗粒或聚合物填充的硅橡胶复合材料，另一种是基于化学交联方法来制备丙烯酸互穿网络型的介电弹性体。

物理共混的方法获得介电弹性体复合材料主要基于以下三种不同的机制。第一种是填充高介电常数的铁电陶瓷颗粒，如钛酸钡、铌镁酸铅（PMN-PT）等。随着填充颗粒含量的增加，复合材料的介电常数和弹性模量均增加 [102]。这种复合材料所需的驱动电场较低，而且在同一电场情况下具有较高的驱动力。第二种是填充高极性的导电颗粒，如碳纳米管等。在外加高电场的作用下，介电弹性体中的颗粒产生极化，提高了聚合物的介电性能，降低了所需要的驱动电场，而且还提高了复合材料的力学性能，如提高弹性模量和驱动力等 [76]。第三种是填充一种高极性的共轭聚合物，如 PHT [poly(3-hexylthiophene)] 等 [100]。与上面所述的颗粒填充型介电弹性体复合材料相比，此种复合材料具有较高的介电常数和较低的弹性模量，更容易产生大的电致变形。

Ha 等制备出性能优异的互穿网络丙烯酸介电弹性体材料 [101]。他们在丙烯酸内部网络中引入了第二种聚合物的网络，可以使薄膜在脱离刚性支撑情况下保持预拉伸状态。常用的第二种聚合物材料包括 TMPTMA（trifunctional trimethylolpropane trimethacrylate）和 HDDA（hexandiol diacrylate）。具体的制备过程是，首先对丙烯酸薄膜进行预拉伸并固定在刚性框架上，然后将 TMPTMA 或 HDDA 均匀地喷洒到薄膜上，溶剂聚合后将在薄膜内形成第二种网络结构。最后，把薄膜从刚性框架取下，形成的第二种网络结构将使其保持在预拉伸状态。经实验验证，互穿网络的介电弹性体复合材料比普通纯净介电弹性体材料的电致变形性能更佳优异。

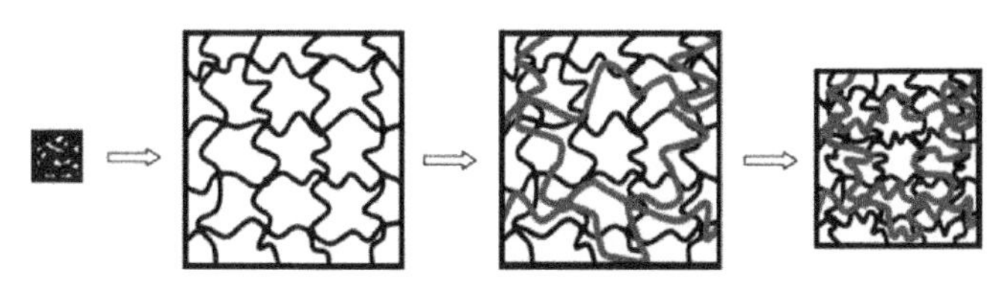

图 7-3-3　互穿网络聚合物薄膜制备过程 [101]

第三节　最近二三十年的研究进展

一、形状记忆聚合物及其复合材料

为了清楚地认识和掌握形状记忆聚合物及其复合材料的形状记忆机制，对形状记忆聚合物及其复合材料的制备和应用起一定的指导作用，建立合理的本构模型来预测其在不同条件下的应力应变响应关系是非常有必要的。目前，国内外学者针对形状记忆聚合物及其复合材料开展了大量的理论、实验研究，取得了丰富的成果。

（一）形状记忆聚合物及其复合材料的理论研究

形状记忆聚合物的形状记忆效应通过热-力学循环过程体现出来 [8,103-104]。其宏观形状记忆循环过程如图 7-3-4 所示；三维的应力-应变-温度的循环过程如图 7-3-5 所示，包括以下步骤：①在一个高于转变温度 T_{trans} 的温度 T_h 下（$T_h > T_{trans}$），形状记忆聚合物试件在外载作用下发生临时变形；②保持外载，降低温度至 T_l ($T_l < T_{trans}$)；③在 T_l 温度卸去外载，材料的应力降为零，材料由于弹性变形产生一个较小的形状回复后，达到预变形状态；④加热材料到 T_h 以上时，材料的预变形几乎完全恢复，只保持少量的残余应变。

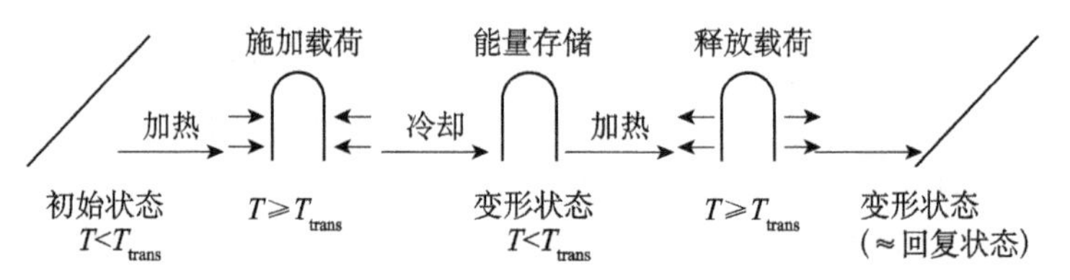

图 7-3-4　热激励形状记忆聚合物在热循环过程中的形状记忆效应 [8]

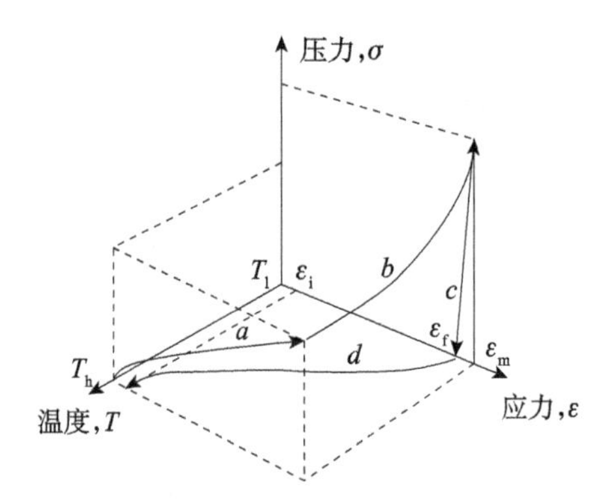

图 7-3-5　应力－应变－温度的三维热－力学循环过程示意图 [8]

由于形状记忆聚合物的变形较大，在对其本构关系研究中，需要考虑材料非线性，故研究难度较大。到目前为止，本构关系的研究模型主要为基于Toubushi 线性与非线性模型和基于相变理论的细观力学模型。

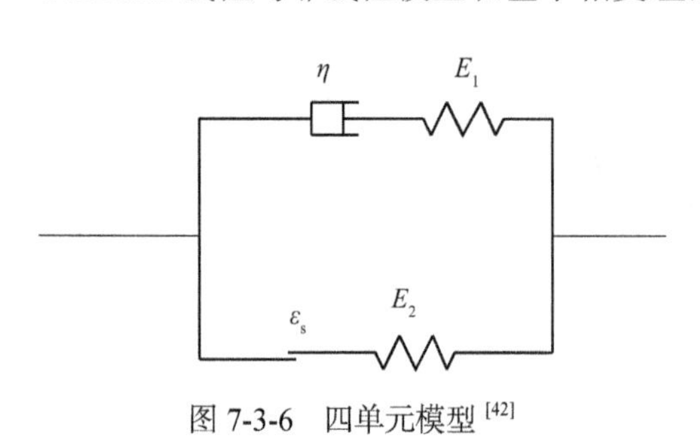

图 7-3-6　四单元模型 [42]

1997 年，Tobushi 等根据线性的黏弹性理论，在传统线性黏弹性三单元模型基础上引入滑移摩擦单元用，以描述形状记忆聚合物的冻结应变，建立了唯象的四单元热力学本构方程（图 7-3-6）[42]。

2001 年 Tobushi 等修正了线性模型，引入了非线性弹性应力和黏性应力，提出非线性模型 [40]，用以描述材料在大应变条件下的本构关系。在变形的加载过程中，如果以常应变率加载，将会产生与时间无关塑性残余应变。使用建立的本构关系，Tobushi 分别得到在 20% 应变水平下应力-应变和应力-温度的变化关系，且预测结果与实验结果基本吻合。然而，由于在模型中没有考虑与时间的有关参数，因此该模型预测的与时间参数有关的结果与实验结果吻合性不好。

在 Tobushi 等建立的黏弹性模型基础上，Zhou 等对其理论进行了相应的扩充，建立了三维的形状记忆聚合物本构模型；同时提出了插值函数，用以表征弹性模量、黏性系数、延迟时间和热膨胀系数等参数随温度的变化规律[105-107]。此外，利用ABAQUS中的UMAT函数对苯乙烯基形状记忆聚合物的力学性能进行了模拟、预测，验证了模型具有较好的有效性及准确性。

Morshedian等将弹簧单元和黏壶单元组合，建立了SMP的响应模型[108]，如图7-3-7所示。该模型中黏壶的黏度系数η_1在高温状态很小，易被拉长；低温状态很大，可使材料变形保持；该模型的形状记忆特性可由弹簧单元E及黏壶单元η_1表示，黏壶单元η_2主要用来表征SMP的黏弹性。Diani等从热力学理论和线性黏弹三单元模型出发，认为SMP在玻璃化转变温度以上，变形能改变主要由熵变化引起；在玻璃化转变温度以下，变形能改变主要由内聚能变化引起。同时，假设材料具有不可压缩性，提出了一个三维有限应变本构方程[109]。Seok等从线性黏弹理论出发，引进松弛模量，得到小变形情况下积分型本构关系[110]。

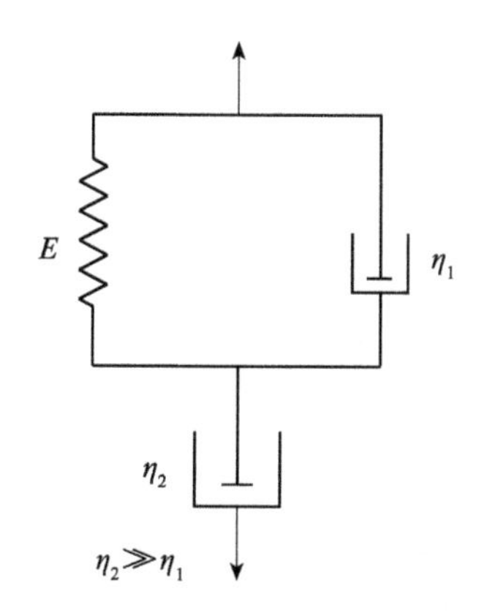

图7-3-7　SMP力学响应模型[108]

为了从分子层次来描述SMP材料的形状记忆行为，Liu等在2006年发展了形状记忆聚合物的一种基于两相转变的微观结构的本构模型[111-112]。该模型能够很好地描述形状记忆聚合物在单轴拉伸载荷作用下产生的小变形行为。在该模型中定义了两个内部状态参数:冻结相和活动相，如图7-3-8所示。在此基础上，定义了材料的总应变分为三部分。该模型假设在低温冷却状态下，材料由于外界力约束所产生的应变将全部转化为应变能储存起来，即为储存应变，且该储存应变是由熵应变所引起的。同时，假设材料中冻结态和活动态所受应力相同，提出了三维、线弹性、小变形的热力学本构方程。

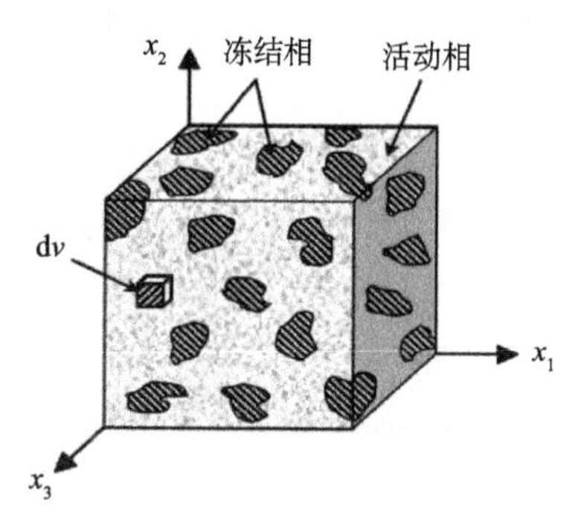

图7-3-8　SMP的相转变模型[111]

基于Liu等的两相理论，Chen和Lagoudas也发展了一种非线性的本构模型，以描述形状记忆聚合物的热力学行为[113-114]。

在该模型中，同样定义了两个内部状态参数：玻璃态相与橡胶态相。该模型应用运动学理论，认为形状记忆聚合物经过一系列晶核成型实现相变。在该本构模型中假设材料变形梯度为一次 First Piola-Kirchhoff 应力及温度的函数。Chen，Lagoudas 模型假设材料为不可压缩 neo-Hookean 且各向同性。因此，应用 Helmholtz 自由能中对于不可压缩 neo-Hookean 材料方程，得用 first Piola-Kirchhoff 应力表示的形状记忆聚合物在橡胶态相和玻璃态相的本构方程。

同样基于相变转换理论，Qi 等也进行了形状记忆聚合物的热-力学性能研究[115]。该模型为三维有限变形模型，假设材料在预变形降温冷却过程中会出现与初始的玻璃态相不同的新的玻璃态相。因此，该模型还有三相结构：橡胶相、初始玻璃态相和冻结玻璃态相。该模型采用一级相变的概念来描述：随着温度的降低，材料由橡胶态的熵弹性过渡到玻璃态的黏弹性的变形机制。假设新形成的冻结玻璃态相并未继承变形橡胶态的变形，则冻结玻璃态相中新的变形可被认为是总的变形的重新分布。该模型可以为具有形状记忆效应的材料建模，并不局限于为弹性体聚合物或者玻璃体聚合物，能准确地模拟材料由高温向低温冷却过程中材料性质的变化。

Barot 对结晶型 SMP 的研究中，认为 SMP 从高温到低温，随着温度的降低，材料内部结晶体增多[116]。同时，假设材料为不可压缩超弹性体，依据能量原理得到结晶型 SMP 的热力学本构方程。

王诗任等认为形状记忆聚合物的形状记忆效应实质为高分子的黏弹性力学行为，同时将材料的形变认为是普弹形变、高弹形变和黏性流动形变三部分的叠加，并建立了一套相应的数学模型[117]。杨青基于高分子材料黏弹性理论，利用非线性有限元分析软件 MSC.Mare 对其黏弹性模块进行二次开发，模拟了材料的形状记忆过程，结果与 Tobushi 和 Liu 的实验与计算结果较好吻合[118, 119]。李郑发等建立了微观力学三维本构模型，很好地解释了材料形状记忆效应的微观机制，考虑其冻结 / 恢复时间延迟效应、应力松弛效应以及热变形效应的影响，对其变形过程进行了理论分析[120]。郭晓岗等基于两相理论，以正态分布函数描述了形状记忆聚合物冻结相与活跃相随温度变化的关系方程，同时首次考虑应力对材料相变的影响，建立了描述形状记忆聚合物热力学行为的三维本构模型[121]。陈建国等研究了环氧基形状记忆聚合物在不同温度下条件下的力学行为，在实验的基础上建立了考虑材料蠕变、松弛行为的三维本构模型，通过与实验结果的对比，验证了模型的准确性[122]。

吕海宝等从聚合物分子热运动所遵循的松弛理论和热力学方程出发，建立了形状记忆聚合物混合体系的形状记忆效应在混合过程中所应遵循的热力学模型[72]，基于聚合物 Flory-Huggins 溶液理论，结合松弛理论的 Eyring 方程和自由能方程，建立了形状记忆聚合物混合体系热力学方程，并进一步描述了在不同受力状态下应力-应变本构方程。

纤维增强形状记忆复合材料具有更高的刚度和回复力。该材料在纵向的最大变形率由纤维决定，一般小于 2%。纤维增强形状记忆复合材料实际应用中多利用层板状纤维增强复合材料的弯曲变形，对该类型材料的力学特性研究，多考虑大挠度弯曲变形条件下的结构展开动力学以及复合材料的屈曲变形特性[123]。

Dow 等根据 Timoshenko 给出的弹性体中一个杆的屈曲解发展了关于单向纤维增强复合材料的经典剪切微屈曲失稳模型[124, 125]。该模型是一个二维简化的单向板，其中纤维层与基体层均匀相间分布。根据该模型描述的结构发生剪切型屈曲变形时的最小临界应力，材料发生剪切型屈曲最小临界应力对应屈曲波长取最大值；而对于拉伸型屈曲，计算发现其最小屈曲应力对应的临界屈曲波长与实验测结果有较大差别。为了实现理论预测结果与实验观察值较为接近，Campbell 考虑单根纤维在弹性软基体中的微屈曲问题，提出了一个针对连续纤维增强形状记忆聚合物复合材料板弯曲变形的纯剪切微屈曲模型[126]。Campbell 基于与纤维一起变形的软基体的剪切变形能等于 Timoshenko 给出的弹簧基础变形能，得到其屈曲临界半波长。在上述工作基础上，Campbell 等进一步考虑多根纤维在弹性基体中的微屈曲变形[127,128]，根据 Timoshenko 的屈曲杆和弹簧基础变形能公式，得到屈曲半波长。上述关于连续纤维增强形状记忆聚合物复合材料板的屈曲理论模型都是基于纯拉伸或纯剪切模式提出的，但实际层合板弯曲变形过程中纤维主要发生剪切 / 拉伸混合屈曲模式。针对连续纤维增强形状记忆聚合物复合材料单向层合板弯曲变形，Wang 等提出了一个简化的二维拉 / 剪耦合屈曲模型，其临界屈曲波长与已有的实验观测结果较为吻合[129]。

关于连续纤维增强形状记忆聚合物复合材料的后屈曲性能分析，Campbell 等假设材料呈双直线本构关系，研究长纤维增强形状记忆聚合物板的弯曲行为，以屈曲面层为界，将板分为屈曲区和非屈曲区。依据纯弯曲条件下其横截面沿板纵向的合力为零，推导出中性层位置系数 α。再根据横截面上的弯矩等于外力偶矩，推导出弯矩与曲率的非线性关系式，从而建立了整体长纤维增强 SMP 板的数值分析模型[130]。Francis 等研究了形状记忆聚合物复合材料在弯曲变形过程中纤维的局部后屈曲变形行为，求得了中性面（应变为零）的偏移距离以及纤维屈曲波长等关键参数的表达式[131]。美国CTD公司的科研人员对材料屈曲的各种情况做了测试和理论推测，利用黏弹性本构关系验证了环氧形状记忆聚合物的回复性能。测试和初步理论验证了纤维增强形状记忆复合材料的失效模式，并观测了其屈曲形貌，进行了弯曲测试并完成了有限元计算验证[132-134]。另外，美国哈佛大学的John Hutchinson教授、Zhigang Suo教授、美国西北大学的Yonggang Huang和John Roger教授都对模量较高的增强体在模量很低的基本中的屈曲问题进行了大量的研究[135-140]。

（二）形状记忆聚合物及其复合材料的应用研究

近年来，随着人们对SMP材料的应用日益关注，智能结构发展较快，极大地推动了航空航天、生物医疗、纺织品、智能模具等领域的发展，一些结构已完成原理性演示验证，并已初步获得应用[3,45,141-144]。

2006年发射的美国冲击号卫星（Encounter Spacecraft）将形状记忆材料用于天线结构的展开。已发射的美国智能微型可操控卫星（DiNO Sat）和Road Running卫星的太阳能电池板帆板应用形状记忆聚合物复合材料铰链进行驱动[144-146]。

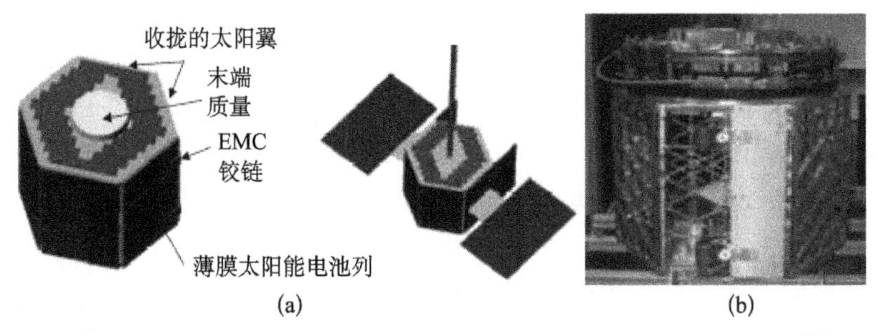

图7-3-9 (a) DiNO Sat卫星的可展开太阳翼；(b) DiNO Sat卫星组装图[145]

形状记忆复合材料可展开梁由三个纵向的可折叠伸缩的形状记忆聚合物复合材料层合板构成，两个形状记忆聚合物复合材料层合板之间间隔120°[115]，如图7-3-10所示。在折叠状态，三个纵向的形状记忆聚合物复合材料层合板以S形折叠收缩；在轨工作后，通电加热层合板，材料变形回复，使可展开梁展开。

哈尔滨工业大学的研究人员还制造了可展开铰链、桁架和天线等，已经实现了原理样机的验证工作，采用电热膜的形式加热使铰链展开，进而驱动桁架或天线结构的整体展开，响应速度稳定[147]。

图 7-3-10　形状记忆复合材料可展开梁[115]

2009 年兰鑫等利用具有高比强度、比刚度的环氧树脂基纤维增强 SMPC 制备了一种新型的铰链结构，该铰链主要包括：铝制材料端部装置，SMPC 复合材料片层和电热薄膜。如图 7-3-11 所示，该铰链具有较轻的重量、简单的机械结构设计，可有效地提高空间可展开结构的展开可靠性。另外，该铰链具有良好的折叠-展开性能，展开速率先慢，进而增加，最后部分再次减慢，可以有效地减少结构展开的振动效应，为空间可展开结构的正常有效工作提供了更好的保障。展开实验如图 7-3-11 所示[147]。

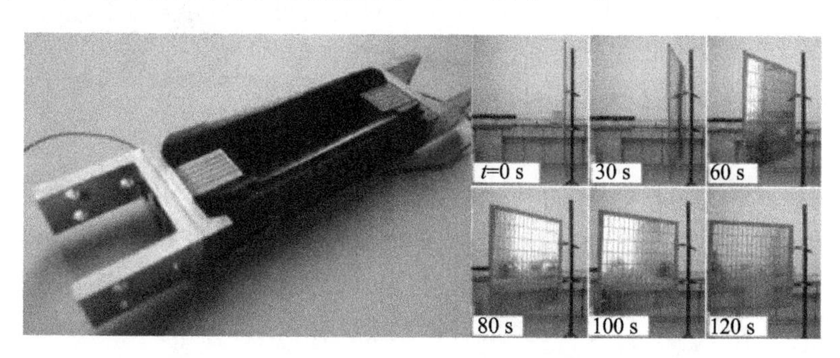

图 7-3-11　基于形状记忆复合材料的铰链结构[147]

为了更有效地研究 SMPC 的可展开性能和展开后的支撑结构能力，2011 年张蕊瑞等利用纤维增强环氧基 SMPC 设计并制备了一种新型的展开梁系，并结合有限元 ABAQUS 的模拟计算得到不同几何形式、材料组分和片层厚度时 SMPC 梁展开过程中的力矩-变形角度、应力-应变关系，为实验结构的制备和优化提供了基本的理论基础。SMPC 梁的组成部分和组成梁系的展开实验如图 7-3-12 所示[148,149]。

基于张蕊瑞等的研究基础，2013 年王通等应用高比强度、比刚度的环氧

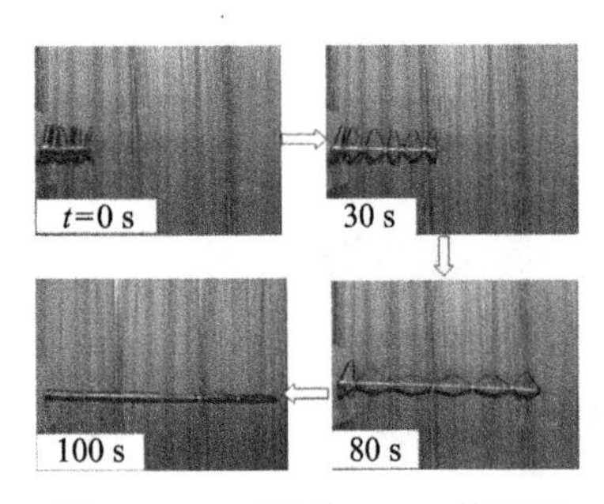

图 7-3-12　环氧基 SMPC 桁架梁展开示意图[148]

树脂基 SMPC，制备了一种新型的三翼可展开智能梁，并测试了该梁系在展开工作时的前 7 阶固有频率，同时，应用有限元软件 ABAQUS，分析模拟了不同 SMPC 片层厚度和角度作用下的动态力学性能，该实验结果可以为 SMPC 空间展开后成为空间结构的支撑结构的动态力学性能分析提供一定的理论依据。制备的三翼梁如图 7-3-13 所示[150]。

桁架梁系 SMP 结构具有良好的可折叠性，该梁系可以折叠为体积较小的压缩状态以便在发射装置中放置。当发射器进入预定的工作轨道时，该梁系结构将被释放压缩变形，成为体积较大的空间可展开结构，如太阳能列阵、空间反射面等。除了上述空间可展开梁系，美国的 CTD 公司发展了一种特殊的螺旋形压缩变形梁系。传统的空间梁系在螺旋折叠时，材料将会产生较大的储存应变能，同时，当变形超过临界状态时，材料表面将会破坏，应用 SMPC 代替原来的金属材料，不仅可以有效地减少压缩过程中储存应变能，而且材料在折叠过程中，材料处在类橡胶的状态，韧性很好，不易破坏。值得注意的是，该种梁系结构可以更有效地降低发射器的储存空间，结构构造简单，制备的空间可展开桁架梁，如图 7-3-14 所示[151]。

图 7-3-13　SMPC 三翼梁桁架结构[150]

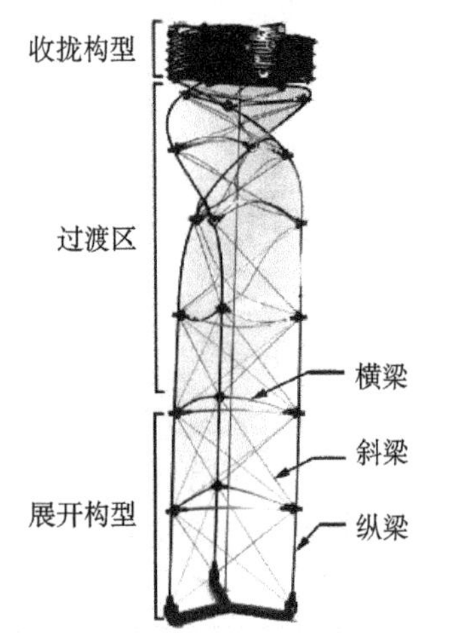

图 7-3-14　螺旋可折叠桁架梁[151]

基于 CTD 公司的研究工作，美国 CRG 公司为满足未来小卫星与地球交流和工作任务，2006 年利用氰酸酯基 SMPC 制备了一个展开状态为直径 6.3cm、高为 72cm 的实验模型桁架梁，该梁折叠的体积仅为长度 6.5cm。该梁在重力作用下的展开实验如图 7-3-15 所示，结果表明该梁系结构可以在硬质塑料管中有效地展开。值得注意的是，由于自身重力的影响，该梁并不能 100% 的展开，如果在顶部固定，梁的展开长度将会略长于初始长度[152]。

图 7-3-15　CRG 公司的螺旋折叠梁展开过程[152]

上述的 SMPC 铰链可以作为空间可展开结构的动力驱动装置，而 SMPC 桁架梁系通常作为空间可展开结构展开以后的支撑结构，用于维持空间展开结构的轨道内正常工作。

CTD 公司发展了一种容易蜷曲并可展开的柔性太阳能列阵，与上述太阳能列阵不同，该太阳能列阵主要由薄膜光伏平面和 SMPC 梁组成。该种太阳能列阵虽然表面能量转化效率不如传统的刚性结构，但是其具有更大的展开面积，更加简单的结构构造，更轻的质量和更低的制造成本，进而可以产生与传统太阳能列阵等量或更大的能量。该结构的展开过程如图 7-3-16 所示[153]。

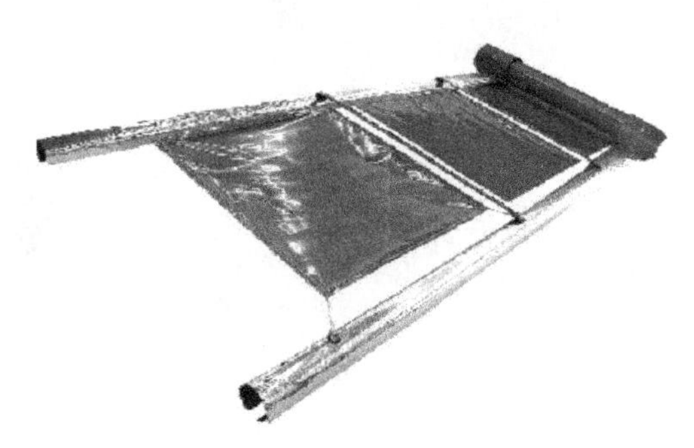

图 7-3-16　柔性可蜷曲太阳能列阵展开过程[153]

低成本的、轻质的、大口径的光学反射面在未来的太空探测中是一个发展趋势，为此，不同的材料制备技术和结构组成被研究和改进，以达到良好的光学性能。如图 7-3-17 所示，基于地面控制的可展开反射镜，应用形状记忆复合材料蜂窝夹层结构作为光学反射器的反射面支撑材料[154]。该反射器的形状记忆复合材料蜂窝夹层结构在工作前为卷曲状态，进入工作状态后形状记忆复合材料逐渐展开，从而对光学反射器提供刚性支撑。

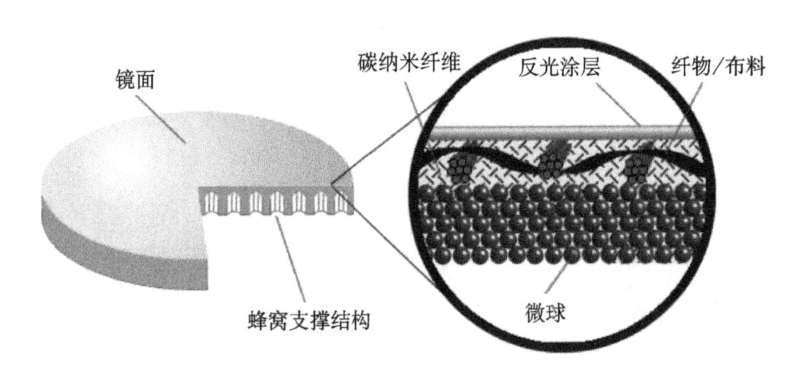

图 7-3-17 形状记忆复合材料可展开光学反射器 [154]

2004 年，Varlese 等发展了一种以 SMPC 为基体的光学反射面，该反射面以碳纤维增强型氰酸酯基 SMPC 为基体，以 0.02mm 厚的镍粉薄膜为表面层。该基体可以为反射面的表面形态提供良好的刚度和强度；该镍表面层又拥有较低的表面应力和柔性的刚度，这个结构在高温下可以人工蜷曲，便于储存，当工作时，在高温的驱使下可以再次自动展开，恢复初始的反射面形态，展开过程如图 7-3-18 所示 [155]。

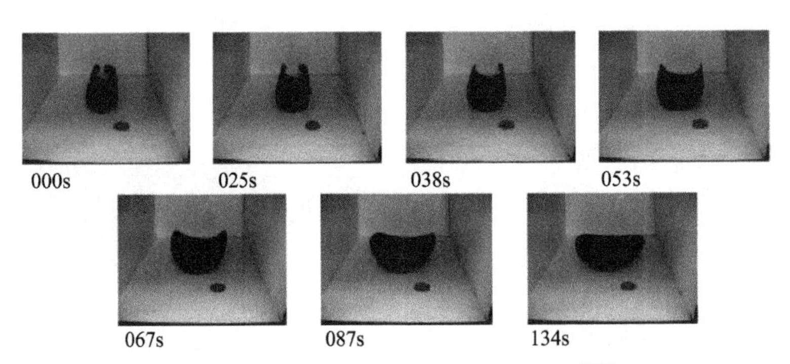

图 7-3-18 SMPC 光学反射面展开过程 [155]

除此之外，2004 年，美国 CTD 与科罗拉多大学应用 SMPC 铰链取代传统的金属铰链，展示了 SMPC 固体表面反射面的工作性能，如图 7-3-19 所示，这个反射面由 6 片子反射面组成，每个反射面底部将用 SMPC 铰链替代原有的金属铰链，制备的新型反射面可以很好地实现自展开 [156]。

大口径充气天线是一类重要的可展开反射器。然而，充气天线的中间部分的变形是非常大的，以至于使充气展开难度增大。最近，形状记忆聚合物

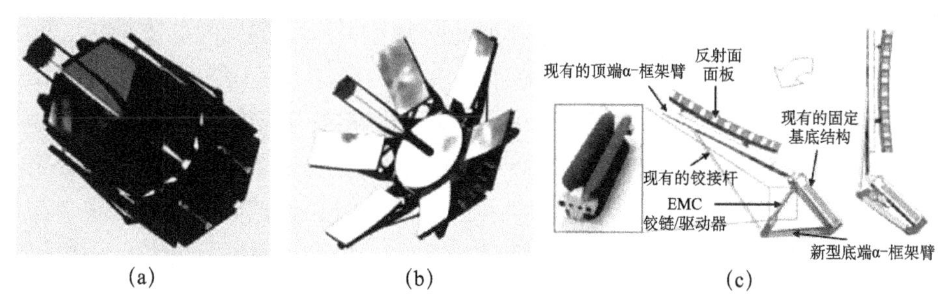

图 7-3-19　SMPC 即铰链驱动的光学反射面折叠和展开状态 [156]

复合材料被设想作为该结构的中间部分，以便实现大幅度展开。在这个设计中，天线的盘面和支撑杆是由形状记忆聚合物复合材料制作的。同时，形状记忆可展开天线可作为旋转抛物面形天线反射面，如图 7-3-20 所示。该天线的反射面可收缩折叠成伞形皱褶状结构 [157]。

图 7-3-20　形状记忆复合材料可展开天线 [157]

基于 SMPC 的变刚度性质，美国 ILC 公司发展了形状记忆充气刚化可展开固体表面发射面，如图 7-3-21 所示，该反射面的边缘部分和底部的支撑部分全由 SMPC 制作，而反射面的表面可以由薄膜材料制备而成。经过分析、设计、制备和性能评估，由实验结果发现，SMPC 的选取是该反射面能否正常工作的一个关键因素 [158]。

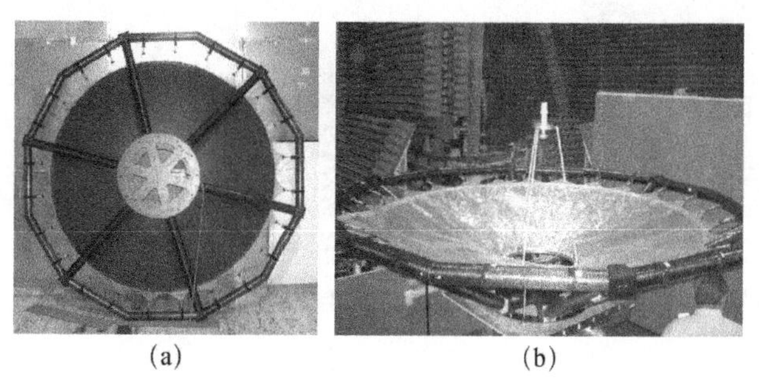

图 7-3-21　混合充气刚化 SMPC 反射面 [158]

此外，形状记忆聚合物复合材料还能应用于可变形机翼的蒙皮结构[159]。洛克希德·马丁公司提出了“折叠机翼”变形的概念，利用无人机作为验证平台，将其机翼设计为折叠式，即内段机翼可以对着机身折叠，折叠接缝部位采用形状记忆聚合物材料，以保证机翼由折叠状态展开后产生平滑的表面，见图 7-3-22[160]。哈尔滨工业大学将弹性纤维加入热固性形状记忆聚合物中，大大增强了形状记忆聚合物的使用安全性，并将其用于可变形后缘机翼结构，制备成无缝舵面，推迟气流分离，提高升阻比，提高隐身性能[161]。针对可变形机翼蒙皮的需求，哈尔滨工业大学还制备了一种基于形状记忆复合材料变刚度管的柔性基体蒙皮，该蒙皮利用形状记忆聚合物的变刚度特性，使蒙皮既具有承载能力又具有变形能力[162]。

图 7-3-22　可变形飞行器的折叠机翼[160]

形状记忆聚合物还可以用来制作各种医疗器械，图 7-3-23 为可生物降解的形状记忆聚合物外科手术缝合线，图 7-3-24 为具有清除血管中血液凝块功能的形状记忆聚合物微驱动器[22,163,164]。形状记忆聚合物用来制作各种医疗器械，主要是应用其形状记忆效应，通过预变形来缩小器械在使用前的体积，之后通过微创手术切口将其植入人体，之后通过外界激励再变形恢复至所需形状，如此可最大限度地减小切口尺寸和创伤。

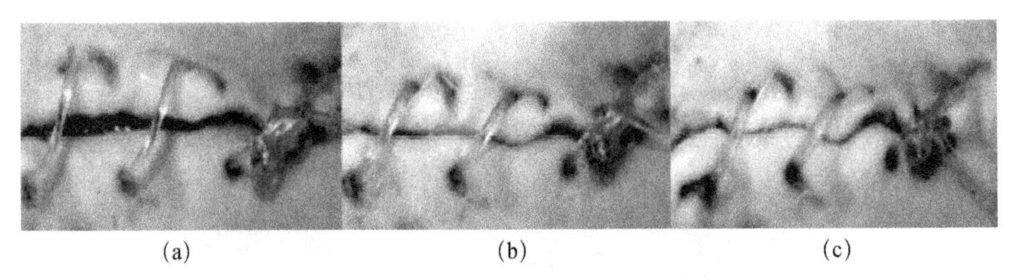

(a)　(b)　(c)

图 7-3-23　形状记忆聚合物缝合线[163]

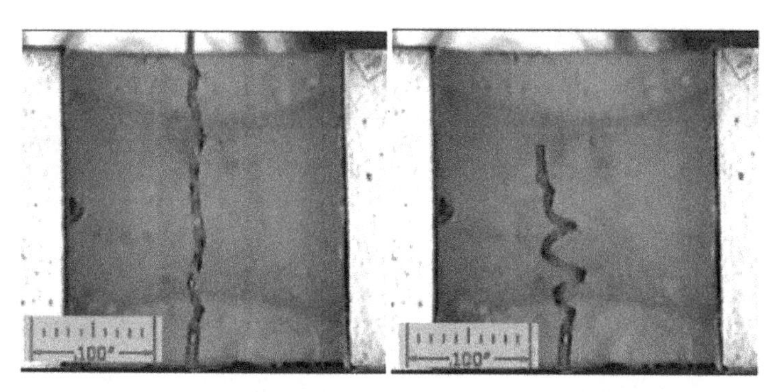

图 7-3-24　形状记忆聚合物螺旋形血栓微型清除器的演示试验 [164]

另外，利用形状记忆聚合物还可以制作复杂轮廓的可变形的智能模具等，如美国 CRG 公司和哈尔滨工业大学的智能材料与结构研究课题组等目前已投入研究 [165-167]。智能模具借助形状记忆聚合物的形状记忆效应，先固化聚合物管；随后将管插入所需制件轮廓金属模具，通过加热胀形聚合物管得到用于工作状态的模具；之后通过缠绕成型在智能模具上成型出所需制件；最后通过加热回复使模具恢复到最初管状，完成脱模过程。图 7-3-25 是 CRG 公司制备的智能模具。

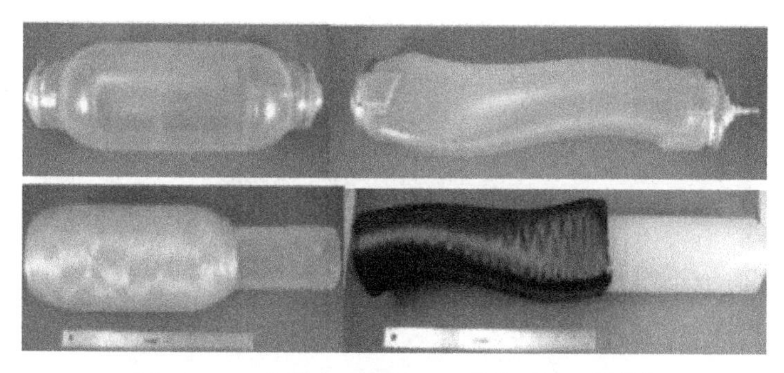

图 7-3-25　最终成型试件以及脱模后的芯模 [166]

哈尔滨工业大学的智能材料与结构研究课题组初步开展了基于形状记忆聚合物的智能模具的研究 [167]。基于苯乙烯形状记忆聚合物研制了一系列不同截面变形率的瓶状智能模具（截面变形率分别为 25%、50%、75% 和 100%），并对其恢复性能进行了测试，分别利用截面变形率为 25% 和 50% 的形状记忆聚合物智能模具进行了纤维缠绕实验，验证了智能模具的脱模效果。图 7-3-26 是利用截面变形率为 25% 和 50% 的智能模具制备的碳纤维瓶状制品。

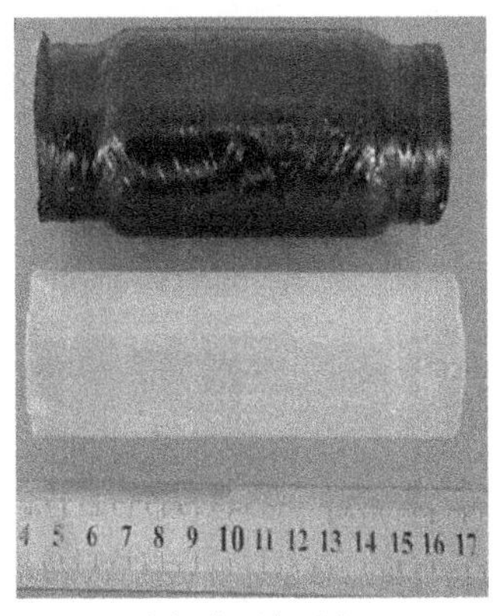

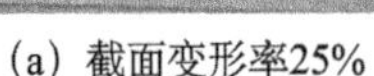
（a）截面变形率25%

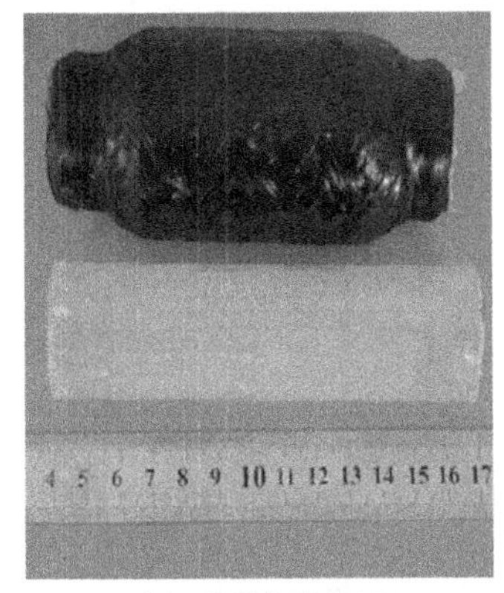

（b）截面变形率50%

图 7-3-26　截面变形率为 25% 和 50% 的碳纤维瓶状模具制品

之后，智能材料与结构研究课题组基于形状记忆聚合物进行了形状更为复杂的飞机进气道异型结构模具的研制。如图 7-3-27 所示，某飞机进气道模型结构模具的形状为一端圆截面、一端矩形截面的 S 形复杂形状异形管，课题组对该模具进行了回复率的测试实验，结果表明其回复率可以达到 99%。

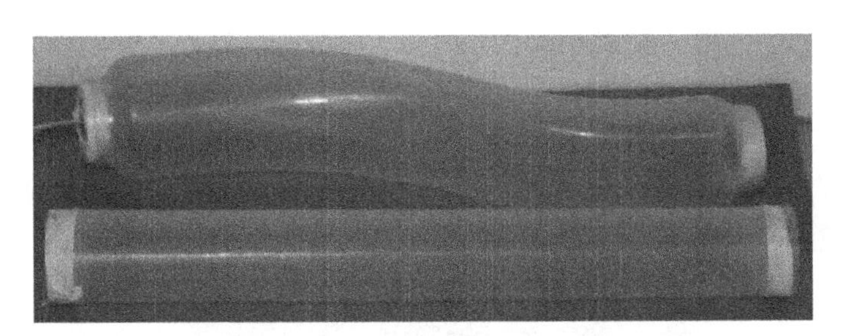

图 7-3-27　变形与未变形的进气道模具样品

二、介电弹性体材料

介电弹性体软质材料具有材料非线性和几何大变形特性，而且其承受的载荷还具有多场耦合特点，如机械力场和电场耦合、机械力场、电场和热场耦合等。另外，虽然介电弹性体在航空航天、机器人等工程领域应用潜力巨大，但商业应用仍然面临着巨大挑战。因此，研究这类材料的力学性能是十分重要和有意义的。介电弹性体的稳定性和大变形两大研究主题引起了研究人员的广泛兴趣。围绕着这两个主题，介电弹性体及其复合材料在多物理场耦合的本构关系、机电稳定性、突跳稳定性、超大电致变形机制、许用区域描绘及其应用器件的失效分析被研究人员深入而广泛的研究[1,168-211]。

（一）介电弹性体材料的理论研究

2000年，Pelrine等建立了一个简单的模型来描述施加电场后的介电弹性体的力学性能。他们把均匀涂覆柔性电极的介电弹性体看成可变电容的平行板电容器，(这是因为施加机械力和电场力后，材料将产生厚度的降低和面积的增加，引起电容的变化)，基于经典的麦克斯韦理论，推导出应力与施加电场和材料介电常数的关系。它揭开了介电弹性体本构理论研究的序幕[168]。基于热力学理论，美国哈佛大学锁志刚教授研究组建立了可变形电介质的热力学理论框架。他们从能量的角度出发，考虑机电耦合效应，建立联合弹性应变能和电场能的系统自由能函数，推导可变形电介质（如介电弹性体）的本构关系，研究其力学行为[169]。他们把自由能直接写成了弹性能和电场能的简单加和形式，没考虑两种能量之间的耦合。这是因为机械响应和电响应的响应时间相差几个数量级。耦合存在于电场能的表达和几何学关系中。在此基础上，刘彦菊等基于超弹性理论和机电耦合理论，并联合材料参数的实验测试，分析了介电弹性体及其复合材料的力学行为，推导出不同条件下的本构关系（比如，约束变形情况下和自由变形情况下）[170]。洪伟建立了黏弹性电介质耗散系统的热力学框架，推导出本构关系，并研究其瞬态的不稳定性[171]。赵选贺等研究了介电弹性体非线性和非平衡态黏弹性力学行为[172]。

介电弹性体在机械力场和电场等物理场耦合作用下将导致系统失稳，影响材料及其应用器件的正常工作。介电弹性体机电稳定性研究开始于2007年[173]。锁志刚等揭示了介电弹性体软质材料从机电稳定到机电不稳定这一过程：施加电场后，由于静电力作用，材料将沿电场施加的方向收缩，厚度变小导致施加的电场强度更高，静电力更大，这一不可逆过程一直持续下去，当超过其临界电场时，介电弹性体被击穿，这就是介电弹性体的机电不稳定[173]。介电弹性体的机电不稳定现象被Plante等在实验中观测到[174]，如图7-3-28所示，在某一特定电压作用下，变形的介电弹性体薄膜一部分区域的状态是平滑的，另一部分区域的状态是褶皱的。薄膜的平滑区比较厚，具有较小的平面拉伸；薄膜的褶皱区比较薄，

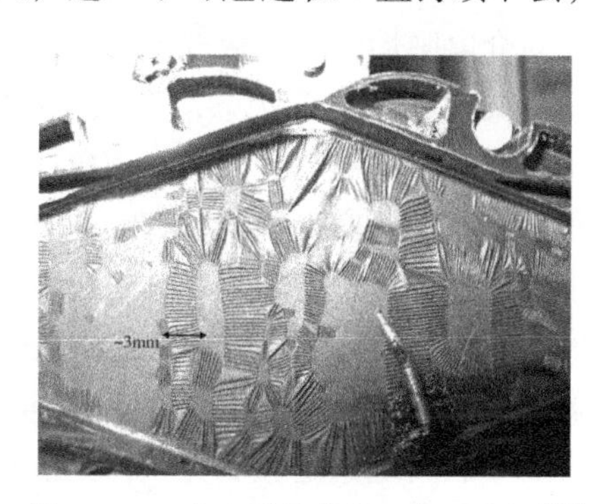

图7-3-28 介电弹性体的机电不稳定[174]

具有较大的平面拉伸。这两种状态共存。而且，实验中也可以明显地观察到介电弹性体薄膜的机电不稳定区域的传播 [175]。

锁志刚等提出可以应用任意的自由能分析介电弹性体机电稳定性 [173]。他们以 neo-Hookean 为例，分别研究了施加等双轴预应力和非等双轴预应力时的性质，描绘了介电弹性体名义电位移和名义电场的关系，理论证明了预拉伸能够显著提高介电弹性体的机电稳定性这一实验现象，计算出的临界击穿电场和实验结果吻合得很好 [173-174]。基于此，介电弹性体机电耦合系统的非线性机电稳定性分析越来越深入和具体 [173,175-186]。刘彦菊等应用两个材料常数的 Mooney-Rivlin 弹性应变能模型分析介电弹性体的机电稳定性行为，引入的材料常数比 k 能够分析不同类型和结构的介电弹性体，得出当 k 增大时，对应不同类型或结构的介电弹性体机电稳定性显著增强。理论结果能够帮助指导介电弹性体驱动器的设计和制造 [176]。进一步，刘彦菊等应用多材料常数的弹性应变能模型进行介电弹性体机电稳定性的分析 [177]。Norrisa 应用 Ogden 弹性应变能分析弹性体的稳定性行为，精确推导出临界真实电场、名义应力与拉伸率的关系，并给出 Ogden 模型简化形式时临界真实电场、名义应力与拉伸率的更为简化的精确关系 [178]。周进雄等研究了介电弹性体的不稳定性传播 [175]。何天虎等研究了介电弹性体经历非均匀变形时的机电稳定性 [180, 181]。Xu 等应用全应力理论进行了介电弹性体的机电稳定性研究 [182]。上述工作中的研究对象是理想介电弹性体。线性或非线性电致伸缩变化的介电弹性体的大变形和机电稳定性被深入研究 [183-187]。介电弹性体经历大变形时的介电常数依赖于自身的变形 [170,187]。赵选贺等提出把介电常数拟合为依赖于拉伸的线性变化函数，研究了介电弹性体的大变形和稳定性行为 [58]。刘彦菊等基于 Gofod 等对丙烯酸的实验研究 [187]，提出介电常数是拉伸的非线性函数，从解析表达和数值模拟两个角度深入研究系统的机电稳定性 [184-186]。在对介电弹性体的电致变形实验中，研究人员观测到它从一个稳定状态突然跳变到另一个稳定状态，这一过程被称为突跳稳定性 [188, 189]。理论结果也表明，介电弹性体有可能在未被电击穿的情况下避免机电不稳定的产生，进而稳定在一个厚度更薄的状态。这是由于介电弹性体达到拉伸极限附近时，硬度突然急剧增大，增加了其抵抗电击穿的能力，避免或者消除了机电不稳定的产生 [190]。Suo 等研究了互穿网络介电弹性体的突跳不稳定性，预测出应用器件的超大电致变形 [189]。李博等进行了经历极化饱和的非线性介电弹性体的大变形和突跳稳定性研究，得到的结论可以指导应用介电弹性体制备大变形高性能的转换器 [190-192]。刘立武等建立了非线性极化饱和介电弹性体的一般理论，应用经历极化饱和的电场能计算出电场力，得出和经典麦克斯韦理论不同的结果，进一步研究突跳不稳定性性能，并预测极化饱和对电致变形的影响[193]。He 等研究了介电弹性体球壳的突跳不稳定性[194]。Rudykh 等研究了电致活性聚合物球体的突跳电致变形驱动性能[195]。朱建等研究了弹性体中气泡的非线性突跳膨胀行为[196]。

实验证明，为了获得介电弹性体大电致变形可以通过以下几种不同方法：预拉伸介电弹性体[168]，互穿网络的介电弹性体[101,197]，应用弹性体在溶剂中的溶胀效应[198]，电荷驱动介电弹性体[199]等。下面列举几个典型的电致变形实验。1998 年，Pelrine 等揭示出对硅橡胶介电弹性体施加电场后可以产生 30% 的电致驱动变形[200]。2000 年，*Science* 报道了经历 300% 的等双轴预拉伸后丙烯酸在施加电场的条件下产生 100% 的电致驱动变形[168]。Ha 等进行的实验证明了在介电弹性体中引入第 2 种网络，使弹性体内部的预拉伸增加，进而增大了其电致驱动变形[197]。另外，应用电荷驱动代替电压驱动介电弹性体可以诱导其产生超过 100% 的电致变形，这是因为应用电荷驱动可以有效地抑制或避免机电不稳定的发生[198]。赵选贺和锁志刚建立了能够产生超大电致变形的介电弹性体理论，根据电压-拉伸曲线和电击穿-拉伸曲线的不同位置关系，提出三种不同机电响应特性的介电弹性体。该理论预测出选择和设计适当的电压变形响应，超过 500% 的超大电致变形是可能达到的[199]。Koh 等理论阐明了预拉伸后的介电弹性体或具有短链的介电弹性体复合材料将产生更大的电致驱动变形，综合考虑预拉伸、应变硬化和聚合物链长的影响，提出了介电弹性体产生超大变形的机制[201]。

介电弹性体的机电相变理论及应用此理论进行更大能量转化的方法被哈佛大学锁志刚教授研究组提出，与物质的相变类似，介电弹性体的厚度改变过程被称为机电相变[202]。他们从理论角度解释了介电弹性体机电相变现象。他们得到介电弹性体在单轴拉伸和施加电压情况下自由能随拉伸率的变化规律：曲线中存在两个局部最小点和一个鞍点，分别对应于介电弹性体拉伸率不同的两个变形稳定状态以及不稳定状态。两个稳定状态对应不同的自由能，对应于介电弹性体的薄厚两个稳定的状态。当机械力及施加电压发生改变时，弹性体将从当前稳定状态跳变到另一个稳定状态。在此基础上，Lu 等[203]研究了介电弹性体管状充气驱动器在电压和内压共同作用下的机电相变及其能量转化，推导出膨胀区域与未膨胀区域共存的条件，并根据电击穿和机械破坏确定了其许用区域。结果表明，膨胀转换可以显著增大机电能量转换，管内通气膨胀区域与未膨胀区域共存时一个机电循环产生的能量是仅有未膨胀区域时的几千倍。

介电弹性体换能器的动态特性以及机电转换能力会受到耗散过程（如黏弹性、介电松弛以及漏电）的影响。介电弹性体受力和电压作用时产生响应是时间相关的耗散过程。聚合物长链间的滑动和单体连接点的旋转可导致黏弹性松弛；电子云畸变和极化群转动可导致介电松弛；电子和离子在介电弹性体内的迁移可导致导电松弛[1]。在外力作用下，介电弹性体在一个特征时间内松弛至一个新的变形状态，这一特征时间即为黏弹性松弛时间；相应地，在外电场作用下，存在介电松弛时间。Foo 等[204]基于非平衡热力学建立了介电弹性体的耗散模型，并对耗散介电弹性体的机电性能进行分析。与耗散有关的变量影响系统的自由能密度函数，但与外载做功无关。他们利用弹簧和黏壶表征黏弹性松弛，利用并联的电容和电阻表征漏电，由此得到的结论与实验结果一致。基于上述模型，Foo 等[205]又对介电弹性体机电换能器性能进行分析。他们对一种机电换能器进行了研究，分析了不同循环速度以及不同变形条件下的换能器性能。李铁风等研究了薄膜充气式能量收集器在考虑非均匀场和黏弹性变形时的能量耗散和循环，指出了能量收集装置在急速加载卸载以及经过预拉伸下具有更优的性能[206]。王惠明等研究了介电弹性体薄膜在气压与电压耦合加载下的黏弹性变形，计算了各种场随时间的变化，结果表明，当外载小时，薄膜会逐渐达到稳定，而外载过大时会随着时间推移发生力电失稳[207]。

在介电弹性体中，每个聚合物链可能由电偶极子单体组成。没有外加电压时，偶极子经历热涨落，取向随机，这种情形和水分子类似。当介电弹性体受电压作用时，偶极子向电场方向转动。当电压足够大时，偶极子和电场完全平行，材料极化达到饱和。这种非线性介电性能可包含在广义理想介电弹性体模型中[1]。李博等研究了极化饱和对介电弹性体的机电不稳定性的影响[208]。他们基于理想弹性体模型，考虑了介电非线性，对等双轴拉伸的介电弹性体膜进行分析，得到了介电弹性体在不发生机电不稳定的情况下获得大变形的几种途径；提出了介电弹性体的不同极化模型，进而研究了当材料失效时的驱动模式，理论证明了极化饱和可以有效抑制或消除机电不稳定，而且驱动模式过渡过程中有助于产生更大电致变形。李博等从分子链角度建立了条件极化模型，通过理论说明了条件极化能够消除介电弹性体的不稳定性[209]。Lallart[210]等考虑了空间电荷分布不均匀性，确立了介电常数-薄膜厚度表达式，并考虑极化饱和的影响，对极化强度进行修正。理论分析及实验结果表明，极化饱和对聚合物的变形有限制作用。在此基础上，他们又对多相电活性聚合物驱动器的变形驱动能力进行了研究[211]，通过理论对影响考虑了极化饱和的两相电活性聚合物系统变形能力的因素（相比，电场、频率、温度等）进行分析，并通过实验进行验证。

（二）介电弹性体材料的应用研究

基于介电弹性体材料的工作原理，介电弹性体可以用来设计和制造不同结构的智能驱动器（图 7-3-29）、人工肌肉（图 7-3-30）、仿生智能机器人（图 7-3-31）、除尘刷（图 7-3-32）、飞艇舵（图 7-3-33）、夹持器（图 7-3-34）、盲文显示装置（图 7-3-35）和扬声器（图 7-3-36）等[212-215,219]。基于此原理逆过程可设计和制造能量收集器（图 7-3-37）[219,231]

图 7-3-29 列举出一些 EAP 驱动器。典型的结构包括三爪夹紧形（a）、四爪夹紧形（b）、半球形（c）、卷形（d）、折叠形（e）、堆栈形（f）和球形（g）等。

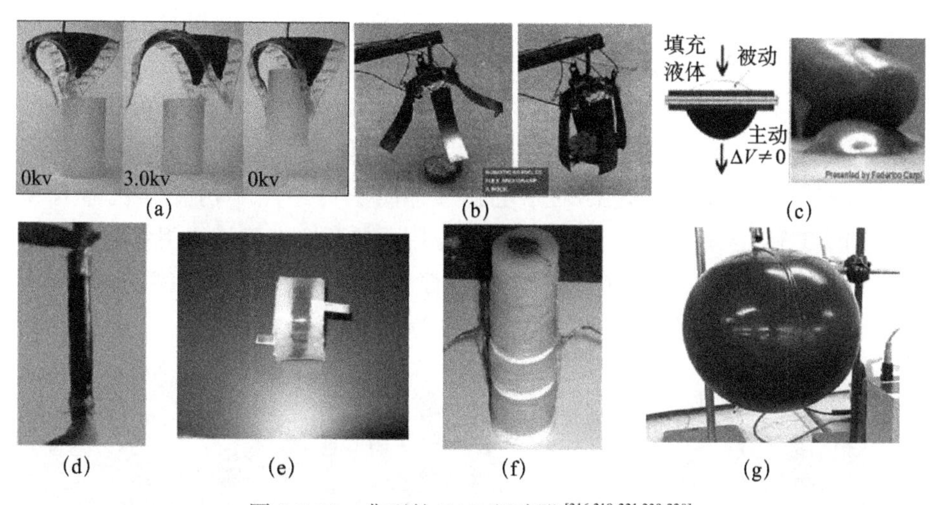

图 7-3-29　典型的 EAP 驱动器[216,219-221,228-230]

图 7-3-30 是基于介电弹性体卷形驱动器驱动的人工肌肉手臂。它与人进行了掰手腕比赛。制造出的人工肌肉手臂能够支撑对抗人类手臂时间长达 26s，这是一个重要的里程碑，显示出介电弹性体在人工肌肉仿生领域的巨大应用潜力[217]。介电弹性体的可塑性和可控制性，将对人造假肢和类人机器人的快速发展起到极大的推动作用。2005 年 11 月在韩国釜山举办的 APEC 最高首脑会议上，由汉森机器人技术公司设计和制造的机器人展示了介电弹性体的在面部表情领域应用的潜力（模拟爱因斯坦面部表情），它可以表现出

快乐、忧伤和愤怒等不同情绪，引起参会人员的广泛关注和极大反响[217]。

图 7-3-30　基于介电弹性体的人工肌肉手臂[222]

科技的迅猛发展有望制造出类昆虫机器人（机器昆虫），它们能够进入隐蔽地区的结构（如飞机的引擎）中执行维修和侦察等任务。可以想象，在不久的将来，电活性聚合物在集成众多学科成就的基础上将完成模拟地球上生物的庞大而崭新的计划，比如，像猫一样软着陆；像蚱蜢一样做远距离跳跃；像鱼一样会游泳；像鸟一样会飞翔；像蚂蚁一样会挖掘和分工协作。为达到理想的仿生学性能，用于制造人工肌肉的材料必须具有类自然肌肉的重要特征，如驱动方式、驱动应力、反应速度和效率等。肌肉功能很多，运动时肌肉通常作为能量的供应者，同时起支撑作用，具有可变的硬度和刚度。具有“人工肌肉”之称的介电弹性体材料已经用于仿生学机器人领域，图 7-3-31 是一些例子，(a) 和 (b) 是应用丙烯酸介电弹性体制备的卷型驱动器驱动的类昆虫机器人，(c) 是南丹麦大学设计和制造的自感知智能机器人，其四肢的传感器是应用介电弹性体材料制造的[217]。

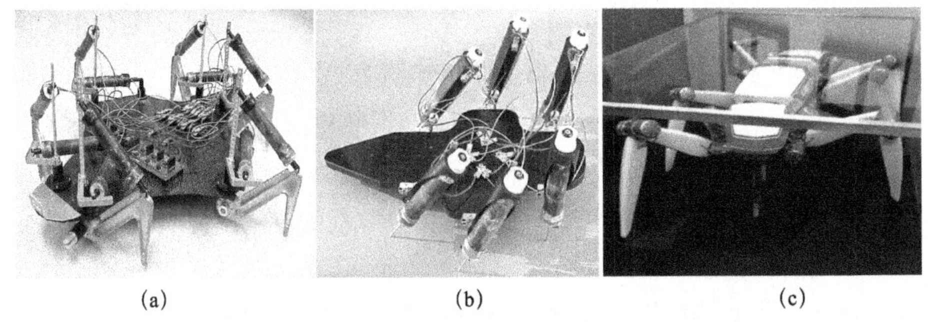

(a)　(b)　(c)

图 7-3-31　介电弹性体仿生机器人[217,227]

介电弹性体由于具有质量轻和柔性好的特点，在航空航天等领域具有广阔的应用前景，美国国家航空航天局、国防部高级计划研究局和欧洲航空航天局等相关机构都积极开展了广泛的探索研究，取得了一系列的研究成果[216]。1995年，美国航空航天局启动了轻质肌肉驱动器的研究计划，由喷气推进实验室Yoseph Bar-Cohen博士研究组开展了介电弹性体驱动机器人手臂的研究。美国斯坦福研究院与日本签署了微型机器人研究计划，并最终选择了介电弹性体材料研制人造肌肉驱动器[217]。原有的空间探测器窗口除尘机械刷质量大、结构复杂，而基于电活性聚合物驱动器，可以制成空间用智能除尘刷［如月球车除尘刷，图7-3-32(a)和(b)分别为示意图与实物图］，具有质量轻、结构紧凑、驱动功率低等优点，可有效减轻空间探测器的重量和节约能源[224]。

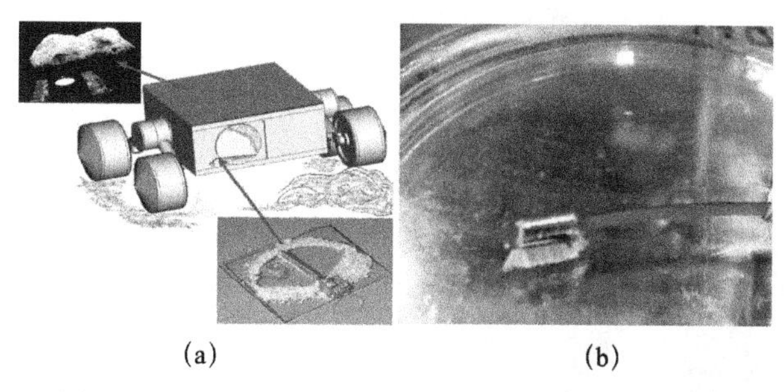

(a)　　(b)

图7-3-32　基于电活性聚合物材料的智能除尘刷[224]

如图7-3-33(a)所示，2007年，瑞士联邦材料测试研究实验室提出采用介电弹性体制作飞艇舵驱动器，能够控制飞艇的自由转向[225]。如图7-3-33(b)所示，从2002年，俄亥俄航天局开始了基于电活性聚合物材料进行的新概念飞行器——固态飞行器（solid state aircraft，SSA）的研究[226]。

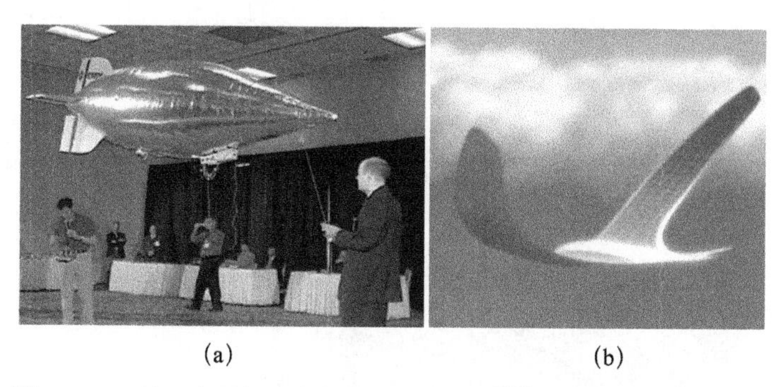

(a)　　(b)

图7-3-33　基于电活性聚合物的(a)飞艇舵[225]和(b)固态飞行器[226]

太空碎片对于运行在地球轨道的航天器已经产生了日益严重的威胁，基于介电弹性体最小能量结构制作的夹持器已经在朝着缓解该问题方面做出了努力[232]。相比于机械臂、网状与非接触式的碎片捕捉机构，所提出的新型夹持器具有质量轻、可重复工作、能耗低等优点。图7-3-34中(a)、(b)与(c)分别是安装有该夹持器的卫星概念图，夹持目标与多节式夹持器实物图。工作过程如图 7-3-34（d）所示，在发射过程中处于卷曲状态（stage 1），入轨后释放，由于弹性能恢复初始状态（stage 2），电压加载后可进一步展开（stage 3），夹持空间碎片后电压卸载可将其抓牢（stage 4 与 5）。经试验测量，已制作出的用于概念验证的夹持器可以实现大于 60°的最大弯曲角与 2.2mN 的最大夹持力，同时结构质量仅有 0.65g。

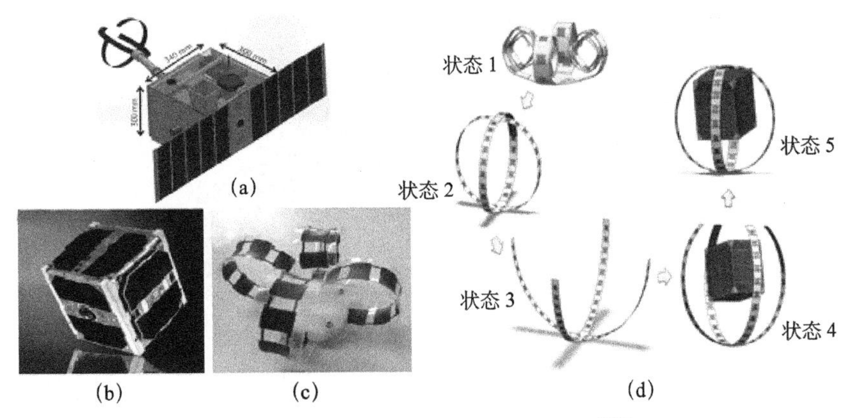

图 7-3-34 卷曲多节式可展开夹持器 [232]

传统的可重复利用的盲文设备都是由压电元件制成的，其结构设计复杂再加上压电盲文元件的昂贵，导致广大视力有障碍的患者对于这种设备望而却步。如图 7-3-35(a) 和 (b) 所示，哈尔滨工业大学冷劲松课题组设计和制造了基于介电弹性体驱动器的盲文触觉显示器。它由控制部分，驱动器和机械传动部分，触摸屏幕及外壳组成。当用户通过人机交互界面输入英文字母或单词时，程序由对应的布莱尔编码关系转换成电信号，通过单片机输出信号，控制 6 个继电器的开断，进而控制驱动器是否变形，而连接在介电弹性体驱动器上的机械传动结构将控制触摸屏幕的规则变化。

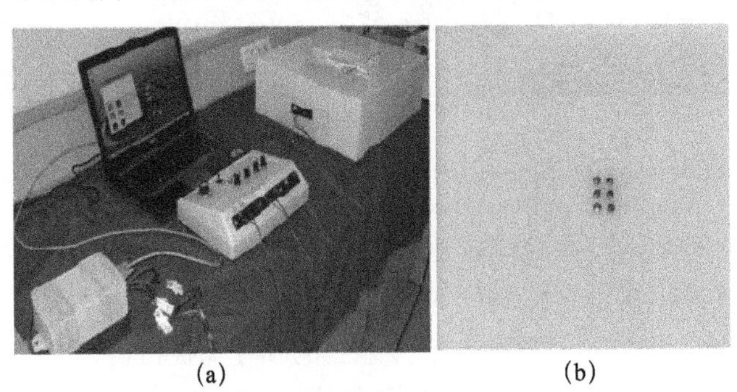

图 7-3-35 基于介电弹性体驱动器的盲文显示器 [223]

由于介电弹性体具有较宽的频带，而且机电响应在毫秒数量级，当对其施加交流电压时，它将以某一频率振动，产生声音。图 7-3-36(a) 和 (b) 分别是介电弹性体扬声器的基本结构和工作原理。面临的一个问题是，声音在传播过程中的失真较为严重。已经有一些初步的研究通过修正声音的非线性部分来提高声音的保真度，取得了一定效果 [229]。

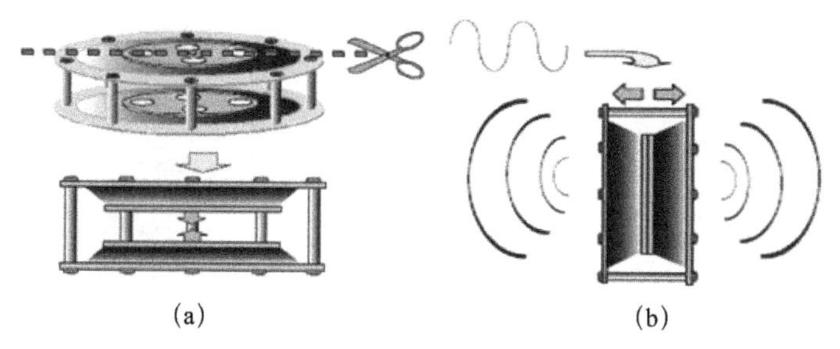

(a)　　(b)

图 7-3-36　介电弹性体扬声器 [229]

基于介电弹性体的机械致电效应，可以设计和制造能量收集器。图 7-3-37(a) 将介电弹性体安装在鞋跟处来收集人类行走时的能量 [218,229]；图 7-3-37(b) 与 (e) 通过水轮式与管式结构收集河水流动产生的机械能并转化成电能 [233,234]；图 7-3-37(c) 展示了一个薄膜充气式能量收集器 [235]；图 7-3-37(d) 通过浮漂式能量收集器将海浪振动的机械能转化为电能 [236]。

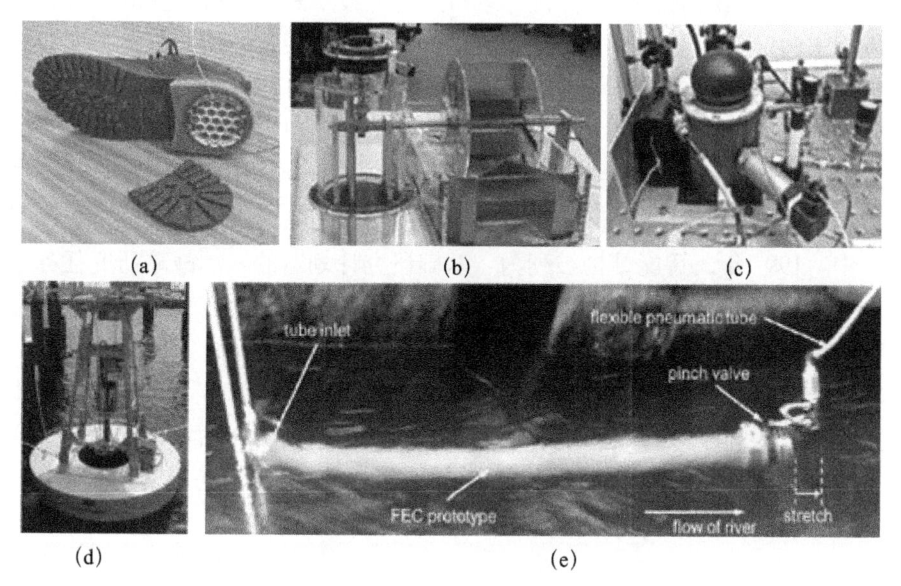

(a)　　(b)　　(c)

(d)　　(e)

图 7-3-37　基于介电弹性体的能量收集器 [218,229,233-236]

介电弹性体材料已经开始商业化。美国3M公司所生产的VHB胶带就是一种性能优异的丙烯酸介电弹性体材料。丹麦Danfoss PolyPower A/S公司2011年成立，并制备出一种新型性能优异的介电弹性体。图7-3-38描述介电弹性体的薄膜的商业生产过程。据报道，Danfoss公司应用此材料制造的驱动器可以驱动10kg的重物。

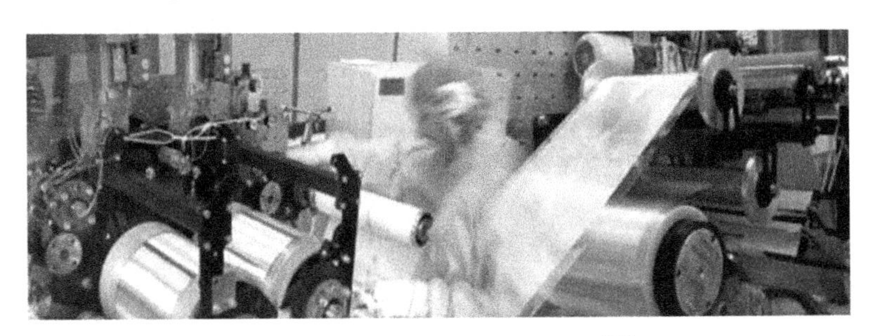

图7-3-38　介电弹性体商业生产[219]

第四节　学科前沿问题

智能材料与结构是一个新兴的前沿学科，所涉及的知识面广，发展潜力巨大，应用前景广阔。智能材料与结构的研究最早是在20世纪70年代末期开展起来的，随后世界各主要发达国家相继开展该领域的研究工作，使智能结构技术得到广泛承认。美国、英国、德国等西方发达国家相继投入巨资开展该领域的研究。进入20世纪90年代，它更是受到高度重视。特别是美国军方和一些政府机构直接参与了研究和开发工作。1995年，白宫科技政策办公室和国家关键技术评审组将智能材料结构技术列入《美国国家关键技术报告》中。1997年，智能结构被列为“基础研究计划”的六项战略研究任务之一。美国各军种、弹道导弹防御局和美国航空航天局以及波音、麦道等大公司都分别制定了研究与发展计划，如弹道导弹防御局的“自适应结构计划”，陆军研究局的“智能材料与结构计划”，空军航天实验室的“智能结构蒙皮计划”等。1980～1991年，英、法、意三国的7家公司在欧共体的支持下完成了欧洲在此领域的第一个合作研究计划“复合材料光学传感计划”。20世纪90年代初英国成立了欧洲这一领域的首家专门研究机构“斯特拉斯立德大学智能材料与结构研究所”。德国正在研究将植入光纤的自诊断智能结构用于可重复使用运载器的损伤探测和评估。日本自1984年即着手空间应用的智能结构的研究，日本航空宇宙研究所等参加了这项工作。

我国对智能复合材料和结构技术的发展十分重视，智能材料已经被列入2020年《国家中长期科学和技术发展规划纲要》“智能材料与结构技术”前沿技术中。国内有关智能材料和结构的研究始于20世纪90年代中期。目前，哈尔滨工业大学、南京航空航天大学、北京航空航天大学、中国科学技术大学、中国科学院金属研究所、兰州大学等单位在形状记忆聚合物、光纤传感器、电致活性聚合物、压电驱动器、形状记忆合金、电/磁流变液体、铁电材料等智能材料和结构等方面进行了初期探索性的研究工作。哈尔滨工业大学从1992年开始在国内较早地开展了智能材料和结构方面的研究，在形状记忆聚合物、光纤传感器、结构健康监测、电致活性聚合物、电（磁）流变液体材料、振动主动控制、压电驱动器、多功能纳米复合材料等研究领域，取得了一定的研究成果。与国外相比，国内有关智能材料和结构的研究虽然取得了一定的研究成果，但是起步较晚，底子较薄，自主创新方面也比较薄弱，尤其是自适应、自感知、智能材料与结构前沿技术的研究才刚刚起步，其在应用过程中面临传感、驱动、自适应、多功能、多场耦合等诸多技术难题和极端复杂环境问题。

以形状记忆材料、电活性材料为代表的新型自感知、自适应、自修复、多功能的智能材料和结构研制及其应用的前沿技术和基础科学问题研究涉及材料、力学、化学、物理及机械等多学科综合交叉，在响应机制和工作环境上具有电、力、热、磁、化学多场耦合的特点，在结构的研究尺度上具有从细观、微观结构到宏观大变形的跨越，是智能材料领域中最具挑战性的前沿和基础研究课题之一。

由于智能材料与结构具有多场耦合和多场调控的性能特点，其应用离不开力-磁-电-热等耦合作用的复杂环境。与此同时，新的智能材料及结构层出不穷，比如，功能软物质结构在力、电、磁等多场耦合下发生大变形，大变形又反过来影响电磁场的分布，而且它还具有湿热和化学环境敏感性，如何表征功能软物质的多场耦合性能成为新的研究方向。同时，由于智能材料及结构的本征特性，它还与声波和电磁波等耦合，可以用来设计电磁场调控声波以及变形调控电磁波传输等功能器件。鉴于智能材料和结构的应用多种多样，不同应用要求的材料维度不一样，有大至毫米、厘米甚至更大尺度，小至微米、纳米尺度的应用。因此，智能材料与结构的多场耦合和动态跨尺度力学研究是固体力学领域的一个迫切和重要的研究方向。

美国哈佛大学George M. Whitesides研究组（化学材料方向）和锁志刚研

究组（材料力学性能方向）在智能软材料研究方面密切合作引领了本领域国际发展前沿。锁志刚研究组就介电弹性体和水凝胶这两类软物质，发展了热力学框架内的大变形力电耦合本构关系，并深入开展了黏弹性大变形力电耦合本构关系研究，在介电弹性体俘能器、制动器、传感器电致变形方面取得了若干开创性工作。美国伊利诺伊大学厄巴纳香槟分校 John A. Rogers 研究组（材料方向）和美国西北大学黄永刚研究组（材料力学性能方向）在柔性电子器件研究方面密切合作取得了重大研究成果，在 *Science*, *Nature*, *Nature Nanotechnology* 上发表了多篇学术论文。

在形状记忆聚合物的本构理论研究上，主要以线性黏弹性本构关系为基础，基于小变形理论和率无关性构造本构模型。美国哈佛大学的 John Hutchinson 教授、锁志刚教授、美国西北大学的黄永刚和伊利诺伊大学香槟分校 John A.Rogers 教授都对模量较高的增强体在模量很低的基本中的屈曲问题进行了大量的研究。针对介电弹性体软材料的相关理论研究，美国哈佛大学锁志刚等基于热力学理论，考虑机电耦合效应，建立系统的自由能函数，推导介电弹性体的本构关系。基于此理论，研究人员进行了大变形、稳定性分析、器件优化设计、黏弹性耗散、动力学响应等一系列分析。对于其他智能聚合物材料，如压电、铁电聚合物等以磁、电场作为激励源的软智能材料的研究，主要以多场耦合条件下材料的电、力学性质为主，并注重观测材料内部的微观结构演化规律与宏观特征之间的跨尺度研究。

从国内外研究文献看，耦合了热场、电场、化学场和力场的智能软聚合物材料的力学性能研究还较少。包含大变形、率相关性以及更多微结构机制的热力学理论研究将成为今后的研究趋势。

第五节　与实际需求结合的重大问题

目前世界各国智能软聚合材料的研究极为活跃，充满了机遇和挑战，新技术、新专利层出不穷。发达国家企图通过知识产权的形式在智能材料领域形成技术垄断，并试图占领中国广阔的市场，这种态势应引起我国的高度重视。结合现有研究成果可以看出，智能材料和结构及其在实际工程应用上应用仍面临着诸多困难，可以总结出其面临的问题与挑战主要包括以下几个方面。

一、变形 / 承载一体化、变刚度技术

蒙皮是保证飞行器气密性和良好气动布局的关键，在传统飞行器上可以采用刚性蒙皮来保证外形形状和承受气动载荷，但是作为变体飞行器的蒙皮不仅仅保证足够的面外刚度承受气动载荷，而且还要有较低的面内刚度能够随结构变形。蒙皮实现变形 / 承载一体化是未来变体飞行器蒙皮结构发展的一个主要方向，变刚度蒙皮结构也是未来研究的重要方向，以形状记忆聚合物、变刚度管等为代表的变刚度材料和结构有望解决此类问题，例如，通过控制液体流动的变刚度管蒙皮结构、基于形状记忆聚合物的变刚度管蒙皮结构、通过气压改变刚度的气动肌纤维蒙皮结构等。

如果采用形状记忆聚合物做的飞机蒙皮，当材料温度低于玻璃化转变温度时，材料呈玻璃态，弹性模量较高，有足够的强度和刚度来满足飞行器承载方面的要求；当材料温度高于玻璃化转变温度时，材料呈橡胶态，弹性模量较低，可以实现 200% 的大变形，满足飞行器变形方面的要求；当温度降低时，材料又会从橡胶态转换成玻璃态，弹性模量提高，可以继续承载。但是，飞行器在航行过程中，一直承受着气动载荷，不可能在变形的时候降低气动载荷，但此时形状记忆聚合物的刚度又较低。因此，材料在低模量时能否承受相应的气动载荷是一个难题。

二、轻质 / 大输出力驱动器技术

结构变形需要驱动器来实现，驱动器既要满足变形结构对大驱动力的需求，同时也要满足轻质、快速响应、高精度、循环响应快等需求，轻质 / 大输出力驱动器是智能材料发展的一个难题。形状记忆合金以其驱动力大的特点已经被广泛使用，但是有驱动循环响应慢、变形小等问题，需要进一步对材料改性，为其进一步应用提供保障。利用压电叠堆陶瓷、磁滞伸缩材料等智能材料设计轻质大输出力的驱动器也是未来驱动器的一个研究方向，此类材料的特点是输出力大、能量密度高、易控制，但是输出位移小。

利用形状记忆聚合物、电活性聚合物等智能软材料和结构的驱动器由于材料本身的特点，已经使驱动器本身满足了轻质的要求。如何高效实现形状记忆聚合物材料的双向驱动，并且加快其响应速度、提高驱动力是形状记忆聚合物在驱动器应用上所面临的问题。介电弹性体材料虽然响应速度快，但是同样存在驱动力小，并且较高的驱动电压也是一个亟须改善的问题。

三、多物理场耦合作用条件下智能材料的力学测试方法

对于智能材料的力学分析需要重点考虑几方面的问题：首先智能材料在力学行为上具有几何与材料非线性的特征；其次在响应机制和工作环境上具有多场耦合的特点；最后在其力学研究的尺度上存在从理解材料微观结构特点到表征宏观力学响应的跨越。此外，对于智能软材料变形稳定性的研究以及材料失效和可靠性的预测也是一个重要问题。

在力学测试方法上面，智能材料的多场耦合加载与测量涉及的器件一般较多，往往会存在器件之间的空间干涉和各物理场之间的场间干扰问题，影响仪器功能实现和测量的准确性。目前国际上有高低温压痕技术、力电耦合的压痕技术，但温度量程和电压量程指标一般较低，缺少力磁耦合微纳米力学测试技术，以及力电磁多场、高低温和气液多氛围联合加载的技术和设备。

第六节　未来5～10年学科发展趋势

针对智能软聚合物材料的自感知、自适应、自修复等问题，需要对智能软聚合物材料进行复合改性设计和多功能设计，研究电、热、光、磁、溶剂等智能聚合物复合材料的记忆效应驱动新机制和形状恢复行为规律；同时针对智能聚合物复合材料材料的新型驱动方式，需要对智能聚合物及其复合材料在温度场、应力场、磁场、电场、化学场、多场耦合等物理场下的响应行为和感应规律进行研究，获得“结构、功能一体化”的智能聚合物形状记忆复合材料及结构，并推动其在实际工程领域的应用。

一、自感知、自适应、自修复智能软聚合物及结构的设计与研制

开展基于光纤传感器、纳米复合材料等嵌入式传感元件的自感知智能软聚合物及多功能、多参数监测机制和方法的研究，实时采集智能软聚合物与结构自身服役状态信息、实时监测与评价其从制备到服役过程中工艺参数及损伤演化规律，评价和评估其安全状况和剩余服役寿命；优化设计与研制形状自适应智能复合材料，进行可变形结构的非线性展开动力学分析，优化设计与研制基于智能复合材料的自适应可变形结构，预报和评估其展开与变形

性能，揭示智能复合材料结构形状自适应变形规律；优化设计与研制适用于多种环境下的高修复效率、高安全性、高可靠性、长使用寿命的多重响应性自修复智能材料，优化设计与研制自修复复合材料的智能复合结构，预报和评估其使用安全性、稳定性与耐久性；研制集结构承载、防雷击及防除冰为一体的纳米纸等纳米复合材料增强的多功能智能复合材料结构，试验研究多功能纳米智能复合材料的防除冰及防雷击性能，评估其使用安全性、稳定性与耐久性。

二、智能软聚合物及结构多场耦合条件下的本构理论建立

新兴智能软聚合物材料及其结构基础科学问题研究涉及力学、物理、材料、化学及机械等多学科综合交叉，在力学上具有几何大变形与材料非线性的特征，在响应机制和工作环境上具有多场耦合的特点，在结构的研究尺度上具有从微观结构到宏观大变形的跨越，需要进一步考虑多学科相互关联的新的力学研究，并且这种研究不仅涵盖了常规固体力学的内容与方法，而且所涉及的非线性、多场耦合和跨尺度的力学建模与性能表征等科学问题是固体力学域中具有挑战性的前沿研究课题。构建能够准确描述热-力-电-磁-化学多物理场耦合作用下智能材料的热力学自由能方程，发展智能材料的三维大变形本构理论，验证和修正智能材料的宏观热力学本构关系。研究材料非线性、几何非线性及力-电-磁-热-化学多物理场耦合条件下对智能材料的应力、应变、应变率、驱动力、变形等性能的影响，分析智能材料的宏观力学行为。

三、智能软聚合物实验方法的发展及其结构力学性能和失效行为表征

随着材料制备技术的发展，智能软聚合物与器件的特征尺寸越来越小，其力学性能存在明显的尺度效应和外场效应，因此建立多场耦合变形和断裂破坏理论模型，发展多场耦合计算方法，提出多场耦合力学性能表征方法，对于研究微纳米尺度材料在外场作用下的力学性能测试具有重要意义，同时也为新的智能材料和结构的设计、制备与应用提供指导，如外场调控的声学器件、基于接触测量的磁场传感器等。智能材料与器件固有的力-磁-电-热耦合特性，使其在力场、磁场、电场和热场及其联合作用下发生微结构（主要是畴域结构）的演化，并进而调控其力、磁、电和热学性能。如何测试和表征智能软聚合物在多场耦合下的微区力学性能对于智能材料的结构与器件

设计和质量评估有重要意义。开发互不干扰的多场多氛围环境耦合加载与屏蔽技术，以及多场多氛围环境微纳米力学测试表征方法尤为重要。测试智能材料在力、电、磁、热多场环境、高低温环境和气液氛围等环境条件的变形曲线，获得杨氏模量、硬度、断裂韧性、蠕变、疲劳等材料参数与多场环境之间的关系，研究外场和氛围引起的材料相变、畸变行为、场致弹性、场致塑性、场致韧性等性能。

致谢： 感谢国家自然科学基金（批准号：11102052, 11272106, 11225211）资助。

刘立武、吕雄飞、刘彦菊、冷劲松（哈尔滨工业大学）

参考文献

[1] Suo Z G. Theory of dielectric elastomers. Acta Mech Solida Sin. 2010, 23(6): 549-578.

[2] 杜善义，冷劲松，王殿富．智能材料系统和结构．北京：科学出版社，2001.

[3] Otsuka K, Wayman C M. Shape Memory Materials. Cambridge University Press, 1999.

[4] Behl M, Lendlein A. Shape-memory polymers. Materials Today, 2007, 10: 20-28.

[5] Rousseau I A. Challenges of shape memory polymers: a Review of the progress toward overcoming SMP's limitations. Polymer Engineering & Science, 2008, 48(11): 2075-2089.

[6] Birman V. Review of mechanics of shape memory alloy structures. Applied Mechanics Reviews, 1997, 50(11): 629-645.

[7] Liu C, Qin H, Mather P T. Review of progress in shape-memory polymers. Journal of Materials Chemistry, 2007, 17(16): 1543-1558.

[8] Leng J S, Lan X, Liu Y J, et al. Shape-memory polymers and their composites: stimulus methods and applications. Progress in Materials Science, 2011, 56(7): 1077-1135.

[9] Huang W M, Liu N, Phee S J. Thermo-moisture responsive polyurethane shape memory polymer and composite: A review. Journal of Materials Chemistry, 2010, 20: 3367-3381.

[10] Baer G, Wilson T S, Mathews D L, et al. Shape-memory behavior of thermally stimulated polyurethane for medical applications. Journal of Applied Polymer Science, 2007, 103(6): 3882-3892.

[11] Xie T. Tunable polymer multi-shape memory effect. Nature, 2010, 464: 267-270.

[12] Kim B K, Lee S Y, Xu M. Polyurethanes Having Shape Memory Effects. Polymer, 1996, 7:

5781-5793.
[13] Karger-Kocsis J. Polypropylene: Copolymers and Blends. London: Chapman & Hall, 1991: 43-47.
[14] Hayashi S, Kondo S, Kapadia P, et al. Room temperature functional shape-memory polymers. Plastic Engineering, 1995, 51(2): 29-31.
[15] Hu J L, Ji F L, Wong Y W. Dependency of the shape memory properties of a polyurethane upon thermomechanical cyclic conditions. Polymer International, 2005, 54(3): 600-605.
[16] Prima M A, Gall K, McDowell D L, et al. Deformation of epoxy shape memory polymer foam: Part II. Mesoscale modeling and simulation. Mechanics of Materials, 2010, 42(3): 315-325.
[17] Chun B C, Cho T K, Chong M H, et al. Structure-property relationship of shape memory polyurethane cross linked by a polyethyleneglycol spacer between polyurethane chains. Journal of Materials Science, 2007, 42(21): 9045-9056.
[18] Li C C, Zhu G, Li Z Y. Effects of rosin-type nucleating agent on polypropylene crystallization. Journal of Applied Polymer Science, 2002, 83(5): 1069-1073.
[19] Li F, Feng J, Zhou R X. Synthesis and characterization of novel biodegradable poly(p-dioxanone-co-ethylethylene phosphate)s. Journal of Applied Polymer Science, 2006, 102(6): 5507-5511.
[20] Lan X, Wang X H, Liu Y J, et al. Fiber reinforced shape-memory polymer composite and its application in a deployable hinge. Smart Materials and Structures, 2009, 18: 024002.
[21] Ji F, Zhu Y, Hu J, et al. Smart polymer fibers with shape memory effect. Smart Materials and Structures, 2006, 15: 1547-1554.
[22] Gall K, Yakacki C M, Liu Y P, et al. Thermomechanics of the shape memory effect in polymers for biomedical applications. Journal of Biomedical Materials Research Part A, 2005, 73A(3): 339-348.
[23] Liu Y J, Du H Y, Liu L W, et al. Shape memory polymers and their composites in aerospace applications: a review. Smart Materials and Structures, 2014, 23(2): 1077-1135.
[24] Hu J L, Zhu Y, Huang H H, et al. Recent advances in shape–memory polymers: structure, mechanism, functionality, modeling and applications. Progress in Polymer Science, 2012, 37(12):1720-1763.
[25] Hornbogen E. Comparison of shape memory metals and polymers. Advanced Engineering Materials, 2006, 8: 101-106.
[26] Leng J S, Lv H B, Liu Y J, et al. Synergic effect of carbon black and short carbon fiber on shape memory polymer actuation by electricity. Journal of Applied Physics, 2008, 104: 104917.

[27] Leng J S, Lv H B, Liu Y J, et al. Electroactivate shape-memory polymer filled with nanocarbon particles and short carbon fibers. Applied Physics Letters, 2007, 91: 144105.

[28] Yakacki C M, Satarkar N S, Gall K, et al. Shape-memory polymer networks with Fe_3O_4 nanoparticles for remote activation. Journal of Applied Polymer Science, 2009, 112(5): 3166-3176.

[29] Small W, Metzger M F, Wilson T S, et al. Laser-activated shape memory polymer microactuator for thrombus removal following ischemic stroke: preliminary in vitro analysis. Quantum Electronics, 2005, 11(4): 892-901.

[30] Leng J S, Wu X L, Liu Y J. Infrared light-active shape memory polymer filled with nanocarbon particles. Journal of Applied Polymer Science, 2009, 114(4): 2455-2460.

[31] Lv H B, Leng J S, Liu Y J, et al. Shape-memory polymer in response to solution. Advanced Engineering Materials, 2008, 10: 592-595.

[32] Ahn S K, Kasi R M, Kim S C, et al. Stimuli-responsive polymer gels. Soft Matter, 2008, 4: 1151-1157.

[33] Lendlein A, Schmidt A M, Schroeter M, et al. Shape-memory polymer networks from oligo(-caprolactone)dimethacrylates. Journal of Polymer Science Part a-Polymer Chemistry, 2005, 43: 1369-1381.

[34] Deng L C, Chun B, Patrick T M. Chemically cross-linked polycyclooctene: Synthesis, characterization, and shape memory behavior. Macromolecules, 2002, 35(27): 9868-9874.

[35] Zhang S F, Feng Y K, Zhang L, et al. Novel interpenetrating networks with shape-memory properties. Journal of Polymer Science Part a-Polymer Chemistry, 2007, 45: 768-775.

[36] Lendlein A, Jiang H Y. light-induced shape-memory polymers. Nature, 2005, 434: 879-882.

[37] Jeong H M, Lee S Y, Kim B K. Shape memory polyurethane containing amorphous reversible phase. Journal of Materials Science, 2000, 35(7): 1579-1583.

[38] Jeong H M, Kim B K, Choi Y J. Synthesis and properties of thermotropic liquid crystalline polyurethane elastomers. Polymer, 2000, 41(5): 1849-1855.

[39] Jeong H M, Ahn B K, Kim B K. Temperature sensitive water vapour permeability and shape memory effect of polyurethane with crystalline reversible phase and hydrophilic segments. Polymer International, 2000, 49(12): 1714-1721.

[40] Tobushi H, Okumura K, Hayashi S, et al. Thermomechanical constitutive model of shape memory polymer. Mechanics of Materials, 2001, 33(10): 545-554.

[41] Tobushi H, Hashimoto T, Ito N, et al. Shape fixity and shape recovery in a film of shape memory polymer of polyurethane series. Journal of Intelligent Material Systems and Structures, 1998, 9(2): 127-136.

[42] Tobushi H, Hashimoto T, Hayashi S, et al. Thermomechanical constitutive modeling in

shape memory polymer of polyurethane series. Journal of Intelligent Material Systems and Structures, 1997, 8(8): 711-718.

[43] Kim B K, Shin Y J, Cho S M, et al. Shape-memory behavior of segmented polyurethanes with an amorphous reversible phase: the effect of block length and content. Journal of Polymer Science Part B-Polymer Physics, 2000, 38(20): 2652-2657.

[44] Kim B K, Lee S Y, Lee J S, et al. Polyurethane ionomers having shape memory effects. Polymer, 1998, 39(13): 2803-2808.

[45] Jang M K, Andreas H, Kim B K. Shape memory polyurethanes cross-linked by surface modified silica particles. Journal of Materials Chemistry, 2009, 19: 1166-1172.

[46] Rule J D, Sottos N R, White S R. Effect of microcapsule size on the performance of self-healing polymers. Polymer, 2007, 48: 3520-3529.

[47] Bertmer M, Buda A, Blomenkamp-Höfges I, et al. Biodegradable shape-memory polymer networks: characterization with solid-state NMR. Macromolecules, 2005, 38(9): 3793-3799.

[48] Voit W, Ware T, Dasari R R, et al. High-strain shape-memory polymers. Advanced Functional Materials, 2010, 20(1): 162-171.

[49] Safranski D L, Gall K. Effect of chemical structure and crosslinking density on the thermo-mechanical properties and toughness of (Meth)acrylate shape memory polymer networks. Polymer, 2008, 49(20): 4446-4455.

[50] Yakacki C M, Shandas R, Safranski D, et al. Strong, tailored, biocompatible shape-memory polymer networks. Advanced Functional Materials, 2008, 18(16): 2428-2435.

[51] Yakacki C M, Shandas R, Lanning C, et al. Unconstrained recovery characterization of shape-memory polymer networks for cardiovascular applications. BioMaterials, 2007, 28(14): 2255-2263.

[52] Xie F, Huang L N, Liu Y J, et al. Synthesis and Characterization of High Temperature Cyanate-Based Shape Memory Polymers with Functional Polybutadiene/Acrylonitrile. Polymer, 2014, 55: 5873-5879.

[53] Wu X L, Zheng H, Liu Y J, et al. Thermomechanical property of epoxy shape memory polymers. International Journal of Modern Physics B, 2010, 24: 2386-2391.

[54] Leng J S, Xie F, Wu X L, et al. Effect of the γ-radiation on the properties of epoxy-based shape memory polymers. Journal of Intelligent Material Systems and Structures, 2014, 25(10): 1256-1263.

[55] Lv H B, Leng J S, Du S Y. A phenomenological approach for the chemo-responsive shape memory effect in amorphous polymers. Soft Matter, 2013, 9(14): 3851-3858.

[56] Monkman G J. Advances in shape memory polymer actuation. Mechatronics, 2000, 10: 489-498.

[57] Wei Z G, Sandstrom R, Miyazaki S. Shape memory materials and hybrid composites for smart systems: Part II shape-memory hybrid composites. Journal of Materials Science, 1998, 33(15), 3763-3783.

[58] Ohki T, Ni Q Q, Ohsako N, et al. Mechanical and shape memory behavior of composites with shape memory polymer. Composites Part A: applied science and manufacturing, 2004, 35(9): 1065-1073.

[59] Gall K, Kreiner P, Turner D, et al. Shape-memory polymers for microelectromechanical systems. Journal of Microelectromechanical System, 2004, 13: 472-483.

[60] Liu Y P, Gall K, Dunn M L, et al. Thermomechanics of shape memory polymer nanocomposites. Mechanics of Materials, 2004, 36: 929-940.

[61] Gall K, Dunn M L. Shape memory polymer nanocomposites. Acta Materialia, 2002, 50: 5115-5126.

[62] Yang B, Huang W M, Li C, et al. Effects of moisture on the thermomechanical properties of a polyurethane shape memory polymer. Polymer, 2006, 47: 1348-1356.

[63] Huang W M, Yang B, An L, et al. Water-driven programmable polyurethane shape memory polymer: demonstration and mechanism. Applied Physics Letters, 2005, 86: 114105.

[64] Leng J S, Lan X, Huang W M. Electrical conductivity of thermoresponsive shape-memory polymer with embedded micron sized ni powder chains. Applied Physics Letters, 2008, 92: 014104.

[65] Leng J S, Huang W M, Lan X, et al. Significantly reducing electrical resistivity by forming conductive Ni chains in a polyurethane shape-memory polymer/carbon-black composite. Applied Physics Letters, 2008, 92: 204101.

[66] Rousseau I A, Xie T. Shape memory epoxy: composition, structure, properties and shape memory performances. Journal of Materials Chemistry, 2010, 20: 3431-3441.

[67] Barrett R, Francis W, Abrahamson E, et al. Qualification of elastic memory composite hinges for spaceflight applications. 47th AIAA/ASME/ASCE/AHS/ASC Structures, Structural Dynamics, and Materials Conference, 2006, 20391-10.

[68] Everhart M C, Nickerson M C, Hreha R D. High-temperature reusable shape memory polymer mandrels. Smart Structures and Materials 2006: Industrial and Commercial Applications of Smart Structures Technologies, Proc. of SPIE, 2006, 6171: 61710K.

[69] Sahoo N G, Jung Y C, Goo N S, et al. Conducting shape memory polyurethane-polypyrrole composites for an electroactive actuator. Macromolecular Materials and Engineering, 2005, 290(11): 1049-1055.

[70] Schmidt A M. Electromagnetic activation of shape-memory polymer networks containing magnetic nanoparticles. Macromolecular Rapid Communication, 2006, 27(14): 1168-1172.

[71] Duncan M J, Metzeger M F, Schumann D, et al. Photothermal properties of shape memory polymers micro-actuators for treating stroke. Lasers in Surgery and Medicine, 2002, 30(1): 1-11.

[72] 吕海宝. 电驱动与溶液驱动形状记忆聚合物混合体系及其本构方程. 哈尔滨: 哈尔滨工业大学博士论文，2010.

[73] Lu H B, Liu Y J, Leng J S, et al. Comment on water-driven programmable polyurethane shape memory polymer: demonstration and mechanism. Applied Physics Letters, 2010, 97(5): 056101.

[74] Chen S J, Hu J L, Yuen C W, et al. Novel moisture-sensitive shape memory polyurethanes containing pyridine moieties. Polymer, 2009, 50(19): 4424-4428.

[75] Pelrine R, Kornbluh R, Pei Q. High-speed electrically actuated elastomers with strain greater than 100%. Science, 2000, 287(28): 836-839.

[76] Brochu P, Pei Q B. Advances in dielectric elastomers for actuators and artificial muscles. Macromolecular Rapid Communications, 2010, 31: 10-36.

[77] Mirfakhrai T, Madden J, Baughman R. Polymer artificial muscles. Materials Today, 2007, 10(4): 30-38.

[78] O′ Halloran A, O′ Malley F, McHugh P. A review on dielectric elastomer actuators, technology, applications, and challenges. Journal of Applied Physics, 2008, 104: 071101.

[79] Madden J D, Vandesteeg N, Madden P G, et al. Artificial muscle technology: Physical principles and naval prospects. IEEE Journal of Oceanic Engineering, 2004, 29: 706.

[80] Bar-Cohen Y. Electroative polymers as artificial muscles capabilities, potentials and challenges. Hanbook on biomimetics. Section 11, in Chapter 8. NTS Inc. 2000, 1-13.

[81] Bar-Cohen Y. Biologically inspired technology using electroactive polymers (EAP). Proc. SPIE, 2006, 6168: 616803.

[82] Bar-Cohen Y. Electroative polymers: current capabili ties and challenges. Proc. SPIE, 2002, 4695: 1-7.

[83] Bar-Cohen Y. Worldwide electroactive polymer (WW-EAP). Newsletter, 2008, 10(2): 1-17.

[84] Bar-Cohen Y. Worldwide electroactive polymer (WW-EAP). Newsletter, 2003, 5(1): 1-15.

[85] Bar-Cohen Y. Worldwide electroactive polymer (WW-EAP). Newsletter, 2004, 6(2): 1-23.

[86] Bar-Cohen Y. Worldwide electroactive polymer (WW-EAP). Newsletter, 2005, 7(1): 1-26.

[87] Bar-Cohen Y. Worldwide electroactive polymer (WW-EAP). Newsletter, 2010, 12(1): 1- 19.

[88] Bar-Cohen Y. Worldwide electroactive polymer (WW-EAP). Newsletter, 2001, 3(2): 1-15.

[89] Bar-Cohen Y. Worldwide electroactive polymer (WW-EAP). Newsletter, 2007, 9(2): 1-15.

[90] Bar-Cohen Y. Worldwide electroactive polymer (WW-EAP). Newsletter, 2004, 6(1): 1- 20.

[91] Bar-Cohen Y. Worldwide electroactive polymer (WW-EAP). Newsletter, 2011, 13(1): 1-13.

[92] Young Yang E E. Phenomenological constitutive models for dielectric elastomer membranes for artificial muscle. The Pennsylvania State University, 2006: 1-144.

[93] Zhao X H. Mechanics of soft active materials. Harvard University. 2009: 1-62.

[94] Kofod G. Dielectric elastomer actuators. The Technical University of Denmark, 2001: 1-83.

[95] Plante J S. Dielectric elastomer actuators for binary robotics and mechatronics. Massachusetts Institute of Technology, 2006: 1-43.

[96] Fox J W. Electromechanical characterization of the static and dynamic response of dielectric elastomer membranes. Virginia Polytechnic Institute and state University, 2007: 1-55.

[97] Leng J S, Lau K T. Multifunctional Polymer Nanocomposites. CRC Press. 2010: 65-136.

[98] Carpi F, De Rossi D, Kornbluh R, et al. Dielectric Elastomers as Electromechanical Transducers. New York: Elsevier, 2008, 12-260.

[99] Bar-Cohen Y. Electroactive Polymer (EAP) Actuators as Artificial Muscles: Reality, Potential, and Challenges. Bellingham: SPIE Press, 2004: 1-765.

[100] Carpi F, Callone G, Galantini F, et al. Silicone-poly(hexylthiophene) blends as elastomers with enhanced electromechanical transduction properties. Advanced Functional Materials, 2008, 18:235-24.

[101] Ha S M, Yuan W, Pei Q, et al. Interpenetrating networks of elastomers exhibiting 300% electrically-induced area strain. Smart Materials & Structures, 2007, 16: S280-S287.

[102] Gallone G, Carpi F, Rossi D D, et al. Dielectric constant enhancement in a silicone elastomer filled with lead magnesium niobate-lead titanate. Materials Science and Engineering C, 2007, 27: 110-116.

[103] Yang B. Influence of moisture in polyurethane shape memory polymers and their electrical conductive composites. Nanyang Technological University, Singapore, 2005.

[104] Tobushi H, Hayahi S, Ikai A, et al. Basic deformation properties of a polyurethane-series shape memory polymer film. Japan Society of Mechanical Engineers: A, 1996, 62: 576-582.

[105] Zhou B, Liu Y J, Leng J S. A macro-mechanical constitutive model for shape memory polymer. Science China Physics, Mechanics and Astronomy, 2010, 53(12): 2266-2273.

[106] Zhou B, Liu Y J, Lan X, et al. A glass transition model for shape memory polymer and its composite. International Journal of Modern Physics B, 2009, 23(6): 1248-1253.

[107] 周博，刘彦菊，冷劲松．形状记忆聚合物的宏观力学本构模型．中国科学：物理学 力学 天文学，2010，(7): 896-903.

[108] Morshedian J, Khonakdar H A, Rasouli S. Modeling of shape memory induction and recovery in heat-shrinkable polymers. Macromolecular Theory and Simulations, 2005, 14(7): 428-434.

[109] Diani J, Liu Y, Gall K . Finite strain 3D thermoviscoelastic constitutive model for shape memory polymers. Polymer Engineering & Science, 2006, 46(4): 486-492.

[110] Seok J H, Wong R Y, Ji H Y, Thermomechanical deformation analysis of shape memory polymers using viscoelastieity. 10th Esaform Conference on Material Forming, 2007: 853-858.

[111] Liu Y P, Gall K, Dunn M L, et al. Thermomechanics of shape memory polymers: uniaxial experiments and constitutive modeling. International Journal of Plasticity, 2006, 22: 279-313.

[112] Liu Y P. Thermomechanical behavior of shape memory polymers. University of Colorado, 2004, 95-120.

[113] Chen Y C, Lagoudas D C. A constitutive theory for shape memory polymers. Part I - large deformations. Journal of the Mechanics and Physics of Solids, 2008, 56(5): 1752-1765.

[114] Chen Y C, Lagoudas D C. A constitutive theory for shape memory polymers. Part II - a linearized model for small deformations. Journal of the Mechanics and Physics of Solids, 2008, 56(5): 1766-1778.

[115] Qi H J, Nguyen T D, Castro F, et al. Finite deformation thermo-mechanical behavior of thermally induced shape memory polymers. Journal of the Mechanics and Physics of Solids, 2008, 56(5): 1730-1751.

[116] Barot G, Rao I J. Constitutive modeling of the mechanics associated with crystallizable shape memory polymers. Zeitschrift für angewandte Mathematik und Physik ZAMP, 2006, 57(4): 652-681.

[117] 王诗任，吕智．热致形状记忆高分子的研究进展．高分子材料科学与工程，2000, 16(1): 1-4.

[118] 黄歆明，郑百林，杨青．形状记忆聚合物热致形状记忆行为的有限元数值模拟．计算机辅助工程，2008, 17(1): 30-32.

[119] 杨青．基于黏弹性理论高分子材料形状记忆过程的有限元数值模拟．上海：同济大学博士论文，2007.

[120] 李郑发，王正道，熊志远．形状记忆聚合物热力学本构方程．高分子学报，2009, 1(1): 23-27.

[121] Guo X G, Liu L W, Liu Y J, et al. Constitutive model for a stress- and thermal-induced phase transition in a shape memory polymer. Smart Materials and Structures, 2014, 23(10): 105019.

[122] Chen J G, Liu L W, Liu Y J, et al. Thermoviscoelastic shape memory behavior for epoxy-shape memory polymer. Smart Materials and Structures, 2014, 23(5): 055025.

[123] 兰鑫．形状记忆聚合物复合材料及其力学基础研究．哈尔滨：哈尔滨工业大学博士

论文，2010.

[124] Dow N F, Rosen B W. Evaluations of filament-reinforced composites for aerospace structural applications. General Electric Co Philadelphia Pa, 1965.

[125] Timoshenko S P, Gere J M. Theory of Elastic Stability. Courier Dover Publications, 2009.

[126] Campbell D, Lake M S, Mallick K. A study of the conpression mechanics of soft-resin composites. In 45th AIAA/ASME/ASCE/AHS/ASC Structures, Structural Dynamics and Materials Conference. California: Palm Springs, 2004.

[127] Campbell D, Maji A. Failure Mechanisms and Deployment Accuracy of Elastic-Memory Composites. Journal of Aerospace Engineering, 2006, 19(3): 184-193.

[128] Campbell D, Maji A. Deployment Precision and Mechanics of Elastic Memory Composites. Proceeding of 44th Structures, Structural Dynamics, and Materials Conference, Norfolk, Virginia, 2003.

[129] Wang Z D, Xiong Z Y, Li Z F, et al. Micromechanism of deformation in EMC laminates. Materials Science and Engineering: A, 2008, 496(1): 323-328.

[130] Campbell D, Lake M S, Mallick K. A study of the bending mechanics of elastic memory composites. AIAA. 45th Structures, Structural Dynamics, and Materials Conference. California: Palm Springs, 2004, 1323-1331.

[131] Francis W H, Lake M S. A review of classical fiber microbuckling analytical solutions for use with elastic memory composites. AIAA Journal, 2006, 21(4): 1764-1776.

[132] Abrahamson E R, Lake M S. Shape memory mechanics of an elastic memory composite resin. Journal of Intelligent Material Systems and Structures, 2003,14(10): 623-632.

[133] Campbell D, Maji A. Failure mechanisms in the folding of unidirectional soft-resin composites. Society for Experimental Mechanics SEM Annual Conference and Exposition of Experimental and Applied Mechanics, 2003: 1-10.

[134] Schultz M R, Francis W H, Campbell D, et al. Analysis techniques for shape-memory composite structures. 48th AIAA/ ASME/ ASCE/ AHS/ ASC Structures, Structural Dynamics, and Materials Conference, Honolulu, Hawaii, AIAA. 2007, 2401: 1-11.

[135] Chen X, Hutchinson J W. Herringbone buckling patterns of compressed thin films on compliant substrates. Journal of Applied Mechanics, 2004, 71: 597-603.

[136] Huang Z Y, Hong W, Suo Z. Nonlinear analyses of wrinkles in a film bonded to a compliant substrate. Journal of the Mechanics and Physics of Solids, 2005, 53 : 2101-2118.

[137] Ryu S Y, Xiao J L, Park W I, et al. Lateral buckling mechanics in silicon nanowires on elastomeric substrates. Nano Letter, 2009, 9(9): 3214-3219.

[138] Xiao J, Ryu S Y, Huang Y, et al. Mechanics of nanowire/nanotube in-surface buckling on elastomeric substrates. Nanotechnology, 2010, 21(8): 1-9.

[139] Ryu S Y, Xiao J L, Park W I, et al. Lateral buckling mechanics in silicon nanowires on elastomeric substrates. Nano Letters, 2009, 9(9): 3214-3219.

[140] Kim D H, Ahn J H, Choi W M, et al. Stretchable and foldable silicon integrated circuits. Science, 2008, 320(25): 507-511.

[141] Ruggiero E J, Inman D J. Gossamer spacecraft: recent trends in design, analysis, experimentation and control. Journal of Space and Rockets, 2006, 43(1): 10-24.

[142] Roybal F A. Development of an elastically deployable boom for tensioned planar structures. 48th AIAA/ASME/ASCE/AHS/ASC Structures, Structural Dynamics, and Materials Conference, Honolulu, Hawaii. AIAA, 2004, 1063: 143-153.

[143] Hazelton C S, Gall K R, Abrahamson E R. Development of a prototype elastic memory composite STEM for large space structures. 44th Structures, Structural Dynamics, and Materials Conference, Norfolk, Virginia, AIAA, 2003, 1977: 1-11.

[144] Taylor R M, Abrahamson E, Barrett R, et al. Passive deployment of an EMC boom using radiant energy in thermal vacuum. 48th AIAA/ASME/ASCE/AHS/ASC Structures, Structural Dynamics, and Materials Conference, Honolulu, Hawaii, AIAA, 2007-2269.

[145] Yee J C H, Soykasap O, Pellegrino S. Carbon fibre reinforced plastic tape springs. 45th AIAA/ASME/ASCE/AHS/ASC Structures, Structural Dynamics, and Materials Conference, Palm Springs, CA, AIAA, 2004-1819.

[146] Ruggiero E J, Inman D J. Gossamer spacecraft: recent trends in design, analysis, experimentation and control. Journal of Space and Rockets, 2006, 43(1): 10-24.

[147] Lan X, Liu Y J, Lv H B, et al. Fiber reinforced shape memory polymer composite and its application in a deployable hinge. Smart Materials and Structures, 2009, 18(2): 024002.

[148] Zhang R R. Design variable stiffness space deployable structure based on shape memory polymer composite. Dissertation for the Master Degree in Engineering, China. Harbin Institute of Technology, 2011.

[149] Zhang R R, Guo X G, Liu Y J, et al. Theoretical analysis and experiments of a space deployable truss structure. Composite Structures, 2014, 112: 226-230.

[150] Wang T. The dynamic behavior research of truss structures based on shape memory polymer composite. Dissertation for the Master Degree in Engineering, China, Harbin Institute of Technology, 2013.

[151] Campbell D, Lake M S, Scherbarth M S, et al. Elastic memory composite material: an enabling technology for Future Furlable Space Structures. 46th AIAA/ASME/ASCE/AHS/ASC Structures, Structural Dynamics and Materials Conference, Austin, TX. April, 2005: 18-21.

[152] Robert M S, Emil J P, Eric W T, et al. Veritex TM struts for antenna application. 47th

AIAA/ASME/ASCE/AHS/ASC Structures, Structural Dynamics and Materials Conference, Newport, Rhode Island. 2006: 1-4.
[153] Taylor R M, Abrahamson E, Barrett R, et al. Passive deployment of an emc boom using radiant energy in thermal vacuum. 48th AIAA/ASME/ASCE/AHS/ASC Structures, Structural Dynamics and Materials Conference, Waikiki, HI, 2007: 23-26.
[154] Arzberger S C, Munshia N A, Lakea M S, et al. Elastic memory composite technology for thin, lightweight space and ground-based deployable mirrors. Optical Materials and Structures Technologies, Proceedings of SPIE, 2003, 5179: 143-154.
[155] Varlese S J, Hardaway L R. Laminated electroformed shape memory composite for deployable light weight optics. Proceedings of SPIE: Earth Observing Systems IX, 2004, 5542: 375-383.
[156] Francis W, Lake M, Hinkle J, et al. Development of an EMC self-locking linear actuator for deployable optics. 45th AIAA/ASME/ASCE/AHS/ASC Structures, Structural Dynamics and Materials Conference, Palm Springs, CA, 2004: 19-22.
[157] Keller P N, Lake M S, Codell D, et al. Development of elastic memory composite stiffeners for a flexible precision reflector. 47th AIAA/ASME/ASCE/AHS/ASC Structures, Structural Dynamics, and Materials Conference, Newport, Rhode Island, AIAA, 2006, 2179: 1-11.
[158] Lin J K H, Knoll C F, Willey C E. Shape memory rigidizable inflatable (Ri) structures for large space systems applications. 47th AIAA/ASME/ASCE/AHS/ASC Structures, Structural Dynamics and Materials Conference, Newport, Rhode Island, 2006: 1-4.
[159] Leng J S, Wu X L, Liu Y J. Effect of linear monomer on thermomechanical properties of epoxy shape memory polymers. Smart Materials and Structures, 2009, 45(4): 52-56.
[160] Murphey T W, Meink T, Mikulas M M. Some micromechanics considerations of the folding of rigidizable composite materials. 42nd AIAA/ASME/ASCBAHS/ASC Structures, Structural Dynamics, and Materials Conference, Seattle, WA, 2001.
[161] Sun J, Liu Y J, Leng J S. Mechanical properties of shape memory polymer composites enhanced by elastic fibers and their application in variable stiffness morphing skins. Journal of Intelligent Material Systems and Structures, 2014, DOI: 10.1177/1045389X14546658.
[162] Chen S B, Chen Y J, Zhang Z C, et al. Experiment and analysis of morphing skin embedded with shape memory polymer composite tube. Journal of Intelligent Material Systems and Structures, 2013, DOI: 10.1177/1045389X13517307.
[163] Lendlein A, Langer R. Biodegradable, elastic shape-memory polymers for potential biomedical applications. Science, 2002, 296(5573): 1673-1676.
[164] Small W, Buckley P R, Wilson T S, et al. Fabrication and characterization of cylindrical light diffusers comprised of shape memory polymer. Journal of Biomedical Optics, 2008,

13(2): 024018.

[165] Everhart M C, Nickerson D M, Hreha R D. High-temperature reusable shape memory polymer mandrels. Smart Structures and Materials. International Society for Optics and Photonics, 2006, 6171: 61710K-1-61710K-10.

[166] Everhart M C, Stahl J. Reusable shape memory polymer mandrels. Smart Structures and Materials. International Society for Optics and Photonics, 2005, 5762: 27-34.

[167] Zhang L, Du H Y, Liu L W, et al. Analysis and design smart mandrels using shape memory polymers. Smart Materials & Structures, 2014, 59: 230-237.

[168] Pelrine R, Kornbluh R, Pei Q B. High-speed electrically actuated elastomers with strain greater than 100%. Science, 2000, 287(28): 836-839.

[169] Suo Z G, Zhao X H, Greene W H. A nonlinear field theory of deformable dielectrics. J. Mech. Phys. Solids, 2008, 56: 467-486.

[170] Liu Y J, Liu L W, Zhang Z, et al. Dielectric elastomer film actuators: characterization, experiment and analysis. Smart Mater. Struct., 2009, 18: 095024.

[171] Hong W. Modeling viscoelastic dielectrics. J. Mech. Phys. Solids., 2011, 59: 637-650.

[172] Zhao X H, Koh S J A, Suo Z G. Nonequilibrium thermodynamics of dielectric elastomers. Int. J. Appl. Mech., 2011, 3: 203-217.

[173] Zhao X H, Suo Z G. Method to analyze electromechanical stability of dielectric elastomers. Appl. Phys. Lett., 2007, 91: 061921.

[174] Plante J S, Dubowsky S. Large-scale failure modes of dielectric elastomer actuators. Int. J. Solids Struct., 2006, 43: 7727-7751.

[175] Zhou J X, Hong W, Zhao X H, et al. Propagation of instability in dielectric elastomers. Int. J. Solids Struct., 2008, 45: 3739.

[176] Liu Y J, Liu L W, Zhang Z, et al. Comment on method to analyze electromechanical stability of dielectric elastomers. Appl. Phys. Lett., 2008, 93: 106101.

[177] Liu Y J, Liu L W, Sun S H, et al. Stability analysis of dielectric elastomer film actuator. Sci. China Ser. E-Tech. Sci., 2009, 52(9): 2715-2723.

[178] Norrisa A N. Comment on Method to analyze electromechanical stability of dielectric elastomers. Appl. Phys. Lett., 2007, 92: 026101.

[179] Stark K H, Garton C G. Electric strength of irradiated polythene. Nature, 1955, 176: 1225-1226.

[180] He T H, Zhao X H, Suo Z G. Equilibrium and stability of dielectric elastomer membranes undergoing inhomogeneous deformation. J. Appl. Phys., 2009, 106: 083522.

[181] He T H, Cui L L, Chen C, et al. Nonlinear deformation analysis of a dielectric elastomer

membrane-spring system. Smart Mater. Struct., 2010, 19: 085017.

[182] Xu B X, Mueller R, Classen M, et al. On electromechanical stability analysis of dielectric elastomer actuators. Appl. Phys. Lett., 2010, 97: 162908.

[183] Zhao X H, Suo Z G. Electrostriction in elastic dielectrics undergoing large deformation. J. Appl. Phys., 2008, 104: 123530.

[184] Leng J S, Liu L W, Liu Y J, et al. Electromechanical stability of dielectric elastomer. Appl. Phys. Lett., 2009, 94: 211901.

[185] Liu Y J, Liu L W, Sun S H, et al. Electromechanical stability of Mooney-Rivlin-type dielectric elastomer with nonlinear variable dielectric constant. Polym. Int., 2010, 59: 371-377.

[186] Liu Y J, Liu L W, Yu K, et al. An investigation on electromechanical stability of dielectric elastomers undergoing large deformation. Smart Mater. Struct., 2009, 18: 095040.

[187] Kofod G, Sommer-Larsen P, Kronbluh R, et al. Actuation response of polyacrylate dielectric elastomers. J. Intell. Mater. Syst. Struct., 2003, 14: 787.

[188] Zhao X H, Hong W, Suo Z G. Electromechanical hysteresis and coexistent states in dielectric elastomers. Phys. Rev. B, 2007, 76: 134113.

[189] Suo Z G, Zhu J. Dielectric elastomers of interpenetrating networks. Appl. Phys. Lett., 2009, 95: 232909.

[190] Li B, Liu L W, Suo Z G. Extension limit, polarization saturation, and snap-through instability of dielectric elastomers. Int. J. Smart. Nano Mater., 2011, 2(2): 59-67.

[191] Li B, Chen H L, Qiang J H, et al. Effect of mechanical pre-stretch on the stabilization of dielectric elastomer actuation. J. Phys. D-Appl. Phys., 2011, 44: 155301.

[192] Li B, Chen H L, Zhou J X, et al. Polarization modified instability and actuation transition of deformable dielectric. Europhysics Letters, 2011, 95: 37006.

[193] Liu L W, Liu Y J, Luo X J, et al. Electromechanical stability and snap-through stability of dielectric elastomers undergoing polarization saturation. Mech. Mater., 2012, 55: 60-72.

[194] He X Z, Yong H D, Zhou Y H. The characteristics and stability of a dielectric elastomer spherical shell with a thick wall. Smart Mater. Struct., 2011, 20(5): 055016.

[195] Rudykh S, Bhattacharya K, Debotton G. Snap-through actuation of thick-wall electroactive balloons. Int. J. Non-Linear Mech., 2012, 47(2): 206-209.

[196] Zhu J, Li T F, Cai S Q, et al. Snap-through expansion of a gas bubble in an elastomer. J. Adhesion, 2011, 87: 466-481.

[197] Ha S M, Yuan W, Pei Q B, et al. Interpenetrating polymer networks for high-performance electroelastomer artificial muscles. Adv. Mater., 2006, 18: 887-891.

[198] Shankar R, Ghosh T K, Spontak R J. Mechanical and actuation nanostructured polymers as tunable actuators. Adv. Mater., 2007, 19: 2218-2223.

[199] Keplinger C, Kaltenbrunner M, Arnold N, et al. Röntgen's electrode-free elastomer actuators without electromechanical pull-in instability. Proc. Natl. Acad. Sci. USA, 2010, 107(10): 4505-4510.

[200] Pelrine R E, Kornbluh R D, Joseph J P. Electrostriction of polymer dielectrics with compliant electrodes as a means of actuation. Sens. Actuator. A-Phys., 1998, 64(1): 77-85.

[201] Koh S J A, Li T F, Zhou J X, et al. Mechanisms of large actuation strain in dielectric elastomers. J. Polym. Sci. Pt B-Polym. Phys., 2011, 49: 504-515.

[202] Huang R, Suo Z G. Electromechanical phase transition in dielectric elastomers. Proc. R. Soc. A-Math. Phys. Eng. Sci., 2012, 468, 1014-1040.

[203] Lu T Q, Suo Z G. Large conversion of energy in dielectric elastomers by electromechanical phase transition. Acta. Mech. Sin., 2011,28, 1106-1114.

[204] Foo C C, Cai S Q, Koh S J A, et al. Model of dissipative dielectric elastomers. J. Appl. Phys., 2012, 111, 034102.

[205] Foo C C, Koh S J A, Keplinger C, et al. Performance of dissipative dielectric elastomer generators. J. Appl. Phys., 2012,111, 094107.

[206] Li T F, Qu S X, Yang W. Energy harvesting of dielectric elastomer generators concerning inhomogeneous fields and viscoelastic deformation. J. Appl. Phys., 2012, 112: 034119.

[207] Wang H M, Lei M, Cai S Q. Viscoelastic deformation of a dielectric elastomer membrane subject to electromechanical loads. J. Appl. Phys., 2013, 113: 213508.

[208] Li B, Liu L W, Suo Z G. Extension limit, polarization saturation, and snap-through instability of dielectric elastomers. Int. J. Smart Nano Mater., 2011, 2(2): 59-67.

[209] Li B, Chen H L, Qiang J H, et al. A model for conditional polarization of the actuation enhancement of a dielectric elastomer. Soft Matter, 2012, 8: 311.

[210] Lallart M, Capsal J F, Kanda M, et al. Guiffard. Modeling of thickness effect and polarization saturation in electrostrictive polymers. Sens. Actuator. B-Chem., 2012, 171-172: 739-746.

[211] Lallart M, Capsal J F, Idrissa A K M, et al. Actuation abilities of multiphasic electroactive polymeric systems. J. Appl. Phys., 2012,112, 094108.

[212] Brochu P, Pei Q B. Advances in dielectric elastomers for actuators and artificial muscles. Macromol Rapid Commun., 2010, 31: 10-36.

[213] Mirfakhrai T, Madden J, Baughman R. Polymer artificial muscles. Mater Today, 2007, 10(4): 30-38.

[214] O'Halloran A, O'Malley F, McHugh P. A review on dielectric elastomer actuators, technology, applications, and challenges. J. Appl. Phys., 2008, 104: 071101.
[215] Madden J D, Vandesteeg N, Madden P G, et al. Artificial muscle technology: physical principles and naval prospects. IEEE J. Ocean. Eng., 2004, 29: 706.
[216] Bar-Cohen Y. Electroative polymers as artificial muscles capabilities, potentials and challenges. Hanbook on Biomimetics. Section 11, in Chapter 8. NTS Inc. 2000, 1-13.
[217] Bar-Cohen Y. Biologically inspired technology using electroactive polymers (EAP). Proc. SPIE, 2006, 6168: 616803.
[218] Bar-Cohen Y. Electroative polymers: current capabili ties and challenges. Proc. SPIE, 2002, 4695: 1-7.
[219] Bar-Cohen Y. Worldwide electroactive polymer (WW-EAP). Newsletter, 2008, 10(2): 1-17.
[220] Bar-Cohen Y. Worldwide electroactive polymer (WW-EAP). Newsletter, 2003, 5(1): 1-15.
[221] Bar-Cohen Y. Worldwide electroactive polymer (WW-EAP). Newsletter, 2004, 6(2): 1-23.
[222] Bar-Cohen Y. Worldwide electroactive polymer (WW-EAP). Newsletter, 2005, 7(1): 1-26 .
[223] Bar-Cohen Y. Worldwide electroactive polymer (WW-EAP). Newsletter, 2010, 12(1): 1-19.
[224] Bar-Cohen Y. Worldwide electroactive polymer (WW-EAP). Newsletter, 2001, 3(2): 1-15.
[225] Bar-Cohen Y. Worldwide electroactive polymer (WW-EAP). Newsletter, 2007, 9(2): 1-15.
[226] Bar-Cohen Y. Worldwide electroactive polymer (WW-EAP). Newsletter, 2004, 6(1): 1- 20.
[227] Bar-Cohen Y. Worldwide electroactive polymer (WW-EAP). Newsletter, 2011, 13(1): 1-13.
[228] Carpi F, De Rossi D, Kornbluh R, et al. Dielectric Elastomers as Electromechanical Transducers. New York: Elsevier, 2008: 12-260.
[229] Bar-Cohen Y. Electroactive Polymer (EAP) Actuators as Artificial Muscles: Reality, Potential, and Challenges. Bellingham: SPIE press, 2004: 1-765.
[230] Kofod G, Wirges W.Energy minimization for self-organized structure formation and actuation.Appl. Phys. Lett., 2007,90: 081916.
[231] Liu Y J, Liu L W, Leng J S. Analysis and manufacture of energy harvester based on Mooney-Rivlin type dielectric elastomer. EPL, 2010, 90: 36004.
[232] Araromi O A, Gavrilovich I, Shintake J, et al. Rollable multisegment dielectric elastomer minimum energy structures for a deployable microsatellite gripper. IEEE-ASME Trans. Mechatron., 2014: 438-446.
[233] Chiba S, Waki M, Kornbluh R, et al. Innovative wave power generation system using electroactive polymer artificial muscles. OCEANS-IEEE. Bremen: IEEE, 2009: 143-145.
[234] Maas J, Graf C. Dielectric elastomers for hydro power harvesting. Smart Mater Struct. 2012, 21: 064006.

[235] Kaltseis R, Keplinger C, Baumgartner R, et al. Method for measuring energy generation and efficiency of dielectric elastomer generators. Appl. Phys. Lett., 2011, 99: 162904.

[236] Chiba S, Waki M, Kornbluh R, et al. Innovative power generators for energy harvesting using electroactive polymer artificial muscles. Proc SPIE. San Diego: SPIE, 2008: 692715.

第四章 场诱导智能软物质材料

第一节 引　　言

软物质是指处于固体和理想流体之间的物质。从物理学角度来看，固体的组成结构是长程有序的，而软物质可以总结为短程有序而长程无序[1]。造成这一区别的根本原因在于两者内部原子动能的不同：软物质中的基本单元（原子或者分子）的动能接近热运动能量 K_bT；而固体的基本单元的动能远小于热运动能量。软物质的组成复杂，它的运动并不由其组成单元中的原子或分子尺度的量子力学作用决定，而主要是热涨落和熵导致了软物质体系复杂物相的变化，这些驱动作用比原子或分子间键能弱得多，表现出“复杂性”、“软”和“易变性”。只要运用得当，一些微弱的刺激就能引起整个软物质系统量的乃至质的改变。

随着时代的发展和科技的进步，人类不再满足于简单地使用原始材料，而是想根据自己的意愿合成制备具有功能或者智能特性的材料。智能材料是指具有可感知外部刺激（如压力、温度、湿度、pH、电场或磁场等）的改变而判断，并处理这些外部刺激的新型功能特性[2]。软物质“小作用、大响应”的特点预示着它对于外部刺激可以具有特定而显著的响应，它的属性中的某些参量可通过一个外部条件而改变，并且这种变化是可逆的，还可以重复多次。所以，如果将软物质的这些特殊的性质加以研究和利用，就可以制备出具有智能特性的软物质材料。

第二节　学科发展背景和现状

智能材料是20世纪80年代中后期由美国和日本科学家先后提出的，是继天然材料、合成高分子材料、人工设计材料之后的第四代材料，是现代高技术新材料发展的重要方向之一。概念形成初期，日本高木俊宜教授和美国Newnhain教授将对环境具有可感知、可响应等功能的新材料定义为智能材料[3, 4]。发展至今，智能材料这一概念不断扩充拓展。

一般来说，根据发展的先后顺序材料分为一般材料、功能材料和智能材料。功能材料是指那些具有优良的电学、磁学、光学、热学、声学、力学、化学、生物医学功能，特殊的物理、化学、生物学效应，能完成功能相互转化，主要用来制造各种功能元器件而被广泛应用于各类高科技领域的高新技术材料。而智能材料是功能材料高级的形式，是新型功能材料，它不仅能够感知环境变化，还能根据这些属性做出相应的响应，以达到某种智能控制的目的。智能材料拥有传感功能、反馈功能、信息识别与积累功能、响应功能、自诊断能力、自修复能力和自适应能力七大功能。由此可见，智能材料是材料领域目前最前沿的研究领域。智能材料的物理机制的研制和大规模应用将导致材料物理科学发展的重大革命。而软物质的基础研究，特别是场诱导软物质的物理研究将大大促进功能材料以及智能材料的发展和应用。

近年来，软物质科学迅速发展，软物质的研究横跨化学、生物、物理三大学科，化学和生物学构成了软物质科学的实验基础，物理学为软物质科学提供理论依据和发展的方向。软物质材料更是成为化学、物理、材料和生物等学科交叉融合的重要领域和天然桥梁，同时又与许多技术和工程问题密切相关。智能材料可以根据其化学成分分为智能金属材料、智能无机非金属材料和智能高分子材料，其中智能高分子材料属于软物质智能材料范畴，如高分子凝胶。高分子凝胶是指三维高分子网络与溶剂组成的体系，其大分子主链或侧链上有离子解离性、极性和疏水基团，类似于生体组织。此类高分子凝胶可因溶剂种类、盐浓度、pH、温度的不同以及电、磁刺激和光辐射而产生可逆的非连续的体积变化[5-7]。

为此研究人员从智能材料接受外界响应和应用范围的不同将场诱导软物质智能材料分为电、磁响应软物质材料、温度响应软物质材料、光响应软物质材料等。

电、磁响应软物质材料也已经应用在我们生活之中了。比如，磁响应软物质——磁流变液和电响应软物质——电流变液，不加外场时，它们表现出类似 Newton 流体行为；加外场时，它们表现出 Bingham 流体特性，随着外场的增强，其屈服应力增大。

一、磁流变液

美国学者 Rabinow 在 1948 年发明了磁流变液（magnetorheological fluid, MRF）。磁流变液是一种包含磁性纳米至微米颗粒的胶体悬浮液（图 7-4-1）[8, 9]。这种胶体悬浮液在外加磁场的作用下，其中的纳米颗粒会产生很强的磁偶极相互作用力，该结构会大大地增加流体的黏度，可由液体变为“固体”，当外磁场去掉之后，磁流变液会迅速地从“固体”状态恢复为一般的液体状态并且是可逆的（图 7-4-1）。

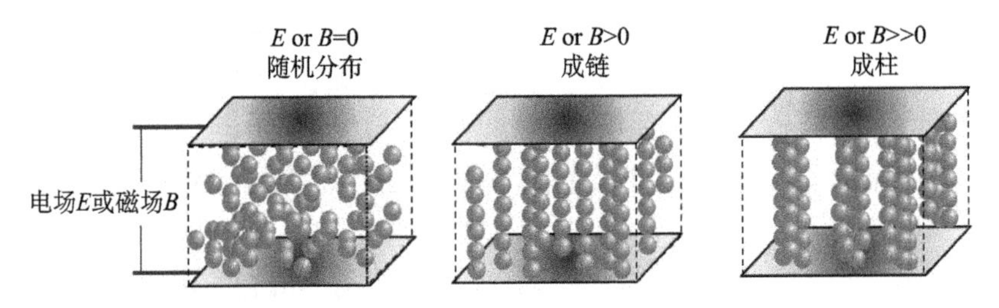

图 7-4-1　磁（或者电）流变液的工作机制：磁流变液，红色颗粒为磁性纳米至微米颗粒；电流变液，红色颗粒为介电纳米至微米颗粒（文后附彩图）

典型的磁流变液由三部分组成：软磁性颗粒、载液和添加剂。

制备磁流变液的软磁性颗粒，一般有羰基铁粉、Fe_3O_4[10]、钴粉[11]、铁钴合金及镍锌合金等。除上述软磁性颗粒之外，最近专家采用以下新技术制备出复合软磁性颗粒。

第一类：用聚合物包覆铁粉。该方法可以减小软磁性颗粒的密度，增加颗粒的表面积，提高所制备磁流变液的沉降稳定性和再分散性[12]。Park 等用聚甲基丙烯酸甲酯（PMMA）包覆羰基铁粉制备出复合磁性颗粒。笔者的研究小组以羰基铁粉为原料，用聚乙二醇包覆羰基铁粉颗粒。

第二类：用软磁性颗粒包覆非金属材料。该方法能减少颗粒的密度，提高所制备磁流变液的沉降稳定性。Jun 等以聚合物为核，以氧化铁颗粒为壳，制备出理想的球形颗粒[14]。

第三类：用金属颗粒包覆软磁性颗粒。该方法可以增强颗粒的磁饱和强度，进而增强所制备磁流变液的屈服强度。John 用化学镀的方法在羰基铁粉表面包覆一层镍粉[15]。

对比以上几种颗粒，羰基铁粉是制备磁流变液常用的软磁性颗粒，目前商品化的磁流变液大多采用普通羰基铁粉制备，但采用聚合物包覆的羰基铁粉是目前各国专家研究的热点。

载液是软磁性颗粒所能悬浮的连续媒介，是磁流变液的重要组成成分，如合成油、矿物油、水等液体[16]都可以作为载液。Bose 报道了用胶体作为载液，制备出一种磁流变弹性体[17]。Fuchs 等以聚合胶体为载液制备出了一种磁流变聚合胶体[18]。

添加剂包括分散剂和防沉降剂等，其作用主要是改善磁流变液的沉降稳定性、再分散性、零场黏度和剪切屈服强度。分散剂主要有：油酸及油酸盐、环烷酸盐、磺酸盐（或酯）、磷酸盐（或酯）、硬脂酸及其盐、单油酸丙三醇、脂肪醇、二氧化硅等。防沉降剂主要有：高分子聚合物、亲水的硅树脂低聚物、有机金属硅共聚物、超细无定形硅胶以及有机黏土和含氢键的低聚物等。此外，Chin 用纳米级的磁性颗粒（$Co\text{-}\gamma\text{-}Fe_2O_3$，$CrO_2$）作为添加剂，提高了磁流变液的沉降稳定性[19]。

磁流变液的制备是工程应用的基础，磁流变液在最初应用过程中遇到了稠化、沉降及磨损等难题，提高包含再分散性和沉降稳定性在内的性能指标将对工程应用产生重要意义。当前应在以下几个制约工程应用的问题上开展深入研究：如何提高磁流变液的再分散性、如何利用表面改性技术提高磁流变液的沉降稳定性、如何制备出高性能的磁流变弹性体以及如何利用纳米级添加剂改善磁流变液的综合特性等。但随着稠化等问题的解决，各工业国竞相展开了对磁流变液及器件的研究，加速了磁流变液的进展。美国 Lord 公司的 Carlson 在磁流变液性能研究和应用开发方面取得了较为突出的成就，该公司先后报道了多种合金制备的磁流变液并有多种商品化磁流变液产品上市[20-23]。在 2002 年的时候，雪佛兰汽车首次将 Lord 公司生产的磁流变液减震器应用于他们的汽车悬挂系统中。到今天，很多高档轿车纷纷使用这种减震器，它的优点是能耗低、反应迅速和连续可调。

由于磁流变液需要磁场来控制，需要电流较大，能源消耗也较大，所以科学家们研究出了另一种能耗非常低的电响应软物质材料——电流变液。电流变液（electrorheological fluid，ERF）与磁流变液工作原理类似，只是悬浮的颗粒为介电性材料，外加电场控制，它们之间的相互作用力为电偶极相互作用（表 7-4-1）。[24]

表 7-4-1　磁流变液与电流变液特性比较

特性	磁流变液	电流变液
最大屈服应力 /kPa	50～100	40～120
电压	2～25V（1～2V）	2～5kV(1～10mA)
场强	～250 kA/m	～4kV/mm
反应时间	毫秒级	毫秒级
密度 /（g/cm^3）	3～4	1～2
稳定性	不受大多数杂质影响	受水和导电杂质影响
温度范围 /°C	-50～+150	-25～+150
耗能 /（J/cm^3）	0.1	0.001

电流变液的黏度会随着电场强度的增大而明显增大，随着电场增大到一个阈值，该液体就会发生由液相向固相的转变，这一过程十分迅速，通常发生在几毫秒的过程内[25]，而且这样的转变过程具有可逆性。这一过程也就意味着电流变液的流变特性会随着电场的变化而发生变化。原本在不加外加电场时，流体呈现出牛顿流体的特性，但是在电场强度足够高的外加电场下，能够转变成黏弹性固体，对外呈现出宾汉流体的性质［图 7-4-2（a）］。

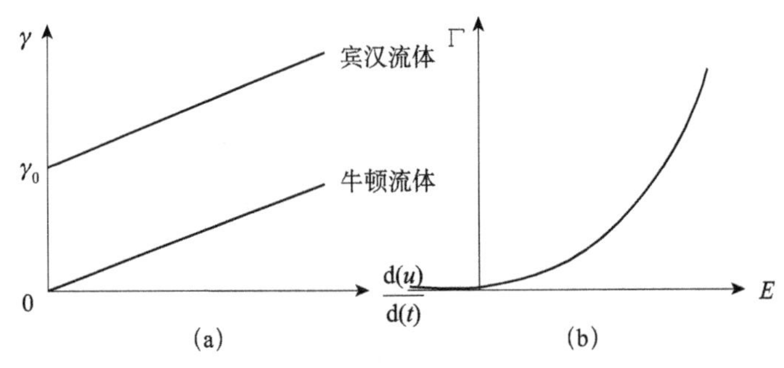

图 7-4-2　(a) ERF 剪切强度随剪切速率的变化；(b) 屈服强度随电场的变化

二、电流变液

（一）电流变液的成分

当前普遍应用的电流变液体主要组分是复杂的悬浊液[26]，但是它的组分大致是由以下三个部分组成的。①组成电流变液分散相的固体颗粒，称为分散相。分散相颗粒具有高介电常数、较强极化能力。目前的分散相颗粒有无机材料，如金属氧化物、金属氢氧化物等，一般来说颗粒尺寸在纳米到微米级别。②用来分散固体颗粒的连续相液体，称为连续相。作为连续相的基础液应具备较低的相对介电常数以及绝缘性能。③改善电流变液体系稳定性以及增强电流变效应的添加剂。

（二）电流变液的微观原理

电流变效应的大致过程[27]：电场作用下分散相粒子发生极化，形成偶极子现象。带偶极矩的粒子产生定向排列，粒子从无序到有序，成链成束或形成某种结构，对外呈现电流变效应。

由于电流变液是一种十分复杂的悬浮液，其构成十分复杂，所以导致电流变效应的原因也十分复杂。但是绝大多数的科研工作者都认为电流变效应产生的原因是来自于极化。表 7-4-2 是四种极化类型[25]。

不难发现，电子和离子极化是一种快速极化过程，其时间在 10^{-12}～10^{-15}s，而偶极子转向极化和双电层极化是一种慢极化过程，其时间在 10^{-2}～10^{-10}s。

实验证明，当施加电场频率增加时电流变效应随之降低，这是由于介电颗粒电偶极子极化弛豫效应引起的，即高频作用下颗粒电极化与电场不能同步而引起电偶极矩相互作用减小。

同时，实验数据表明高频时电流变效应远远高于低频时，这说明偶极子转向极化和双电层极化是产生电流变效应的主要原因。

表 7-4-2　四种极化形式的比较

极化形式	具有这种极化形式的电解质	极化所用时间 /s	与温度的关系
电子位移式极化	气体、液体、固体	10^{-14}～10^{-15}	在气体中温度上升极化削弱
离子位移式极化	离子式结构固体介质	10^{-12}～10^{-13}	温度升高极化增强
电偶极子转向极化	极性电介质	10^{-2}～10^{-10}	在某些温度出现最大值
双电层极化	固液两相悬浮液	10^{-2}～10^{-4}	温度升高极化削弱

1. 纤维理论

1947 年，美国学者 Winslow 第一次使用分散微粒与基液形成的悬浮物制成电流变液，并提出了电流变效应的纤维结构理论[28, 29]。该理论认为，原本无序的颗粒在电场的作用下会有序地排列形成纤维结构。该理论的基础是基于颗粒之间的相互作用力，但是这种作用力显然远低于研究人员在实验中测量到的实际值，所以不能很好地解释电流变效应。

2.“水桥”理论

“水桥”理论由 Stangroom 在 1983 提出[30]。首先介绍“水桥”现象。“水桥”现象就是当我们给两杯加满水的杯子通电后，慢慢把两个杯子分开，两个杯子之间溢出的水不会由于重力作用向下流，而是在两个杯子之间形成一个“水桥”。

“水桥”理论认为，对于具有水的电流变液体来说，电流变效应产生的原因主要是基础液里的水分子之间的相互作用。当加上一个外加电场后，原本在颗粒的空隙中自由流动的离子向空隙的两端运动，而在这些离子的周围就汇聚了许多水分子，离子与离子之间聚集的水分子就产生类似于“水桥”的结构，而正是这种“水桥”结构促使悬浮在液体中的小颗粒产生了紧密的联系。当外加电场消失之后，原本聚集的水分子又快速散开，电流变效应随之消失。但是这种理论有一个致命缺点，就是完全没有办法解释疏水性的半导体颗粒作为分散相的电流变液的电流变效应。

3. 双电层极化理论

双电层极化理论由 Klass 在 1967 年提出[31]。在电流变悬浮液中，由于大量的固体粒子和连续相基础液相接处，在它们接触的表面上都会带上电荷，从而形成带点粒子。当然，固体粒子带电可能是因为本身的电离，其他离子在固体粒子表面的吸附也可能是固体粒子表面离子的溶解。带电的固体粒子从而可以吸引基础液中的异性离子，排斥同性离子，使同性离子远离颗粒在基础液中扩散，所以在颗粒表面形成正或者负电荷而与基础液中被吸引的负或者正电荷形成双电层。在没有加一个外电场时，这个双电层会均匀地分布在颗粒的表面；当加上一个外电场时，原本均匀分布的双电层开始发生变化。颗粒上吸附的反离子受到电场的作用而发生定向的偏移，产生类似于离子的位移极化，形成类偶极子结构，而这个类偶极子结构在电场下开始定向移动，有序排列。目前很多学者都认为这是引起电流变效应的主要原因。

4. 介电理论

在电流变液机制的研究过程中，早期的双层理论和“水桥”理论可以在一定程度上解释含水的电流变液机制。但是，1985 年 Block 等制备出无水的电流变液后[32]，上述的理论就不能很好地解释了。介电理论认为电流变效应的主要原因是电流变液中极性分子在强电场下发生诱导极化，且分散相粒子有远大于连续相液体的介电常数[33]。但此阶段研制的电流变材料力学性能、悬浮稳定性和温度使用范围距实际应用仍有较大差距，难以满足工业和工程应用的实际需求。

介电理论认为由于分散相颗粒具有较大的介电常数，正负电荷分布不均匀，在较强的电场作用下发生诱导极化，正电荷向负电极一方移动而负电荷向正电极一方移动，形成偶极子，相邻偶极子之间由于静电力相互吸引形成链状结构，随着电场的增大链变成柱状结构，对外呈现出较强的屈服强度。马红孺等运用第一性原理计算出介电型电流变液屈服强度的理论极限大约为10kPa，其屈服强度随电场的变化呈现二次关系［图 7-4-2（a）］。

三、巨电流变液的重大突破

（一）巨电流变液的材料

实验发现，如果电流变液中的介电颗粒含水，其电流变效应将会有显著提高。而由于水的易挥发性，这种电流变液难以实用，但这启发了研究人员通过设计一种具有大分子电偶极矩的材料来增强电流变效应。研究人员把具有巨电流变（giant electrorheological，GER）效应的纳米颗粒电流变液称为巨电流变液，此类电流变液在外加电场作用下所表现出的剪切强度远远超过了通常理论所预测到的“上限”，达到了 100kPa 以上，且响应时间小于 10ms，同时还具有温度稳定性好、介电常数大、电流密度低和不沉淀等优点[34]。

巨电流变流体由表面包裹尿素薄层的钛氧基草酸钡［化学式为 $BaTiO(C_2O_4)_2$］纳米颗粒与硅油混合而成。尿素薄层的存在改变了纳米颗粒的表面特性。图 7-4-3 为外层含有尿素薄层的 $BaTiO(C_2O_4)_2$ 纳米颗粒的透射电子显微镜（TEM）图像。从图中可以看出，纳米颗粒的尺寸在 30～70nm 范围内，表面尿素薄层的厚度在 2～5nm 范围内。当有外加电场作用于巨电流变液时，纳米颗粒便沿电场方向排列成柱状结构，如图 7-4-3(b) 所示。图 7-4-3(c) 为某一柱状结构的放大图，可以看到两颗粒的接触界面（contact areas）趋于平整，说明尿素薄层具有一定的柔软度（softness）。

（二）巨电流变液的机制

研究发现，巨电流变液剪切强度与外加电场呈线性变化关系，而不是通常的二次方关系。如图 7-4-4 所示，随着电场强度的线性增加，巨电流变液的剪切强度也随之呈线性增长。对于浓度为 30% 的巨电流变液，当外加 4kV/mm

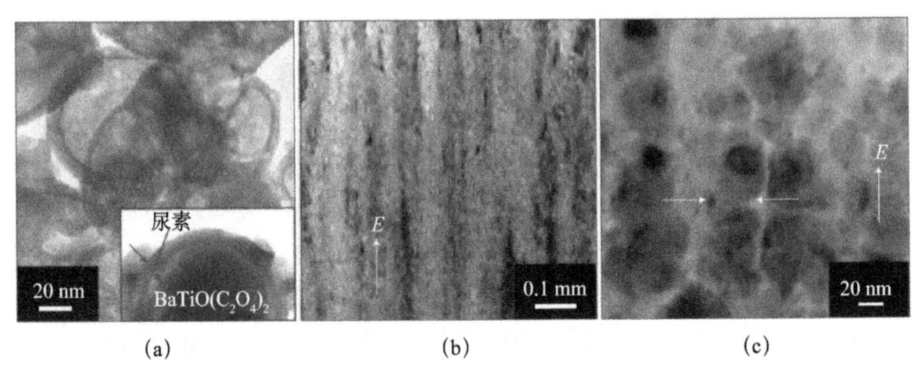

图 7-4-3 (a) 表面包裹尿素薄层的 $BaTiO(C_2O_4)_2$ 纳米颗粒的 TEM 图像；(b)、(c) 加上电场后巨电流变液的结构，(c) 箭头所示为某一平整接触界面

的电场强度时，其剪切强度大于 100kPa。图中圆圈所示是实验测得的在线性增长的外加电场下，巨电流变液的剪切强度随电场变化的关系图像，实线为有限元模拟结果。由此看出，有限元模拟结果与实验测量结果十分吻合。对于此类电流变液，无论是剪切强度的大小还是剪切强度随外加电场的变化规律，都与普通电流变液有着本质的不同。利用传统的介电理论模型已经无法解释其作用机制，其中必定有传统电流变模型未考虑到的因素在起作用。

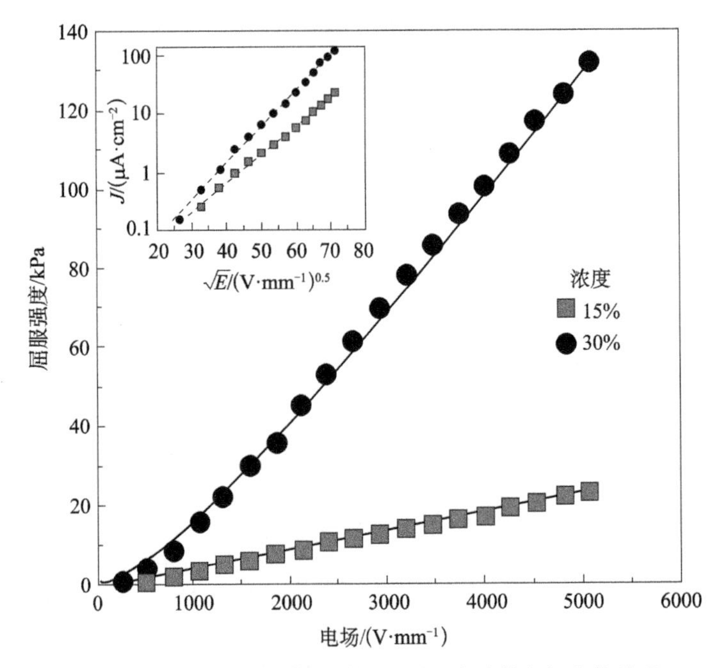

图 7-4-4 巨电流变流体的剪切屈服强度随外电场变化关系

为解释巨电流变液的作用机制，研究人员提出了“表面极化饱和”模型。当有外加电场作用于巨电流变液时，纳米颗粒先被极化，沿电场方向排列成有序的结构。当颗粒间相互接触，且电场增大到某个阈值时，这些颗粒便会在接触部分形成饱和的极化层，如图 7-4-5 所示。饱和极化层之间的相互作用，使得巨电流变液的剪切强度得到大大的提高。基于此模型，研究人员通过数值模拟得到其中的静电场能量为

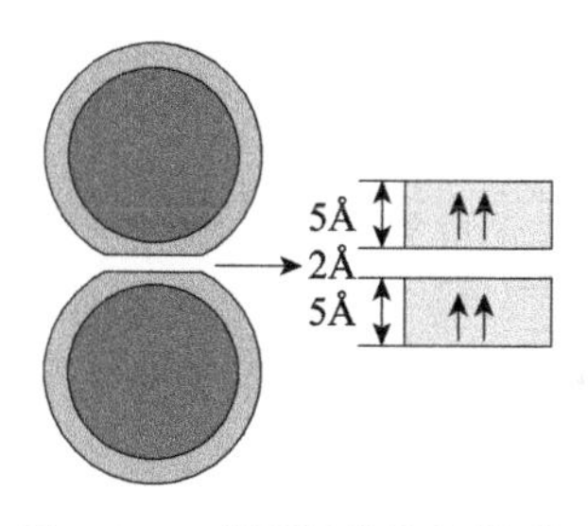

图 7-4-5 表面极化饱和模型，图右侧所示为饱和极化层

$$w_{\mathrm{es}}=-\frac{1}{8\pi}\int_{V_0}D\cdot E\mathrm{d}V-\frac{1}{4\pi}\int_{V_s}\mathrm{d}V\int_0^E D\cdot\delta E$$

颗粒接触时的弹性相互作用能为

$$w_{\mathrm{el}}=(\Delta L)^{5/2}\frac{2}{5D}\left(\frac{R}{2}\right)^{1/2}$$

由这两种能量的作用导出相应电场下剪切强度的大小，结果如图 7-4-5 中实线所示，计算结果与实验结果十分匹配。表面极化饱和模型的提出很好地解释了巨电流变液的作用机制。

巨电流变液的发现引起了国内外的广泛关注，英国出版的《新科学家》杂志发表了题为“‘硬’液体很快就应用于汽车刹车上”的文章。美国福特汽车公司高级工程研究部的物理学家 John Ginder 称“这一技术为智能的电流变液体的应用找到了一条途径”，标志着电流变材料研究的第 3 次浪潮的到来。美国联邦科学工程和技术协调会的报告中，将电流变液研究列为一个重要领域，美国能源部《关于电流变液研究需求估量的最终报告》中指出，“电流变液有潜力成为电气-机械转换中能源效率最高的一种，而且价格合理、结构紧凑、响应快速、经久耐用以及动态范围可变，这些特性是任何其他电气-机械转换方法都无法做到的”。巨电流变液的发现向人们展示了电流变液的巨大应用价值。一些西方国家的国防及工业部门也投入数以亿计的资金，进行电流变材料及其阻尼器件的研究。电流变液和磁流变液都可以应用在减震器、汽车离合器、机器人甚至健身器材等领域。如果将来的减震器都是使用基于电流变液或者磁流变液的减震器，则其市场前景无疑是巨大的，将带来巨大的价值。

（三）温度响应性软物质

含有醚键，取代的酰胺、羟基等功能团的高分子凝胶是温度响应软物质中最具代表性的一种，如聚（N-异丙基烯酰胺）（PNIPAm）、聚氧化烯醚（PEO）、聚乙烯吡咯烷酮（PVP）等。其中PNIPAm由于其广阔的应用前景，成为当前温度响应软物质材料研究的热点。其大分子链上同时具有亲水性的酰胺基和疏水性的异丙基，使线型PNIPAm的水溶液及交联后的PNIPAm水凝胶呈现出温度敏感特性。在常温下，线型PNIPAm溶于水中形成均匀的溶液，当温度升高至30～35℃时，溶液发生相分离，表现出最低临界溶液温度（LCST）。在NIPAm聚合过程中加入交联剂或经处理产生化学交联后，就成为PNIPAm水凝胶。它在室温下溶胀，而在33℃左右发生体积相变而收缩。这种由温度敏感性而引起高聚物产生的智能型和记忆效应引起了人们很大的兴趣。在对PNIPAm的研究中，人们最关心的一个问题是PNIPAm产生这种热敏特性的机制，这也是当前对PNIPAm研究的一个重点。目前较容易被人接受的观点是：PNIPAm分子内具有一定比例的疏水和亲水基团，它们与水在分子内、分子间会产生相互作用。在低温时，PNIPAm与水之间的相互作用主要是酰胺基团与水分子间氢键的作用。PNIPAm分子链在LCST以下溶于水时，由于氢键及范德瓦耳斯力的作用，大分子链周围的水分子将形成一种由氢键连接的、有序化程度较高的溶剂化壳层。随着温度上升，PNIPAm与水的相互作用参数突变，其分子内及大分子间疏水相互作用加强，形成疏水层，部分氢键被破坏，大分子链疏水部分的溶剂化层被破坏，水分子从溶剂化层的排出表现为相变，产生体积收缩即温度响应性的体积变化。基于以上软物质材料设计的智能膜可以广泛用于智能玻璃。这种玻璃能够感应和响应外部环境的变化，在周围光、热或者电场的作用下，智能玻璃能够从最初完全透明的状态转变为半透明或者不透明的状态[35, 36]。这种透光度改变的过程是可逆的，并且透光度是可以随着温度、光强或者电场的变化而变化。由于智能玻璃可以任意可逆地调节光的透过性，所以它在工业和日常生活中的应用正受到越来越多的重视。

软物质智能材料所包含的内容相当宽广，而在实际中也已经有相当广泛的应用，比如液晶，它在显示器中有着不可取代的作用。从20世纪70年代开发出第一台液晶显示器开始，它经历了动态散射模式到旋转向列场效应模式的发展。液晶显示器有很多优点，比如说机身薄节省空间、省电不产生高温、低辐射以及画面柔和不伤眼睛。在软物质中，有一类材料在光照作用下会发生一系列的物理的或者化学的变化，研究人员称之为具有光响应软物质材料，最具有代表性的就是光致变色材料。光致变色材料在实际应用中有广泛的用途，可以用来制备低能耗的显示器，可调波长滤光片和智能楼宇的变色玻璃等。最近在光控智能软物质材料方面，利用溶胶凝胶制备钒系氧化物复合物的方法，提高溶胶凝胶过程中钒钨氧化物纳米材料的络合能力，可以大大增强电子迁移概率，提高光电转换效率。在光电致变过程中，可产生过氧化氢。负载有二氧化钛和钒酸盐的氧化物会均匀地分散在溶胶中，其中在紫外线照射之前和之后的吸收率变化明显，在可见光照射下表现出高的活性，从而用于抗细菌、抗肿瘤和光催化降解有机污染物方面，且功能明显[37,38]。

第三节　软物质智能材料的前沿问题

软物质智能材料发展呈现出复杂性和跨学科性，需要各个学科的协同发展，才可能取得突破性进展。软物质智能材料已经发展了 30 多年，但是大部分研究成果只是停留在实验室，未能推向工业化应用，主要原因是软物质智能材料在应用过程中遇到的问题未能得到解决，需要制备出高性能的软物质智能材料，这是当前软物智能材料的前沿研究。

例如，电流变液的制备不仅涉及分散质的选择，还需要选择合适的分散相以及添加剂。特别是高性能巨电流变液的制备，需要综合考虑各方面的因素，电流变液之前的研究注重分散质即介电颗粒材料的选择，较少关注介电颗粒表面极化分子、分散相和添加剂的作用。然而，分散相以及添加剂会影响到纳米颗粒表面饱和极化，不同的分散剂和添加剂会有不同电流变效应，目前大部分的实验结果是用硅油作为分散剂，并没有系统地研究不同种类的分散剂对电流变液的影响，需要实验上测试不同分散相对巨电流变液的影响，同时也需要对它的影响进行进一步科学分析和解释，探讨固-液相的最佳匹配机制。

介电颗粒表面的极性分子在电场作用下的极化是巨电流变液机制基础。目前研究人员利用分子模拟的方法，发现巨电流变液的极性分子在电场的作用下能够在分散相中形成按照电场取向的分子链，这进一步描述了极性分子的具体作用及其微观机制，但是仍然需要实验的证实[40]。

电流变液的纳米介电材料的制备是当前电流变液的另一个研究前沿。研究人员证实了介电材料的尺寸效应，即尺寸越小，电流变效应越大。研究人员进一步制备除了中空多层多孔的纳米介电颗粒，发现其电流变效应更好，而且大大减小其密度和沉淀效应，为制备性能更强、稳定性更好的电流变液开辟了一条新的道路[41]。

PNIPAm 的水凝胶热敏性相转变是由交联网络的亲水性/疏水性平衡受外界条件变化而引起的，是大分子链构象变化的表现。虽然人们对温度响应性的机制已有初步的认识，但就疏水基团相互作用机制及其与相转变温度的关系而言，在定量方面尚有许多问题有待澄清，重复性也有待提高。

介电弹性体是目前已经商业化了的软物质智能材料，使用电场能够控制弹性体的压缩形变。这种材料的优点是分散剂是固体，比磁/电流变液和高分子水凝胶更加稳定和重复性更好，不会出现沉淀等问题，其缺点就是需要极高的电场（～100kV/mm）。将聚苯胺导电聚合物颗粒加入聚二甲基硅氧烷弹性体里面，发现能够降低其所需电压。其物理机制有待建立和研究[42-43]。

软物质智能材料在应用过程中遇到许多实际的问题需要解决，总结起来包括合成简单、成本低、稳定性好、响应快速、性能增强、可重复使用、无环境污染等方面。在实际应用中科研人员需要提出一些简单的合成方法以制备有一定智能特性的软物质材料，因为对于工业应用来说，简单的合成方法将制备出合格率高的和廉价的材料，方便应用于现实生活中；软物质本身的特性——液体溶液的流动性决定了它的稳定性要弱于固体，而它的稳定性主要集中体现在化学物质的抗氧化、抗腐蚀、化学降解以及沉降性等方面，于是如何提高软物质材料在功能上面的稳定性是一个非常重要的问题；对部件的腐蚀性、与容器/密封圈的相容性等也是软物质材料实际应用需要解决的问题；智能材料会感知并响应环境的变化，在特殊的条件下，例如高速列车上，快速响应是非常重要的条件，在一般的应用中也需要比传统材料更短的响应时间；在工业应用中，除了成本是一个重要的阻碍因素之外，另外一个就是软物质材料的性能，由于这是一个新兴的领域，许多材料的性能跟传统的材料相比还有很大的提升空间。例如，在电流变液发展的初期，它的强度不能达到工业应用的最低 30kPa 的要求，而磁流变液在当时是有这一强度的，于是，在汽车减震器的应用中，磁流变液的使用更早，但是在 2003 年巨电流变液发明之后，电流变液的强度经过短时间的发展，现在已经超过了磁流变液的最高强度（巨电流变液可以达到 250kPa 以上）[24]。因此，材料性能的提高也是软物质材料发展道路上亟待解决的问题；材料的多次重复使用也需要在今后的科研中解决，因为不能多次重复使用会提高使用成本，同时也会引入新的不稳定性；无环境污染是对现代材料提出的新要求，在以前粗放式的发展模式中，环境承受了太大的压力，导致很多地区污染严重，近年来我国非常重视环境保护，对工业企业提出了严格的要求。软物质材料中的有机污染物、重金属以及纳米颗粒等可能对环境造成的污染需要予以重视。

第四节 未来5～10年发展趋势

先进材料是我国制造业强国战略的物质基础，改革开放以来我国科技事业取得巨大进展，为国家经济发展提供了基础和驱动力。目前我国正在推进将中国制造全面升级为中国创造，主攻方向是智能制造。智能制造的实现离不开材料的创新与发展。其中软物质智能材料的发展与应用具有非常重要的特性，特别是面向工业化应用，把材料与器件集成化，做成可以实际应用的智能化器件。同时，智能材料的研究一直以来都是国家高精尖领域的热门项目，而软物质的智能特性又是新兴发展起来的方向，它的发展必将推动整个工程应用领域的许多技术革命及产品的升级换代。例如，巨电流变液被认为是现在最有希望在高铁、国防、军工等方面获得广泛应用的软物质智能材料，它可以用来制作性能优良的新一代智能减震器，而在这三个领域中，智能减震是一个科研上的难点，而巨电流变液正好符合装置简单、响应速度快、减震效果好等要求。除了电、磁作用智能软物质材料，具有热、光及其他作用的智能软物质也是今后材料的作用发展领域（图7-4-6）。

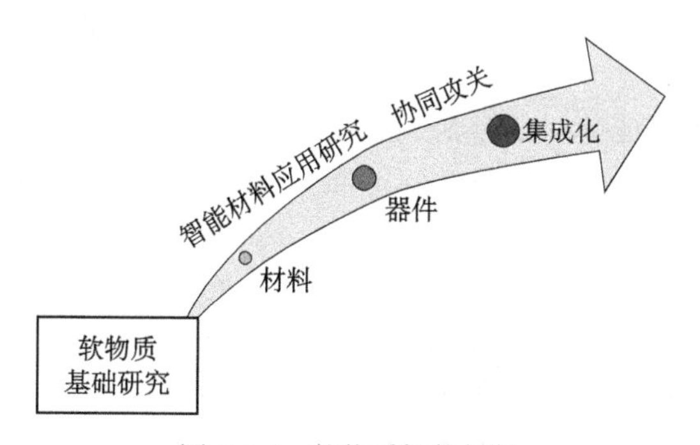

图7-4-6 软物质智能材料

当然，软物质智能材料在智能制造的应用仍然遇到许多问题，其中包括基础科学问题和工程技术问题。未来5～10年，软物质智能材料学学科重点发展包括：

（1）电、磁流变体固-液相的匹配及分散相的选择多样性。其应用与环境保护问题。

（2）纳米颗粒材料的表面修饰及结构对材料影响，以及添加剂对电（磁）流变液性能改变的基础物理机制[43]。

（3）固相纳米材料颗粒沉淀及极限应用下物理特性评价。

（4）软物质智能材料在不同的应用条件过程中的服役、失效问题。

随着社会经济的迅速发展，人们对新材料的需求日益增加，世界各国在材料的研发方面都投入了巨大的人力、物力和财力。虽然我国在软物质智能材料研究领域取得了很大进步，但是我国与发达国家相比在软物质智能材料的种类、数量和性能上是有一定差距的。希望集众多研究者的智慧，促进我国软物质和材料科学在今后的发展过程中牢牢把握住软物质智能材料这一新的领域，结合国家发展战略需求，研发出引领材料科学前沿的响应性更多、更快、重复性更好的软物质智能材料。

温维佳（香港科技大学），巫金波（上海大学）

参考文献

[1] 德热纳 P G, 巴杜 J. 软物质与硬科学 . 长沙：湖南教育出版社，2000.

[2] 蒋成保，赵晓鹏，王树彬，等 . 智能材料 . 中国新材料产业发展报告 . 航空航天材料专辑 , 2006.

[3] 姚康德，许美萱 . 智能材料 . 材料导报，1993,05.

[4] Takagi T.A concept of intelligent materials. Journal of Intelligent Material System Structure, 1990, 1(1): 149-156.

[5] Osasa Y, Matsuda A. Shape memory hydrogels. Nature, 1995, 376: 219.

[6] 陈莉 . 智能高分子材料 . 北京：化学工业出版社，2004.

[7] Meng Z Y. M. A. Sc University of Toronto, Canada, 2002, ISBN 0612738167.

[8] Lemaire E, Grasselli Y, Bossis G, et al. J. Phys. II (France), 1992, 2: 359; Grasselli Y, Bossis G, Lemaire E, et al. J. Phys. II (France), 1994, 4: 253.

[9] Rabinow. The magnetic fluid clutch. AIEE Transactions, 1948,67:1308-1315.

[10] Lopez-Lopez M T, Kuzkir P, Meunier A, et al. Synthesis and magnetorheology of

suspensions of cobalt particles with tunable particle size. Journal of Physics: Conference Series, 2009, 149: 1-4.

[11] Carlgon J D, Weiss K D. Magnetorheological Materials Based on Alloy Particles: US, 5382373. 1995.

[12] Choi H J, Park B J, You J L, et al. Preparation and magneto-rheological characterization of polymer coated-carbonyl iron suspensions //faramarz gerdaninejad. Proceedings of the 10thInternational Confemnee on Electrorheeloglcal Fluids and MagnetorheoloicalSuspensions. Lake Tahoe, USA: World Scientific, 2006: 121-127.

[13] Park B J, Park C W, Yang S W, et al. Coreshell typed polymer coated-earbonyl iron suspensions and their magnetorheology. Journal of Physics: Conference Series, 2009, 149(1): 1-5.

[14] Jun J, Uhm S, Ryu J, et a1. Synthesis and characterization of monodisperse magnetic composite particles for magnetorheologica-fluid materials. Colloids and Surfaces A Physicechemical Engineering Aspects, 2005, 260(1-3): 157-164.

[15] Ulicny J C, Mance A M. Evaluation of electroless nickel surface treatment for iron powder used in MR fluids. Materials Science and Engineering: A, 2004, 369(1/2): 309-313.

[16] Barber D E, Carlson J D. Performance characteristics of prototype MR engine mounts containing LORD glycol MR fluids. Journal of Physics: Conference Series, 2009, 149: l-4.

[17] Bose H, Roder R. Magnetorheological elastomers with high variability of their mechanical properties. Jourual of Physics: Conference Series, 2009, 149: 1-6.

[18] Fuchs A, Gerdaninejad F, Wang X J, et a1. Development and characterization of hydrocarbon polyol polyurethane and silicone magnetorheological polymer gels. Journal of Applied Polymer Science, 2004, 92(2): 1176-1182.

[19] Chin B D, Park J H, Kwon M H, et a1. Rheological properties and dispersion stability of magnetorheological(MR)suspensions. Bheol. Acta., 2001, 40(3): 211-219.

[20] Carlson J D. Catanzarite D M, Clair K A. Commercial magneto-rheological fluid devices, and associated technolgoies. In: Proc of the 5thIntConf on ER Fluids and MR Suspensions and Associated Technology, Singapore: World Scientific, 1996: 20-28.

[21] Carlson J D. Low-cost MR fluid sponge devices. Intelligent Material Systems and Structures, 2000, 10(8):589-594.

[22] Carlson J D. John C. Controllable Brake. US Patent, 5842547, 1998.

[23] Carlson J D. Clair K A St, Chrzan M J. Controllable vibration apparatus. US Patent, 5878851, 1999.

[24] Halsey T C. Science, 1992, 258: 761-766.

[25] Winslow.Methods and means for transmitting electrical impulses into mechanical force. US

Patent 2417850, 1974。
[26] 魏宸官．电流变技术：机制，材料，工程应用．北京：北京理工大学出版社，2000.
[27] 龚烈航，崔占山．电流变液机制及其研究现状．润滑与密封，2002, (1): 66-68.
[28] 陆坤权，沈容，王学昭，等．极性分子型电流变液．物理，2007, 36(10): 742-749.
[29] 张东恒．电流变液及其在润滑油领域的应用． 润滑油．2011, 26(S1): 9-15.
[30] Winslow W M. Induced fibration of suspensions. Journal of applied physics, 1949, 20(12): 1137-1140.
[31] Stangroom J E. Electrorheological fluids. Physics in Technology, 1983, 14(6): 290.
[32] Klass D L, Martinek T W. Electroviscous fluids. I. Rheological properties. Journal of Applied physics, 1967, 38(1): 67-74.
[33] Block H, Kelly J P. Electro-rheology. Journal of Physics D: Applied Physics, 1988, 21(12): 1661.
[34] Wen W, Huang X, Yang S, et al. The giant electrorheological effect in suspensions of nanoparticles. Nature materials, 2003, 2(11): 727.
[35] Gong X, Li J, Chen S, et al. Copolymer solution-based "smart window" . Applied Physics Letters, 2009, 95(25): 251907.
[36] Li J, Gong X, Yi X, et al. Facile fabrication, properties and application of novel thermo-responsive hydrogel. Smart Materials and Structures, 2011, 20(7): 075005.
[37] Wang C, Zhou B P, Zeng X P, et al. Enhanced photochromic efficiency of transparent and flexible nanocomposite films based on PEO–PPO–PEO and tungstate hybridization. Journal of Materials Chemistry C, 2015, 3(1): 177-186.
[38] Wang C, Gao X, Gao Y, et al. Controlled H_2O_2 release via long-lived electron–hole separation mediated to induce tumor cell apoptosis. Journal of Materials Chemistry B, 2015, 3(41): 8115-8122.
[39] Chen S, Huang X, van der Vegt N F A, et al. Giant electrorheological effect: a microscopic mechanism. Physical review letters, 2010, 105(4): 046001.
[40] Lee S, Lee J, Hwang S H, et al. Enhanced electroresponsive performance of double-shell SiO_2/TiO_2 hollow nanoparticles. ACS nano, 2015, 9(5): 4939-4949.
[41] Pelrine R, Kornbluh R, Pei Q, et al. High-speed electrically actuated elastomers with strain greater than 100%. Science, 2000, 287(5454): 836-839.
[42] Opris D M, Molberg M, Walder C, et al. New silicone composites for dielectric elastomer actuator applications in competition with acrylic foil. Advanced Functional Materials, 2011, 21(18): 3531-3539.
[43] Xu Z, Hong Y, Zhang M, et al. Performance tuning of giant electrorheological fluids by interfacial tailoring. Soft matter, 2018, 14(8): 1427-1433.

第五章 基于生物矿化构建的“生物-材料”复合体

生物矿化使生物体利用无机物质，并在进化中起到重要作用。受到鸡蛋与硅藻的保护性外壳的启发，科学家们通过人工仿生的手段使更多生命体拥有功能性的材料，最终制备出生物-材料复合物，提升生物体的生存能力，并赋予其新的功能。本章主要介绍最近功能材料外壳对生物体改进的相关进展以及有关应用，包括细胞保护与储存、热稳定性改进、生物掩蔽、光合作用与生物催化等。不同于以往对生物矿化的理解，本章着重探讨功能材料对生物体的调控与改进，通过材料科学技术实现人工生物进化。合理地设计材料可以解决很多世界性难题，如能源危机、环境污染、生物医学以及复杂材料的合成。此外，仿生构建功能化的“生物-材料”复合体可以将生物学与材料学更紧密地结合在一起。

第一节 引 言

在自然界，生物体拥有生产复杂结构与形貌的材料，如骨骼、牙齿、珍珠和贝壳都是常见的例子[1]。值得注意的是，生物体总是倾向于选择自然界含量丰富的简单无机物（如磷酸钙、碳酸钙或二氧化硅等）作为其材料的成分[2]。不同于人工合成材料，生物矿物在有机体的精确调控下总是拥有复杂的多级结构，其中胶原及多聚糖等有机基质在无机结晶过程中起到了关键调控作用[1,3]。一般认为生物有机基质能够有效调控无机晶体的成核、生长、相态和有序度以及组装过程[4]。鉴于生物材料的优异特性并受到生物矿化的启发，科学家开始通过仿生策略去设计合成各类功能材料[5]。如今，人们已经通过仿生矿化实现了牙齿的修复[6]、抑制肾结石发病[7]以及减少各种异位病理性矿化的发生[8]。很多仿生矿化材料也被成功制备甚至部分材料已经商业化，例如羟基磷灰石与Ⅰ型胶原的有机-无机复合物已被用于骨修复材料[9]。此外，人们还通过仿生矿化制备了各类氧化物、金属及半导体功能材料[10]。

我们还注意到，生物矿化在自然进化过程中也起到十分重要的作用。生命体正是通过这种方法将无机物加以利用，实现对自身的保护以及功能的改进[11]。其中，鸡蛋壳是一个很好的例子[12]；以碳酸钙为主要成分的硬蛋壳可以为柔软的细胞提供额外的保护，避免细菌入侵，同时还起到保持水分、调节与外界的气体分子交换以及为胚胎细胞提供钙离子等作用[13]。另一个例子则是以二氧化硅作为外壳的硅藻；正是由于硅外壳的存在，硅藻比其他藻类拥有更好的生存能力[2]。除了保护作用，生物矿化材料还可以为生物体提供基于材料学基础的生物功能拓展，例如，磁细菌可以利用 Fe_3O_4 或者 Fe_3S_4 晶体作为生物磁感应器[14]。此外，存在于内耳中以碳酸钙为主要成分的耳石，可被生物体用于重力感应器[15]。然而，通过生物进化获得生物矿化能力往往需要漫长的时光，并且这也和生物自身种类有着很大的相关性。幸运的是，通过对生物矿化的研究，我们可以通过仿生的方法赋予生物体原本没有的功能材料，这甚至可以认为基于人工材料的生物功能进化。过去十年间，许多科学家尝试赋予生物体各类人工材料，以此赋予其新的功能[13,16]，并开始逐步建立受生物矿化启发的基于材料的生物改造（图 7-5-1）。

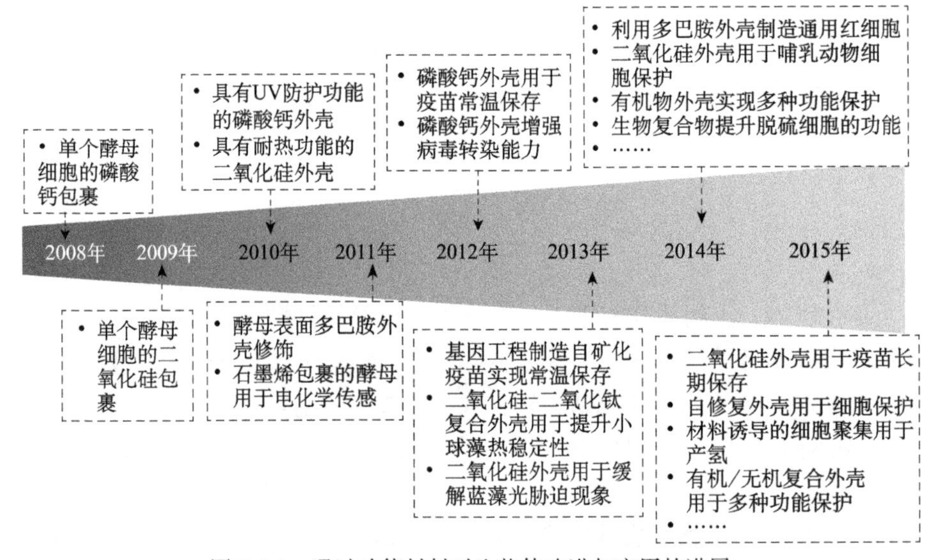

图 7-5-1　通过功能材料对生物体改进与应用的进展

第二节 矿化与调节

在探讨材料与生物之间的相互结合之前，我们回顾一下仿生矿化的基本原理。分子层面上的生物矿化调控可以控制纳米材料的生成、组装、尺寸以及形貌[4,17]。例如，骨中的羟基磷灰石晶体 ($Ca_{10}(PO_4)_6(OH)_2$) 是片状的，但是在牙釉质中却是针状的。它们之间的差别主要是由于不同有机基质参与调控的结果。

一般来说，与生物矿化相关的有机基质具有特殊调控基团。例如，在钙矿物成矿中体现出活性的生物分子往往具有一个共同的特征，即其往往拥有大量负电荷的磷酸根、羧基或者硫酸根等[3, 4, 18]。这些有机基团可以螯合钙离子并以此提高局部离子浓度，进而能够促进钙矿物在有机分子周围的异相成核。除了钙矿化植物，硅矿化是自然界中另一种常见的生物矿化现象。和钙矿化不同，具有硅矿化诱导能力的生物大分子则拥有大量带正电的氨基残基，此类基团在质子化后可以富集硅酸根并催化硅矿化过程[19]。正是由于存在各种各样的有机大分子，许多具有不同精细多级结构的生物矿化材料可以在温和的生物环境中得以形成。因此，许多包含相关基团的有机大分子已经被人们用来调控仿生矿化的过程并且已经取得了很大的进展[20]。然而，在生物界中的生物矿化的本源是受到基因表达的调控，这也是生物矿化与仿生矿化的明显差异[21]。自然界的生物矿化是从分子、基因、蛋白、基质到细胞逐级调控的过程，例如分子 - 分子作用控制着纳米尺寸晶体的有序性，而细胞-细胞则调节微观尺寸晶体的组装[22]。以脊椎动物牙生长为例，生物体首先从上皮细胞和成骨细胞演化出成釉细胞，并分泌胞外基质为矿化做好相关准备[23]，其中通过对胞外基质蛋白分泌基因的调控可以构成钙结合磷酸蛋白基因家族（secretory calcium-binding phospho-protein gene family），而这个家族蛋白又是通过骨粘连蛋白（osteonectin）演化而来的[24]。此外，科学家也成功地在基因与蛋白层面上研究了生物硅矿化的调控[25]。正是有着对生物矿化调控的准确理解，科学家可以在不同层面上仿生矿化，从而为在生物体上原位矿化提供了可行的方法。

第三节　生物－材料复合体的制备

在自然界中，一些生物通过矿化手段获得材料外壳用以提高自身的生存能力[2c,12,21]。由于矿化蛋白的缺失，许多生物体不能自发矿化。人为地向生物体表面引入矿化相关分子可以赋予生物体矿化的能力，以此通过仿生矿化生成材料外壳。通过化学或生物修饰，科学家已经可以成功地在生物体表面添加“成核位点”，以此来增强生物体的矿化能力，进而制备材料外壳[13,16]。表 7-5-1 总结了近期相关领域的代表性成果。

表 7-5-1　生物体功能外壳与改进

生物体	方法	有机基质	目的	参考文献
酵母	LBL assemble	PDADMAC(+)/PAA(−), PAH(+)/PSS(−)	CaP 外壳生长	[27]
酵母，蓝藻	LBL assemble	PDADMAC(+)/PSS(−), PEI(+)/PSS(−)	二氧化硅外壳生长	[36,37]
酵母	LBL assemble	PAH(+)/PSS(−)	高岭石吸附	[38]
斑马鱼卵	LBL assemble	壳聚糖 (+)/PAA(−)	磷酸镧外壳生长	[87]
酵母	LBL assemble	$GO\text{-}COO^-/GO\text{-}NH_3^+$	磁性纳米粒子生长	[31]
酵母	LBL assemble	PAH(+)/ β-cyclodextrinsulphate(−)	负载分子	[39]
酵母	LBL assemble	单宁酸 /$FeCl_3$	可循环外壳	[41]
绿藻	LBL assemble	20-mer peptide/titanium bis(ammonium lactate) dihydroxide	二氧化钛外壳生长	[42]
酵母	LBL assemble	PDADMAC/ 硅酸	二氧化硅外壳生长	[43]
酵母	Ion binding	柠檬酸，Ca^{2+}	金纳米外壳生长	[47]
酵母	Ion binding	GR sheet, Ca^{2+}	石墨烯外壳生长	[48]
日本脑炎病毒（JEV），腺病毒疫苗 5 (Ad5)	Ion binding	病毒表面负电蛋白	CaP 外壳生长	[50]
芽孢杆菌	Ion binding	细胞壁表面负电蛋白	钯纳米外壳生长	[51]
酵母	Self-formation	PDA	PDA 外壳生长	[55]
酵母	Self-formation	Poly-(norepinephrine),PEI	二氧化硅外壳	[56]
脱硫细菌，蓝细菌	Self-formation	α-Lysine/Au NPs,L-cysteine/ Au NPs	生物复合外壳	[60,61]

续表

生物体	方法	有机基质	目的	参考文献
酵母	Self-formation	Peptide, $R_4C_{12}R_4$ (R: arginine;C: cysteine)	二氧化硅外壳	[64]
M13 病毒	Genetic engineering	金特异性识别蛋白	金-氧化钴外壳生长	[67]
人肠道病毒疫苗 71(EV71)	Genetic engineering	磷酸 / 钙螯合蛋白	磷酸钙外壳	[69]
酵母	Disulfide bond	含巯基蛋白	量子点标记	[72]
大肠杆菌	Spray drying	硅溶胶，脂质体	二氧化硅外壳生长	[73]

一、层层自组装

酵母是真核细胞的重要模式生物，其自身不具备矿化能力[26]。层层自组装（LBL）是 2008 年首先被应用于增强酵母矿化能力的方法，并且也是目前最普遍的一种生物-材料复合手段[27]。LBL 是通过交替沉积具有相反电荷或分子作用力的分子来生成超薄膜的技术[28]。酵母细胞壁是由带有微弱负电荷的聚多糖、甘露糖和 N-乙基葡萄糖胺等具有非矿化能力的有机基质构建所成[29]。PDADMAC（聚二烯丙基二甲基氯化铵）是一种在生理条件下带有正电基团的聚合物，可以吸附在带负电的酵母细胞壁表面形成一个带正电的表面层。然后，将带相反电荷的 PAA(聚丙烯酸) 吸附于 PDADMAC 上来形成带负电的外表面。通过交替添加这两种电荷相反的聚合物，一个富含官能团的有机超薄膜便可以在酵母表面生成并可以用于矿化能力的增强[27,30]。对于钙矿化过程，富含羧基的 PAA 可作为最外层结构，以便于有效地与钙离子相互作用。通过在细胞表面对溶液钙离子的富集则能诱导磷酸钙（CaP）在细胞表面的原位沉积［图 7-5-2（a）］，最终生成具有核-壳结构的酵母-磷酸钙复合体[27]。因此，LBL 方法可以将酵母从非矿化细胞转变为矿化细胞[31]。

利用自组装功能的枝装聚合物分子作为 LBL 的单元分子也有类似的细胞修饰功能，它们不仅能和内层相反电荷的聚合物连接，同时又可增加最外层的电荷 / 官能团［图 7-5-2（b）］[32]。通过控制 LBL 的循环次数可以调节最外层的矿化活性残基的密度或对应的矿化能力。其他一些特殊的聚合物也可以扮演“桥梁”的角色，并能进一步控制溶液中细胞的聚集能力［图 7-5-2（c）］[33]。例如，具有嵌段结构的丝蛋白（SF）聚合物，可以分别修饰给予聚-L-赖氨酸（SF-PLL）、聚-L-谷氨酸（SF-PGA）以及聚乙二醇（PEG）侧链。实验表明，带聚阳离子氨基酸基团的聚合物可以显著促进细胞聚集并构建出细胞－材料复合聚集体。相比之下，富含 PEG 的以氢键为主的外壳则趋向于生成稳定的单细胞悬浮液。此项研究表明，聚电解质可以被用于调控细胞的聚散，并选择性地在单细胞或细胞聚集体上诱导形成材料外壳[33,34]。

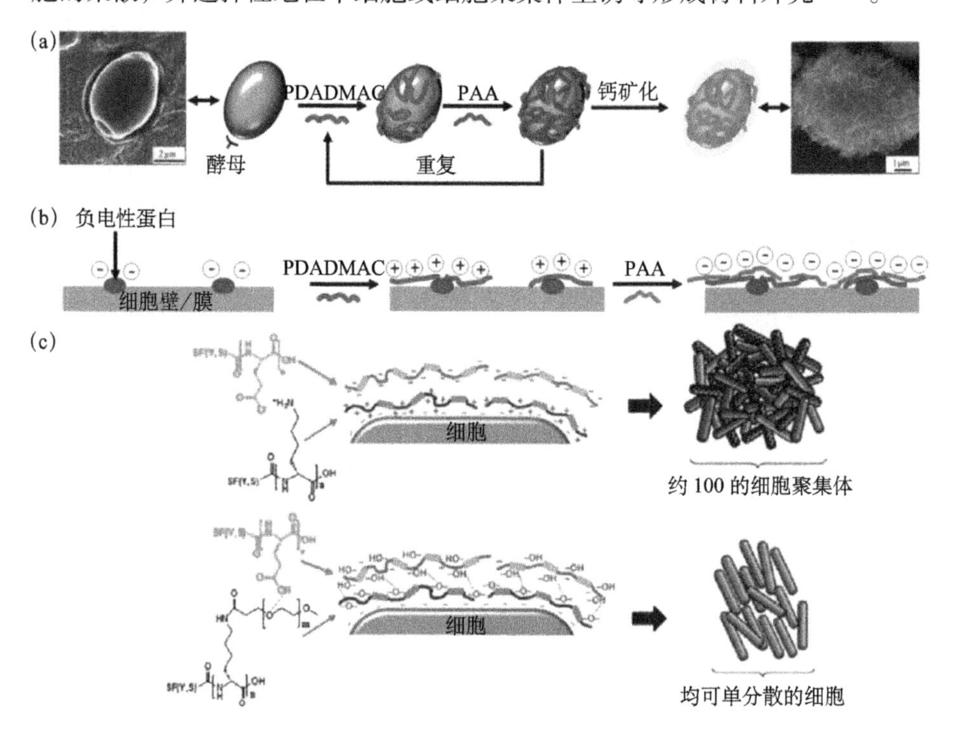

图 7-5-2　LBL 对细胞的修饰

(a) 在酵母表面实现人造 CaP 外壳。(b)LBL 方法可以增加表面结合位点或电荷密度。(c) 通过不同丝蛋白共聚物实现细胞聚集调控。分子内作用力影响细胞聚集行为：电荷结合（上图）与氢键结合（下图）Copyright 2015, American Chemical Society from Ref. [33]

对于 LBL 自组装薄膜，最外层的组装物质性质决定矿化所能沉积的材料。CaP 与其他钙类矿物趋向于在带负电的聚合物表面沉积，然而二氧化硅（SiO_2）材料则更易于在带正电的聚合物表面生长[35]。PDADMAC 拥有大量的正电荷氨基，可以增强细胞的硅矿化能力[36]。与聚苯乙烯磺酸钠（PSS，一种生物相容的带负电的聚合物）共同可以在蓝藻表面上构建出 LBL 薄膜（PDADMAC/PSS）$_6$–PDADMAC，其中最外层 PDADMAC 上的氨基作为硅矿化的活性位点［图 7-5-3(a)］[37]。通过 LBL 自组装也可以帮助细胞实现对纳米材料颗粒的选择性吸附[38]。此外，细胞-材料的功能协同也可以通过聚合物上的残基实现。例如，环糊精基团可以让细胞表面产生主客体组装行为，进而能通过环境响应性外壳调节细胞活性[39]。通过聚合物与无机相的LBL组装则可以直接制备出有机-无机复合材料外壳[40]。这种方法的优势在于材料外壳的厚度可以通过LBL组装层数来控制［图7-5-3(b)］。利用天然有机多酚（单宁酸，TA）与Fe^{3+}之间存在的非共价作用力，人们通过LBL技术完美地在活细胞表面制备出有机-无机仿生功能外壳[41]。此外，二氧化钛/多肽[42]和二氧化硅/PDADMAC[43]复合外壳也已经被成功制备，这些意味着无机纳米颗粒也可以作为细胞LBL修饰中的组装单元。

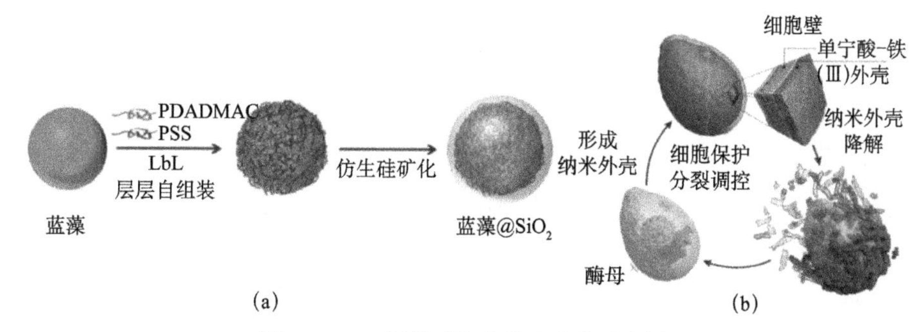

图7-5-3 二氧化硅包裹单个蓝藻示意图

(a)通过LBL在细胞表面沉积PDADMAC和PSS来生成二氧化硅包裹外壳。Copyright 2013, the Royal Society of Chemistry from Ref. [37]. (b)TA–Fe^{III}外壳在单个酵母细胞表面的生成与降解。Copyright 2014, Wiley-VCH from Ref. [41]

LBL技术是最广泛使用的细胞表面修饰方法。然而，在LBL处理中涉及去除多余聚合物的离心、分离与纯化等步骤。这些步骤不仅操作相对比较复杂，更严重的是其很容易导致细胞损伤。因此，LBL方法目前仅适用于有完整细胞壁结构的生物，如酵母与藻类等。然而，对于环境特别敏感并且相对脆弱的哺乳动物细胞，这种方法并不适用。

二、离子键合

在某些特定情况中，细胞表面与特定矿物离子间的电荷作用可以直接用于细胞-材料复合物的制备。钙离子是大多数生物体必需的元素，尤其在生物信号通路上有着重要的作用[44]。一些细胞能通过其表面的离子受体吸收并富集溶液中的钙离子（Ca^{2+}）[45]。事实上，这些表面Ca^{2+}受体作为离子键合位点也可以诱导钙矿物材料在细胞表面的原位成核生长[46]。通过离子键合的思路，人们将柠檬酸修饰的金纳米颗粒（Au NPs）与氯化钙溶液共混，可以将金纳米颗粒在酵母细胞表面形成一层纳米金外壳[47]。以此类推，将石墨烯与钙离子-金纳米颗粒（Ca-Au NPs）共混同样可以在酵母表面形成基于石墨烯材料的功能外壳［图 7-5-4(a)］。这些实验中的细胞-材料间的相互识别与组装都是通过钙离子介导与完成的[48]。

此外，一些病毒颗粒表面也拥有高密度的负电荷[49]。因此，它们可以直接吸附溶液中矿物离子并在表面富集，进而提高过饱和度并促进材料矿化材料在其表面的沉积结晶。以日本脑炎病毒（JEV）为例，这种病毒可在生物培养液中大量吸附钙离子导致其原位矿化并获得了 CaP 外壳［图 7-5-4(b)］[50]。同时在细胞表面引入特定的晶种促进材料的异相成核也可以为细胞表面提供与材料复合的途径。例如，在 *Bacillus* 细胞表面首先吸附钯纳米粒子（Pd NPs）就可以很好地促进磁性 CoNiP 外壳的生成[51]。

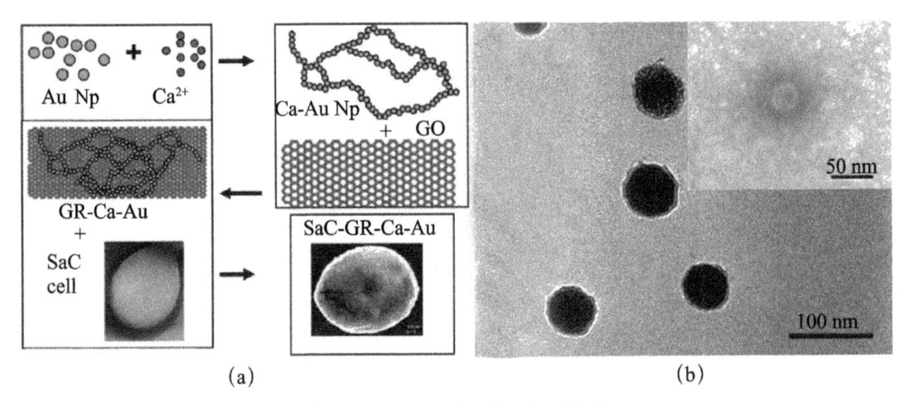

图 7-5-4　基于离子的表面修饰

(a) Au 纳米粒子通过钙离子作用后形成链状聚集体。进一步，Ca-Au 纳米可以在片状 GO 表面沉积并形成 GO-Ca-Au 聚集体，并进一步还原为 GR-Ca-Au。接下来将酵母细胞（SaC）与片状 GR-Ca-Au 通过离子键合修饰在一起。Copyright 2011, American Chemical Society from Ref. [48]. (b) JEV 病毒因其表面拥有大量负电荷，可以自发地在富钙体系中矿化。插入图是未染色的 JEV。Copyright 2012, Wiley-VCH from Ref. [50]

不同于 LBL 修饰方法，离子键合方法可以更简便地生成材料外壳构建生物-材料复合体。但这种方法仅仅局限于自身表面带有高密度电荷的生物体系，通用性并不广泛。

三、自生长

不同于酵母细胞，哺乳动物细胞不具有细胞壁结构，而它们的细胞膜对材料结合显得极其敏感并且十分脆弱。探寻一种有效的哺乳动物细胞矿化方法目前仍然是一项重大挑战。目前，一种简化的 LBL 修饰方法可以在哺乳动物细胞上通过一次性组装实现仿生矿化构建。Choi 的课题组通过聚电解质辅助的方法在 Hela 细胞表面实现了人工硅矿化[52]。在他们的实验中，细胞首先与带正电的聚乙烯酰亚胺（PEI, 硅矿化模板）在溶液中混合，之后加入硅矿化前驱体（正硅酸甲酯）促进硅外壳在细胞表面的原位生成。其原因是 PEI 拥有极高的正电荷密度，因此有利于二氧化硅的沉积，但也正是 PEI 的高正电荷性能够导致细胞毒性[53]。尽管硅外壳可以生成，但被包裹的细胞也仅仅只能存活 12 小时。因此，寻找与开发一种生物友好的聚电解质来用于哺乳动物细胞矿化功能修饰变得更为紧迫。

自聚合（self-polymerization）是另一种促进生物体矿化能力的有效方法。聚多巴胺（PDA) 是一种受蚌类黏附蛋白启发的分子，并发展成为一种可靠的广泛用于在基底表面功能化的方法，特别适用于生物体表面[54]。其修饰的过程需要在具有弱碱性环境的多巴胺溶液中。人们认为 PDA 会通过多酚基团及氨基与基底形成共价键与非共价键。利用这个反应过程，酵母细胞可以被包裹在 PDA 外壳之中［图 7-5-5(a)］[55]。此外，PDA 上的氨基与巯基也存在一定的反应活性，能为卵白素在内的多种物质化学修饰提供可能[55]。事实上，PDA 已经被成功用于红细胞（RBC）的表面修饰，而红细胞又拥有与其他哺乳动物类似的细胞膜结构[57]。聚甲肾上腺素（PN）是一种多巴胺衍生物，由于其拥有更多的羟基官能团，因此被认为是更合适的表面修饰分子。实验已经表明，PN 外壳可以更好地保持细胞活性并体现出很好的生物相容性［图 7-5-5(b)］[56]。

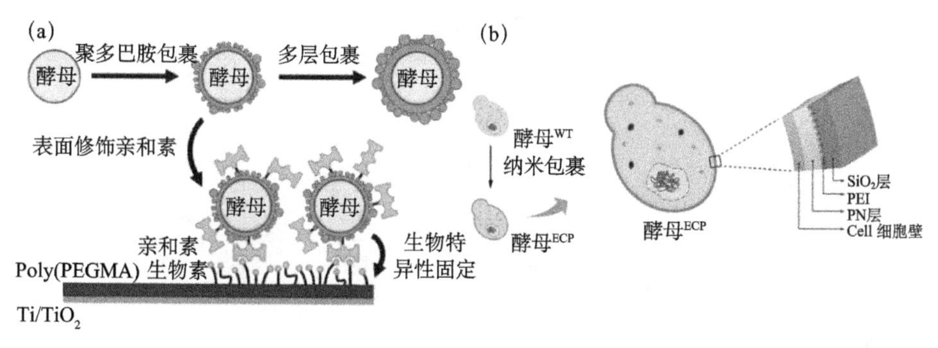

图 7-5-5 (a) PDA 包裹的酵母及其功能化。Copyright 2011, American Chemical Society from Ref. [55]. (b) 由 PN 和二氧化硅外壳组成的人造外壳（yeast^{WT}: 普通酵母；yeast^{ECP}: 包裹后的酵母）。Copyright 2015, the Royal Society of Chemistry from Ref. [56]

自组装是一种在温和环境下用于细胞修饰的重要方法。自组装被定义为通过非共价键、分子或纳米粒子自发并可逆的有序化的过程[58]。对于有机-无机复合物的构建，选择合适的有机基团与无机纳米颗粒作用可以产生可观的非共价作用[59]。例如，L-半胱氨酸是一种有巯基的生物分子，并且可以与金纳米颗粒产生作用力。因此，L-半胱氨酸和金纳米颗粒可以自发地形成拥有纳米结构的生物复合物，并在生物体周围形成材料外壳[60]。此外，L-赖氨酸也可以替代L-半胱氨酸与金纳米粒子形成纳米聚集，从而诱导细胞表面的材料壳化[61]。由于是非共价键驱动外壳的生成，这种生物复合物可以在细胞分裂的同时实现自我修复功能[60]。选择一种功能化的纳米颗粒参与以上的自组装甚至可以直接赋予生物体更多的功能[62]。

还有一种简化的LBL组装方法是利用高度功能化的分子实现生物-材料复合。半胱氨酸中的巯基可以模仿硅蛋白-α中的丝氨酸残基，并用于催化加速硅酸四乙酯的水解过程[63]，而在多肽中的带正电的精氨酸基团则可以使其吸附在细胞的表面上。利用这两种氨基酸的相互协同，人们可以设计出一种生物相容的多肽分子，用于促进酵母表面自发的硅矿化过程（图7-5-6）[64,65]。此类方法可以大幅度提高细胞修饰效率并减少对细胞的伤害。事实上，通过对很多生物分子的多功能特征研究有利于矿化的开展。例如，幽门螺杆菌（*Helicobacter pylori*）可以分泌尿素酶来获得碱性的微环境[66]。以此，通过设计含有尿素酶的分子并吸附在生物体表面，可以直接将酸碱敏感的矿化功能材料构建在细胞表面。

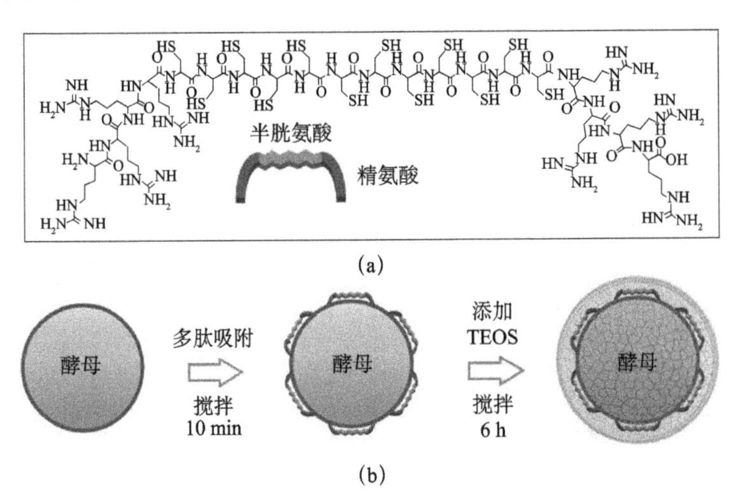

图7-5-6　(a)$R_4C_{12}R_4$的分子结构；(b)二氧化硅包裹单细胞示意图。Copyright 2015, the Royal Society of Chemistry from Ref. [64]

四、基因工程

在自然界中，生物通过基因调控分泌出特定的功能蛋白实现自矿化[21]。需要指出的是，之前的化学修饰方法都是暂时性的，也就是说当细胞分裂之后需要重复同样的修饰方法以保持新细胞的矿化能力。基因表达具有遗传性，因此重新编码生物体的基因赋予细胞可遗传的矿化能力是一种比化学修饰更好的途径。

到目前为止，科学家通过基因工程尝试过对生物体可遗传矿化能力的改造[67]。例如，溶菌酶表达的酵母细胞（*Pichiapastoris*）和硅蛋白-α 表达的大肠杆菌（*E.coli*）已经被应用于细胞硅外壳的生成[68]；其中溶菌酶和硅蛋白-α 在细胞质中先被合成，然后通过细胞壁渗透到胞外并和溶液中的二氧化硅前驱体反应。此外，病毒也可以通过基因工程而获得生物矿化的基因并最终实现自发矿化[69]。通过基因修饰，CaP 成核多肽可以在病毒表面表达从而赋予其 CaP 的矿化能力。通过基因重组技术，我们课题组将磷酸螯合蛋白（N6p）或钙螯合蛋白（NWp and W6p）重组于 A12 系 EV71 病毒的 cDNA 之中［图 7-5-7(a)］[69]。被基因改性的病毒可以携带成核多肽在体外转染人源细胞。这些基因修饰过的细胞有着和普通 EV71 细胞类似的生物性能。通过病毒结构建模发现新插入的多肽会在天然病毒蛋白 1 中表达，并且 60 个成核多肽均匀地在病毒表面表达，基因工程修改后的病毒拥有在近生理条件下自发矿化的能力［图 7-5-7(b)］[69]。

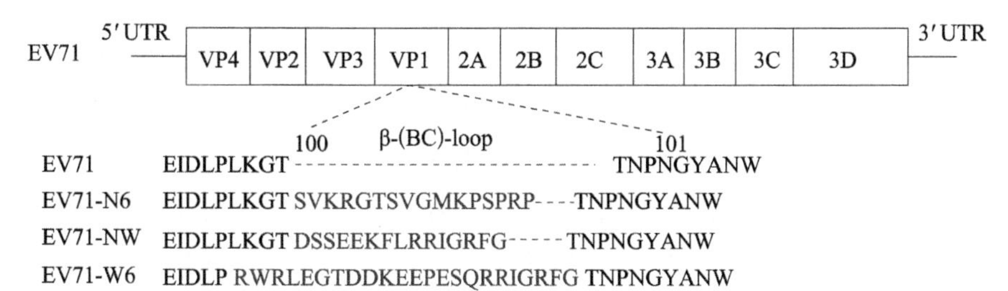

(a)

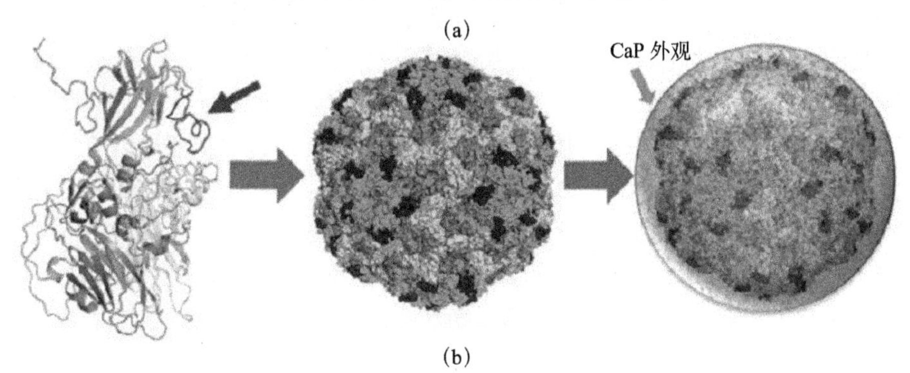

(b)

图 7-5-7　含有结合蛋白的 EV71 设计思路

(a) EV71 的基因组和 VP1 上 β-(BC)-loop 的插入位点。(b) 病毒蛋白建模结果。EV71 外壳蛋白 VP1、VP2 和 VP3 分别以青色、黄色和橙色表示。内嵌蛋白以蓝色标记，60 段复制体在表面均匀分布，这些蛋白可以诱导 CaP 外壳（灰色）的原位生成。Copyright 2013, the National Academy of Science from Ref. [69]（文后附彩图）

基因工程是一项先进的生物体改造技术，尤为突出的就是其修饰具有可遗传性。然而，基因工程很大程度上受制于基因组学和蛋白组学。尽管对病毒的基因改造技术已经比较成熟，但对真核生物的改性仍在研究之中。

五、其他方法

分子识别是指两个分子之间通过非共价作用来实现特定结合的过程，也是生物体系中实现受体—配体（抗原—抗体，蛋白—蛋白，糖—外凝聚素等）相互作用的重要基础[70]。合成拥有特定配体的分子也可以用于为生物体选择性制备材料外壳。半乳糖和甘露糖修饰的量子点可以特定与糖类和外凝聚素作用，可以将量子点作为特定的标记复合在酵母细胞表面[71]。此外，生物体系中二硫键也是一种常见的键合，通常是通过两个巯基相连。三（2-羧乙基）磷盐酸盐可以将酵母表面含有二硫键的膜蛋白断裂，从而形成自由的巯基，此时如果加入含有巯基的量子点则可以形成新的二硫键，从而实现量子点和细胞的复合[72]。

通过喷雾干燥技术则可以规模化地实现对生物体的人工矿化[73]。由于具有使用方便、前驱液体丰富、高效以及连贯性等优点，喷雾干燥技术被大规模地用于制药及食品工业。在人造矿化外壳与喷雾干燥联用方面还需要结合真空诱导的自组装技术[74]。以二氧化硅为例，溶剂蒸发会浓缩溶解在溶液（通常为水-乙醇体系）中的二氧化硅和表面活性剂，从而导致胶束的形成并自发组装成拥有三维组装尺寸的二氧化硅——表面活性剂介观相。通过这种策略，选用与细胞膜相同的磷脂作为表面活性剂的替代品来提高生物相容性，可以实现细胞介导的材料自组装[73]。在这个过程中，两性磷脂充当结构引导剂的作用来引导硅酸前驱体有序化，并在旋涂干燥的过程中形成高度有序化的脂质—二氧化硅介观相。在加入细菌、酵母或者其他生物单元后，在真空环境下浓缩可以在生物体周围形成均匀和连续的脂质—二氧化硅纳米外壳[73]。将这种细胞引导的自组装过程引入喷雾干燥之中已经可以成功包裹多种细胞，且二氧化硅外壳可以调控（图 7-5-8）[73]。不同于其他方法，这种方法可以实现大规模地赋予生物体功能化的外壳材料，具有很好的工业实用性。

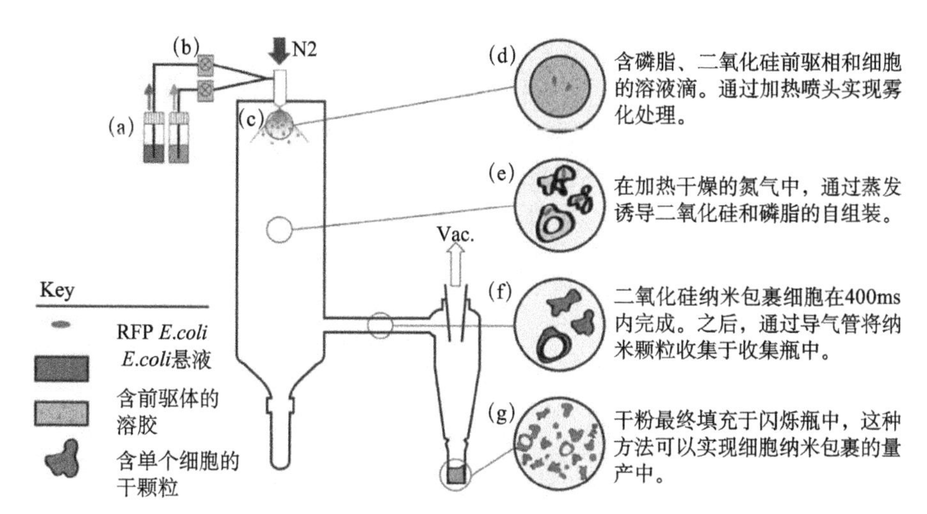

图 7-5-8 通过喷雾干燥实现脂质体－二氧化硅外壳的生成。分散于脂质体的细胞悬液与二氧化硅前驱体在闪烁瓶中混合 (a) 之后，通过蠕动泵在 Y 形管快速混合后在喷嘴中喷出 (b) 混合物，之后被加热的氮气雾化 (c)，生成的液滴直径为 10～100μm，由细胞、脂质体、二氧化硅前驱体和溶剂组成 (d)。通过氮气流速和液体流速可以控制液滴尺寸 (e) 脂质体促使二氧化硅前驱体生成有序的纳米结构，同时伴随着溶剂挥发 (g) 在气旋分裂前，颗粒已完全干燥 (f) 并通过气旋分离器进入收集仓。真空泵（Vac.）会提供真空环境以除去氮气。Copyright 2015, American Chemical Society from Ref. [73]

第四节 基于功能材料外壳的生物体改性

生物体通过自然选择获得各种成熟的生物功能，而我们可以利用其中一些功能来解决很多世界性难题，如温室气体循环 [75]、太阳能转化 [76]、污染物降解 [77] 以及疾病治疗 [78] 等。尽管对于生物体而言有些功能是现成的，但由于生物自身对环境敏感以及脆弱等原因的限制，我们事实上并不能直接利用这些功能。因此，通过为生物体合理设计功能材料外壳可以保护并提升生物体相关功能并方便人们应用。

一、细胞保护

受鸡蛋壳的启发，酵母首次被通过 LBL 方法在 2008 年被赋予了矿物材料外壳 [27]。最终形成的 CaP 外壳拥有几百纳米的厚度，并能保护内部的活细胞免受外部侵害 [27,79]。例如，溶菌酶可以降解酵母的细胞壁，导致酵母因渗透压而破裂[80]。普通的酵母细胞会因溶菌酶的存在而破裂死亡。然而，拥有CaP外壳的酵母可以隔离酵母和溶菌酶，从而有效保障被包裹细胞的结构完整性［图7-5-9(a)］。例如，其他矿化材料（如二氧化硅构建的外壳）同样也有保护细胞的功能［图7-5-9(b)］[56]。类似的外壳也被用于保护哺乳动物细胞，增强其对胰酶的抵御能力[52,81]。值得注意的是，由于只有纳米尺度的材料外壳可以保证包括营养交换在内的正常细胞代谢过程，因此被包裹的细胞可以持久地保持活性[27,56,82-83]。

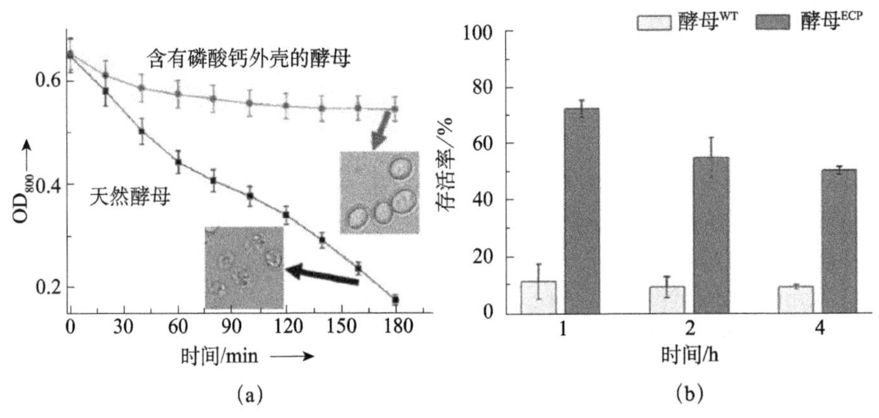

图7-5-9 (a)在酵母裂解酶存在条件下酵母的存活率变化。裂解过程通过光学测量手段在低盐含酶条件下测定。细胞裂解伴随着光密度的下降。两幅内嵌图代表实验末期光学显微镜照片。可见酶会降解细胞壁，从而使细胞失活，但是在有CaP外壳存在下，活性可以保持。Copyright 2008, Wiley-VCH from Ref. [27]。(b)裂解酶处理后，普通酵母和二氧化硅包裹酵母活性变化.（yeast^{WT}：普通酵母；yeast^{ECP}：包裹酵母）。Copyright 2015, the Royal Society of Chemistry from Ref. [56]

全球环境由于人类大量活动而已经产生了明显的变化，其中最严重的问题之一就是臭氧空洞[84]。这造成了大量的中波紫外射线（UVB radiation,280-320 nm)抵达地球表面，许多生物体结构及发育过程由于过量的UVB射线而被损害[85]。受到球石藻[86]（一种拥有滤光功能外壳生物）的启发，我们为细胞设计制备了能吸收紫外线的磷酸镧外壳，并构建在斑马鱼受精卵表面(图7-5-10)[87]。通过对斑马鱼卵发育形态的观察，大量天然胚胎（～86%）在紫外线照射下停止发育，大概只有7%的胚胎可以存活3天。相比之下，超过半数（～51%）磷酸镧外壳包裹的胚胎在同样条件下仍可以正常发育，最终形成的幼鱼能够突破无机矿物外壳而完成整个孵化过程[87]。这为强辐射下生物发育及自然生态的维持提供了基于材料学的新策略。

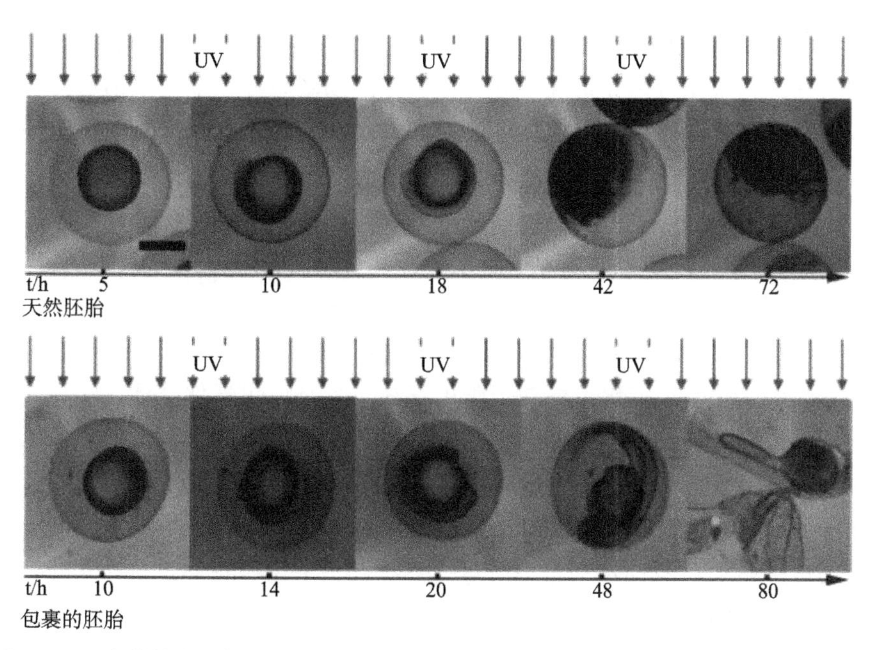

图 7-5-10 在紫外光照射下普通（上）和磷酸镧包裹的（下）斑马鱼卵发育过程。标尺，500 mm。 Copyright 2010, Public Library of Science from Ref. [87]

此外，材料外壳更可以保护细胞免于外部机械力的破坏。通过喷雾干燥制备的二氧化硅外壳包裹的大肠杆菌可以极大地提升细胞耐受的机械强度。一般来说，有细胞壁的细胞有着 150～200MPa 的强度[73]。但矿化包裹后的生物体可以有约 13.0GPa 的杨氏模量以及 1.4GPa 的强度。我们还注意到，正是由于生物体的存在及复合，材料的强度已经显著超过了生物硅和介孔硅等材料的强度。另外，有报道说表面材料外壳可以有效地保护细胞免于渗透压冲击的影响[88]。渗透压冲击是当胞外溶液浓度突变，导致水快速通过细胞膜的过程[89]。这个过程会导致细胞因渗透压不平衡而裂解[90]。但是通过在酵母表面组装石墨烯（拥有～0.3TPa 的强度）可以极大地优化生物体的机械性能，以及在渗透压不平衡环境下的生存能力。而这主要归功于石墨烯材料的韧性，从而控制细胞由渗透压造成的体积变化。

二、细胞储存

目前的细胞储存往往需要将细胞在含 10% 二甲亚砜的培养基溶液中冷冻，并最终转移到液氮之中[91]。而这一冷冻过程会对细胞造成永久损伤，例如细胞会因冰晶的形成而造成体积变化和破裂。此外，凋亡也是一种冷冻过程中细胞死亡的原因[13]。因此，是否有可能实现在常温中储存细胞？正如我们所知，鸡蛋壳可以在常温下几周内保证内部胚胎的稳定和生物活性。此外，仙女虾卵因其外壳可以在高温、冷冻及脱水环境下生存多年并仍然保持活性[92]。这些例子表明生物储存中外壳的重要性。真核细胞的分裂是通过特定的生长因子与营养调控的，当缺少这些物质的时候细胞便会进入非分裂时期（G0 期）[93]。其中酵母就可以通过包裹上 CaP 外壳而进入 G0 期［图 7-5-11(a)］[27]。因此，材料外壳可以延长酵母细胞的寿命。实验发现，大部分（＞80%）未处理的酵母细胞会在一个月内于水溶液中死亡，相比之下，大部分（～85%）用 CaP 矿化包裹的细胞可以在一个月后仍然保持活性［图 7-5-11(b)］。其主要原因是材料外壳可以减少细胞与周围环境的物质交换和生物交流，通过改变微环境分子梯度而使细胞强制进入休眠期。如果将矿化外壳溶解，则内部的酵母细胞又可以被重新激活分裂（G1 期）。从实验中可见，重新复活后细胞的增殖曲线与普通酵母无异，说明生物功能不会因外壳的去除而产生影响［图 7-5-11(a)］[27]。相比之下，被外壳包裹的细胞则在培养基中一直处于休眠期（G0）并不分裂［图 7-5-11(c)］[27]。因此，通过材料外壳控制的细胞储存与重激活可以为细胞生命周期调控提供新思路。这种特征为接下来的细胞常温储存研究提供了可能性。

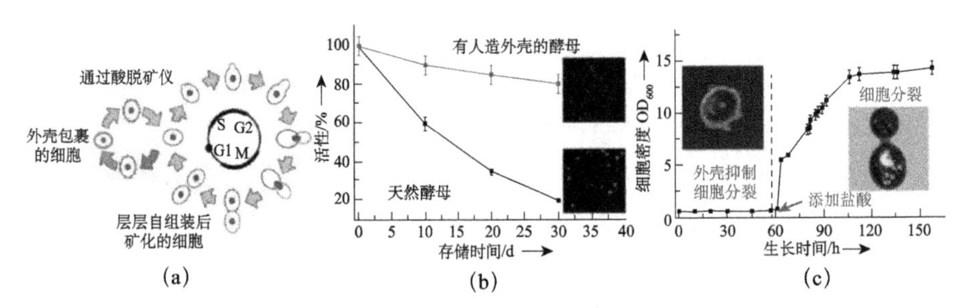

图 7-5-11 (a) 酵母细胞与包裹后酵母的生物周期 (b) 通过曲线可知活细胞数量在清水中会随着时间而减少。内嵌的图是通过死活细胞染色剂（FUN1）处理后，30 天后的荧光图。红色表示细胞是活的，绿色表示细胞死亡。(c) 普通与包裹后酵母的生长曲线。细胞密度是通过浊度法测定的（OD 600）。本实验中带壳的细胞是在细胞活性实验最后收集的，所以所有细胞都是被完整包裹的。1mM HCl 在 60h 时被加入以用于溶解矿物外壳。左侧插图：荧光图像说明细胞是活的；右侧插图：TEM 图展示体系中的分裂细胞。Copyright 2008, Wiley-VCH from Ref. [27]（文后附彩图）

通过喷雾干燥在大肠杆菌上制备的二氧化硅-脂质体复合物外壳也拥有与 CaP 外壳类似的性能。研究表明，这种复合物外壳阻碍细胞生长并强制其进入非分裂休眠期[73]。另一种生物相容的矿化外壳由两层外壳组成：海藻酸-二氧化硅和钙-海藻酸外壳，并用于 HepG2 细胞的包裹[94]。用这种复合物包裹的哺乳动物细胞在超过六个月之后仍保持较高的存活率，也暗示着哺乳动物细胞的非冷冻保存存在可能性。细胞分裂是由母细胞分裂成两个或更多子细胞的重要生理过程[95]。这个过程实际上可以通过材料外壳所造成的空间限定进行抑制[96]。用儿茶酚修饰的 PEI 和透明质酸用于酵母表面的 LBL 组装后再交联聚电解质外壳能够很好地实现对内部细胞的空间束缚。通过增强聚电解质外壳的物理强度，细胞分裂可以被抑制且程度取决于聚电解质的层数即材料外壳的强度[96]。

三、热稳定性

生物质热稳定性（thermal stability）是生物储存研究和应用中的重要问题。一般情况下，高温总会促使蛋白质失活变性并最终失去其功能[97]，因此提高生物体和蛋白质的热稳定性是很有必要的。基于硅外壳及蛋白聚集体可以显著提高被包裹生物体在高温下的稳定性[98]。例如，酵母表面的二氧化硅外壳可以在 49～53°C 下依旧保持内部细胞的活性。对于普通的酵母，在 52°C 条件下的存活率不足 50%，但以二氧化硅包裹的酵母在相同条件下可达到 70% 的存活率[98]。在显微镜观察下发现，拥有二氧化硅外壳的酵母细胞在高温下不会有显著的结构变化，但是普通细胞可见明显的细胞壁坍缩。这个现象表明硅外壳可以像“绷带”一样稳定住细胞结构，从而使酵母在高温下稳定[98]。

对于蛋白及酶类的储存，往往比细胞保存更重要。受生物矿化的启发，为蛋白复合上无机矿化材料外用于热稳定增强是一种可行的策略。二氧化硅[99]、CaP[100]、聚合物[101]以及有机金属框架（MOF）[102]均被作为外壳用于蛋白及酶类的保护。据报道，由锌离子和 2-甲基咪唑组成的名为 ZIF-8 的金属有机框架可以用于蛋白、酶和 DNA 的热保护［图 7-5-12(a)］[102]。如果材料拥有越小的孔道，则在干燥环境中对生物分子赋予更好的稳定性，同时越紧密的复合也可以实现更好的生物热稳定性增强。最近，我们课题组研究发现，关于无定形矿物能在蛋白周围形成稳定的结合水，通过稳定的氢键结合降低蛋白本身在加热过程中的变性［图 7-5-12(b)］[100]。这些方法都表明通过材料稳定有机体必须有紧密结合的外部作用力，其中无论是利用氢键还是电荷作用均是可行的。

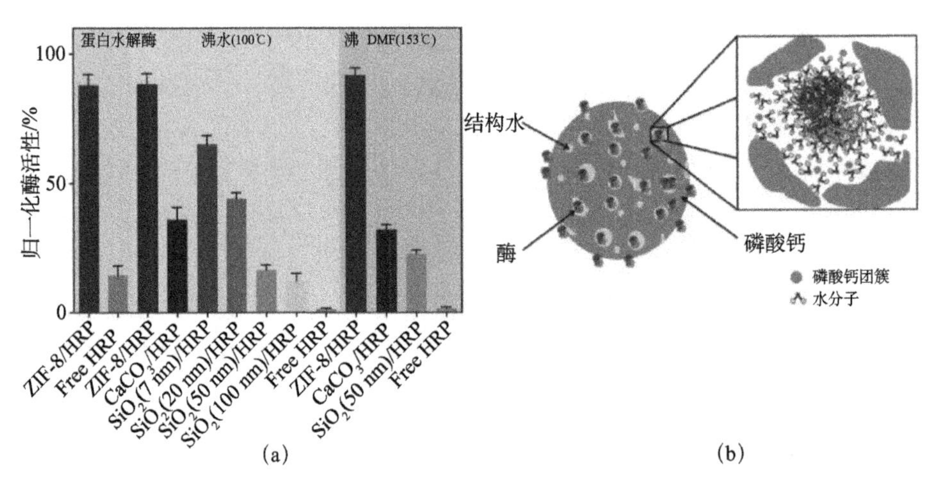

图 7-5-12 (a) ZIF-8 包裹于辣根过氧化物酶表面可以实现热保护作用。对 HRP，ZIF-8 包裹的 HRP(ZIF-8/HRP)，碳酸钙包裹的 HRP($CaCO_3$/HRP)，二氧化硅包裹的 HRP(SiO_2/HRP, SiO_2 平均尺寸在 7nm, 20nm, 50nm 及 100nm)，在蛋白水解酶（胰酶）的存在下，分别在沸水下处理 1h，以及在 153℃沸腾的二甲基甲酰胺（DMF）下处理 1h 后酶活性的结果。数据以常温下 HRP 的活性为基准。Copyrights 2015, Macmillan Publishers Limited. from Ref. [102]。(b) CaP-酶纳米复合物的示意图。Copyright 2015, the Royal Society of Chemistry from Ref. [100]

四、无须冷藏的疫苗

疫苗为人类提供有效的免疫治疗手段并保护数十亿人类免受疾病危害[103]。然而，每年仍然有 1700 万人受感染而死亡，其中大多数是因为未能受到疫苗有效保护而失去生命的[104]。目前在贫困国家，疫苗的使用仍然没有普及。主要原因是疫苗对热十分敏感，需要有效的冷链。但是，这些地区由于各种原因并没有拥有可靠的冷链[105]。针对这些情况，*Science* 和比尔·盖茨基金提出了一个重要的全球健康挑战，即如何制备无须冷藏的疫苗[106]。目前已经有许多方法用于解决疫苗的热稳定性问题，如利用明胶、蛋白、重水、氯化镁以及多糖等来稳定疫苗[107]。此外，蛋白工程也是一种增加疫苗热稳定性的方法[108]。然而，以上所涉及的方法往往比较复杂且容易造成病毒的失活。

通过材料复合实现疫苗热稳定性提高是一个十分重要的发现。我们的研究表明，活的日本脑炎病毒（JEV）可以被直接应用于仿生矿化并构建出 CaP 外壳[50]。体外实验表明，矿化后的 JEV 疫苗（B-JEV）总是表现出比普通疫苗高出 3 倍的热稳定性，并且在常温下拥有超过一周的热稳定性。由于病毒在被细胞内吞后能够自发地去除外壳，因此 B-JEV 拥有和普通 JEV 疫苗相同的生物体内行为和特征，也就是说矿化疫苗可以像普通疫苗那样方便使用。体内实验证明，常温保存一周之后 B-JEV 均可以像正常 JEV 一样诱导中和抗体和γ-干扰素（代表免疫反应能力），表现出十分有效的免疫功能。

然而并非所有疫苗都能直接自发矿化生成外壳，所以我们通过基因工程为疫苗表面接入“成核多肽”，从而实现在生理条件下触发疫苗自矿化[69]。人肠道病毒疫苗 71（EV71）被编码表达钙结合多肽（W6p）后就拥有了可遗传的自矿化能力［图 7-5-13(a)］[69]。体内免疫学实验表明，具有自矿化能力的 EV71 具有更好的免疫能力和热稳定性。即便在 37°C 高温条件下存放 5 天后，其免疫能力和新制备的 EV71 具有相同的功效［图 7-5-13(b) 及图 7-5-13(c)］[69]。更多实验结果表明，将基因技术与生物矿化技术结合有望为发展中国家提供新的热稳定疫苗－材料复合体。

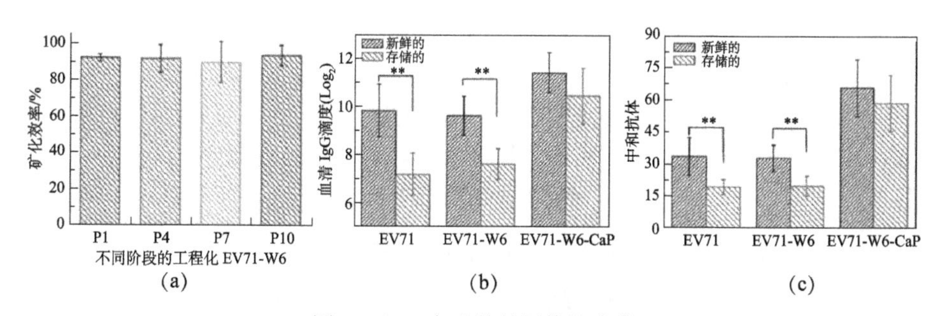

图 7-5-13 自矿化基因修饰疫苗

(a) 通过 qRT-PCR 实验定量 RNA 可以实现子代 EV71-W6 的矿化能力测定。对储存疫苗的动物实验结果，其中用 EV71、EV71-W6 和 EV71-W6-CaP 在 37°C 储存 5 天后，IgG titers (b) 和小鼠中性抗体 titers(c) 在免疫 4 周之后测定。Copyright 2013, the National Academy of Science from Ref [69]

为进一步改进疫苗的热稳定性，我们发现二氧化硅材料是疫苗材料外壳更理想的选择。人们发现在早期的地球上存在嗜热，甚至超级嗜热的微生物，而它们往往都拥有着二氧化硅外壳[109]。从温泉菌[110]到深海海绵[111]，从低级的单细胞硅藻[112]到高等的多细胞大米植物[113]，都拥有硅矿化能力，能够保护生物体免受外部高温的侵害。这些现象说明生物硅矿化过程可能是生物进化出的增强耐热性的材料学策略，由此我们为疫苗设计制备了二氧化硅的外壳[114]。体外实验表明，硅外壳将疫苗的热稳定性显著提高大约 10 倍。在室温下，硅矿化的疫苗可以拥有超过一个月的储存期限，这个时间相当于普通疫苗在 4°C 条件下的存放时间（图 7-5-14）[114]。相比将热稳定性只能提高 3 倍的 CaP 外壳，这个结果说明 SiO_2 作为功能材料在热保护领域的优越性。从结果看来，拥有二氧化硅外壳的疫苗最符合“无须冷藏的疫苗”这个概念。这种策略同样被用于脊髓灰质炎（小儿麻痹症）疫苗上，经过硅矿化处理后该疫苗可以在常温条件下储存近 2 个月[114]。

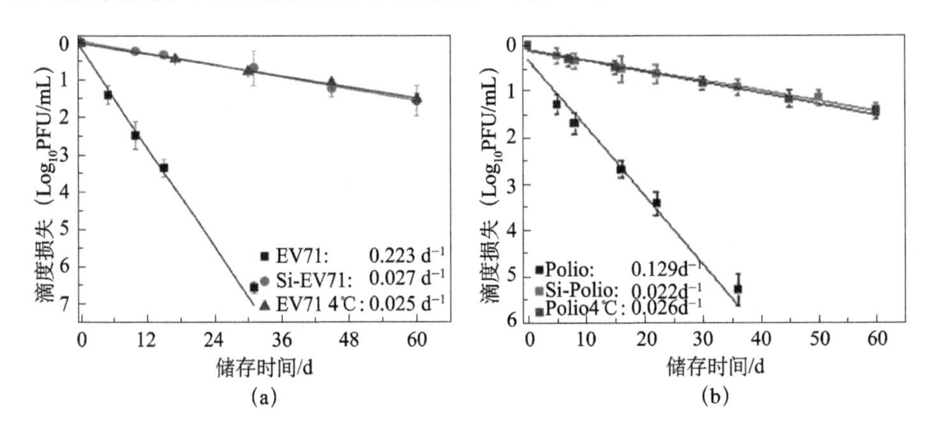

图 7-5-14 硅矿化病毒的热稳定实验：(a) EV71 和 Si-EV71 在 25℃下的热失活动力学，同时提供普通 EV71 在 4℃条件下的活性作为对比（$n \geq 4$）。右下角是计算出的平均失活速率。(b) Polio 和 Si-Polio 在 25℃下的热失活动力学，同时提供普通 EV71 在 4℃条件下的活性作为对比（$n \geq 4$）。右下角是计算出的平均失活速率。Copyright 2015, American Chemical Society from Ref. [114]

通过与生物材料的复合是提高疫苗热稳定性的有效方法，这种基于仿生矿化的途径具有快速、高效和廉价等优点。这项发明对保障全球疫苗免疫计划，尤其是对缺乏冷链的贫困地区免疫效果提升具有十分重要的意义。

五、生物隐身

分子识别可以帮助生物系统传递信号以及抵御外源物质的侵害[70]。然而，一些识别过程反而会不利于治疗过程。例如，在基因治疗中病毒作为载体被广泛研究[115]。但是，在体内实验中发现，这些病毒载体往往会被人体免疫系统识别为外源物质，从而被快速清除[116]，极大地限制了病毒载体在生物医学上的应用。由于识别过程是通过与病毒表面的特定分子作用，所以通过设计材料外壳来掩盖表面识别分子会是一种有效的方法。

腺病毒 5 型（Ad5）通常作为病毒载体及临床疫苗的物种[117]，因此，Ad5 被选择用来作为 CaP 矿化复合的模型病毒[118]。由于材料外壳的保护作用[119]，被矿化的 Ad5 可以在生物体内有效阻止特定抗体对其的识别。与对照组相比，抗体可以快速识别 Ad5 病毒粒子，但矿化后的 Ad5 则不被识别清除[118]。从这个实验可以说明，矿化后的病毒可以拥有在生物系统内潜伏的能力。Ad5 病毒的感染也是通过特定识别实现的，一般是通过腺病毒受体（CAR）进入细胞的，因此对于没有 CAR 的细胞，Ad5 不具有感染或者基因治疗能力。但是，对于矿化后的细胞，则对没有 CAR 的细胞仍具有感染性，也说明矿化病毒拥有了新的感染通路。这种矿化后的 Ad5 病毒就像特洛伊木马（图 7-5-15）[118] 扩展病毒的感染能力，进而提高矿化病毒在基因治疗等方面的应用价值 [120]。

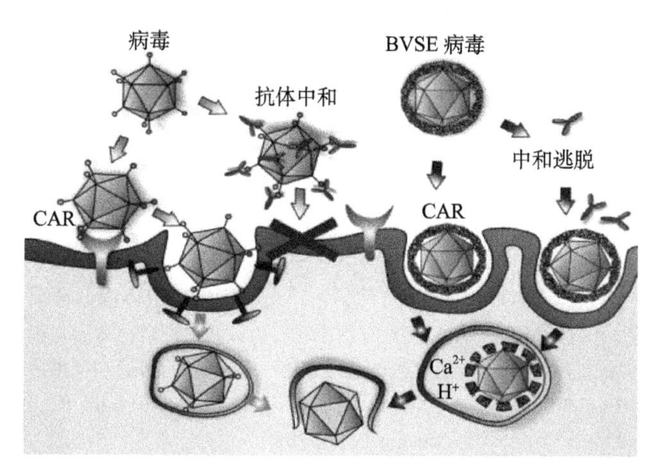

图 7-5-15 基于生物矿化的病毒外壳工程（biomineralization-based virus shell engineering,BVSE）用于生物潜入的示意图。Reproduced by Permission of Wiley-VCH from Ref. [118]

六、抗原掩蔽

如前所述，材料外壳可以掩蔽有机体本质特征，包括有机体上的抗原，或特定的多肽以及功能基团 [118,119]。在输血方面，长期困扰人们的问题是每年大量的血液需求量，其中血型不匹配尤为突出 [121]。通过用 PDA 对红细胞（RBC）表面的修饰可以成功地掩蔽其表面抗原，避免因血型不同而引起的凝血过程 [57]。通过对结构功能分析发现，PDA 包裹后的 RBC 在输氧功能上没有任何影响，同时在体内血液替换的实验上有着优异的表现 [57]。与一般的血液相比，PDA-RBC 可以拥有相似的血液循环周期。通过抗原掩蔽后的红细胞在输血后可以保持 40 天以上的体内循环，与一般的红细胞有相似的循环周期［图 7-5-16(a)］[57]。免疫原性试验发现，重复对小鼠注射 PDA-RBC 不会产生免疫系统对这种修饰红细胞的排异作用，这些红细胞在三次注射后仍可在体内保存，最多可占用原血液体积的 60%［图 7-5-16(b)］[57]。此外，注射过 PDA 复合的 RBC 小鼠未见明显的体重及生理功能缺失。因此，这种 PDA 材料复合的红细胞可以有效避免血型不匹配带来的问题，同时又不失去红细胞本身的活性。

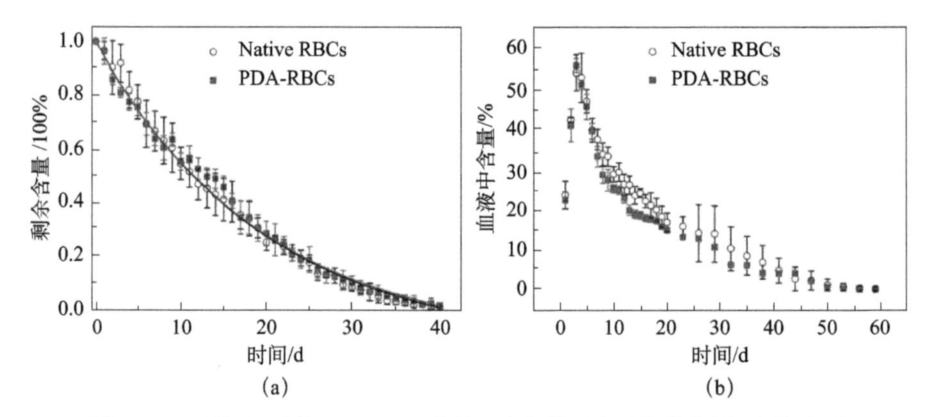

图 7-5-16　用标记后的 PDA-RBC 注射入小鼠模型中进行体内循环周期实验
(a) 一次注射之后的保留时间。结果表明在存活率、行为以及生理活性方面与正常红细胞没有显著差异，也表明未包裹的细胞没有显著的毒性。(b) 三次注射后普通和包裹后的红细胞含量数据表明，PDA-RBC 可以像“万能血”一样使用。Copyright 2014, the Royal Society of Chemistry from Ref. [57]

七、光合系统改进

蓝藻作为一种常见的陆生和水生物种，占据了地球上 20%～30% 的光合作用产量并且以约 450 TW 的功率将太阳能转化为生物质能 [122]。但微生物光合作用效率往往会受自然环境的胁迫而降低，其中光胁迫就是其中之一 [123]。过量的光可以显著地抑制蓝藻的光合作用，甚至引起对光合细胞器的光氧化损伤作用 [124]。

受硅藻外壳的启发，我们通过 LBL 技术将蓝藻细胞复合上二氧化硅外壳，并保持其光合作用的能力 [37]。该二氧化硅外壳可以减少进入蓝藻细胞的光强度，从而减弱光胁迫效应。通过对光合系统 II（PSII）的表征，正常蓝藻 PSII 的活性随光强增强而显著减弱，但是这种趋势效在用二氧化硅复合的蓝藻中并不明显（图 7-5-17）。这说明材料外壳的存在提高了蓝藻的光合作用效率，减弱了环境的光胁迫 [37]。

植物和微生物的生长经常受外部因素的影响，在没有特定适宜培养环境的条件下，光合系统将不会有理想的效率 [125]。然而，这个问题可以通过对生物体加保护性材料外壳来解决。类囊体是一种由双层膜组成的植物细胞器，也是叶绿体中的光合作用核心。能产生光合作用的生物体可以拥有近 100% 的量子产率，但是对于分离出来的光合细胞器则没有那么稳定。之前，苏宝连课题组提出了将二氧化硅基质复合在类囊体上的方法 [126]。材料复合后的类囊体活性可以保持超过 40 天，相比之下未复合的类囊体在 3 天之后就失去产氧能力。因此，二氧化硅材料复合的类囊体可以像“光合反应器”那样长期生产氧气。类似的方法也被用于植物细胞 [127]、蓝藻 [128] 以及绿藻上 [129]，并且这些材料复合的细胞往往可以维持数周以上的生物活性。

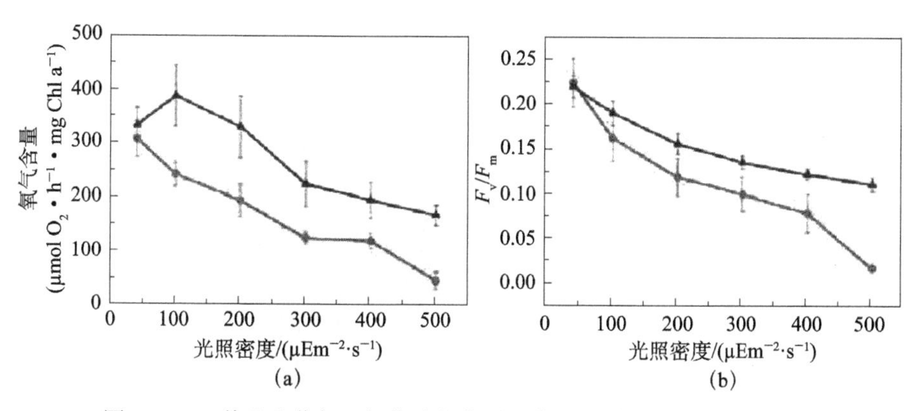

图 7-5-17 普通蓝藻与二氧化硅包裹后蓝藻的不同光合作用参数对比

(a) 以 1,4- 苯醌为电子受体的光合产氧变化过程。(b) PSII 的最大量子产率（F_v/F_m），红线表示普通绿藻，蓝线表示二氧化硅包裹后的绿藻。Copyright 2013, the Royal Society of Chemistry from Ref. [37]（文后附彩图）

八、生物产氢

氢气被认为是能够替代化石燃料的未来绿色能源 [130]，但目前的氢气来源仍然主要依赖石油产品的裂解。如何利用光能实现生物产氢是一个重大的能源挑战，事实上自然界的绿藻具有利用光及氢酶裂解水产生氢气的能力，但是仅仅在明暗转换的瞬间可以发生 [131]。这主要是因为氢酶会在氧气的存在下失去活性 [132]。在从暗转换到明的瞬间，光电子会在光合系统 II（PSII）中通过光氧化水而获得（$H_2O \rightarrow 2H^+ + \frac{1}{2}O_2 + 2e^-$），之后被传递给氢酶用于氢气生成（$2H^+ + 2e^- \rightarrow H_2$）[133]。然而，在光合过程中氧气的产生会使氢酶快速失活，从而失去了产氢能力 [134]。最近我们发现，有二氧化硅引起的细胞聚集体，在聚集体核心处可以形成一个无氧的环境，用于氢气的生成 [135]。*Chlorella pyrenoidosa* 是一种商业化的绿藻，当它们和二氧化硅复合后就可通过 100μm 大小的聚集体按大约 0.20μmol $H_2 \cdot h^{-1} \cdot mg$ 叶绿素$^{-1}$ 的速率实现光合产氢，这相当于自然界生物质-燃料转化速率的 1.75 倍［图 7-5-18(a)］[135]。正是由于二氧化硅材料的复合，矿化后的绿藻才可以聚集从而诱导产生空间功能分化［spatial–functional differentiation，图 7-5-18(b)］：在细胞-材料聚集体表层的绿藻细胞和正常细胞一样进行光合作用并放出氧气，但内核处的细胞由于处于封闭环境，其光合氧气被细胞的呼吸作用消耗并构建出一个缺氧的动态平衡环境，因此这些细胞内的氢酶可以被激活实现光合产氢。这个研究扩展了我们对细胞材料外壳的理解，为开发新生物的新功能提供了思路[135]。

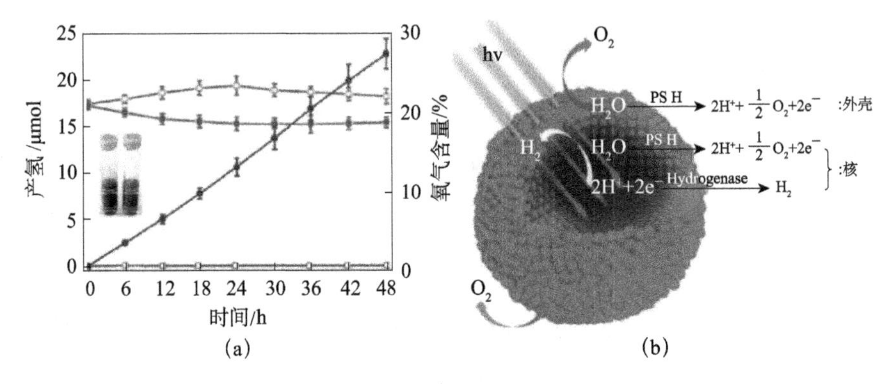

图 7-5-18　在有氧环境中通过聚集小球藻实现生物产氢

(a) 在 100 $\mu Em^{-2} \cdot s^{-1}$ 光照强度下，不同时期的产氢量和氧含量。红线：氧含量；蓝线：氢气生成量。空心方格：普通小球藻。实方格：聚集的小球藻。(b) 聚集小球藻空间功能化示意图。Copyright 2015, Wiley-VCH from Ref. [135]（文后附彩图）

九、生物催化

微生物拥有多种酶来催化各种不同的反应，并且这种生物转化过程对应用也很有帮助[136]。很多生物体及酶在有机溶剂中往往会失去稳定性，所以通过材料外壳的赋予是一种有效的保护方法[137]。例如，胶体粒子可以同时稳定乳液并在两相溶剂的界面处催化反应进行，这个过程称为 Pickering 界面催化[138]。利用细胞进行的相界面生物催化也是一种很有价值的催化方法，但是细菌对有机溶剂敏感，不仅很容易失活而且还不能形成有效的稳定乳液。所以，通过材料外壳可以实现并提升这一生物催化过程[139]。例如，*Alcaligenesfaecalis* ATCC 8750 细胞在复合 CaP 之后再掺杂磁性纳米粒子，然后通过十二烷基磷酸钠的表面修饰可以增强 Pickering 乳液界面的稳定性

（图 7-5-19）[139]。这种合理设计的生物材料外壳显著地提升了界面处的催化效能。通过超过 30 次的循环实验发现，这些经过材料修饰改造后的细胞仍能具有较高的选择催化活性、细胞存活率及分裂能力，同时也说明材料外壳能很好地保护内部细胞免于长期的有机溶剂侵害[139]。

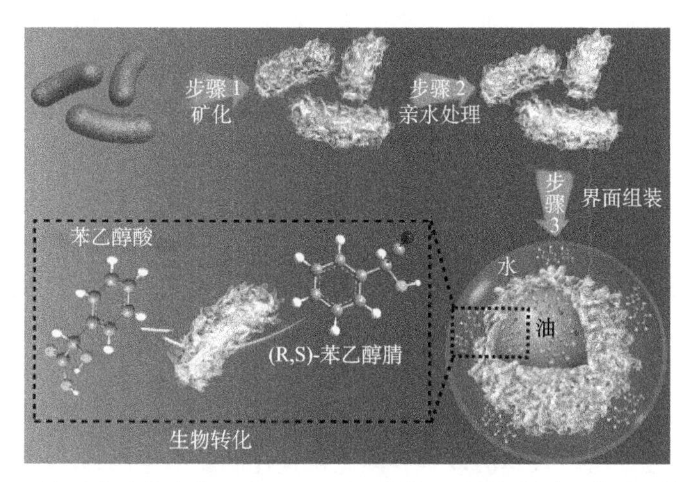

图 7-5-19 通过细胞包裹技术实现 Pickering 界面催化，以及相转移立体选择转化。步骤 1，掺杂 Fe_3O_4 纳米颗粒的 CaP 外壳在细菌表面生长。步骤 2，通过吸附十二烷基磷酸钠实现矿物外壳的浸润。步骤 3，包裹后的细菌在 Pickering 界面水解将油相中疏水的（R,S）-苯乙醇腈转化为亲水的 R-(-)- 苯基乙醇酸。Copyright 2015, Wiley-VCH from Ref. [139]

第五节 展望与挑战

受生物矿化启发，我们提出了通过功能材料实现生物体改进的新策略。传统生物矿化研究强调有机物质对无机结晶过程的调控[3-4,10]；反之，材料也可以干预生物过程。基于材料的生物体改进则强调了材料对生物体系的调控。这些研究加深我们了对材料在生物体中作用的理解，也成功地将材料科学和生物功能优化相互结合在一起。通常我们认为生物体是“活的”，而材料是“不活的”。但是基于材料的生物体改性却将活物与非活物的界限变得不那么明显。由于生物体的功能可以通过功能材料来改进，所以也表明材料在生物中的重要性。在某种程度上，我们也可以将其理解为通过材料促进生物的人工进化。这是一个全新的研究领域并已经取得了重要的进展。

伴随着材料科学的飞速发展，许多新型功能材料已经被开发，在将来的发展中将这些材料与生物体结合会是这个学科的发展方向。将来会有更多智能材料应用到外壳包裹研究中，实现更精确高效的细胞控制与改进。例如，之前的研究表明，外壳材料可以有效地选择性透过生物质，从而对内部细胞的生长实现调控[27]。如我们所知，细胞内离子以及大分子对细胞中很多重要过程起到调控作用[70,140]。那么将来必然会有更多有特色的材料用于对相关离子及分子的调控：金属-有机框架，多孔二氧化硅以及沸石，在修饰之后能实现特定分子的液相及气相分离[141]；响应型离子泵以及纳米孔道均可以控制离子的通透[142]。通过将这些材料与生物体结合，必然会实现更智能化地调节细胞功能，实现更高效实用的生物体功能改进。

此外，对于这个学科在发展的同时也不能忽略一些重要的问题与挑战。其一是如何批量并友好地制备生物体表面的材料外壳[143]。众所周知，硅藻能生成二氧化硅外壳全都是归功于其基因的调控，并且具有遗传性。我们认为通过基因工程引发的自矿化过程将会是一种可行的策略，来简化外壳制备方法。由于基因修饰后的细胞具有生物可遗传性，这种生物学上的改进将会是永久的，也会消耗较少的成本。因此，这样制造的生物体-材料复合物将会是方便而且经济的。尽管通过基因工程在病毒表面实现自矿化已经成功实现[69]，对于其他如真核细胞在内的生物体仍然缺乏探索。此外，对于基因修饰可能造成的生物安全问题也是必须要注意的。

其二是去寻找拥有现成功能的生物体和合适的材料来优化并使用这个功能。这种自然生物和人造外壳的结合可以用于解决很多国际性问题，如能源危机、环境污染及医疗等方面。在自然生物通过数亿年的选择与进化后，我们已经可以通过利用自然的智慧来解决许多问题。例如，通过细菌，古生菌和绿藻可以用于生物制氢来解决能源问题[144]；利用乙酸菌和甲烷菌的厌氧降解可以有效地降解有机废料[145]；利用微生物，哺乳动物细胞以及植物细胞可以用于生物制药[146]。不幸的是，大部分生物体及生物制剂会对环境敏感，并不利于存在于工业环境，这降低了它们在商业化及批量生产方面的应用价值。但是，通过材料技术为生物体制备合适的材料外壳可以显著提升它们的应用可行性。此前我们已经阐述过通过材料外壳改进光合生物的方法，这也是改进其他生物体的可行策略。当然，选择合适的材料和生物将会使这类研究更富有创意，也是十分重要的一点。此外，在材料外壳处理后依旧保留生物的活性也是需要注意的，如果提升材料外壳的生物相容性以及制备外壳的方法，尤其是哺乳动物细胞，也是需要考虑的方面。

其三是自然界中的生物外壳通常都具有多级结构，例如硅藻表面的二氧化硅外壳在微观上可以具有周期性的纳米结构，同时宏观上又具有3D结构的[2]。然而，目前人为制造的外壳从结构上与自然外壳相比要简单很多。需要指出的是，许多拥有独特功能的天然矿物往往是由多级结构造成的[1,2,22,147]。因此，拥有多级结构的外壳很有可能更好地改善内部生物体的功能。尽管已经有很多制备多级结构生物材料的例子[59,148]，如何在生物体表面制备这种多级结构的外壳仍具有挑战性。或许通过有机物多步精确调控仿生外壳的生长将会是十分必要的，因此对生物矿化有更深入的理解也是必须的。

总之，基于生物矿化构建的“生物-材料”复合体研究是一项新兴的同时也十分有潜质的科学。通过材料技术对生物体的改性可以理解为人为的生物功能进化策略，并且相关研究也可以促进生物学和材料科学的结合，为新的交叉学科建立提供思路，也同时为软物质研究发展创造了新的天地。

刘昭明、唐睿康（浙江大学化学系）

参考文献

[1] a) Cusack M, Freer A. Biomineralization: elemental and organic influence in carbonate systems. Chem. Rev., 2008, 108: 4433-4454; b) Palmer L C, Newcomb C J, Kaltz S R, et al. Biomimetic systems for hydroxyapatite mineralization inspired by bone and enamel. Chem. Rev., 2008, 108: 4754-4783.

[2] a) Porter S M. Seawater chemistry and early carbonate biomineralization. Science, 2007, 316: 1302-1302; b) Wang L, Nancollas G H. Calcium orthophosphates: crystallization and dissolution. Chem. Rev., 2008, 108: 4628-4669; c) Hildebrand M. Diatoms, biomineralization processes, and genomics. Chem. Rev., 2008, 108: 4855-4874.

[3] Arias J L, Fernández M. Polysaccharides and proteoglycans in calcium carbonate-based biomineralization. Chem. Rev., 2008, 108: 4475-4482.

[4] a) George A, Veis A. Phosphorylated proteins and control over apatite nucleation, crystal growth, and inhibition. Chem. Rev., 2008, 108: 4670-4693; b) Cölfen H, Antonietti M. Mesocrystals: inorganic superstructures made by highly parallel crystallization and controlled alignment. Angew. Chem. Int. Ed., 2005, 44: 5576-5591.

[5] a) Xu A W, Ma Y, Cölfen H. Biomimetic mineralization. J. Mater. Chem., 2007: 17, 415-449; b) Nudelman F, Sommerdijk N A. Biomineralization as an inspiration for materials chemistry.

Angew. Chem. Int. Ed., 2012, 51: 6582-6596.

[6] a) Cai Y, Tang R. Calcium phosphate nanoparticles in biomineralization and biomaterials. J. Mater. Chem., 2008, 18: 3775-3787; b) Sun J.et al.Biomimetic promotion of dentin remineralization using L-glutamic acid: Inspiration from biomineralization proteins. J. Mater. Chem. B, 2014, 2: 4544-4553; c) Wang J, et al. Remineralization of dentin collagen by meta-stabilized amorphous calcium phosphate. CrystEngComm, 2013, 15: 6151-6158; d) Li L. et al. Bio-Inspired Enamel Repair via Glu-Directed Assembly of Apatite Nanoparticles: an Approach to Biomaterials with Optimal Characteristics. Adv. Mater., 2011, 23: 4695-4701; e) Li L. et al. Repair of enamel by using hydroxyapatite nanoparticles as the building blocks. J. Mater. Chem., 2008, 18: 4079-4084.

[7] Coe F L, Parks J H, Asplin J R. The pathogenesis and treatment of kidney stones. N. Engl. J. Med., 1992, 327: 1141-1152.

[8] Sage A P, Tintut Y, Demer L L.Regulatory mechanisms in vascular calcification. Nature Reviews Cardiology, 2010, 7: 528-536.

[9] Qiu Z Y, et al. Mineralized Collagen: Rationale, Current Status, and Clinical Applications. Materials, 2015, 8: 4733-4750.

[10] Dickerson M B, Sandhage K H, Naik R R. Protein-and peptide-directed syntheses of inorganic materials. Chem. Rev., 2008, 108: 4935-4978.

[11] a) Weiner S, Addadi L. Design strategies in mineralized biological materials. J. Mater. Chem., 1997, 7: 689-702; b) Smith B L, et al. Molecular mechanistic origin of the toughness of natural adhesives, fibres and composites. Nature, 1999, 399: 761-763.

[12] Nys Y, Gautron J, Garcia-Ruiz J M. Avian eggshell mineralization: biochemical and functional characterization of matrix proteins. Comptes Rendus Palevol, 2004, 3: 549-562.

[13] Wang B, Liu P, Tang R. Cellular shellization: Surface engineering gives cells an exterior. Bioessays, 2010, 32: 698-708.

[14] a) Bazylinski D A, Frankel R B. Magnetosome formation in prokaryotes. Nature Reviews Microbiology, 2004, 2: 217-230; b) Faivre D, Schuler D. Magnetotactic bacteria and magnetosomes. Chem. Rev., 2008, 108: 4875-4898.

[15] Sahney S, Wilson M V. Extrinsic labyrinth infillings imply open endolymphatic ducts in Lower Devonian osteostracans, acanthodians, and putative chondrichthyans. J. Vert. Paleontol., 2001, 21: 660-669.

[16] a) Meunier C F, Rooke J C, Léonard A, et al. Living hybrid materials capable of energy conversion and CO 2 assimilation. Chem. Commun., 2010, 46: 3843-3859; b) Park J H, et al. Nanocoating of single cells: from maintenance of cell viability to manipulation of cellular activities. Adv. Mater., 2014, 26: 2001-2010.

[17] a) Jiang S, Pan H, Chen Y, et al. Amorphous calcium phosphate phase-mediated crystal nucleation kinetics and pathway. Faraday Discuss, 2015;b) Chen Y, Gu W, Pan H, et al. Stabilizing amorphous calcium phosphate phase by citrate adsorption. CrystEngComm, 2014, 16: 1864-1867; c) Ding H, Pan H, Xu X, et al.Toward a detailed understanding of magnesium ions on hydroxyapatite crystallization inhibition.Crystal Growth & Design, 2014, 14: 763-769; d) Chu X, et al. Unique roles of acidic amino acids in phase transformation of calcium phosphates. The Journal of Physical Chemistry B, 2010, 115: 1151-1157; e) Hu Q, et al. Preparing nano-calcium phosphate particles via a biologically friendly pathway. Biomedical Materials, 2010, 5: 041001; f) Tao J, Zhou D, Zhang Z, et al. Magnesium-aspartate-based crystallization switch inspired from shell molt of crustacean. Proc. Natl. Acad. Sci. U. S. A., 2009, 106: 22096-22101.

[18] a) Addadi L, Weiner S. Interactions between acidic proteins and crystals: stereochemical requirements in biomineralization. Proceedings of the National Academy of Sciences, 1985, 82: 4110-4114; b) Tsuji T, Onuma K, Yamamoto A, et al. Direct transformation from amorphous to crystalline calcium phosphate facilitated by motif-programmed artificial proteins. Proceedings of the National Academy of Sciences, 2008, 105: 16866-16870.

[19] a) Kröger N, Deutzmann R, Bergsdorf C, et al. Species-specific polyamines from diatoms control silica morphology. Proceedings of the National Academy of Sciences, 2000, 97: 14133-14138; b) Sumper M, Kröger N. Silica formation in diatoms: the function of long-chain polyamines and silaffins. J. Mater. Chem., 2004, 14: 2059-2065; c) Pohnert G. Biomineralization in Diatoms Mediated through Peptide-and Polyamine-Assisted Condensation of Silica. Angew. Chem. Int. Ed., 2002, 41: 3167-3169.

[20] a) Tao J, Pan H, Zeng Y, et al. Roles of amorphous calcium phosphate and biological additives in the assembly of hydroxyapatite nanoparticles. The Journal of Physical Chemistry B, 2007, 111: 13410-13418; b) Xu X R, Pan H H, Tang R K, et al. Additive-dependent morphogenesis of oriented calcite crystals on mica. CrystEngComm, 2011, 13: 6311-6314; c) Xu X R, et al. The roles of water and polyelectrolytes in the phase transformation of amorphous calcium carbonate. J. Cryst. Growth, 2008, 310: 3779-3787; d) Cai A, et al. Direct synthesis of hollow vaterite nanospheres from amorphous calcium carbonate nanoparticles via phase transformation. The Journal of Physical Chemistry C, 2008, 112: 11324-11330; e) Shi Z, Huang X, Cai Y, et al. Size effect of hydroxyapatite nanoparticles on proliferation and apoptosis of osteoblast-like cells. Acta Biomater., 2009, 5: 338-345.

[21] Bäuerlein E, Behrens P, Epple M. Handbook of biomineralization.Wiley-VCH, 2007.

[22] a) Hall B K.Bones and cartilage: developmental and evolutionary skeletal biology. Cambridge : Academic Press, 2005; b) Weiner S, Dove P M. An overview of

biomineralization processes and the problem of the vital effect. Reviews in Mineralogy and Geochemistry, 2003, 54: 1-29.

[23] Ten Cate A, Nanci A. Structure of the oral tissues. Ten Cate’ s Oral Histology, 6th edn. Mosby, St Louis, 2003: 1-16.

[24] a) Kawasaki K, Weiss K M. Mineralized tissue and vertebrate evolution: the secretory calcium-binding phosphoprotein gene cluster. Proc. Natl. Acad. Sci. U. S. A., 2003, 100: 4060-4065; b) Kawasaki K, Suzuki T, Weiss K M. Genetic basis for the evolution of vertebrate mineralized tissue. Proc. Natl. Acad. Sci. U. S. A., 2004, 101: 11356-11361; c) Kawasaki K, Weiss K M. Evolutionary genetics of vertebrate tissue mineralization: the origin and evolution of the secretory calcium-binding phosphoprotein family. Journal of Experimental Zoology Part B: Molecular and Developmental Evolution, 2006, 306: 295-316.

[25] Round F E, Crawford R M, Mann D G.The diatoms: biology & morphology of the genera. Cambridge : Cambridge University Press, 1990.

[26] Mell J C, Burgess S M. Yeast as a model genetic organism. eLS, 2001.

[27] Wang B, et al. Yeast cells with an artificial mineral shell: Protection and modification of living cells by biomimetic mineralization. Angew. Chem. Int. Ed., 2008, 47: 3560-3564.

[28] a) Ariga K, Hill J P, Ji Q. Layer-by-layer assembly as a versatile bottom-up nanofabrication technique for exploratory research and realistic application. PCCP, 2007, 9: 2319-2340; b) Wang Y, Angelatos A S, Caruso F. Template Synthesis of Nanostructured Materials via Layer-by-Layer Assembly†. Chem. Mater., 2007, 20: 848-858.

[29] Cabib E, Roh D H. Schmidt M, et al. The yeast cell wall and septum as paradigms of cell growth and morphogenesis. J. Biol. Chem., 2001, 276: 19679-19682.

[30] Diaspro A, Silvano D, Krol S, et al. Single living cell encapsulation in nano-organized polyelectrolyte shells. Langmuir, 2002, 18: 5047-5050.

[31] a) Kadowaki K, Matsusaki M, Akashi M. Control of cell surface and functions by layer-by-layer nanofilms. Langmuir, 2010, 26: 5670-5678; b) Yang S H, et al. Interfacing living yeast cells with graphene oxide nanosheaths. Macromol. Biosci., 2012, 12: 61-66.

[32] Schönhoff M.Self-assembled polyelectrolyte multilayers. Current opinion in colloid & interface science, 2003, 8: 86-95.

[33] Drachuk I, et al. Silk Macromolecules with Amino Acid-Poly (Ethylene Glycol) Grafts for Controlling Layer-by-Layer Encapsulation and Aggregation of Recombinant Bacterial Cells. ACS nano, 2015, 9: 1219-1235.

[34] a) Konnova S A, et al. Functional artificial free-standing yeast biofilms. Colloids Surf. B. Biointerfaces, 2011, 88: 656-663; b) Louzao I, Sui C, Winzer K, et al. Cationic polymer

mediated bacterial clustering: Cell-adhesive properties of homo-and copolymers. Eur. J. Pharm. Biopharm, 2015; c) Lee D W, Kim T, Park I S, et al. Multivalent nanofibers of a controlled length: regulation of bacterial cell agglutination. J. Am. Chem. Soc., 2012, 134: 14722-14725.

[35] a) Bellomo E G, Deming T J. Monoliths of aligned silica-polypeptide hexagonal platelets. J. Am. Chem. Soc., 2006, 128: 2276-2279; b) Tomczak M M, et al. Polypeptide-templated synthesis of hexagonal silica platelets. J. Am. Chem. Soc., 2005, 127: 12577-12582.

[36] a) Yang S H, et al. Biomimetic encapsulation of individual cells with silica. Angew. Chem. Int. Ed., 2009, 48: 9160-9163; b) Yang S H, Ko E H, Jung Y H, et al. Bioinspired Functionalization of Silica-Encapsulated Yeast Cells. Angew. Chem., 2011, 123: 6239-6242; c) Lee J, Choi I S, Yang S H. Cytocompatibility Optimization for Silica Coating of Individual HeLa Cells. Bull. Korean Chem. Soc., 2015, 36: 1278-1281.

[37] Xiong W, et al. Alleviation of high light-induced photoinhibition in cyanobacteria by artificially conferred biosilica shells. Chem. Commun., 2013, 49: 7525-7527.

[38] Konnova S A, et al. Biomimetic cell-mediated three-dimensional assembly of halloysite nanotubes. Chem. Commun., , 2013, 49: 4208-4210.

[39] Mathapa B G, Paunov V N. Fabrication of viable cyborg cells with cyclodextrin functionality. Biomaterials Science, 2014, 2: 212-219.

[40] Ejima H, et al. One-step assembly of coordination complexes for versatile film and particle engineering. Science, 2013, 341: 154-157.

[41] a) Park J H, et al. A Cytoprotective and Degradable Metal-Polyphenol Nanoshell for Single-Cell Encapsulation. Angew. Chem. Int. Ed., 2014, 53: 12420-12425; b) Li W, Bing W, Huang S, et al. Mussel Byssus-Like Reversible Metal-Chelated Supramolecular Complex Used for Dynamic Cellular Surface Engineering and Imaging. Adv. Funct. Mater, 2015.

[42] Yang S H, Ko E H, Choi I S. Cytocompatible encapsulation of individual Chlorella cells within titanium dioxide shells by a designed catalytic peptide. Langmuir, 2011, 28: 2151-2155.

[43] Lee H, et al. Layer-by-Layer-Based Silica Encapsulation of Individual Yeast with Thickness Control. Chemistry-An Asian Journal, 2015, 10: 129-132.

[44] Berridge M J. Calcium signalling and cell proliferation. Bioessays, 1995, 17: 491-500.

[45] Dunn T, Gable K, Beeler T. Regulation of cellular Ca^{2+} by yeast vacuoles. J. Biol. Chem., 1994, 269: 7273-7278.

[46] Clapham D E. Calcium signaling. Cell, 1995, 80: 259-268.

[47] Maheshwari V, Fomenko D E, Singh G, et al. Ion mediated monolayer deposition of gold nanoparticles on microorganisms: discrimination by age. Langmuir, 2009, 26: 371-377.

[48] Kempaiah R, Chung A, Maheshwari V. Graphene as cellular interface: electromechanical coupling with cells. ACS nano, 2011, 5: 6025-6031.
[49] Wang X, et al. Functional Single-Virus-Polyelectrolyte Hybrids Make Large-Scale Applications of Viral Nanoparticles More Efficient. Small, 2010, 6: 351-354.
[50] Wang G, et al. Eggshell-Inspired Biomineralization Generates Vaccines that Do Not Require Refrigeration. Angew. Chem., 2012, 124: 10728-10731.
[51] Zhang X, Yu M, Liu J, et al. Bioinspired metal-cell wall-metal sandwich structure on an individual bacterial cell scaffold. Chem. Commun., 2012, 48: 8240-8242.
[52] Lee J, et al. Cytoprotective silica coating of individual mammalian cells through bioinspired silicification. Angew. Chem. Int. Ed., 2014, 53: 8056-8059.
[53] Putnam D, Gentry C A, Pack D W, et al. Polymer-based gene delivery with low cytotoxicity by a unique balance of side-chain termini. Proceedings of the National Academy of Sciences, 2001, 98: 1200-1205.
[54] Lee H, Dellatore S M, Miller W M, et al. Mussel-inspired surface chemistry for multifunctional coatings. Science, 2007, 318: 426-430.
[55] Yang S H, et al. Mussel-inspired encapsulation and functionalization of individual yeast cells. J. Am. Chem. Soc., 2011, 133: 2795-2797.
[56] Hong D, et al. Organic/inorganic double-layered shells for multiple cytoprotection of individual living cells. Chemical Science, 2015, 6: 203-208.
[57] Wang B, et al. Antigenically shielded universal red blood cells by polydopamine-based cell surface engineering. Chemical Science, 2014, 5: 3463-3468.
[58] a) Ulman A. Formation and structure of self-assembled monolayers. Chem. Rev., 1996, 96: 1533-1554; b) Love J C, Estroff L A, Kriebel J K, et al. Self-assembled monolayers of thiolates on metals as a form of nanotechnology. Chem. Rev., 2005, 105: 1103-1170.
[59] a) Berry V, Saraf R F. Self-assembly of nanoparticles on live bacterium: an avenue to fabricate electronic devices. Angew. Chem., 2005, 117: 6826-6831; b) Zhai H, et al. Self-Assembled Organic-Inorganic Hybrid Elastic Crystal via Biomimetic Mineralization. Adv. Mater., 2010, 22: 3729-3734.
[60] Jiang N, et al. “Self-repairing” nanoshell for cell protection. Chemical Science, 2015, 6: 486-491.
[61] Jiang N, et al. Amino acid-based biohybrids for nano-shellization of individual desulfurizing bacteria. Chem. Commun., 2014, 50: 15407-15410.
[62] Drachuk I, et al. Cell Surface Engineering with Edible Protein Nanoshells. Small, 2013, 9: 3128-3137.
[63] Roth K M, Zhou Y, Yang W, et al. Bifunctional small molecules are biomimetic catalysts for

silica synthesis at neutral pH. J. Am. Chem. Soc., 2005, 127: 325-330.

[64] Park J H, Choi I S, Yang S H. Peptide-catalyzed, bioinspired silicification for single-cell encapsulation in the imidazole-buffered system. Chem. Commun., 2015, 51: 5523-5525.

[65] a) O' Leary L E, Fallas J A, Bakota E L, et al. Multi-hierarchical self-assembly of a collagen mimetic peptide from triple helix to nanofibre and hydrogel. Nat. Chem., 2011, 3: 821-828; b) Kar K, et al. Self-association of collagen triple helic peptides into higher order structures. J. Biol. Chem., 2006, 281: 33283-33290.

[66] Weeks D L, Eskandari S, Scott D R, et all. A H+-gated urea channel: the link between Helicobacter pylori urease and gastric colonization. Science, 2000, 287: 482-485.

[67] Nam K T, et al. Virus-enabled synthesis and assembly of nanowires for lithium ion battery electrodes. Science, 2006, 312: 885-888.

[68] Guan C, et al. Bioencapsulation of living yeast (Pichia pastoris) with silica after transformation with lysozyme gene. J. Sol-Gel Sci. Technol., 2008, 48: 369-377.

[69] Wang G, et al. Rational design of thermostable vaccines by engineered peptide-induced virus self-biomineralization under physiological conditions. Proceedings of the National Academy of Sciences, 2013, 110: 7619-7624.

[70] a) Lockett M R, et al. The binding of benzoarylsulfonamide ligands to human carbonic anhydrase is insensitive to formal fluorination of the ligand. Angew. Chem., 2013, 125: 7868-7871; b) Breiten B, et al. Water networks contribute to enthalpy/entropy compensation in protein-ligand binding. J. Am. Chem. Soc., 2013, 135: 15579-15584.

[71] Coulon J, Thouvenin I, Aldeek F, et al. Glycosylated quantum dots for the selective labelling of Kluyveromyces bulgaricus and Saccharomyces cerevisiae yeast strains. Journal of fluorescence, 2010, 20: 591-597.

[72] Chouhan R S, Qureshi A, Niazi J H. Determining the fate of fluorescent quantum dots on surface of engineered budding S. cerevisiae cell molecular landscape. Biosens. Bioelectron., 2015, 69: 26-33.

[73] Johnson P E, et al. Spray-Dried Multiscale Nano-biocomposites Containing Living Cells. ACS nano, 2015, 9: 6961-6977.

[74] Baca H K, et al. Cell-directed assembly of lipid-silica nanostructures providing extended cell viability. Science, 2006, 313: 337-341.

[75] a) Sumida K, et al. Carbon dioxide capture in metal–organic frameworks. Chem. Rev., 2011, 112: 724-781; b) Liu Z, et al. Ionization controls for biomineralization-inspired CO_2 chemical looping at constant room temperature. PCCP, 2015, 17: 10080-10085.

[76] Arico A S, Bruce P, Scrosati B, et al. Nanostructured materials for advanced energy conversion and storage devices. Nature materials, 2005, 4: 366-377.

[77] Xiang Q, Yu J, Jaroniec M. Graphene-based semiconductor photocatalysts. Chem. Soc. Rev., 2012, 41: 782-796.

[78] Jemal A, et al. Global cancer statistics. CA Cancer J. Clin., 2011, 61: 69-90.

[79] Wang B, et al. Biomimetic construction of cellular shell by adjusting the interfacial energy. Biotechnol. Bioeng., 2014, 111: 386-395.

[80] Kitamura K, Kaneko T, Yamamoto Y. Lysis of viable yeast cells by enzymes of Arthrobacter luteus. Arch. Biochem. Biophys., 1971, 145: 402-404.

[81] Na H K, et al. Cytoprotective effects of graphene oxide for mammalian cells against internalization of exogenous materials. Nanoscale, 2013, 5: 1669-1677.

[82] Kim B J, et al., Cytoprotective Alginate/Polydopamine Core/Shell Microcapsules in Microbial Encapsulation. Angew. Chem. Int. Ed., 2014, 53: 14443-14446.

[83] Yang S H, et al. Cytocompatible In Situ Cross-Linking of Degradable LbL Films Based on Thiol-Exchange Reaction. Chemical Science, 2015.

[84] Prather M J, McElroy M B, Wofsy S C. Reductions in ozone at high concentrations of stratospheric halogens. Nature, 1984, 312: 227-231.

[85] McKenzie R L, Aucamp P J, Bais A F, et al. Changes in biologically-active ultraviolet radiation reaching the Earth's surface. Photochemical & Photobiological Sciences, 2007, 6: 218-231.

[86] Quintero-Torres R, Aragón J, Torres M, et al. Strong far-field coherent scattering of ultraviolet radiation by holococcolithophores. Physical Review E, 2006, 74: 032901.

[87] Wang B, et al. Guarding embryo development of zebrafish by shell engineering: a strategy to shield life from ozone depletion. PLoS One, 2010: 5.

[88] Kempaiah R, Salgado S, Chung W L, et al. Graphene as membrane for encapsulation of yeast cells: protective and electrically conducting. Chem. Commun., 2011, 47: 11480-11482.

[89] a) Neu H C, Heppel L A. The release of enzymes from Escherichia coli by osmotic shock and during the formation of spheroplasts. J. Biol. Chem., 1965, 40: 3685-3692; b) Klipp E, Nordlander B, Krüger R, et l. Integrative model of the response of yeast to osmotic shock. Nat. Biotechnol., 2005, 23: 975-982.

[90] Liu Z, et al. Calcium phosphate nanoparticles primarily induce cell necrosis through lysosomal rupture: the origination of material cytotoxicity. Journal of Materials Chemistry B, 2014: 3480-3489.

[91] a) Montanari M, et al. Long-term hematologic reconstitution after autologous peripheral blood progenitor cell transplantation: a comparison between controlled-rate freezing and uncontrolled-rate freezing at 80℃. Transfusion (Paris), , 2003, 43: 42-49; b) Galmés A, et al. Long-term storage at −80℃ of hematopoietic progenitor cells with 5 ercent dimethyl

sulfoxide as the sole cryoprotectant. Transfusion (Paris), 1999, 39: 70-73.

[92] Plodsomboon S, Maeda-Martínez A M, Obregón-Barboza H, et al. Reproductive cycle and genitalia of the fairy shrimp Branchinella thailandensis (Branchiopoda: Anostraca). J. Crustacean Biol., 2012, 32: 711-726.

[93] a) Werner-Washburne M, Braun E, Johnston G, et al. Stationary phase in the yeast Saccharomyces cerevisiae. Microbiol. Rev., 1993, 57: 383-401; b) Herman P K. Stationary phase in yeast. Curr. Opin. Microbiol., 2002, 5: 602-607.

[94] Dandoy P, Meunier C F, Michiels C, et al. Hybrid shell engineering of animal cells for immune protections and regulation of drug delivery: towards the design of "artificial organs". PLoS One, 2011, 6: e20983.

[95] Hartwell L H, Culotti J, Pringle J R, et al. Genetic control of the cell division cycle in yeast. Science, 1974, 183: 46-51.

[96] Lee J, et al. Chemical Control of Yeast Cell Division by Cross-Linked Shells of Catechol-Grafted Polyelectrolyte Multilayers. Macromol. Rapid Commun., 2013, 34: 1351-1356.

[97] Jiskoot W, et al. Protein instability and immunogenicity: roadblocks to clinical application of injectable protein delivery systems for sustained release. J. Pharm. Sci., 2012, 101: 946-954.

[98] a) Wang G, et al. Extracellular silica nanocoat confers thermotolerance on individual cells: a case study of material-based functionalization of living cells. ChemBioChem, 2010, 11: 2368-2373; b) Ko E H, et al. Bioinspired, cytocompatible mineralization of silica-titania composites: thermoprotective nanoshell formation for individual chlorella cells. Angew. Chem., 2013, 125: 12505-12508.

[99] a) Cao A, et al. A facile method to encapsulate proteins in silica nanoparticles: encapsulated green fluorescent protein as a robust fluorescence probe. Angew. Chem. Int. Ed., 2010, 49: 3022-3025; b) Wang S, et al. Mesosilica-coated ultrafine fibers for highly efficient laccase encapsulation. Nanoscale, 2014, 6: 6468-6472.

[100] Yang Y, et al. The effect of amorphous calcium phosphate on protein protection against thermal denaturation. Chem. Commun., 2015, 51: 8705-8707.

[101] Yan M, Ge J, Liu Z, Ouyang P. Encapsulation of single enzyme in nanogel with enhanced biocatalytic activity and stability. J. Am. Chem. Soc., 2006, 128: 11008-11009.

[102] Liang K, et al. Biomimetic mineralization of metal-organic frameworks as protective coatings for biomacromolecules. Nature communications, 2015: 6.

[103] Rappuoli R, Miller H I, Falkow S. The intangible value of vaccination. Science, 2002, 297: 937-939.

[104] a) Chen X, et al. Improving the reach of vaccines to low-resource regions, with a needle-free vaccine delivery device and long-term thermostabilization. J. Controlled Release,

2011, 152: 349-355; b) Clemens J, Holmgren J, Kaufmann S H, et al. Ten years of the Global Alliance for Vaccines and Immunization: challenges and progress. Nat. Immunol., 2010, 11: 1069-1072.

[105] Das P. Revolutionary vaccine technology breaks the cold chain. The Lancet infectious diseases, 2004, 4: 719.

[106] Varmus H, Klausner R, Zerhouni E, et al. Grand challenges in global health. Science, 2003, 302: 398.

[107] a) Alcock R, et al. Long-term thermostabilization of live poxviral and adenoviral vaccine vectors at supraphysiological temperatures in carbohydrate glass. Sci. Transl. Med., 2010, 2: 19ra12-19ra12; b) Milstien J B, Lemon S M, Wright P F. Development of a more thermostable poliovirus vaccine. J. Infect. Dis., 1997, 175: S247-S253.

[108] Zhang J, et al. Stabilization of vaccines and antibiotics in silk and eliminating the cold chain. Proceedings of the National Academy of Sciences, 2012, 109: 11981-11986.

[109] Orange F, et al. Experimental silicification of the extremophilic Archaea Pyrococcus abyssi and Methanocaldococcus jannaschii: applications in the search for evidence of life in early Earth and extraterrestrial rocks. Geobiology, 2009, 7: 403-418.

[110] Walker J J, Spear J R, Pace N R. Geobiology of a microbial endolithic community in the Yellowstone geothermal environment. Nature, 2005, 434: 1011-1014.

[111] Sundar V C, Yablon A D, Grazul J, et al. Fibre-optical features of a glass sponge. Nature, 2003, 424: 899-900.

[112] Hamm C E, et al. Architecture and material properties of diatom shells provide effective mechanical protection. Nature, 2003, 421: 841-843.

[113] Neethirajan S, Gordon R, Wang L. Potential of silica bodies (phytoliths) for nanotechnology. Trends Biotechnol., 2009, 27: 461-467.

[114] Wang G, et al. Hydrated Silica Exterior Produced by Biomimetic Silicification Confers Viral Vaccine Heat-Resistance. ACS nano, 2015, 9: 799-808.

[115] a) Wu Z, Asokan A, Samulski R J. Adeno-associated virus serotypes: vector toolkit for human gene therapy. Mol. Ther., 2006, 4: 316-327; b) Blömer U, et al. Highly efficient and sustained gene transfer in adult neurons with a lentivirus vector. J. Virol., 1997, 71: 6641-6649.

[116] Thomas C E, Ehrhardt A, Kay M A. Progress and problems with the use of viral vectors for gene therapy. Nature Reviews Genetics, 2003, 4: 346-358.

[117] a) Leopold P L, Crystal R G. Intracellular trafficking of adenovirus: many means to many ends. Adv. Drug Del. Rev., 2007, 59: 810-821; b) St George J. Gene therapy progress and prospects: adenoviral vectors. Gene Ther., 2003, 10: 1135-1141.

[118] Wang X, et al. Biomineralization-Based Virus Shell-Engineering: Towards Neutralization Escape and Tropism Expansion. Advanced healthcare materials, 2012, 1: 443-449.

[119] a) Chen W, et al. Overcoming cisplatin resistance in chemotherapy by biomineralization. Chem. Commun., 2013, 49: 4932-4934; b) Chen W, et al. Nano Regulation of Cisplatin Chemotherapeutic Behaviors by Biomineralization Controls. Small, 2014, 10: 3644-3649; c) Chen W, Liu X, Xiao Y, et al. Overcoming Multiple Drug Resistance by Spatial-Temporal Synchronization of Epirubicin and Pooled siRNAs. Small, 2015, 11: 1775-1781.

[120] Hu Q, et al. Engineering Nanoparticle-Coated Bacteria as Oral DNA Vaccines for Cancer Immunotherapy. Nano Lett., 2015, 15: 2732-2739.

[121] Shao C P. Transfusion of RhD-positive blood in "Asia type" DEL recipients. N. Engl. J. Med., 2010, 362: 472-473.

[122] Waterbury J B, Watson S W, Guillard R R, et al. Widespread occurrence of a unicellular, marine, planktonic, cyanobacterium. Nature, 1979, 277: 293-294.

[123] a) Blankenship R E, et al. Comparing photosynthetic and photovoltaic efficiencies and recognizing the potential for improvement. Science, 2011, 332: 805-809; b) Chen T H, Murata N. Enhancement of tolerance of abiotic stress by metabolic engineering of betaines and other compatible solutes. Curr. Opin. Plant Biol., 2002, 5: 250-257.

[124] a) Kok B. On the inhibition of photosynthesis by intense light. Biochim. Biophys. Acta, 1956, 21: 234-244; b) Powles S B. Photoinhibition of photosynthesis induced by visible light. Annual Review of Plant Physiology, 1984, 35: 5-44.

[125] Hellwig S, Drossard J, Twyman R M, et al. Plant cell cultures for the production of recombinant proteins. Nat. Biotechnol., 2004, 22: 1415-1422.

[126] a) Meunier C F, Van Cutsem P, Kwon Y U, et al. Thylakoids entrapped within porous silica gel: towards living matter able to convert energy. J. Mater. Chem., 2009, 19: 1535-1542; b) Meunier C F, Van Cutsem P, Kwon Y U, et al. Investigation of different silica precursors: design of biocompatible silica gels with long term bio-activity of entrapped thylakoids toward artificial leaf. J. Mater. Chem., 2009, 19: 4131-4137.

[127] Meunier C F, Rooke J C, Léonard A, et al. Design of photochemical materials for carbohydrate production via the immobilisation of whole plant cells into a porous silica matrix. J. Mater. Chem., 2010, 20: 929-936.

[128] a) Léonard A, et al. Cyanobacteria immobilised in porous silica gels: exploring biocompatible synthesis routes for the development of photobioreactors. Energy Environ. Sci., 2010, 3: 370-377; b) Jiang N, et al. A Stable, Reusable, and Highly Active Photosynthetic Bioreactor by Bio-Interfacing an Individual Cyanobacterium with a Mesoporous Bilayer Nanoshell. Small, 2015, 11: 2003-2010.

[129] Desmet J, et al. Green and sustainable production of high value compounds via a microalgae encapsulation technology that relies on CO_2 as a principle reactant. Journal of Materials Chemistry A, 2014, 2: 20560-20569.

[130] a) Kruse O, Hankamer B. Microalgal hydrogen production. Curr. Opin. Biotechnol., 2010, 21: 238-243; b) Srirangan K, Pyne M E, Chou C P. Biochemical and genetic engineering strategies to enhance hydrogen production in photosynthetic algae and cyanobacteria. Bioresour. Technol., 2011, 102: 8589-8604.

[131] Gaffron H, Rubin J. Fermentative and photochemical production of hydrogen in algae. The Journal of General Physiology, 1942, 26: 219-240.

[132] Stripp S T, et al. How oxygen attacks [FeFe] hydrogenases from photosynthetic organisms. Proc. Natl. Acad. Sci. U. S. A., 2009, 106: 17331-17336.

[133] a) Melis A, Zhang L, Forestier M, et al. Sustained Photobiological Hydrogen Gas Production upon Reversible Inactivation of Oxygen Evolution in the Green AlgaChlamydomonas reinhardtii. Plant Physiol., 2000, 122: 127-136; b) Ghirardi M L, et al. Microalgae: a green source of renewable H_2. Trends Biotechnol., 2000, 18: 506-511.

[134] Eroglu E, Melis A. Photobiological hydrogen production: recent advances and state of the art. Bioresour. Technol., 2011, 102: 8403-8413.

[135] Xiong W, et al. Silicification-Induced Cell Aggregation for the Sustainable Production of H2 under Aerobic Conditions. Angew. Chem. , 2015.

[136] Schmid A, et al. Industrial biocatalysis today and tomorrow. Nature, 2001: 409, 258-268.

[137] Klibanov A M. Why are enzymes less active in organic solvents than in water? Trends Biotechnol., 1997, 15: 97-101.

[138] a) Crossley S, Faria J, Shen M, et al. Solid nanoparticles that catalyze biofuel upgrade reactions at the water/oil interface. Science, 2010, 327: 68-72; b) Zapata P A, Faria J, Ruiz M P, et al. Hydrophobic zeolites for biofuel upgrading reactions at the liquid–liquid interface in water/oil emulsions. J. Am. Chem. Soc., 2012, 134: 8570-8578; c) Zhou W J, et al. Tunable Catalysts for Solvent-Free Biphasic Systems: Pickering Interfacial Catalysts over Amphiphilic Silica Nanoparticles. J. Am. Chem. Soc., 2014, 136: 4869-4872.

[139] Chen Z, et al. Individual Surface-Engineered Microorganisms as Robust Pickering Interfacial Biocatalysts for Resistance-Minimized Phase-Transfer Bioconversion. Angew. Chem. Int. Ed., 2015, 54: 4904-4908.

[140] a) Gadsby D C.Ion channels versus ion pumps: the principal difference, in principle. Nat. Rev. Mol. Cell Biol. , 2009, 10: 344-352; b) Reyes N, Gadsby D C. Ion permeation through the Na+, K+-ATPase. Nature, 2006, 443: 470-474; c) Catterall W A. Structure and function of voltage-sensitive ion channels. Science, 1988, 242: 50-61; d) D J X, Lavan A. Designing

artificial cells to harness the biological ion concentration gradient. Nat. Nanotechnol., 2008, 3: 666-670.

[141] a) Li J, Sculley J, Zhou H. Metal–organic frameworks for separations. Chem. Rev., 2011, 112: 869-932; b) Pera-Titus M. Porous Inorganic Membranes for CO_2 Capture: Present and Prospects. Chem. Rev., 2013, 114: 1413-1492.

[142] a) Zhang H, Hou X, Zeng L, etal.Bioinspired artificial single ion pump. J. Am. Chem. Soc., 2013, 135: 16102-16110; b) Liu Q, Xiao K, Wen L, etal. Engineered ionic gates for ion conduction based on sodium and potassium activated nanochannels. J. Am. Chem. Soc., 2015, 137: 11976-11983; c) Jiang Y, Liu N, Guo W, et al. Highly-efficient gating of solid-state nanochannels by DNA supersandwich structure containing ATP aptamers: a nanofluidic IMPLICATION logic device.J. Am. Chem. Soc, 2012, 134: 15395-15401.

[143] Chen W, Wang G, Tang R. Nanomodification of living organisms by biomimetic mineralization. Nano Research, 2014, 7: 1404-1428.

[144] Demirbas A. Biohydrogen: For Future Engine Fuel Demands. New York: Springer, 2009.

[145] Mata-Alvarez J, Mace S, Llabres P. Anaerobic digestion of organic solid wastes. An overview of research achievements and perspectives. Bioresour. Technol., 2000, 74: 3-16.

[146] Walsh G. Biopharmaceutical benchmarks 2010. Nat. Biotechnol., 2010, 28: 917-924.

[147] a) Weiner S, Wagner H D. The material bone: structure-mechanical function relations. Annu. Rev. Mater. Sci., 1998, 28: 271-298; b) Katti K S, Katti D R. Why is nacre so tough and strong? Materials Science and Engineering: C, 2006, 26: 1317-1324.

[148] a) Zhai H, Chu X, Li L, et al. Controlled formation of calcium-phosphate-based hybrid mesocrystals by organic-inorganic co-assembly. Nanoscale, 2010, 2: 2456-2462; b) Li Y, et al. Biomimetic graphene oxide-hydroxyapatite composites via in situ mineralization and hierarchical assembly. RSC Advances, 2014, 4: 25398-25403; c) Zhai H, et al. Spontaneously amplified homochiral organic-inorganic nano-helix complexes via self-proliferation. Nanoscale, 2013, 5: 3006-3012; d) Quan Y, Zhai H, Zhang Z, et al. Lamellar organic–inorganic architecture via classical screw growth. CrystEngComm, 2012, 14: 7184-7188.

关键词索引

C

D

E

F

G

H

J

K

L

M

N

P

Q

R

S

T

W

X

Y

Z

其他

彩　图

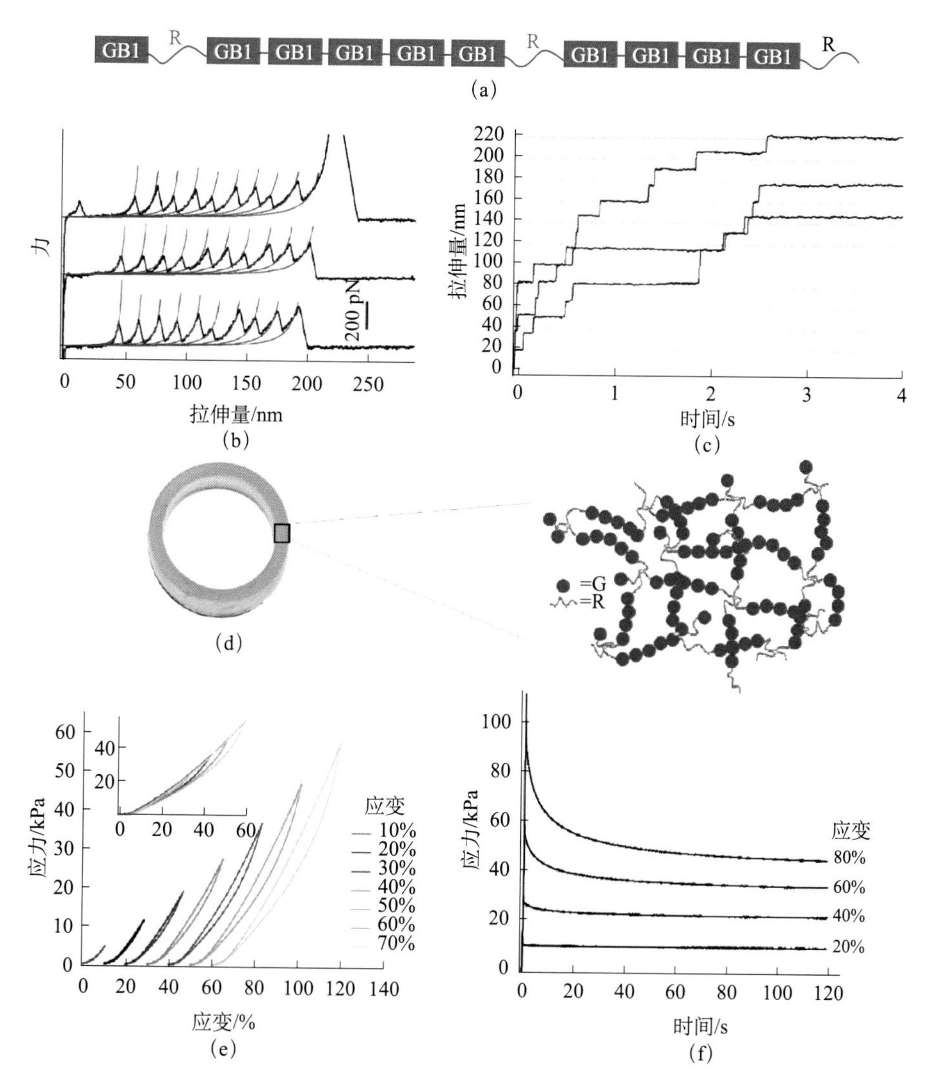

图 3-2-10　通过自下而上的方法由蛋白质工程方法得到的基于弹性蛋白的生物材料

（a）模仿巨肌蛋白设计的弹性蛋白 GRG_5RG_4R。G 代表 GB1 结构域，R 代表 resilin 序列。（b）力谱曲线 GRG_5RG_4R，前面没有力的部分表示 resilin 的拉伸过程，后面锯齿样图案为 GB1 解折叠过程。（c）恒力拉伸曲线。（d）由 GRG_5RG_4R 制作的水凝胶环的照片。（e）该材料的应力-应变曲线以及应力-松弛曲线。（f）应变不变的情况下，应力随时间的变化（图片来自文献 [65]）

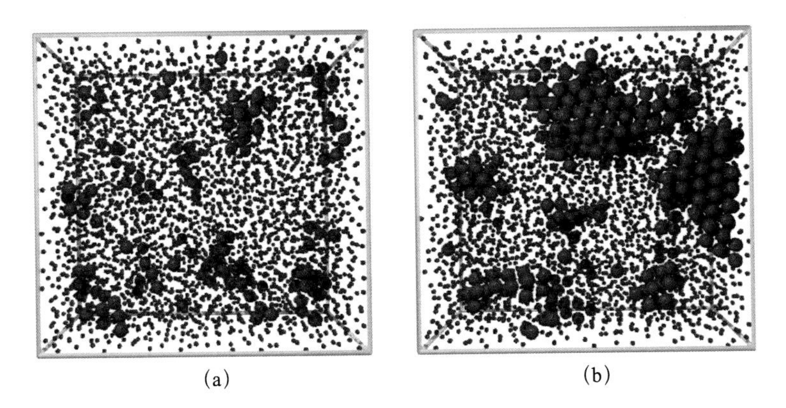

图 3-4-9　晶核的形成和生长过程。使用共聚焦显微镜观察胶体粒子结晶成核，红色粒子表示结晶区域，而在液体区域由更小的蓝色粒子表示（图片来自文献 [11]）

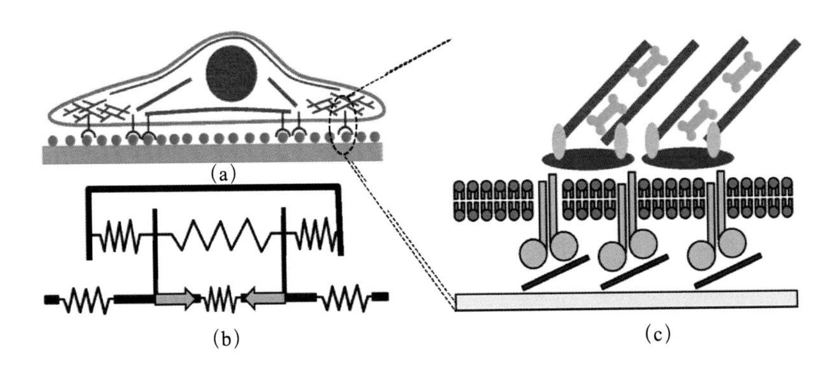

图 3-7-4　细胞在物理表面的黏附示意图及其受力模型

（a）在平面上铺展开的细胞呈典型的煎蛋样，细胞边缘是由肌动蛋白纤维向外撑开的盘状伪足，中间是由肌动蛋白交联形成束状的收缩应力纤维。其受力模型可以用（b）中耦合的弹簧及一对向内收缩的力偶来模拟。（c）黏着斑的主要组成模式：跨膜的整联蛋白二聚体（绿色），一端连接沉积在胞外的胶原蛋白（褐色），另一端通过一些结合蛋白（如 Talin, Vinculin 等），连接到由可伸缩的肌浆球蛋白（绿色）串联组装起来的肌动蛋白纤维（红色）细胞骨架上

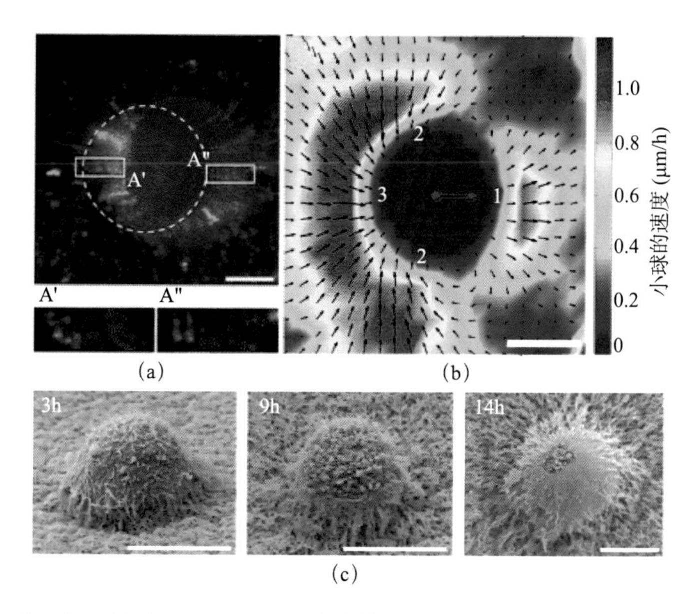

图 3-7-11　乳癌细胞单个 MDA-MB-231 在胶体 matrigel 中移动动态行为的研究。细胞移动引起的周围网络的改变，而环境网络的改变是通过当中掺杂的荧光小球（绿色）位置的变化反映的（b）

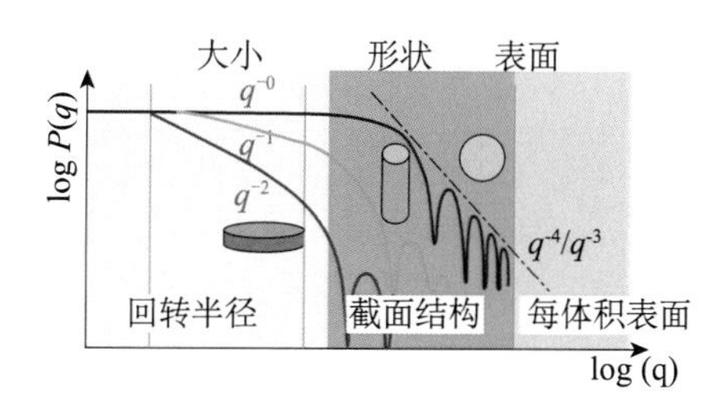

图 4-1-17　形状因子 P 与散射矢量长度 q 的双对数关系（摘自文献 [72]，Anton Paar GmbH，2013 年）

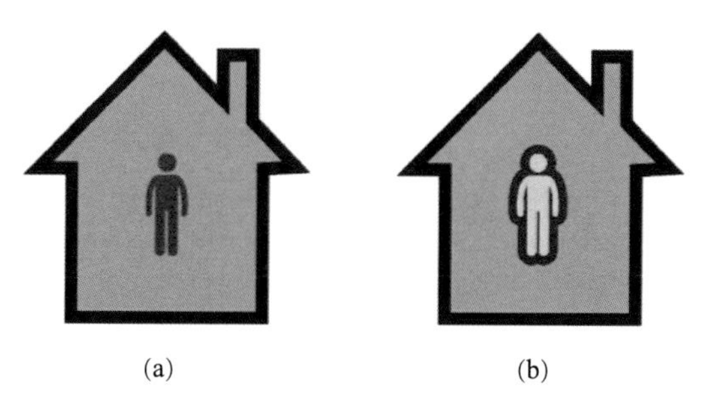

图 4-4-1　变换热学简要示意图

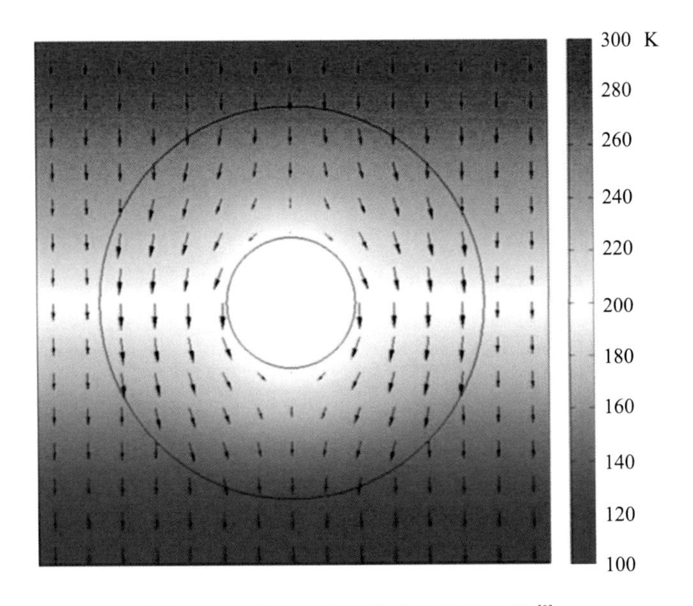

图 4-4-4　稳态二维热隐身衣模拟结果[9]

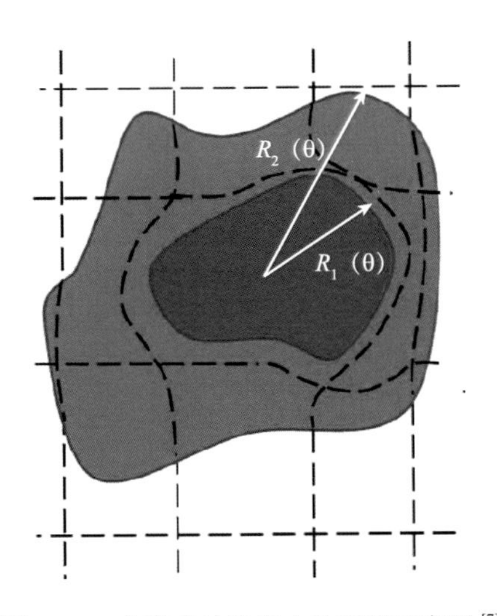

图 4-4-5　非稳态热隐身衣的原理示意图[7]

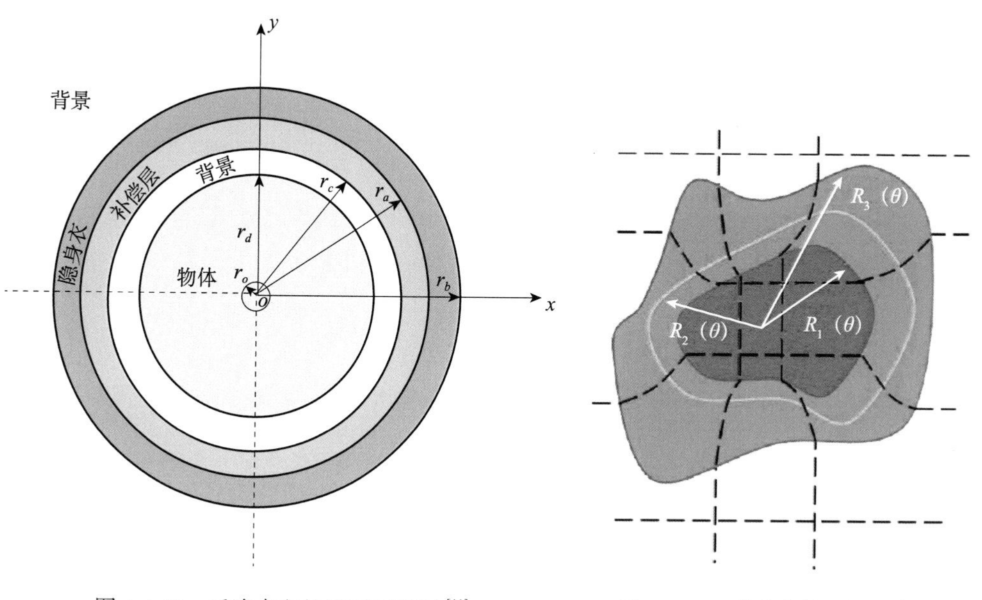

图 4-4-10　反隐身衣的原理示意图 [11]

图 4-4-12　热聚集的原理示意图

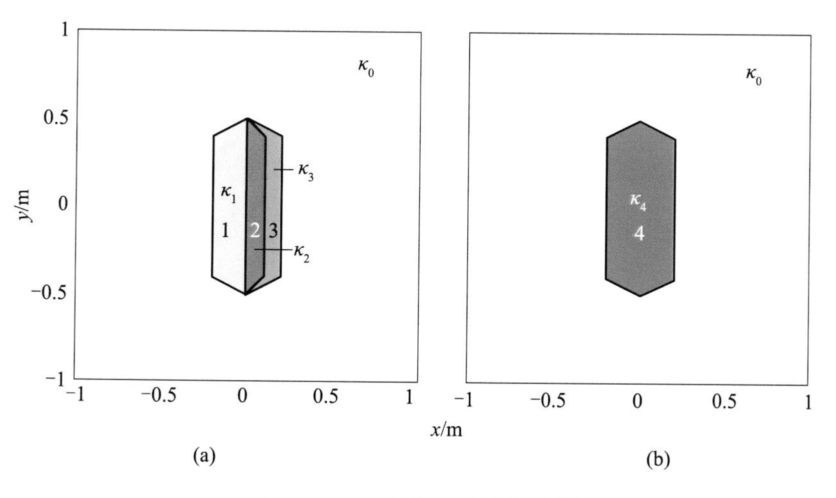

图 4-4-15　热幻像装置示意图 [12]

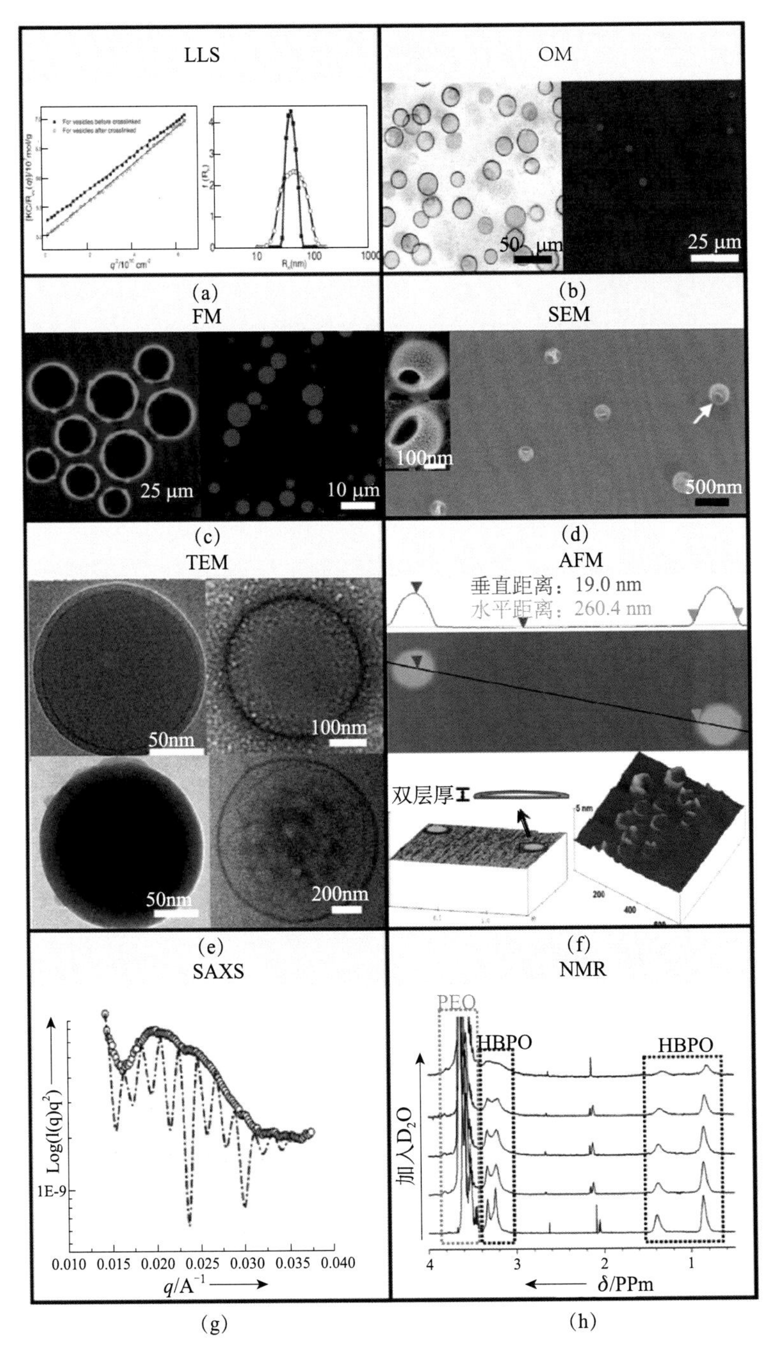

图 4-6-11　囊泡的表征手段

（a）LLS;（b）OM;（c）FM;（d）SEM;（e）TEM;（f）AFM;（g）SAXS;（h）NMR[84]

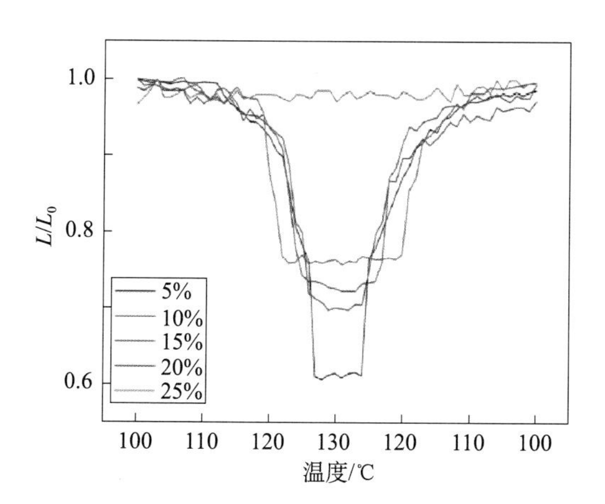

图 4-7-5　热致液晶弹性体形变量随交联度提高而降低

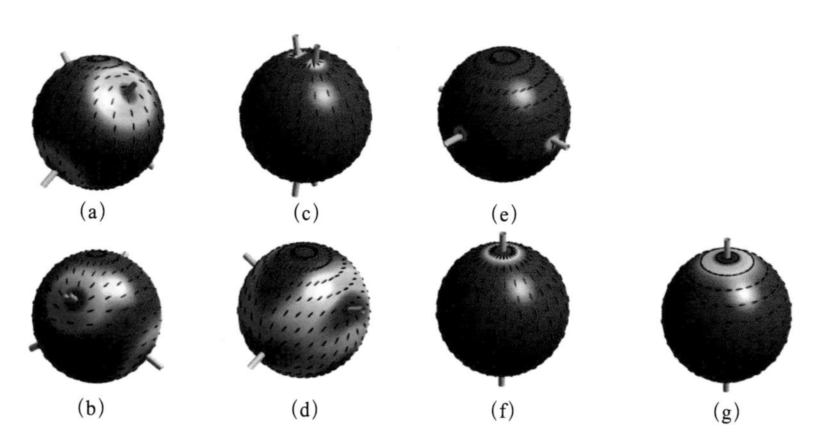

图 5-1-7　文献 [92] 给出的基于自洽场模型的稳态和亚稳态

(a)A 类四面体结构(THA),(b)B 类四面体结构(THB),(c)A 类网球结构(TA),(d)B 类网球结构(TB),(e)旋切结构(CR),(f)发散结构(LS),(g)弯曲结构(LB)。图中所示颜色为序参量从大(红)到小(蓝)。其中每个奇点位置被小圆柱标注

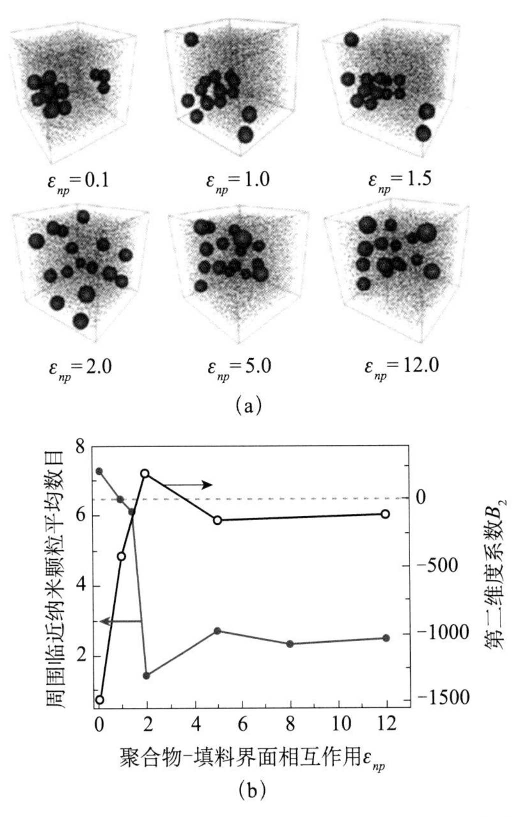

图 5-3-4 （a）不同界面相互作用下纳米颗粒分散的 snapshot 图；（b）周围临近纳米颗粒的平均数目与第二维里系数随界面相互作用的变化

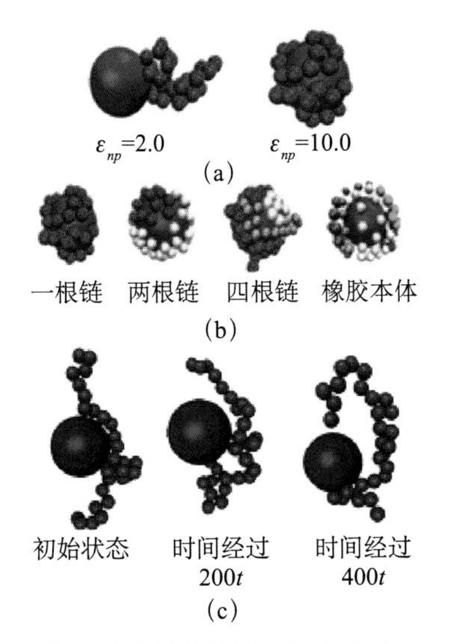

图 5-3-5 （a）不同界面相互作用下单根分子链在纳米颗粒表面吸附状态；（b）不同分子链数目在纳米颗粒表面吸附状态，不同颜色代表不同分子链；（c）类似氢键界面作用下分子链吸附-解吸附过程

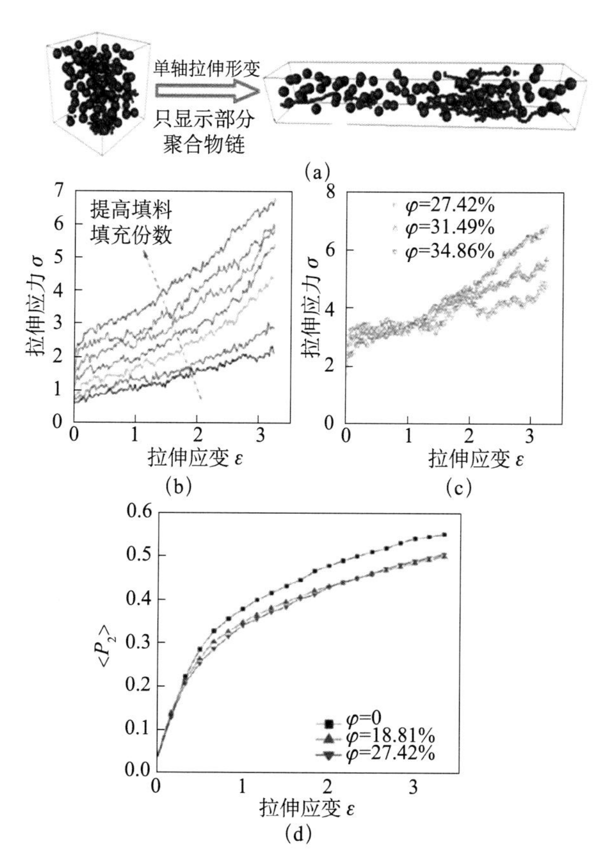

图 5-3-7 （a）球状填充体系单轴拉伸示意图；（b）低填充分数与（c）高填充分数下应力应变曲线；（d）玻璃态下不同填充分数下分子链键取向随拉伸应变变化示意图

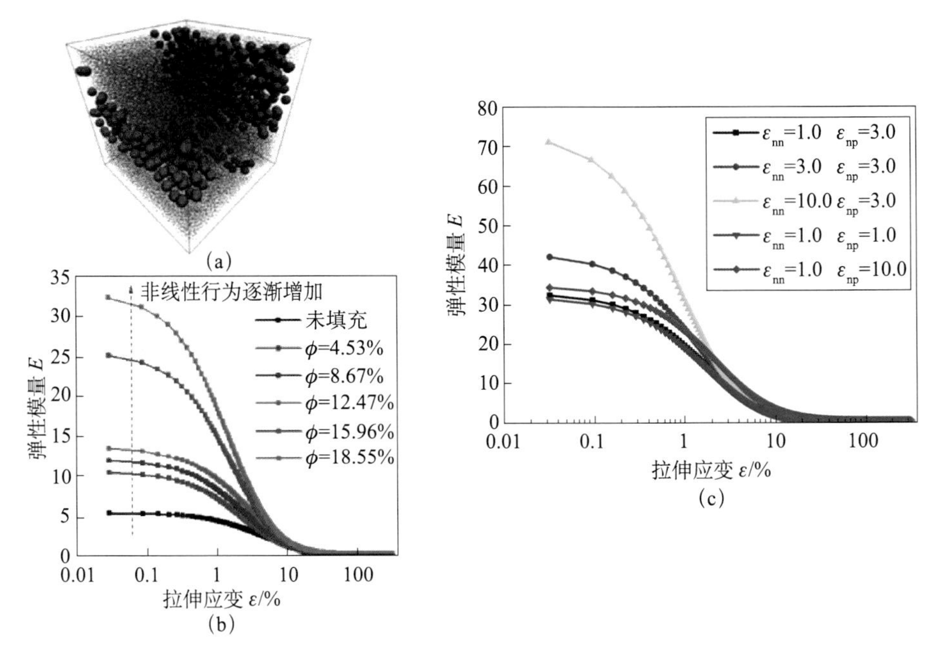

图 5-3-8 （a）初始状态为球状颗粒聚集的体系；非线性行为随着（b）填充分数与（c）颗粒-颗粒相互作用能增加而逐渐增大

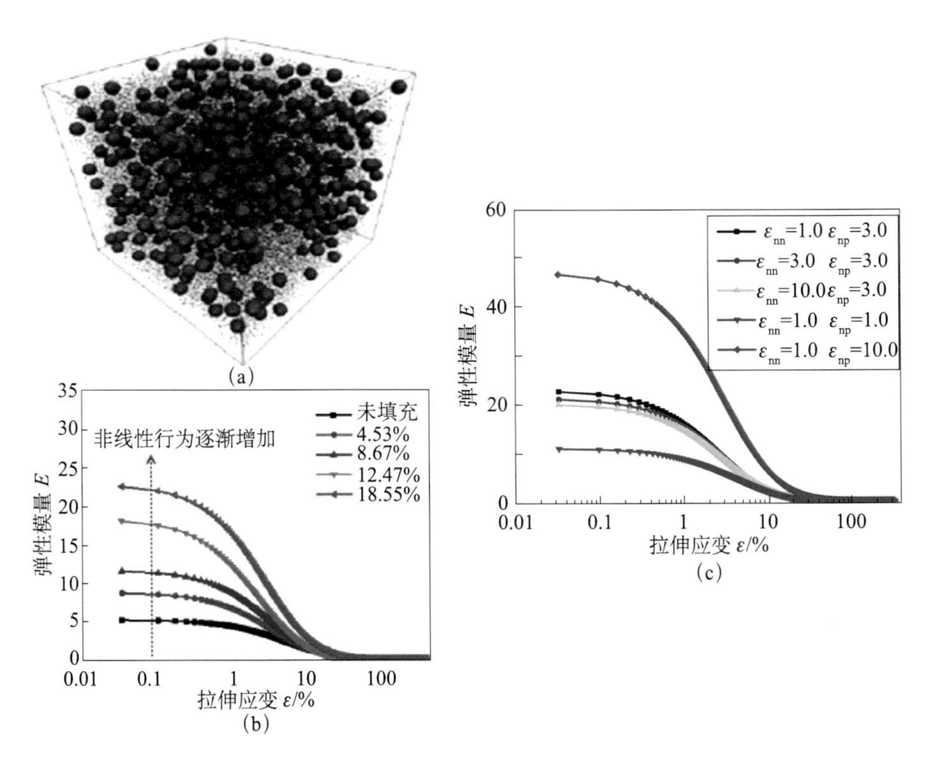

图 5-3-9 （a）初始状态为球状颗粒分散的体系；非线性行为随着（b）填充分数与（c）颗粒-聚合物相互作用能增加而逐渐增大

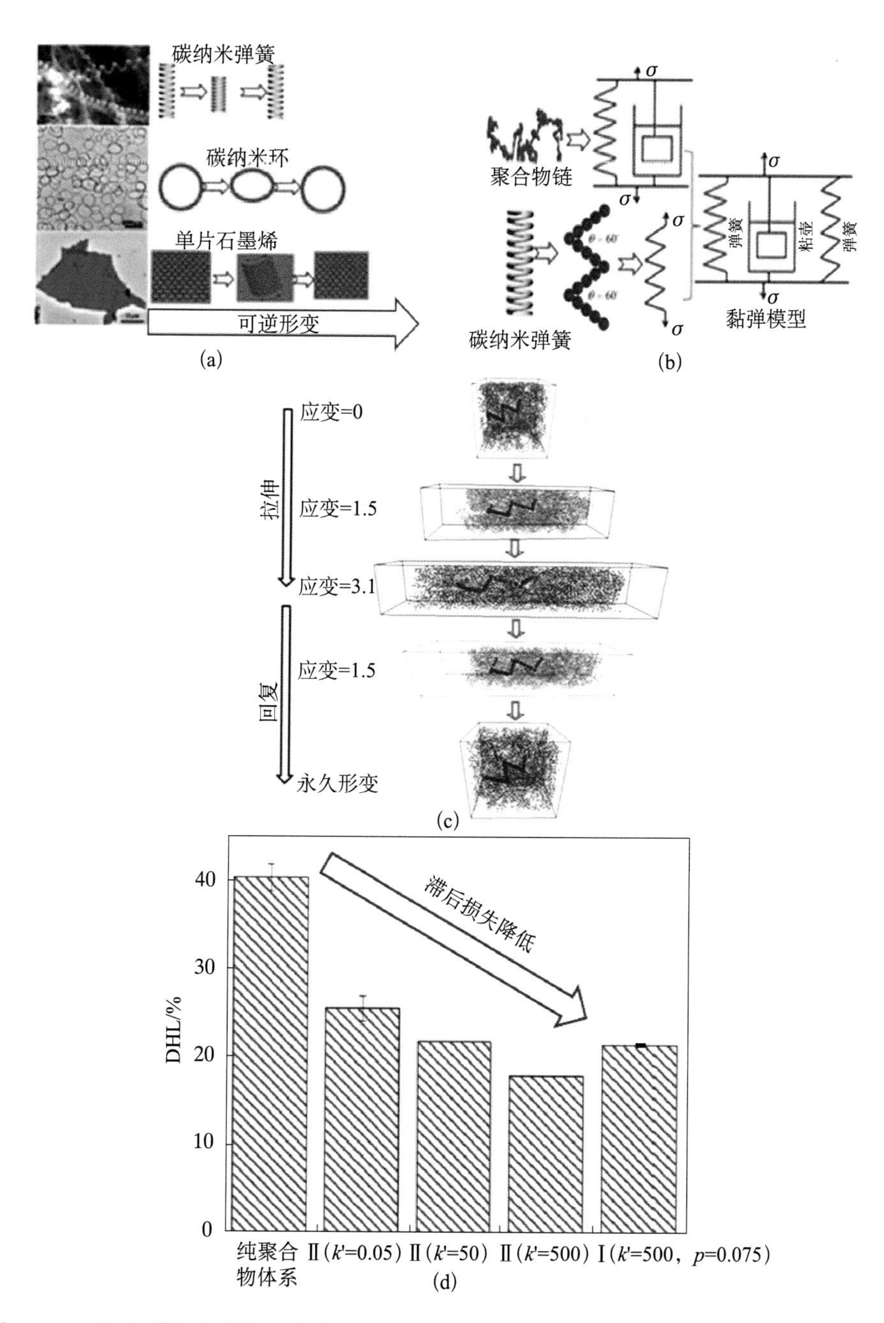

图 5-3-10 （a）碳纳米弹簧、碳纳米环与单片石墨烯力学可回复变形示意图；（b）碳纳米弹簧填充橡胶体系黏弹性模型示意图；（c）碳纳米弹簧填充橡胶体系拉伸-回复过程示意图；（d）动态滞后损失随着碳纳米弹簧弹性体系数逐渐增大而减小

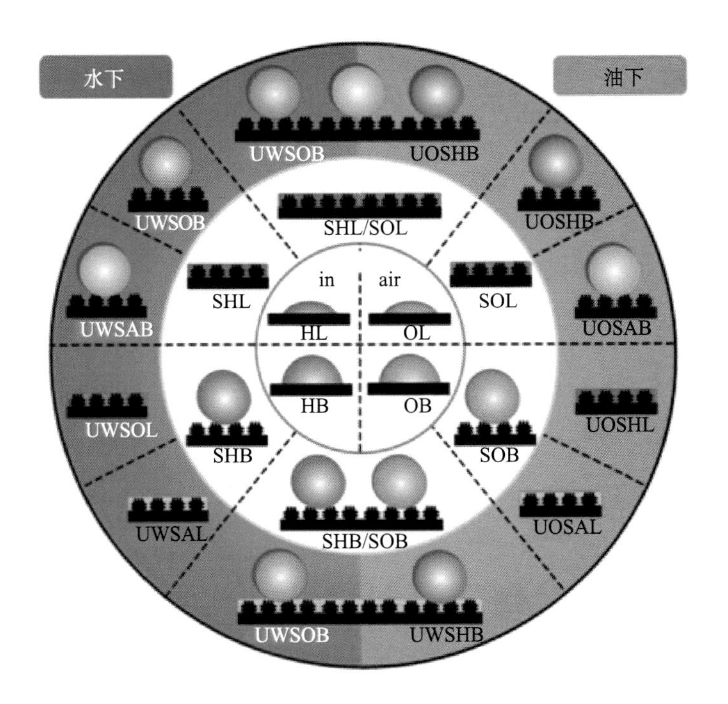

图 5-4-2　超浸润系统示意图

中间部分：空气中平面基体上的，亲水（HL），疏水（HB），亲油（OL），疏油（OB）性质。超亲水（SHL）、超疏水（SHB）、超亲油（SOL）以及超疏油（SOB）的状态可以通过引入微 / 纳尺度的二元粗糙度结构获得。左侧蓝色的半圆形区域分别是：粗糙基底上的水下超亲油（UWSOL），水下超疏油（UWSOB），水下超亲气（UWSAL）和水下超疏气（UWSAB）性质。右侧橙色的半圆形区域分别是：粗超基体上的油下超亲水（UOSHL），油下超疏水（UOSHB），油下超亲气（UOSAL），油下超疏气（UOSAB）性质

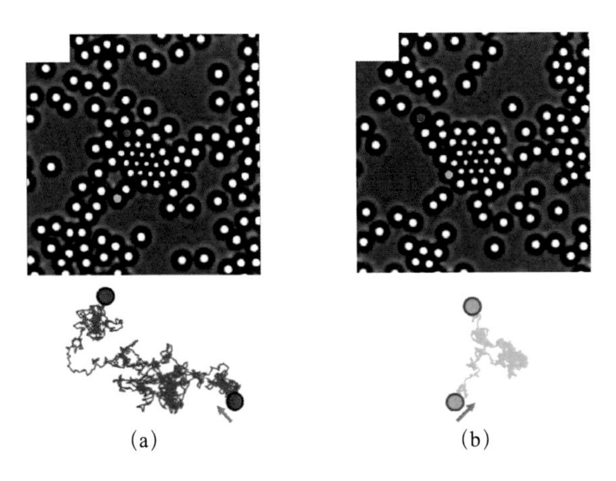

图 5-7-1　胶体颗粒的运动和观察

通过和显微镜连接的数码相机和粒子追踪技术，可以记录还原胶体颗粒在运动过程中的实时位置和轨迹。（a）（b）展示了在结晶相变过程中，胶体颗粒的运动。红、绿两个胶体颗粒在晶核生长过程中的轨迹表明晶核与周围的气相处于平衡状态

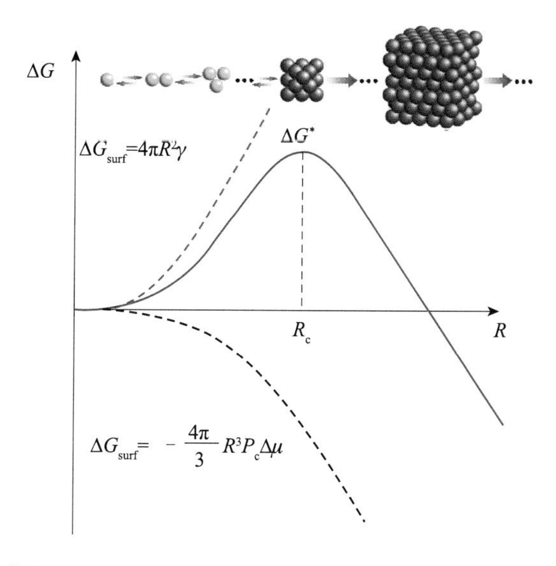

图 5-7-2　经典成核理论。成核是一个体自由能减少和表面自由能增加两种趋势相互竞争的过程。成核势垒和临界尺寸的存在正是这一竞争的结果

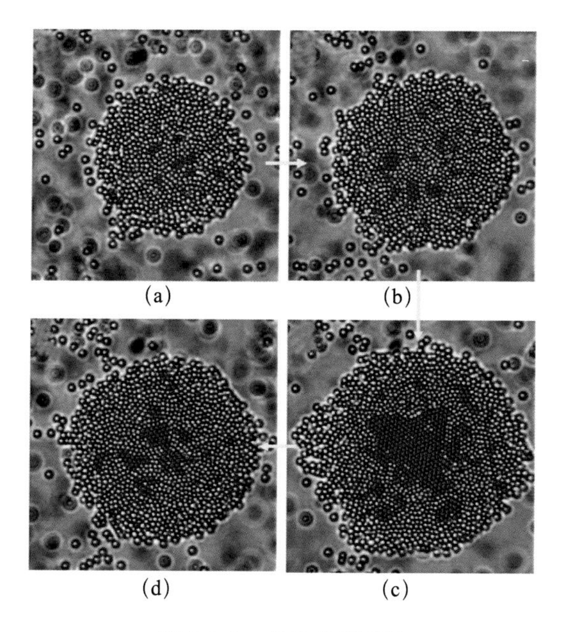

图 5-7-3　多步成核过程

（a）～（d）代表了一个多步成核过程。红色标出了有序结构区域。（a）无序的液态结构首先形成；（b）、（c）局部不稳定的有序结构开始出现；（d）稳定的有序结构形成

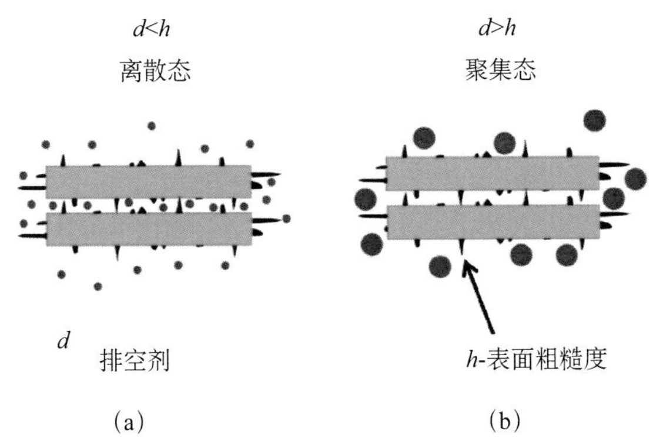

图 5-8-3　表面粗糙度对排空吸引的影响

（a）当表面粗糙度大于排空剂（红色圆球）尺寸时，片状颗粒之间的排空吸引弱，颗粒处于离散态；（b）当表面粗糙度小于排空剂尺寸时，颗粒之间的排空吸引与光滑颗粒之间的排空吸引相当，颗粒处于聚集态

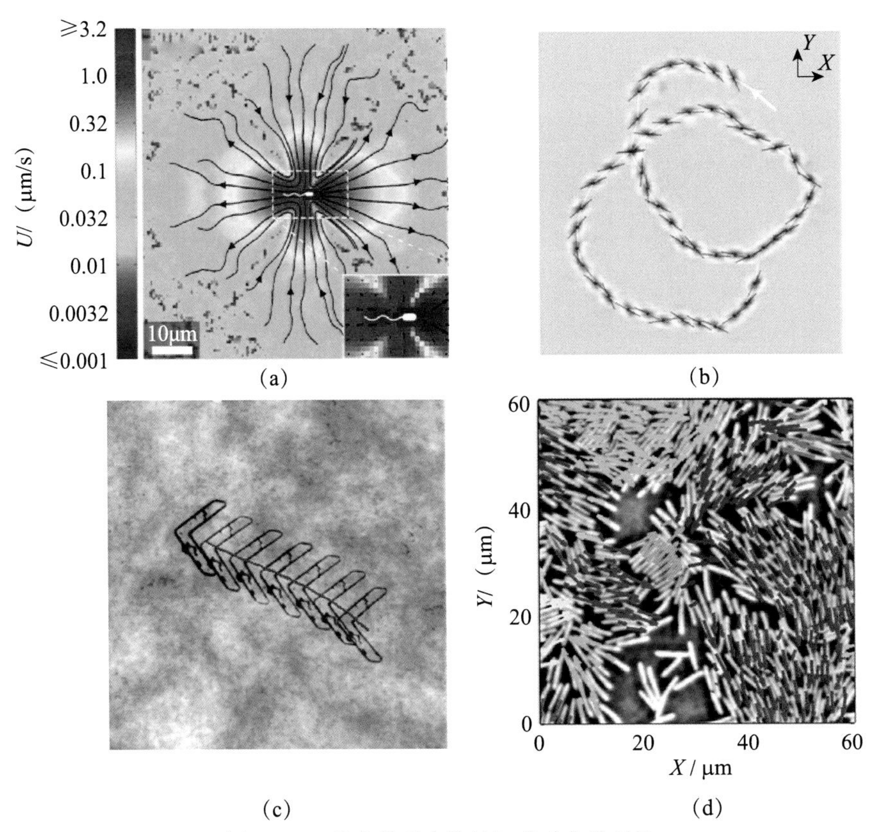

图 5-9-3　从个体到多体的细菌动力学现象

（a）细菌周围流场的实验测量结果，颜色代表流速大小，黑线为流线（参考文献［73］）；（b）细菌在气液界面做逆时针运动，黑色物体为细菌瞬时位置，白线为细菌轨迹，白色箭头为细菌初始位置（参考文献［71］）；（c）细菌悬浊液主动输运三角形物体，红色为物体轨迹，物体边长为 262μm（参考文献［78］）；（d）细菌菌落中集体运动，背景中高亮度杆状物体为细菌，箭头代表瞬时速度，带有同样颜色箭头的近邻细菌属于同一动态团簇，不属于任何团簇的细菌未被标注速度箭头（参考文献［82］）

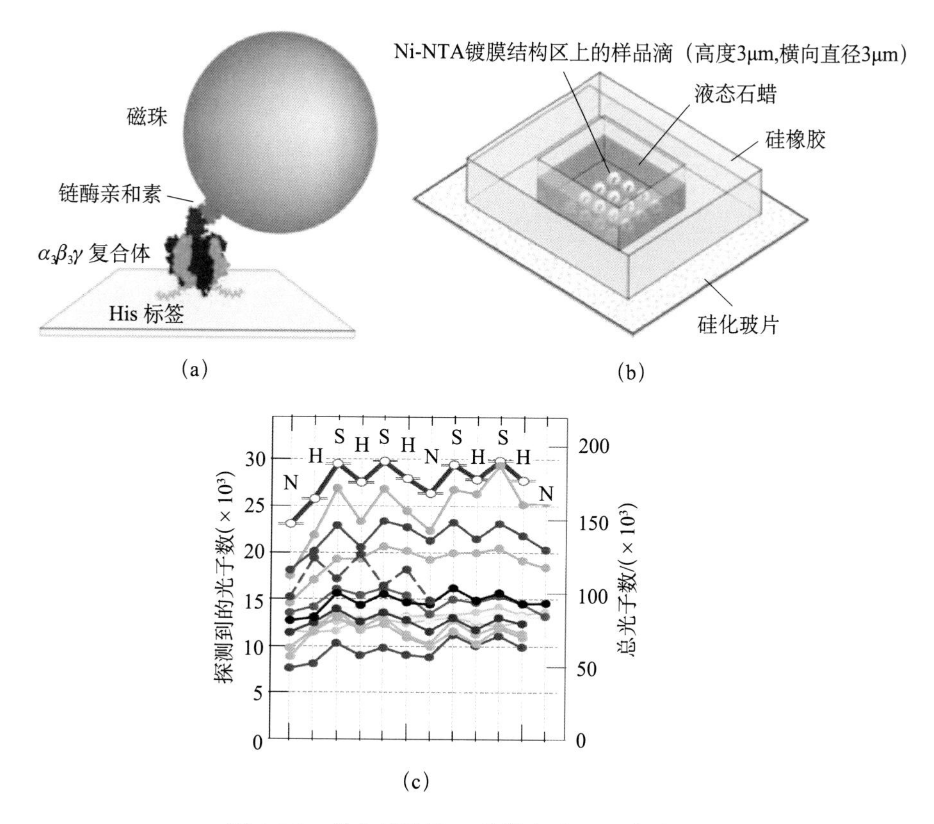

图 6-5-7　单分子操控 F_1 体外合成 ATP 实验

(a)“定子”$\alpha_3\beta_3$ 环固定在底板上，“转子”γ 亚基顶端结合磁珠；(b)16 个隔室内各置入一个马达，外置的旋转磁场可以操控马达合成或水解 ATP，同时利用荧光激素酶测量每个隔室的荧光强度变化表征 ATP 的合成或水解；(c) 12 个有效隔室的荧光强度变化（对应 12 条不同颜色）与马达转向的关系，N：不转；S：10Hz 合成转速；H：10Hz 水解转速。最高处品红粗线代表所有 12 个隔室的荧光总和，数据表明 ATP 总量的变化与马达转向正相关。图片摘自文献 [46]

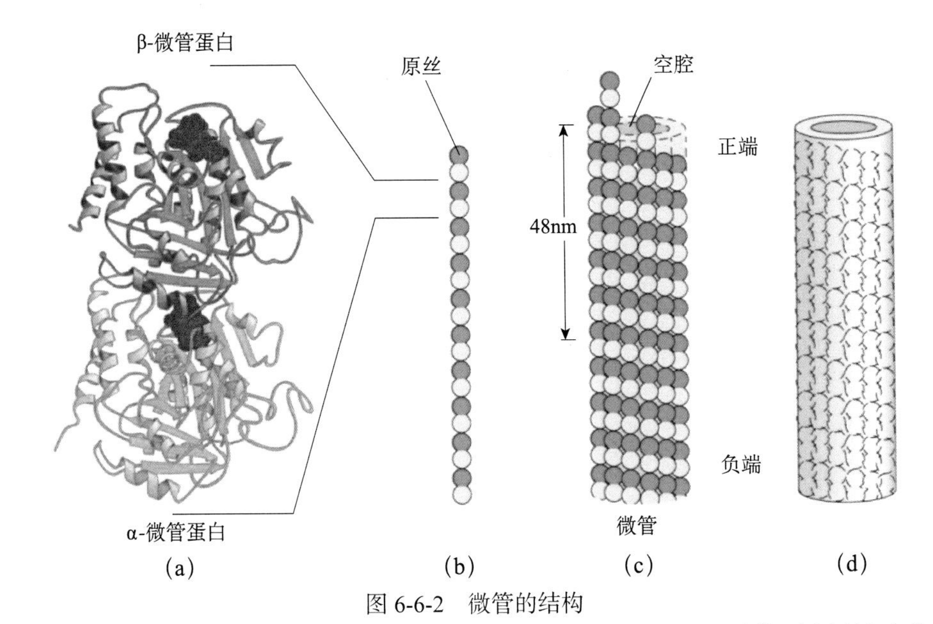

图 6-6-2　微管的结构

(a) 微管蛋白亚基重复单元的结构图，它是由 α 和 β 微管蛋白亚基形成的二聚体。图中的红色块体为它们黏结的 GTP 分子所在的位置。(b) α 和 β 微管蛋白亚基以首尾相连的方式形成单根原丝。(c) 微管是由 13 根原丝排列形成的中空圆柱体。微管蛋白亚基重复单元本身具有极性，从而由它组装形成的微管也具有极性，其中 β 微管蛋白永远存在于微管的正端，而 α 微管蛋白存在于微管的负端。(d) 微管通常被认为是一个直径为 25 nm 的空心圆柱体（摘自 Alberts 编写的 Molecular Biology of the Cell，第六版，Garland Science，2014）

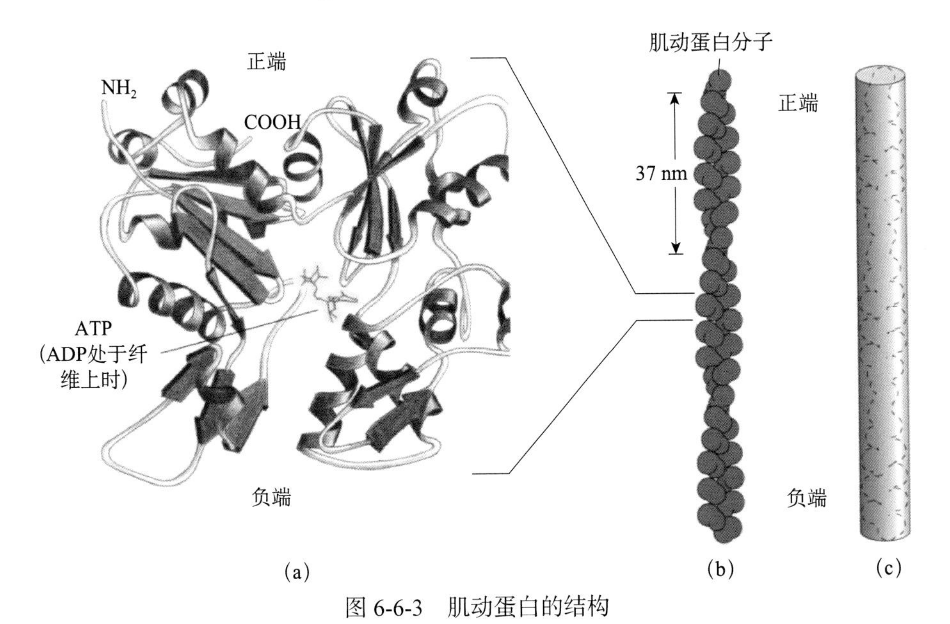

图 6-6-3　肌动蛋白的结构

(a) 单个肌动蛋白分子的结构图，它是直径为 5.4nm 的球状蛋白。图中黄色部分为它们黏结的 ATP(ADP) 分子所在位置。(b) 肌动蛋白纤维是由 2 根原丝组成的，它们形成的螺旋结构的螺距为 37nm。所有的肌动蛋白亚基指向相同的方向。每新增一个肌动蛋白单体，纤维的长度增加 2.7nm。(c) 肌动蛋白纤维可近似看成直径为 8nm 的圆柱体（摘自 Alberts 编写的 Molecular Biology of the Cell，第六版，Garland Science，2014）

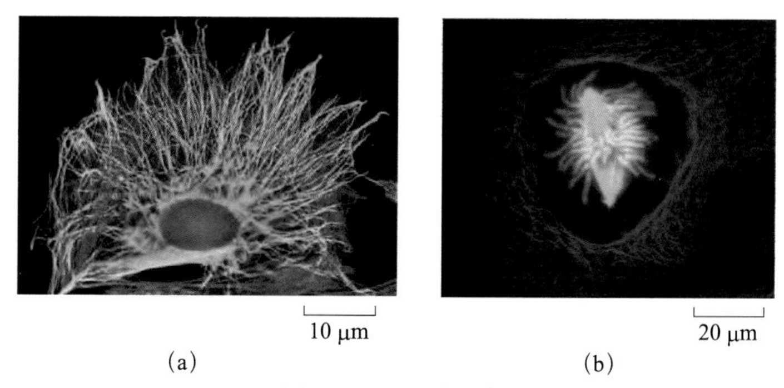

图 6-6-5　细胞骨架

(a) 细胞内细胞质阵列：微管（绿色）和肌动蛋白纤维（红色）。(b) 分裂细胞内组成纺锤体的微管（绿色）和周围的中间纤维（红色）在所有的细胞中 DNA 标记为蓝色（Albert Tousson, Conly Riender 惠赠）

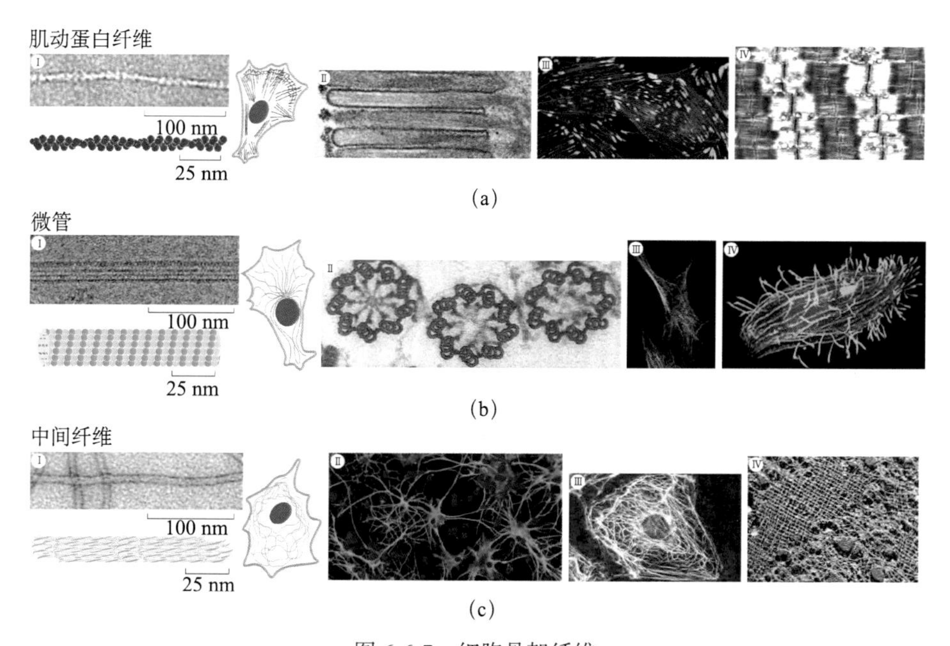

图 6-6-7　细胞骨架纤维

(a) 肌动蛋白纤维：它是由肌动蛋白亚基组成的螺旋状微丝。它的直径约为 8nm 且具有柔性结构，它能组装形成束状结构、二维的网状结构和三维的凝胶结构。肌动蛋白纤维分布于细胞内，主要集中分布于细胞的顶端和质膜下方。(Ⅰ) 单根肌动蛋白纤维，(Ⅱ) 微绒毛，(Ⅲ) 细胞局部黏结时形成的应力纤维，(Ⅳ) 肌肉内部纤维。(b) 微管：它是直径为 25 nm 的中空圆柱体。它比肌动蛋白纤维的持久长度要高。微管的一端通常黏结于微管组织中心（MTOC，中心体）。(Ⅰ) 单根微管，(Ⅱ) 纤毛内部微管三聚体的截面，(Ⅲ) 中间相微管阵列（绿色）和有机体（红色），(Ⅳ) 纤毛原生动物。(c) 中间纤维：它是直径为 10 nm 的绳子状纤维。一种中间纤维形成网状结构主要分布在细胞核内核纤层，另外一种中间纤维主要分布在细胞质内提供力学强度。在上皮细胞组织，它们能连接细胞以提供上皮组织的力学强度。(Ⅰ) 单根中间纤维，(Ⅱ) 神经元内的中间纤维（蓝色），(Ⅲ) 上皮细胞，(Ⅳ) 核纤层（R. Craig, P. T. Matsudaria 惠赠）

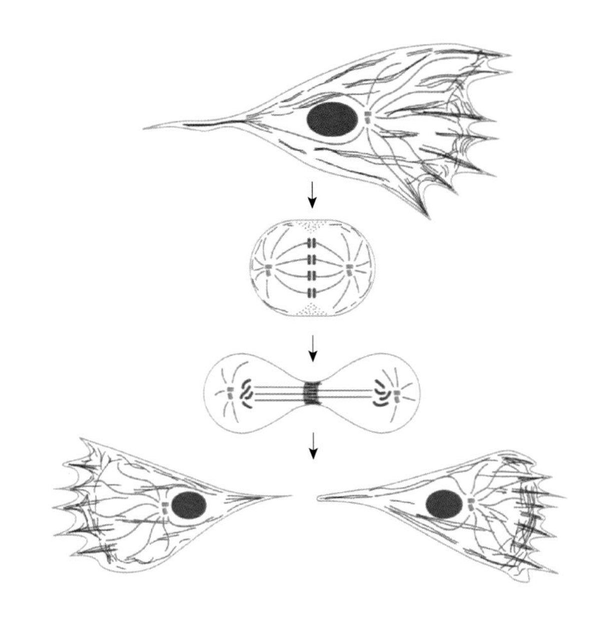

图 6-6-8　细胞分裂过程中细胞骨架的重构示意图

爬行的成纤维细胞具有极性、动态的肌动蛋白细胞骨架（红色）。在板状伪足和成纤维细胞内形成的肌动蛋白细胞骨架推动板状伪足和成纤维细胞向前移动。微管细胞骨架（绿色）对这些肌动蛋白细胞骨架提供支持作用。这些微管细胞骨架在细胞核内以一个中心点发散形成微管细胞骨架。当细胞分裂时，微管细胞骨架首先重组形成双极性纺锤体。纺锤体主要为染色体（棕色）分离提供作用力。在染色体分离后，肌动蛋白细胞骨架在细胞中间形成动态收缩环，它提供限制作用力将细胞一分为二。在细胞完全分裂后，子细胞将继续重组微管和肌动蛋白细胞骨架，从而子细胞继续爬行等待下一次分裂（摘自 Alberts 编写的 Molecular Biology of the Cell，第五版，Garland Science，2008）

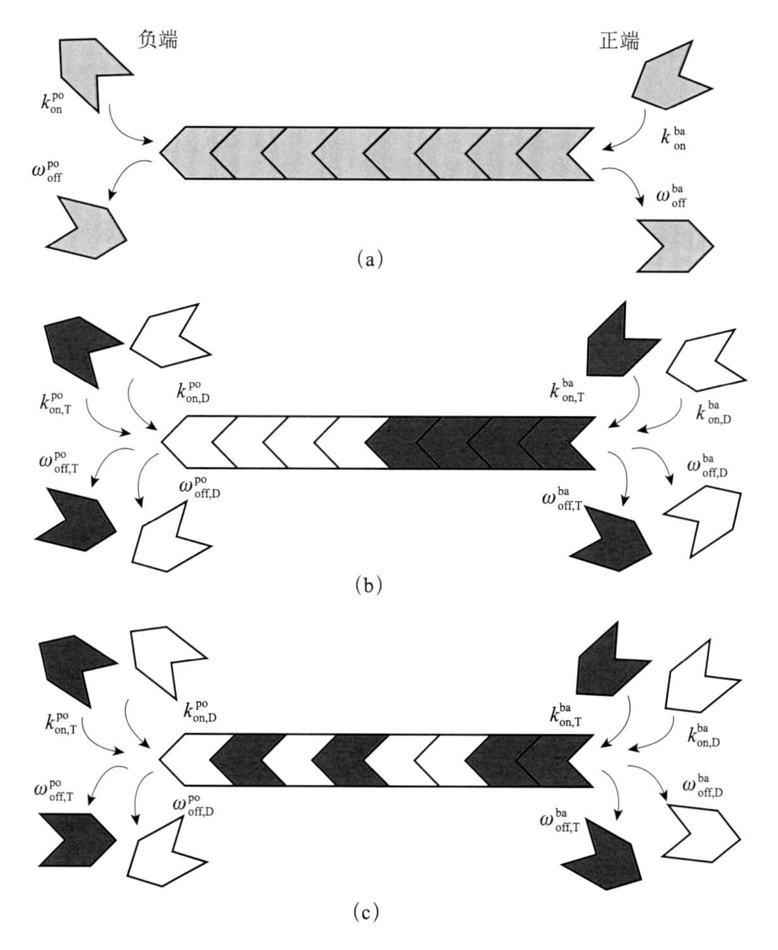

图 6-6-10　细胞骨架聚合的复杂性增加层次示意图，其中包括不对称速率和水解

(a) 由于肌动蛋白纤维的结构不对称性，它的两端具有不同的聚合和解聚速率，分别称为正端和负端，对于肌动蛋白纤维又称为倒钩端和尖端。(b) 矢量水解下，纤维两端以不同的速率聚合和解聚。(c) 随机水解下，纤维两端以不同的速率聚合和解聚。其中，红色表示 ATP- 肌动蛋白单体；白色表示 ADP- 肌动蛋白单体

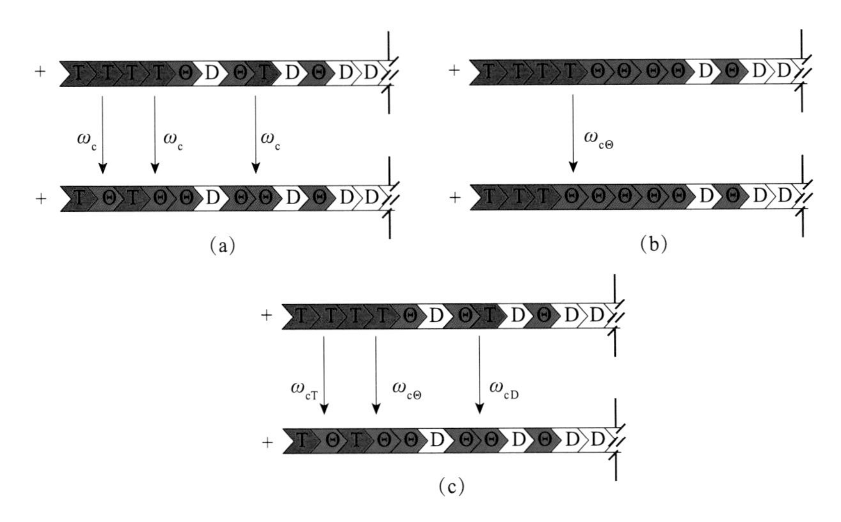

图 6-6-18 (a)ATP- 肌动蛋白原聚体（T，红色）以分裂速度 ω_c 随机分裂成 ADP/Pi- 肌动蛋白原聚体（Θ，蓝色），该过程不依赖于邻近原聚体的核苷酸状态；(b)ATP- 肌动蛋白原聚体（T，红色）以分裂速度 $\omega_{c\Theta}$ 矢量分裂成 ADP/Pi- 肌动蛋白原聚体，该分裂只局限于 TΘ 或者 TD 的边界处；(c) 协同分裂，该分裂速率依赖于邻近单体的核苷酸状态，具有三个分裂速率 ω_{cT}、$\omega_{c\Theta}$ 和 ω_{cD}。关于随机、矢量和协同无机磷酸 Pi 释放过程也有类似的定义

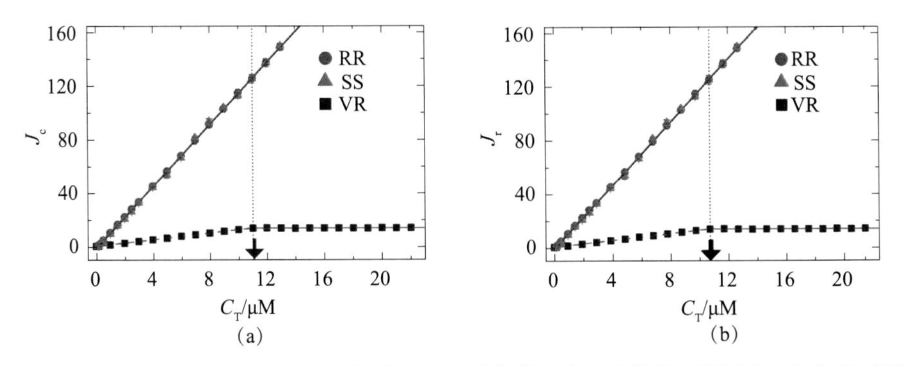

图 6-6-22 ATP 总分裂速率 J_c 和无机磷酸 Pi、总释放速率 J_r（单位：原聚体 /s）和肌动蛋白单体浓度 c_T 的依赖关系。圆、正三角和方块分别表示水解过程 RR、SS 和 VR。黑色的箭头表示矢量分裂的临界浓度，$c_{c,T}$=11.0μM

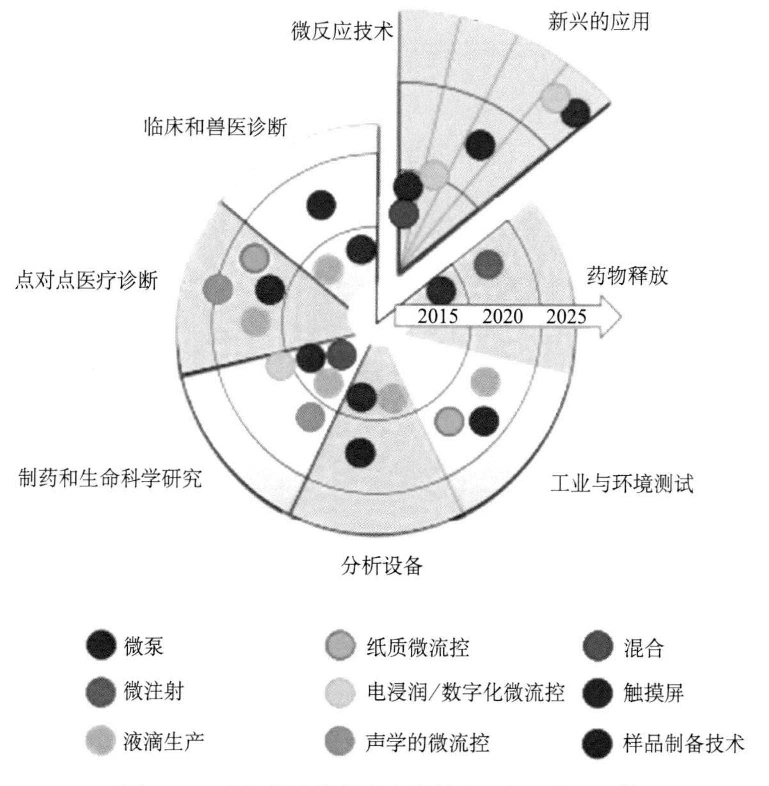

图 7-1-6　麦姆斯咨询的微流控技术 / 应用路线图 [7]

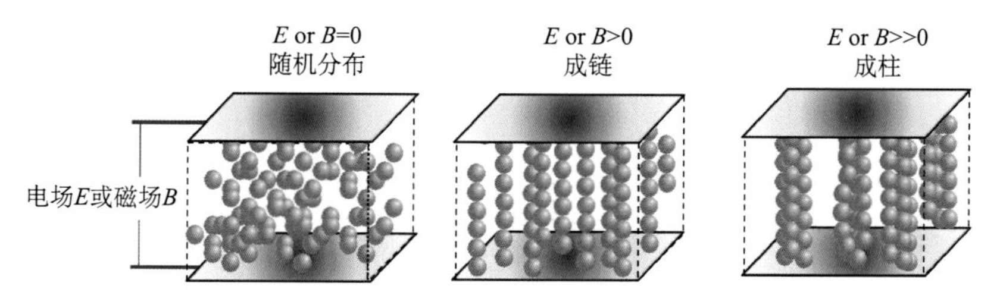

图 7-4-1　磁（或者电）流变液的工作机制：磁流变液，红色颗粒为磁性纳米至微米颗粒；电流变液，红色颗粒为介电纳米至微米颗粒

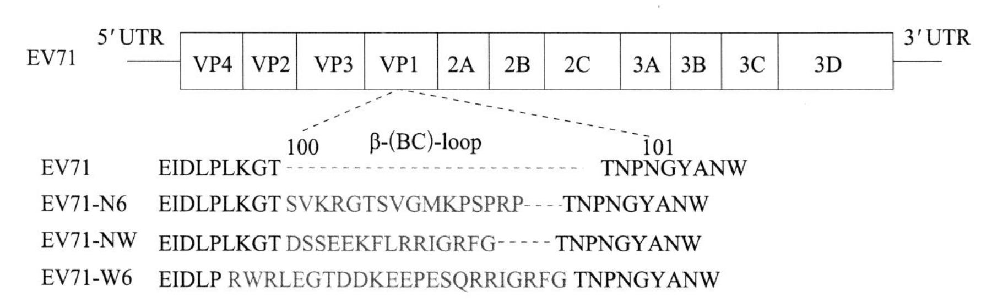

(a)

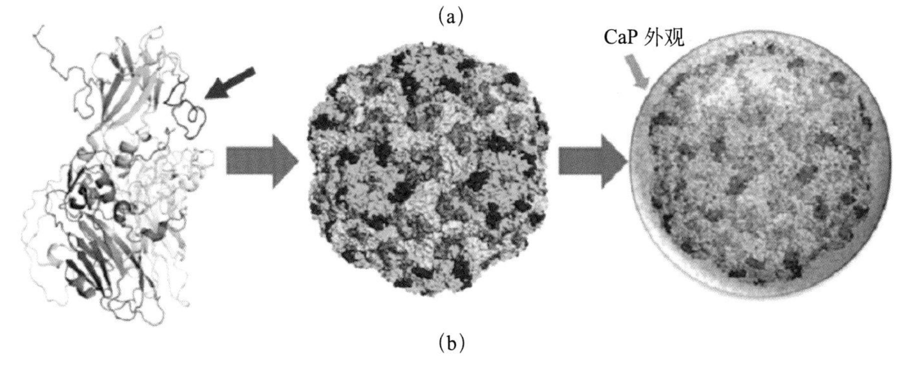

(b)

图 7-5-7　含有结合蛋白的 EV71 设计思路

(a) EV71 的基因组和 VP1 上 β-(BC)-loop 的插入位点。(b) 病毒蛋白建模结果。 EV71 外壳蛋白 VP1、VP2 和 VP3 分别以青色、黄色和橙色表示。内嵌蛋白以蓝色标记，60 段复制体在表面均匀分布，这些蛋白可以诱导 CaP 外壳（灰色）的原位生成。Copyright 2013, the National Academy of Science from Ref. [69]

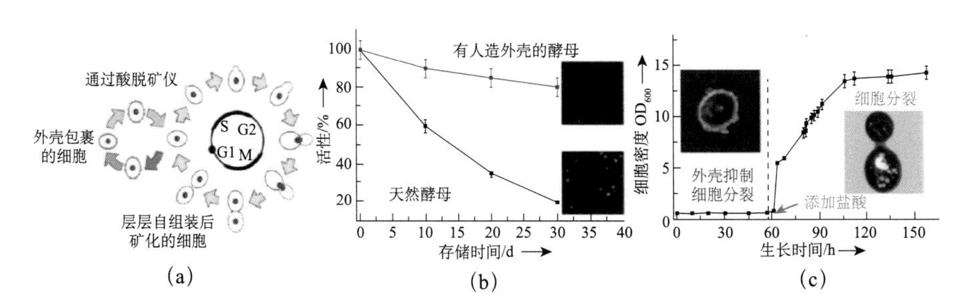

图 7-5-11　(a) 酵母细胞与包裹后酵母的生物周期 (b) 通过曲线可知活细胞数量在清水中会随着时间而减少。内嵌的图是通过死活细胞染色剂（FUN1）处理后，30 天后的荧光图。红色表示细胞是活的，绿色表示细胞死亡。(c) 普通与包裹后酵母的生长曲线。细胞密度是通过浊度法测定的（OD 600）。本实验中带壳的细胞是在细胞活性实验最后收集的，所以所有细胞都是被完整包裹的。1mM HCl 在 60h 时被加入以用于溶解矿物外壳。左侧插图：荧光图像说明细胞是活的；右侧插图：TEM 图展示体系中的分裂细胞。Copyright 2008, Wiley-VCH from Ref. [27]

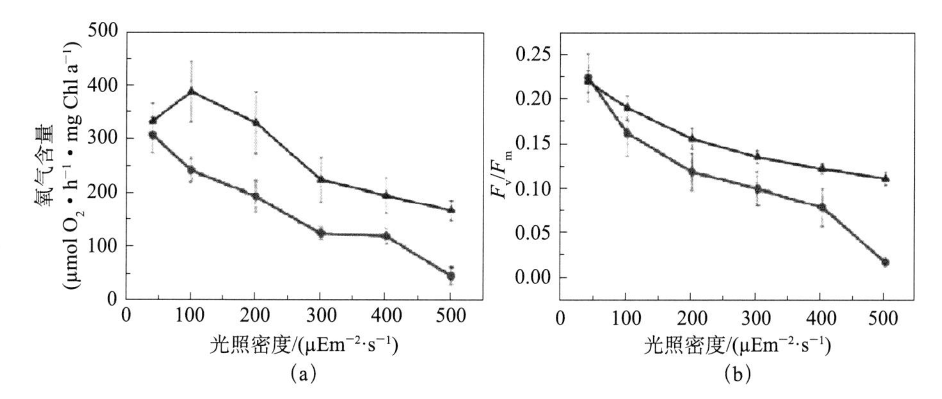

图 7-5-17　普通蓝藻与二氧化硅包裹后蓝藻的不同光合作用参数对比

(a) 以 1,4- 苯醌为电子受体的光合产氧变化过程。(b) PSII 的最大量子产率（F_v/F_m），红线表示普通绿藻，蓝线表示二氧化硅包裹后的绿藻。Copyright 2013, the Royal Society of Chemistry from Ref. [37]

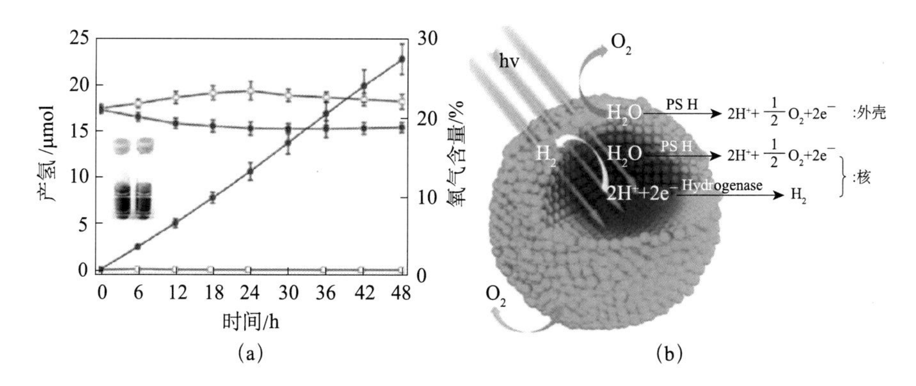

图 7-5-18　在有氧环境中通过聚集小球藻实现生物产氢

(a) 在 100 $\mu Em^{-2} \cdot s^{-1}$ 光照强度下，不同时期的产氢量和氧含量。红线：氧含量；蓝线：氢气生成量。空心方格：普通小球藻。实方格：聚集的小球藻。(b) 聚集小球藻空间功能化示意图。Copyright 2015, Wiley-VCH from Ref. [135]